Developmental Biology

Developmental Biology

Edited by Leonard Roosevelt

New York

Published by Syrawood Publishing House,
750 Third Avenue, 9th Floor,
New York, NY 10017, USA
www.syrawoodpublishinghouse.com

Developmental Biology
Edited by Leonard Roosevelt

International Standard Book Number: 978-1-68286-520-0 (Hardback)

Cataloging-in-Publication Data

Developmental biology / edited by Leonard Roosevelt.
p. cm.
Includes bibliographical references and index.
ISBN 978-1-68286-520-0
1. Developmental biology. 2. Biology. I. Roosevelt, Leonard.

QH491 .D48 2018
571.8--dc23

TABLE OF CONTENTS

PREFACE

Developmental biology studies hereditary relationships between species by analyzing embryonic developments occurring at the molecular level. This book on developmental biology presents researches and studies performed by experts across the globe. Theoretical advancements and technological changes that have occurred in this field as well as in molecular genetics have been included in it. While understanding the long-term perspectives of the topics, the book makes an effort in highlighting their impact as a modern tool for the growth of the discipline. It will act as a guide for students and academicians in the fields of evolutionary biology, deep homology and genetics.

Significant researches are present in this book. Intensive efforts have been employed by authors to make this book an outstanding discourse. This book contains the enlightening chapters which have been written on the basis of significant researches done by the experts.

Finally, I would also like to thank all the members involved in this book for being a team and meeting all the deadlines for the submission of their respective works. I would also like to thank my friends and family for being supportive in my efforts.

Editor

1

Division of Labor in the Hyperdiverse Ant Genus *Pheidole* Is Associated with Distinct Subcaste- and Age-Related Patterns of Worker Brain Organization

Mario L. Muscedere*¤, James F. A. Traniello

Department of Biology, Boston University, Boston, Massachusetts, United States of America

Abstract

The evolutionary success of ants and other social insects is considered to be intrinsically linked to division of labor among workers. The role of the brains of individual ants in generating division of labor, however, is poorly understood, as is the degree to which interspecific variation in worker social phenotypes is underscored by functional neurobiological differentiation. Here we demonstrate that dimorphic minor and major workers of different ages from three ecotypical species of the hyperdiverse ant genus *Pheidole* have distinct patterns of neuropil size variation. Brain subregions involved in sensory input (optic and antennal lobes), sensory integration, learning and memory (mushroom bodies), and motor functions (central body and subesophageal ganglion) vary significantly in relative size, reflecting differential investment in neuropils that likely regulate subcaste- and age-correlated task performance. Worker groups differ in brain size and display patterns of altered isometric and allometric subregion scaling that affect brain architecture independently of brain size variation. In particular, mushroom body size was positively correlated with task plasticity in the context of both age- and subcaste-related polyethism, providing strong, novel support that greater investment in this neuropil increases behavioral flexibility. Our findings reveal striking levels of developmental plasticity and evolutionary flexibility in *Pheidole* worker neuroanatomy, supporting the hypothesis that mosaic alterations of brain composition contribute to adaptive colony structure and interspecific variation in social organization.

Editor: Deborah M. Gordon, Stanford University, United States of America

Funding: This work was funded by the following National Science Foundation (www.nsf.gov) grants: Graduate Research Fellowship to ML Muscedere; IOB 0725013 to JFA Traniello; IOB 0724591 to Wulfila Gronenberg (U. of AZ). The funders had no role in study design, data collection and analysis, decision to publish, or preparation of the manuscript.

Competing Interests: The authors have declared that no competing interests exist.

* E-mail: mario@bu.edu

¤ Current address: Program in Neuroscience, Boston University, Boston, Massachusetts, United States of America

Introduction

Comparing neuroanatomical variation among individuals within and between invertebrate and vertebrate species has proven fruitful for exploring links between brain, behavior, ecology, life history and evolution [1]. Studies typically derive from the fundamental axiom that neural tissue is energetically expensive and thus relative investment in functionally distinct brain areas is shaped by natural selection on the sensory capabilities and cognitive behaviors supported by these regions [2]. Analyses have correlated brain structure and organization with diet, olfactory and visual acuity, group structure, and cognitive traits such as memory capacity in taxa as varied as birds, mammals and insects, providing important insights into the forces shaping animal nervous systems [3–18]. Sociality has received particular attention as a selective agent potentially favoring the evolution of larger brains and more elaborate higher-level integration regions [3].

Ants are exemplars of social life, and their extraordinary diversity and abundance is frequently attributed to the division of labor and collective behavior that underscore colony operations [19]. Although the question of how miniaturized ant brains generate adaptive behavior was posed at least as early as the 19th century, when Darwin mused on the "extraordinary mental activity" of such an "extremely small absolute mass of nervous matter" [20], explorations of the association of sociality, ecology and neuroanatomical variation in ants and other social insects are still nascent [17,18,21–24]. The hyperdiverse ant genus *Pheidole* offers an ideal system in which to investigate these relationships because of its remarkable adaptive radiation, striking degree of intra- and interspecific variation in worker behavior, and strong subcaste morphological differentiation and division of labor, which is considered key to the diversification of the genus [25]. As in many eusocial hymenopterans, *Pheidole* workers develop behaviorally and change task performance patterns as they age: only older individuals work outside the nest, where they encounter a more complex and heterogeneous sensory environment than younger workers residing within dark nest chambers below ground or within wood. In *Pheidole dentata*, minor workers undergo behavioral maturation during the first few weeks of adult life by adding outside-nest tasks such as foraging to their existing within-nest activities [26]. During this maturational period, *P. dentata* minors become more efficient at some tasks [27] and their cephalic musculature concurrently matures [28]. Additionally, their brains experience synaptic remodeling [29], changes in biogenic amine content [30], and increased levels of serotonergic immunoreactivity [31]. Division of labor in *Pheidole* based on morphology involves subdivision of workers into discrete physical subcastes (Figure 1A), which often have distinct behavioral roles. More numerous minor workers attend to a broad range of tasks including brood care, nest maintenance, and foraging, while major workers ("soldiers"), which are larger in size and have disproportionately large heads, typically specialize on colony defense and seed milling [25], compromising their task plasticity [32]. Lastly, there is considerable variation among the ca. 1,100 *Pheidole* species in colony size, nesting ecology, diet, body size, subcaste investment, and in behavioral traits such as foraging mode, worker aggressiveness, and the breadth and plasticity of major worker behavior [25,32,33].

Is the diversity of division of labor among *Pheidole* workers of different species associated with variation in brain structure?

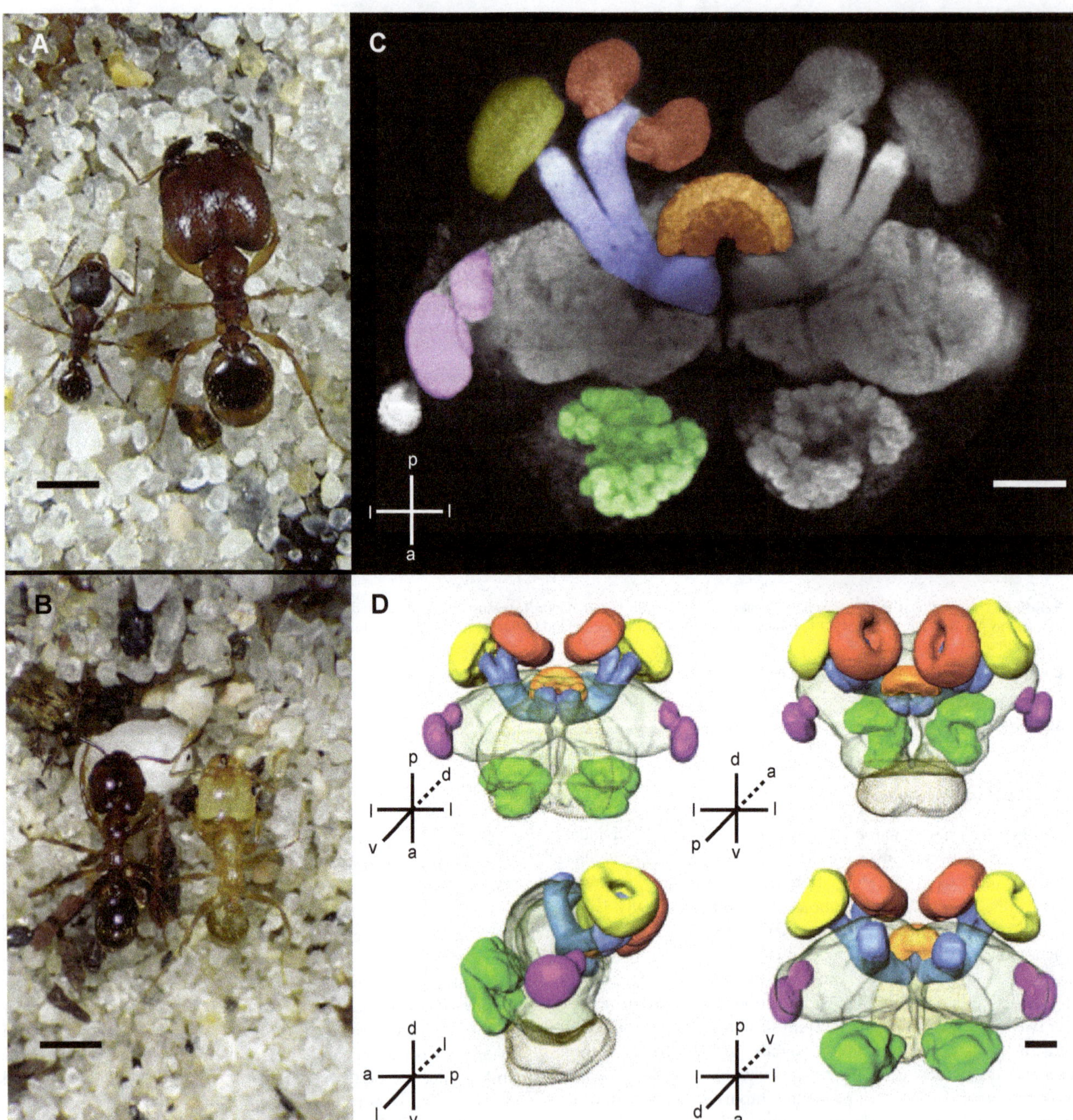

Figure 1. *Pheidole* worker morphology and neuroanatomy. (**A**) Mature *P. pilifera* minor (left) and major (right) workers. (**B**) *P. dentata* mature (left) and young (right) major workers, identifiable by cuticle colour. (**C**) Confocal micrograph of a *Pheidole* worker brain in horizontal section showing regions of neuropil. Central brain subregions visible in this section are pseudocolored. Section is slightly oblique, only one optic lobe is visible in the plane of sectioning. (**D**) 3D reconstructions of the brain shown in (**C**). Central brain functional subregions are rendered solidly, other central brain neuropils are transparent grey, and the subesophageal ganglion is a point cloud. Key: a, anterior; p, posterior; d, dorsal; v, ventral; l, lateral; green, antennal lobe; orange, central body; red, mushroom body medial calyx; yellow, mushroom body lateral calyx; blue, mushroom body peduncle & lobes; purple, optic lobe. Scale bars are 1 mm in (**A**) and (**B**) and 50 μm in (**C**) and (**D**).

Investigations of the neuroanatomical correlates of insect behavior often focus solely on the mushroom bodies (MBs), a sensory integration and processing region of the brain considered vital for learning, memory, and behavioral flexibility [24,34–37]. Evidence supporting the role of the MBs as neural substrates important for behavioral flexibility in social hymenopterans includes correlations between age-related changes in task performance and MB development. Older workers generally have larger MBs than younger workers, presumably reflecting the greater sensory processing and behavioral plasticity required for efficient outside-nest work [38–41]. However, chronological age and prior task experience, when experimentally uncoupled [38,40,42–44], both affect MB size. MB growth in adult social hymenopterans thus likely reflects both "experience-expectant" canalized developmental processes that precede worker behavioral transitions and "experience-dependant" plastic responses to task performance [43,45,46]. MB size also tracks reproductive division of labor: the MBs of behaviorally flexible worker ants are relatively larger compared with the MBs of conspecific sexuals specialized on reproduction [47]. The volume of the MBs increases in locusts when they transition from solitary to gregarious phases, which are associated with increased dietary generalism [7]. Comparative studies arguing that structurally complex MBs have evolved due to the cognitive demands of a generalist diet in coleopterans [10], and parasitoid life history in hymenopterans [11] further support a general coupling of behavioral plasticity and MB size in insects. While studied less frequently, age- and task-related size variation has also been documented in the antennal lobes and other neuropil compartments in social insects [17,18,40]. Worker behavioral differentiation, therefore, is probably not generated solely by differences in MB size or organization, but by adaptive patterns of brain organization that reflect the integrated functional roles of all brain subregions in task performance. Studies of the relative investment by workers in multiple functionally distinct brain subregions – which globally can be considered a worker's neural phenotype – may thus provide a more holistic and detailed understanding of the interplay between behavior, ecology, evolution, sociality, and neuroanatomy.

Here we provide the first comprehensive scaling analysis of multiple, functional brain subregions and the first description of the associations between brain structure, age, and body size in a social insect. Using multivariate methods and a robust sample of *Pheidole* brains collected from workers that vary in age, subcaste, and species, we first determined if brain organization differed among behaviorally differentiated worker groups. We predicted that selection imposed by variation in species ecology and social environments of workers of different subcaste and age favored distinctive patterns of brain structure. Second, we analyzed how individual brain subregions scaled with brain size; we predicted differences in neural phenotype among worker groups could not be explained solely by straightforward developmental scaling relationships, but would illustrate adaptive patterns of interspecific variation and intraspecific developmental plasticity. Finally, we determined whether variation in brain structure was correlated with behavioral differences among worker subcaste and age groups in our focal species. We predicted that plasticity in the sizes of sensory input (optic and antennal lobes) and processing structures (MBs) would be associated with species-typical patterns of division of labor. Examining this prediction integrated the analysis of two classical correlates of division of labor – age and morphology – with neuroethological concepts of task plasticity that suggest variation in behavioral flexibility and specialization among workers are the result of differential elaboration of sensory input and processing regions, including the MBs.

Materials and Methods

Ethics statement

Colonies were maintained in the lab in compliance with conditions specified in USDA APHIS PPQ permit number P526-11-01197, and earlier issues of the permit. Culture methods maximized colony growth and health. No specific permits were required to collect colonies in the field and none of our focal species are endangered or protected. Colonies were collected from unprotected public land with no measurable habitat disturbance.

Species selection, ecology and natural history of focal species, and ant husbandry

We selected three North American *Pheidole* species (*P. dentata*, *P. morrisi* and *P. pilifera*) whose sociobiology, ecology, and life history encompass the broad variation in colony structure and division of labor identified in this genus [25]. The most salient interspecific differences concern colony size (*P. morrisi*>*P. dentata*>*P. pilifera*), diet (*P. morrisi* and *P. dentata* are scavengers, *P. pilifera* is primarily granivorous), and major worker investment and behavioral plasticity (*P. morrisi*>*P. dentata*>*P. pilifera*). *Pheidole dentata* is a small-bodied, ground- and wood-nesting ant species native to the mid-Atlantic to southeastern United States and west to Illinois, Texas and northern Mexico. Colonies often contain more than 1,000 workers and forage on live and dead insect prey. Typically, majors are uninvolved in most colony tasks, although they leave the nest to provide resource and colony security, including enemy-specific defensive behavior [48]. *Pheidole morrisi* is distributed along the east coast of the U.S. from Florida to Long Island, New York.

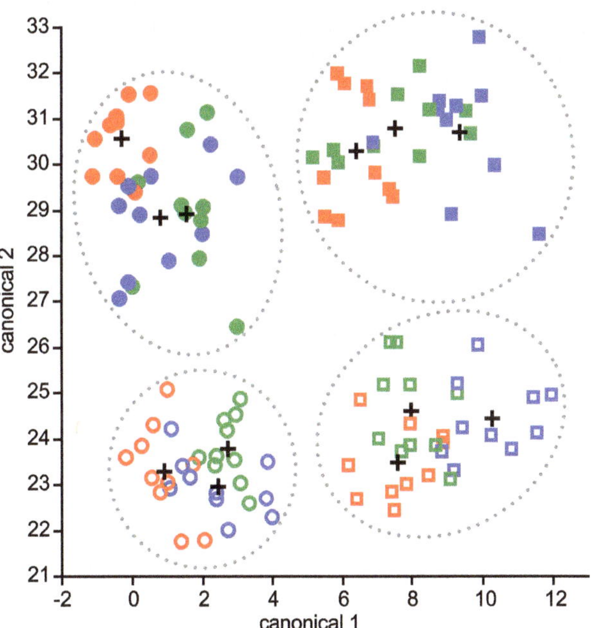

Figure 2. *Pheidole* subcastes and age groups have neural phenotypes that are well separated in multivariate space. Plot shows the separation of all 12 worker groups on the first two canonical variables generated by discriminant analysis (accounting for 87% of the total variance explained by the analysis). Four non-overlapping clusters (dotted ellipses) are generated by very strong discrimination among young minors, old minors, young majors, and old majors. Key: red, *P. dentata*; green, *P. morrisi*; blue, *P. pilifera*; open, young age cohort; closed, mature age cohort; circles, minor workers; squares, major workers; crosses, group centroids.

Pheidole morrisi colonies are also insectivorous, ground-nesting scavengers, but are larger (often >10,000 workers) and competitively dominant, with numerous foraging workers that aggressively defend the nest and compete with sympatric ant species. Colonies contain a relatively high proportion of major workers that are unusually active in some tasks typical of *Pheidole* minors [49], including foraging and other outside-nest work. *Pheidole pilifera* is widely distributed throughout the continental United States. Colonies nest in the ground, are comparatively small (~600–800 workers), and characterized by relatively small numbers of "shy" foragers that either flee or feign death in response to danger. *Pheidole pilifera* colonies have a low proportion of large, docile majors that rarely leave the nest and have a repertoire restricted largely to seed milling and blocking nest passages. *Pheidole pilifera* colonies scavenge insect prey but predominantly collect, store, and feed on seeds. Major workers are typically found in nest chambers where seeds are cached.

Queenright colonies were collected in Centereach, NY (*P. pilifera* and *P. morrisi*), Rocky Point, NY (*P. morrisi*), Concord, MA (*P. pilifera*), and Alachua County, FL (*P. dentata*), and maintained in the laboratory using established protocols [26,27]. Newly eclosed callows (0–2 days old) and fully mature workers from all three species were identified by cuticular coloration [26,27,29–31]. Callow workers in all three species have uniform light yellow cuticles that darken predictably over the first few weeks of adult life (Figure 1B), a process that takes 16–20 days in *P. dentata* minor workers.

Immunohistochemistry and confocal microscopy

Individual worker brains were dissected from the head capsule in cold insect Ringer's solution, quickly transferred to cold Dent's fixative (4:1 absolute methanol:dimethyl sulfoxide [DMSO]), and fixed for 12 hours at −20°C (with one change into fresh fixative after ~30 min). Brains were washed 1 time in absolute methanol, then stored in absolute methanol at −20°C until further processed. Brains were rehydrated through a graded methanol series (95%, 70%, 50%, 30%, PBS, for 5 min each), washed (6×10 min) in 0.01 M PBS with 1% bovine serum albumin and 1% DMSO (PAD), incubated for 1 hour in a blocking solution of 10% normal goat serum in PAD (PADN), and incubated for 60 hours at room temperature in the primary antiserum SYNORF1, diluted 1:50 in PADN. We initially obtained SYNORF1 as a gift from Dr. E. Buchner, University of Würzburg; SYNORF1 was later obtained from the Developmental Studies Hybridoma Bank (developed under the auspices of the NICHD and maintained by The University of Iowa, Department of Biology, Iowa City, IA 52242). After incubation, brains were washed (6×10 min) in PAD and incubated in the dark for 36 hours in a solution of Alexa Fluor 568 goat anti-mouse secondary antibody diluted 1:200 in PADN. Stained brains were washed (3×10 min) in PAD, washed (3×10 min) in 0.01 M PBS, and dehydrated through a graded methanol series (30%, 50%, 70%, 95%, 100%, 100%, for 5 min each). Dehydrated brains were transferred to a 30% methyl salicylate/70% methanol solution for 5 min, then transferred to a dish of 100% methyl salicylate and allowed to fully clear. Cleared brains were whole-mounted in 100% methyl salicylate using custom-made double-sided steel well slides.

Brains were imaged using an Olympus Fluoview BX50 laser-scanning confocal microscope with a 20× objective (N.A. = 0.5). Labeled structures were excited with a Krypton laser at 568 nm and their fluorescence detected using a 605/45 nm band-pass filter. Brains were optically sectioned in their entirety through the horizontal plane at 0.6 μm increments. To correct for axial shortening along the z-axis introduced by the refractive index mismatch between air and methyl salicylate [50], we multiplied the nominal section thickness (0.6 μm) by a correction factor of 1.59 to obtain the true section thickness of approximately 1 μm (0.954 μm). We stained and imaged 10 brains from each of 12 worker groups (young minors, young majors, mature minors, and mature majors from *P. morrisi*, *P. pilifera*, and *P. dentata*), for a total $n = 120$ worker brains.

Table 1. Results of discriminant function classification of worker brains to the *a priori* groups listed.

data subset	*a priori* groups (*n*)	classification scheme	correct classifications	Q	p
full dataset	species, age, subcaste (12)	full	92.5%	1.11×10^{-4}	*<0.0001*
		jackknifed	79.2%	788	*<0.0001*
full dataset	age, subcaste (4)	full	100%	360	*<0.0001*
		jackknifed	99.2%	352	*<0.0001*
young minors	species (3)	full	96.7	54.2	*<0.0001*
		jackknifed	63.3%	12.2	*0.0005*
mature minors	species (3)	full	90%	43.4	*<0.0001*
		jackknifed	70.0%	18.2	*<0.0001*
young majors	species (3)	full	100%	60.0	*<0.0001*
		jackknifed	76.7%	25.4	*<0.0001*
mature majors	species (3)	full	100%	60.0	*<0.0001*
		jackknifed	73.3%	21.6	*<0.0001*
young workers	species, subcaste (6)	full	100%	300	*<0.0001*
		jackknifed	80.0%	173	*<0.0001*
mature workers	species, subcaste (6)	full	91.7%	243	*<0.0001*
		jackknifed	80.0%	173	*<0.0001*

Each row indicates classification outcomes using either the full discriminant model or the jacknifed (leave-one-out) model for the given analysis. Significant Press' Q statistics indicate that workers in a given analysis were classified to the correct *a priori* group at greater-than-chance levels.

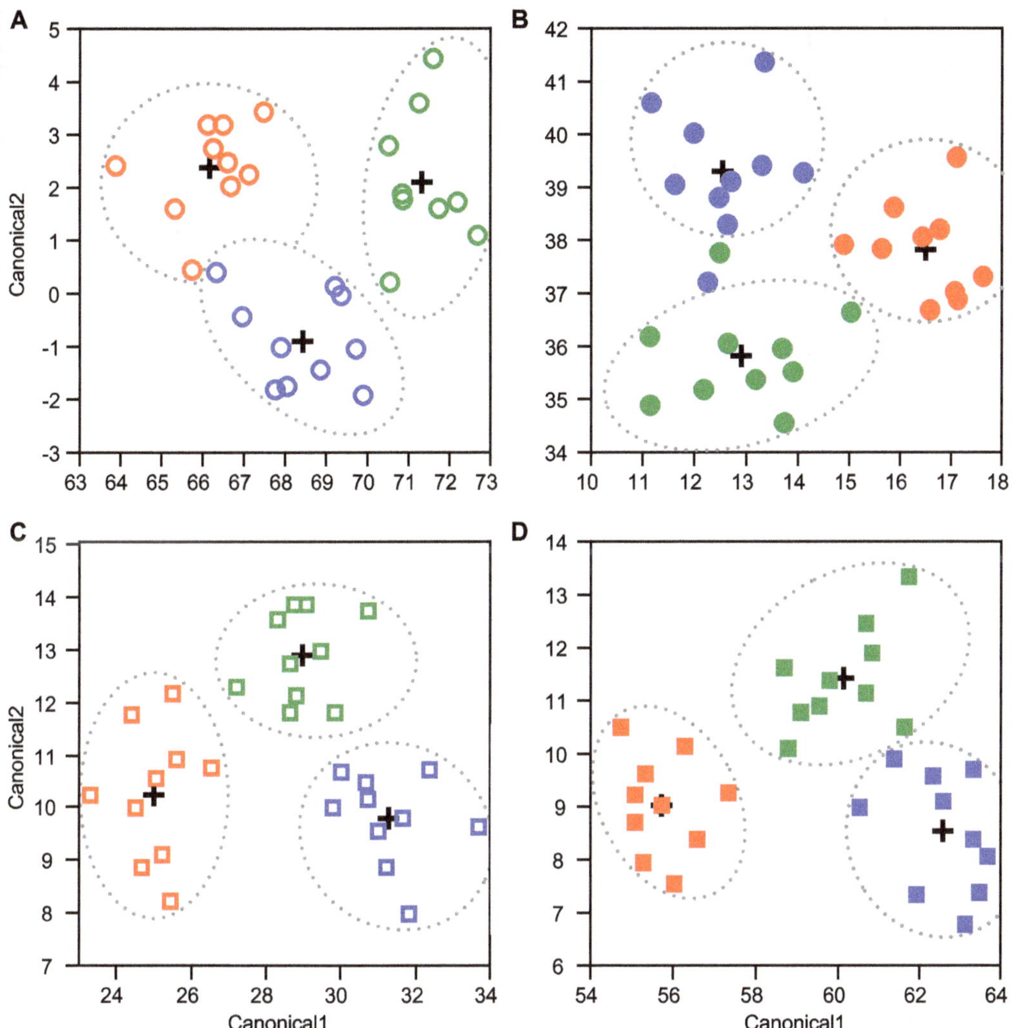

Figure 3. Interspecific variation in *Pheidole* multivariate neural phenotypes. Separate discriminant function analyses reveal significant differentiation of neural phenotype among species for (**A**) young minors, (**B**) old minors, (**C**) young majors, and (**D**) old majors. Key: red, *P. dentata*; green, *P. morrisi*; blue, *P. pilifera*; open, young age cohort; closed, mature age cohort; circles, minor workers; squares, major workers; crosses, group centroids.

Neuroanatomical measurements

We used Amira v3.1 and ImageJ v1.41d to measure the volume of one neuropil hemisphere and seven of its functional subregions from each brain (Figure 1C,D): the lobula and medulla of the optic lobe (OL), the antennal lobe (AL), the mushroom body (MB) medial calyx, the MB lateral calyx (both calyces together denoted as MB-C), the MB peduncle, including the vertical and medial lobes (MB-PL), the central body (CB), and the subesophageal ganglion (SEG). Because the SEG neuropil is fused and completely contiguous with the central brain (supraesophageal ganglion) neuropil in *Pheidole*, we delineated this structure from the central brain by assigning all neuropil ventral to the esophageal foramen to the SEG. To capture shape variation among brains in addition to volume differences, we made six more linear measurements in horizontal section where well-defined anatomical landmarks could be identified: MB medial calyx maximal width, MB lateral calyx maximal width, MB peduncle maximal width at its point of entry into the protocerebral lobe, CB maximal height and width, and SEG width.

Statistical analyses

Multivariate discriminant analyses were conducted to assess overall variation in neural phenotype among groups. All size and shape variables were size-corrected and included in these analyses (as central brain subregion volume/remainder-of-central brain volume [ROCBV, calculated separately for each brain subregion], SEG volume/central brain volume [CBV], or linear shape measurement/maximal width of central brain neuropil across the protocerebrum). Scaling between individual brain subregions and brain size was analyzed using standardized major axis (SMA) regression with the program (S)MATR v. 2.0 [51]. The contributions of between-groups effects (i.e., effects of age or subcaste) to scaling differences identified with (S)MATR were assessed with traditional ANCOVA. Finally, we tested for differences among worker groups in the absolute volumes of the OLs, ALs, and MBs, and for differences in relative brain size among mature worker groups, using ANOVA. Full statistical details are presented as Supporting Information S1.

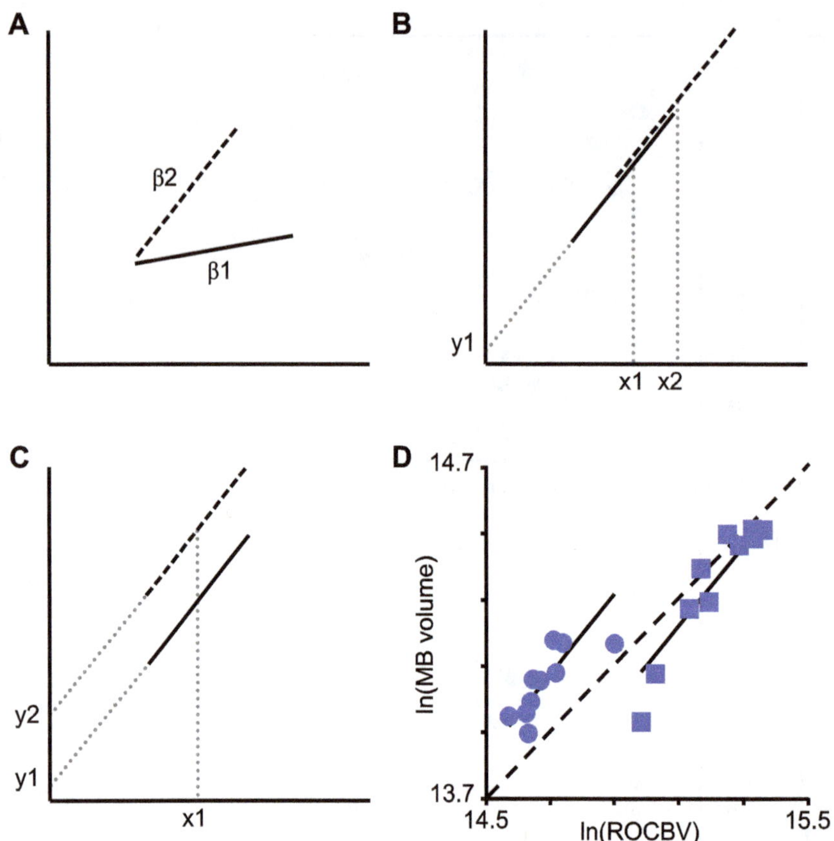

Figure 4. Examples of altered brain scaling relationships. Solid and dashed lines in (**A–C**) indicate hypothetical standardized major axis (SMA) scaling relationships of brain region volume (y variable) with brain size (x variable). (**A**) Altered slopes (β1 and β2), which preclude further scaling analyses using these methods. (**A**). Groups share a common scaling slope and y-intercept (y1) but average brain size differs between groups (×1 and ×2), producing an x-axis shift. (**C**) Groups share a common scaling slope and average brain size (×1) but y-intercepts (y1 and y2) are shifted, producing a grade shift. (**D**) Solid and dashed lines represent actual SMA scaling regressions of mushroom body volume versus remainder-of-central-brain volume (ROCBV) for mature *P. pilifera* minors (circles) and mature *P. pilifera* majors (squares), representative of a grade shift accompanying a shift along the x-axis. Dashed line in (*d*) indicates slope of 1 (isometry) for reference.

Results and Discussion

Pheidole worker subcastes, age groups, and species have distinct brain architectures

Multivariate discriminant analysis revealed highly significant gross neuroanatomical differentiation among our 12 worker groups (Wilks' $\Lambda = 4\times10^{-4}$, $p<0.0001$). The model precisely classified brains to correct species, age cohort, and subcaste in 92.5% of cases (Figure 2), significantly exceeding chance expectations (8.3%; Press's $Q = 1.11\times10^{-4}$, $p<0.0001$). All incorrectly classified brains were correctly identified to age group and subcaste but not species. Reanalysis using only age cohort and subcaste as the four *a priori* groups enhanced discrimination, correctly classifying all 120 brains (Wilks' $\Lambda = 8.92\times10^{-3}$, $p<0.0001$; 100% correct, $Q = 360$, $p<0.0001$). However, separate discriminant analyses of data subsets using only species identity as *a priori* categories confirmed significant interspecific differentiation (Table 1, Figure 3) for young minors (Wilks' $\Lambda = 0.049$, $p = 0.0002$), mature minors (Wilks' $\Lambda = 0.067$, $p = 0.0010$), young majors (Wilks' $\Lambda = 0.039$, $p<0.0001$), and mature majors (Wilks' $\Lambda = 0.037$, $p<0.0001$). More conservative jackknifed (leave-one-out) classification designs, in which each brain was categorized by discriminant functions derived using only the other *n*-1 samples, also had high correct classification rates that significantly exceeded chance levels in all cases (Table 1). When using only age cohort and species as *a priori* groups, jackknifed classification was nearly as accurate as whole-model classification (99.2% correct), emphasizing that worker age cohorts and subcastes have strikingly differentiated neural phenotypes (Table 1).

Task experience can lead to neuroanatomical changes; differences among worker age cohorts in our analysis could therefore reflect experience-dependent and/or independent effects on brain development [38,40,42–46]. We tested whether discrimination among subcastes and species was enhanced among mature individuals, which have more diverse task repertories and greater task experience than recently eclosed workers, and thus greater potential for experience-related divergence in neural phenotype. Resulting whole-model classification of worker brains to subcaste and species was in fact slightly more accurate in young workers (Wilks' $\Lambda = 0.0035$, $p<0.0001$; 100% correct, $Q = 300$, $p<0.0001$) than in mature workers (Wilks' $\Lambda = 0.0052$, $p<0.0001$; 91.7% correct, $Q = 243$, $p<0.0001$), and the accuracy of jackknifed classification was equivalent for both age cohorts (80%, Table 1). This suggests that much of the neuroanatomical differentiation among subcastes and species is generated by alternative developmental scaling of brain regions prior to adult emergence, notwithstanding additional plastic responses to worker experience during adult life.

Table 2. Brain subregions scale differently with brain size across *Pheidole* worker groups.

y	x	scaling slope (β)	β different from 1?	grade shifts?	x-axis shifts?
SEG volume	CBV	0.99	no	yes	yes
		(0.89, 1.10)	(χ^2 = 0.075,	(W^2 = 324,	(W^2 = 405,
			p = 0.8)	p<0.0001)	p<0.0001)
AL volume	ROCBV	0.97	no	yes	yes
		(0.88, 1.09)	(χ^2 = 0.305,	(W^2 = 137	(W^2 = 227,
			p = 0.6)	p<0.0001)	p<0.0001)
MB volume	ROCBV	1.20	yes	yes	yes
		(1.10, 1.33)	(χ^2 = 14.8,	(W^2 = 518,	(W^2 = 220,
			p<0.0001)	p<0.0001)	p<0.0001)
MB-C volume	ROCBV	1.25	yes	yes	yes
		(1.15, 1.36)	(χ^2 = 23.5,	(W^2 = 267,	(W^2 = 188,
			p<0.0001)	p<0.0001)	p<0.0001)
MB-PL volume	ROCBV	1.15	yes	yes	yes
		(1.04, 1.29)	(χ^2 = 6.71,	(W^2 = 711,	(W^2 = 287,
			p = 0.01)	p<0.0001)	p<0.0001)
CB volume	ROCBV	worker groups differ	yes for two groups	n/a	n/a
OL volume	ROCBV	worker groups differ	yes for five groups	n/a	n/a

We tested whether worker groups shared common scaling slopes, whether there were deviations from isometric scaling (β = 1), and whether there were significant shifts among worker groups in scaling elevation (grade shifts) or position along the x-axis (Figure 4, [51]). Values in parentheses indicate 95% confidence intervals of slopes (column 3) or statistical details (χ^2-tests, column 4; Wald tests, columns 5 and 6). All variables ln-transformed before analysis. CBV, central brain volume; ROCBV, remainder-of-central-brain volume.

Differentiated brains result from mosaic alterations of subregion scaling relationships

We conducted bivariate scaling analyses [51] of individual brain subregions to explore differences in brain composition that could lead to the highly differentiated brains identified by our discriminant models. These analyses revealed scaling relationships among brain subregions and total brain size that included differences among *Pheidole* worker groups in log-log slope, y-intercept, and position of scaling relationships along the x-axis (Figure 4, Table 2). Slopes (β) differed significantly among worker groups (Figure 4A) for the scaling of OL volume (χ^2 = 22.3, d.f. = 11, p = 0.02) and CB volume (χ^2 = 21.4, d.f. = 11, p = 0.03) with brain size, precluding further analyses. For scaling of other substructures with brain size, slopes did not significantly differ among groups. These relationships were characterized by different levels of nonlinearity, from isometry between SEG volume and CBV (β = 0.99) to positive allometry between MB-C volume and ROCBV (β = 1.25, Table 2). Allometric scaling of brain subregions could lead to the altered brain proportions among worker groups identified by our discriminant analyses simply because the groups differed significantly in central brain volume (ANOVA: $F_{11,108}$ = 20.5, p<0.0001), resulting in significant shifts

Table 3. Between-group grade shifts in *Pheidole* brain subregion scaling are indicated by significant ANCOVA effects.

brain region	effect	p	description of grade shift (at equivalent brain size)
SEG	age	<0.0139	larger in mature workers
	species × subcaste	<0.0089	larger in majors; subcaste difference is greater in *P. pilifera*
AL	age	<0.0001	larger in mature workers
	species × subcaste	<0.0002	larger in minors; subcaste difference is greatest in *P. dentata*, smallest in *P. morrisi*
MB	species	<0.0043	largest in *P. dentata*, smallest in *P. pilifera*.
	age	<0.0001	larger in mature workers
	subcaste	<0.0001	larger in minors
MB-C	subcaste	<0.0001	larger in minors
	species × age	<0.0247	larger in mature workers; age difference is greatest in *P. pilifera*, smallest in *P. dentata*
MB-PL	species × subcaste	<0.0289	larger in minors; subcaste difference is smallest in *P. morrisi*
	species × age	<0.0094	larger in mature workers; age difference reduced in *P. morrisi*

Factor main effects are not listed for significant interaction terms. Grade shift descriptions are based on inspection of marginal means. Full ANCOVA model details are given in Table S1.

in the position of worker groups along the x-axis (Figure 4B) for all scaling comparisons with common slopes (Table 2 column 6). Nevertheless, these scaling relationships all exhibited significant grade shifts (different y-intercepts, Figure 4*c*) among worker groups (Table 2 column 5, Table 3, Table S1), which had strong effects on brain proportions independent of brain size differences. Grade shifts are particularly meaningful in brain scaling comparisons (e.g. [7,52]) because they describe cases in which a structure is systematically larger or smaller in one group, and thus clearly indicate differential patterns of neural investment unrelated to differences in total brain size.

Intraspecfic variation in sensory neuropil investment correlates with worker task performance

We interpreted the ubiquity and magnitude of grade shifts in our dataset as evidence that the distinct patterns of neuropil investment of *Pheidole* worker groups are a functionally relevant outcome of adaptive developmental processes, rather than a by-product of scaling constraints among brains of different sizes. We therefore predicted that neural phenotypes should underscore variation in task performance among worker groups. Within *Pheidole* colonies, behavioral differences primarily exist along two non-mutually exclusive axes defined by subcaste (minors vs. majors) and age (mature vs. young). In both cases, groups working more frequently outside the nest (mature workers, minors) are faced with challenging navigational and sensory discrimination tasks, and have a larger behavioral repertoire than other groups that leave the nest rarely if ever (young minors) or less frequently (majors). We thus predicted that investment in neuropils that receive and integrate sensory stimuli (the OL, AL, and MBs), which are likely critical for worker task performance, would vary in similar ways along the subcaste and age axes. Patterns of investment fit this prediction (Table 4). Controlling for all other variables, mature workers (relative to young workers) and minors (relative to majors) had proportionally larger ALs and MBs (Table 3). OL scaling could not be analyzed using ANCOVA to hold all other variables constant because slopes differed among worker groups. However, averaging across all worker groups, OL differences also conformed to this prediction, being proportionally smaller in mature workers and minors (young vs. old: $t_{118} = 3.71$, $p = 0.0003$, 2.9±0.08% vs. 2.5±0.07%; major vs. minor: $t_{118} = 10.0$, $p<0.0001$, 3.1±0.07% vs. 2.2±0.05%).

These patterns are consistent with a central role of olfaction in generating behavioral flexibility in *Pheidole*. Larger olfactory input regions (AL glomeruli) are associated with increased olfactory sensitivity [18,53], and comparatively large MBs are associated with behavioral flexibility and heterogeneous sensory environments [7,10,11] in several taxa. In social hymenopterans, MB and AL growth in adults is correlated with age-related behavioral maturation [38,40,41,54]. Our results agree with these findings and extend these analyses beyond age-related development to behaviorally differentiated worker subcastes. The fact that *Pheidole* minors have relatively larger MBs and ALs than majors supports the hypothesis that the size of these neuropils is related to worker task diversities and proficiencies and validates the linkage of MB size and behavioral plasticity in insects.

Olfaction is the dominant sensory modality of *Pheidole*; workers have small eyes with few facets (minors: ca. 50 ommatidia per eye; majors: ca. 75 ommatidia per eye), and foraging, defense, and colony emigration are organized predominantly by pheromones [19]. It is therefore not surprising that worker OLs are very small (grand mean across all worker groups: 2.7±0.06% of central brain volume), reflecting the limited role of vision in task attendance [24]. Smaller proportional OL sizes in workers more likely to perform tasks outside the nest (mature workers and minors), while seemingly counterintuitive, may simply reflect a disproportionate investment in odor-processing neuropils rather than a reduction in visual ability *per se*. In support of this interpretation, absolute OL volumes did not significantly differ between age cohorts (contrasts of least-square means from factorial ANOVA [Table S2]: minors, $F_{1,112} = 3.04$, $p = 0.08$, $1.01\pm0.05\times10^{-4}$ vs. $8.77\pm0.05\times10^{-5}$ mm^3 for young vs. mature workers, respectively; majors, $F_{1,112} = 3.81$, $p = 0.054$, $1.62\pm0.05\times10^{-4}$ vs. $1.77\pm0.05\times10^{-4}$ mm^3), while other neuropil regions including the ALs (minors, $F_{1,112} = 5.75$, $p = 0.0181$, $5.40\pm0.2\times10^{-4}$ vs. $5.94\pm0.2\times10^{-4}$ mm^3; majors, $F_{1,112} = 77.0$, $p<0.0001$, $5.51\pm0.2\times10^{-4}$ vs. $7.47\pm0.2\times10^{-4}$ mm^3) and MBs (minors, $F_{1,114} = 14.9$, $p = 0.0002$, $1.25\pm0.04\times10^{-3}$ vs. $1.49\pm0.04\times10^{-3}$ mm^3; majors, $F_{1,114} = 101$, $p<0.0001$, $1.23\pm0.04\times10^{-3}$ vs. $1.86\pm0.04\times10^{-3}$ mm^3) grew significantly larger in absolute volume in mature workers. Thus for age cohorts, differences in relative OL size seem to be associated largely with the growth of olfactory processing brain regions, which consequently alter the proportional sizes of other brain neuropils including the OLs. Similarly, differences in OL size between worker subcastes are likely due in large part to differences in eye size caused by the allometric scaling of head size that generates dimorphic subcastes in *Pheidole*.

Motor control subregions of the *Pheidole* central nervous system, which would not be predicted to affect worker behavioral flexibility in the same way as sensory input and processing regions, did not vary similarly in size along the subcaste and age axes (Table 4). As with the OL, CB scaling could not be analyzed with ANCOVA due to heterogeneity of scaling slopes among worker groups. Across all worker groups, the CB was relatively larger on average in young workers, but did not differ in relative size among subcastes (young vs. old: $t_{118} = 7.89$, $p<0.0001$, 0.86±0.01% vs. 0.71±0.01%; major vs. minor: $t_{118} = 0.3$, $p = 0.76$, 0.78±0.02% vs. 0.77±0.02%). As CB function is not well characterized in any ant, the consequences of variation in CB size are speculative. Differences in SEG volume relative to CBV were also dissimilar along the young-mature and major-minor axes (Table 4): the SEG was proportionally larger in

Table 4. Patterns of behavioral and sensory neuropil variation are concordant between worker age cohorts and subcastes.

		outside-nest	brain sensory subregions			brain motor subregions	
worker groups	**behavioral repertoire size**	**task frequency**	**relative MB size**	**relative AL size**	**relative OL size**	**relative CB size**	**relative SEG size**
age cohorts	mature>young	mature>young	mature>young	mature>young	mature<young	mature<young	mature>young
subcastes	minor>major	minor>major	minor>major	minor>major	minor<major	minor = major	minor<major
concordance?	*yes*	*yes*	*yes*	*yes*	*yes*	*no*	*no*

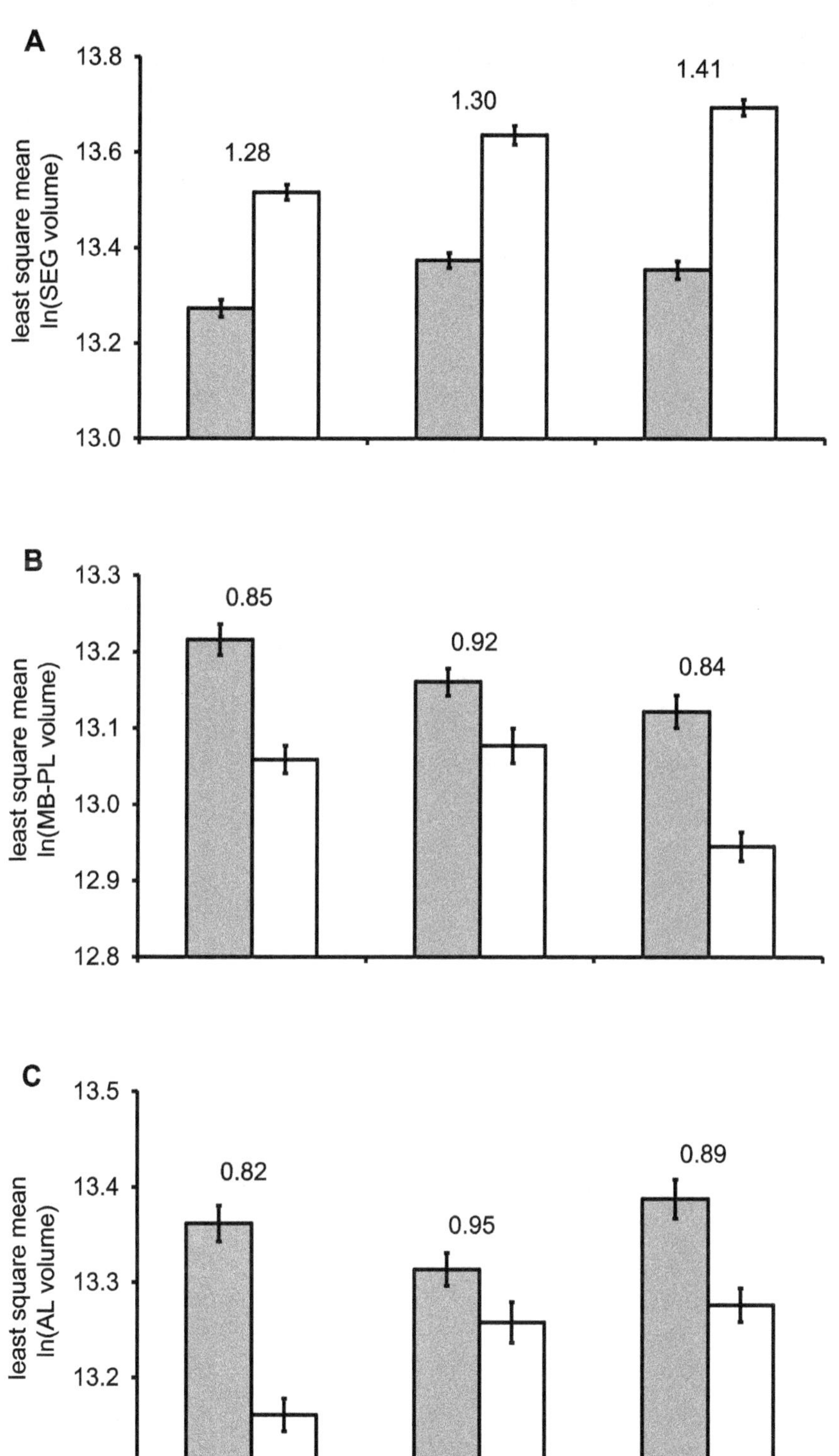

Figure 5. Interspecific variation in subcaste brain structure scaling. Least-square mean (± s. e. m.) brain subregion sizes from ANCOVA model (Table 3, Table S1) illustrating significant species×subcaste interactions for ln-transformed (**A**) subesophageal ganglion volume, (**B**) mushroom body peduncle and lobes volume, and (**C**) antennal lobe volume. Numbers above pairs of bars are major-to-minor-worker ratios of (untransformed) least-square means for each species. Key: grey bars, minor workers; open bars, major workers.

mature workers and in majors (Table 3). The SEG is associated with the motor control of the mandibles, the principal tools used by ants to work, and with the control of other mouthparts as well as integration of their afferent sensory information. The mandibular muscles occupy most of the head capsule in *Pheidole* workers and undergo significant growth after eclosion as *P. dentata* workers age

[28]. The same may be true in other species. Furthermore, majors of all three species, which have enlarged heads and mandibular muscles relative to conspecific minors, had much larger SEGs relative to CBV (Table 3). This pattern mirrors that shown for several species of termites with defensive biting soldiers, which have larger SEGs than conspecific pseudergates ("workers") [55]. Together, these results suggest differences in relative SEG size among worker groups may at least in part represent underlying relationships between mandibular size, musculature, and sensory/motor innervation of these structures.

Interspecific variation in subcaste brain organization correlates with life history

The three taxa chosen characterize a significant range of variation in behavioral and morphological divergence among *Pheidole* species and subcastes. We therefore examined whether the significant species×subcaste interactions in our ANCOVA analysis (SEG, AL, and MB-PL, Table 3) revealed relationships between the degree of major worker specialization and species life history. Majors of *P. dentata* and *P. morrisi* had SEGs approximately 28–30% larger, controlling for CBV and worker age, than conspecific minor workers (Figure 5A). While the degree of morphological difference among majors and minors is similar in *P. dentata* and *P. morrisi*, *P. pilifera* majors are proportionally much larger than their conspecific minors and have large heads with blunt mandibles used to grind seeds. In *P. pilifera*, SEGs were approximately 41% larger in majors than in minors after controlling for central brain volume and worker age. Although it is unclear whether or how SEG size scales with the size and/or closing power of the mandibular musculature, enlarged SEGs in *P. pilifera* majors could reflect sensory/motor specializations for the use of their mouthparts. Significant species×subcaste interactions were also evident for the ALs and one component of the MBs, the pedunculi and lobes (MB-PL). Of our three focal species, *P. morrisi* minors and majors are notable for their convergence in behavioral diversity: majors are active and engage in many of the same tasks as conspecific minors, including foraging [25,49]. After correcting for ROCBV and worker age, the differences in MB-PL and AL sizes between subcastes were smallest in *P. morrisi* (Figure 5B,C), mirroring the behavioral similarities between subcastes and suggesting similarity in the neural substrates that regulate task performance in *P. morrisi* minor and major workers.

Brain size/body size relationships differ among mature workers

Among mature workers, which we assume have completed most or all neural development, brain volume relative to body size (estimated as 2×half central brain volume/head width) differed significantly among worker groups (Figure 6; ANOVA: $F_{5,54} = 39.3$, $p<0.0001$). There were significant main effects of species identity ($F_{2,54} = 36.9$, $p<0.0001$) and subcaste ($F_{1,54} = 121$, $p<0.0001$), which together explained most of the variation among mature workers in relative brain size ($R^2 = 0.78$). There was no

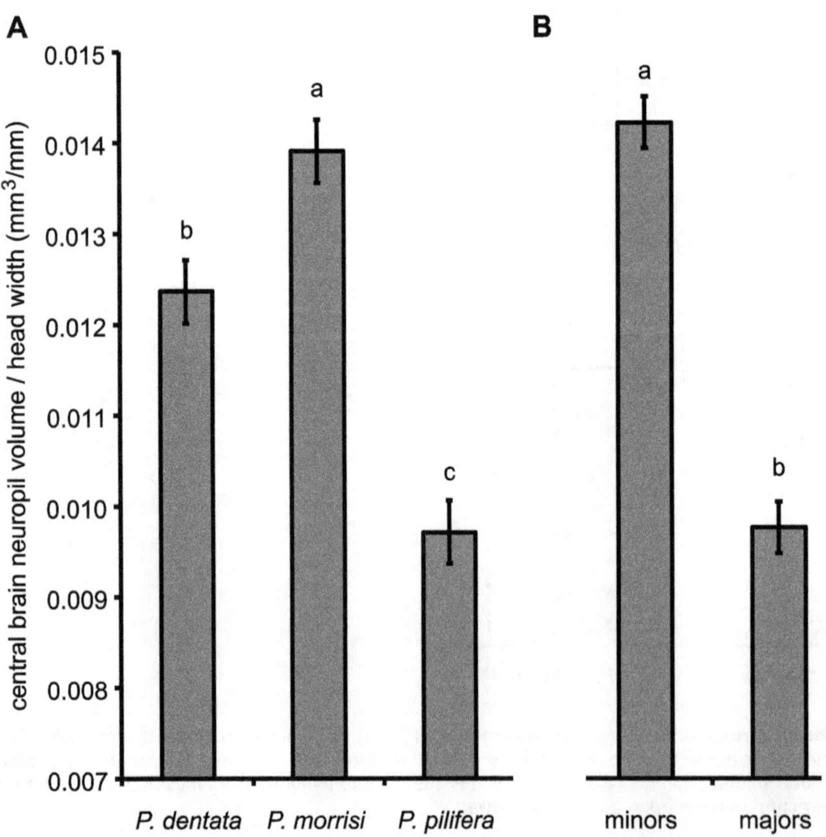

Figure 6. Brain-to-body-size scaling in *Pheidole* workers. Least-square mean (± s. e. m.) relative brain volumes from ANOVA model (mature workers only) indicating significant (**A**) species and (**B**) subcaste effects. Levels not sharing the same letter are significantly different ([**A**]: Tukey-Kramer *post hoc* comparisons of least-square means; [**B**]: main effect of subcaste).

significant species×subcaste interaction effect ($F_{2,54} = 0.653$, $p = 0.5$). Therefore, the variation in brain composition we describe exists in addition to differences in the brain-to-body-size scaling relationships of developmentally mature workers.

Conclusions

Differences in brain size and divergent patterns of brain subregion scaling among *Pheidole* workers result in size-, age-, and species-specific patterns of investment in functional neuropil regions, reflecting the preeminent colony-wide division of labor and interspecific sociobiological variation characteristic of this species-rich genus. We believe our results have significant implications for understanding brain evolution in *Pheidole*, and in ants in general.

First, patterns of division of labor in *Pheidole* are associated with brain composition, particularly the relative sizes of olfactory input (AL) and sensory processing (MB) regions. Outside-nest work and increased behavioral flexibility correlate with greater investment in these regions. This is evident from comparisons of behaviorally differentiated workers of different ages and different subcastes, suggesting olfaction is a key determinant of task performance in *Pheidole*. Variation in subcaste brain structure is not simply due to uniform scaling of total brain size: major brains display increased relative investment in the SEG versus the central brain, but decreased investment in the ALs and MBs relative to the remainder of the central brain, likely affecting their patterns of task attendance. Interspecific variation in brain organization, which we document in addition to age- and subcaste-related effects, could explain the puzzling result that *Pheidole* major worker behavioral specialization does not appear to be correlated with their external morphology or degree of morphological specialization relative to conspecific minors [32]. While we recognize our volumetric analysis of brain subregions does not reveal details of the neuronal circuits that comprise brain neuropils, the volumetric differences we identify likely reflect fine-scale developmental differences in processes such as dendritic arborization [46] and formation of synaptic complexes (e.g. MB-C microglomeruli, [56,57]), which presumably have more proximate impacts on the functions of behaviorally relevant neural circuits [58]. Differences in worker neural phenotype are thus relevant to understanding species-, subcaste-, and age-related division of labor. Events that occur during pupal development may produce the species- and subcaste-specific neural phenotypes evident even in the newly eclosed workers in our study, while processes that continue into the adult stage, and integrate both adaptive developmental trajectories and plastic responses to task experience, likely underlie behavioral maturation. Overall, our results support the hypothesis that division of labor in *Pheidole*, as well as in ants and other social insects, is mediated largely by neurobiological and associated upstream physiological and genetic factors [59,60].

Second, *Pheidole* colonies, like those of most social insects, have decentralized decision-making systems [61]. Recent work has generally accentuated the "complexity" of emergent group behaviors while emphasizing "simplicity" at the individual level [62–65]. However, we have shown that brain structure, which underlies the ability of *Pheidole* workers to perceive social signals and cues and thus organize and participate in collective decisions and actions, varies dramatically in ways that likely affect how individuals of different subcastes and ages perceive, integrate and respond to sensory stimuli cueing behavior. Distinguishing between individual and group "complexity" may therefore be less important than understanding social and ecological factors acting on worker phenotypes, including neurobiological mechanisms regulating how workers detect stimuli and interact to produce emergent behavior. Swarm intelligence need not result simply from decision-making by large numbers of behaviorally and neuroanatomically similar individuals. The efficiency of group actions, both in insect societies [66] and vertebrate social groups [67], may well be enhanced by the presence of behaviorally sophisticated individuals whose differential task and sensory abilities can alter global patterns.

Supporting Information

Supporting Information S1 Detailed statistical methods.

Table S1 Full details of ANCOVA models used to assess between-group grade shifts in brain subregion scaling for brain structures that share common scaling slopes among worker groups.

Table S2 Full details of ANOVA models used to assess differences among groups in absolute OL, AL, and MB volume.

Acknowledgments

We thank Stefan Cover for assisting with field collection and sharing his insights on *Pheidole* ecology. Jeff DaCosta, Ysabel Giraldo, Wulfila Gronenberg, J. Frances Kamhi, Adina Rusakov, and several anonymous reviewers provided helpful discussion and critical readings of the manuscript.

Author Contributions

Conceived and designed the experiments: MLM JFAT. Performed the experiments: MLM. Analyzed the data: MLM. Contributed reagents/materials/analysis tools: JFAT. Wrote the paper: MLM JFAT.

References

1. Sherry DF (2006) Neuroecology. Annu Rev Psychol 57: 167–197.
2. Niven J, Laughlin SB (2008) Energy limitation as a selective pressure on the evolution of sensory systems. J Exp Biol 211: 1792–1804.
3. Dunbar RIM, Shultz S (2007) Evolution in the social brain. Science 317: 1344–1347.
4. Suga N, Jen PH (1976) Disproportionate tonotopic representation for processing CF-FM sonar signals in the mustache bat auditory cortex. Science 194: 542–544.
5. Byrne RW, Corp N (2004) Neocortex size predicts deception rate in primates. Proc R Soc Lond B 271: 1693–1699.
6. Leingartner A, Thuret S, Kroll TT, Chou SJ, Leasure JL, et al. (2007) Cortical area size dictates performance at modality-specific behaviors. Proc Natl Acad Sci U S A 104: 4153–4158.
7. Ott SR, Rogers SM (2010) Gregarious desert locusts have substantially larger brains with altered proportions compared with the solitarious phase. Proc R Soc Lond B 277: 3087–3096.
8. Shultz S, Dunbar RIM (2010) Social bonds in birds are associated with brain size and contingent on the correlated evolution of life-history and increased parental investment. Biol J Linnean Soc 100: 111–123.
9. Pitnick S, Jones KE, Wilkinson GS (2006) Mating system and brain size in bats. Proc R Soc Lond B 273: 719–724.
10. Farris SM, Roberts NS (2005) Coevolution of generalist feeding ecologies and gyrencephalic mushroom bodies in insects. Proc Natl Acad Sci U S A 102: 17394–17399.
11. Farris SM, Schulmeister S (2011) Parasitoidism, not sociality, is associated with the evolution of elaborate mushroom bodies in the brains of hymenopteran insects. Proc R Soc Lond B 278: 940–951.
12. Devoogd TJ, Krebs JR, Healy SD, Purvis A (1993) Relations between song repertoire size and the volume of brain nuclei related to song - comparative evolutionary analyses amongst oscine birds. Proc R Soc Lond B 254: 75–82.
13. Roth TC, Pravosudov VV (2009) Hippocampal volumes and neuron numbers increase along a gradient of environmental harshness: a large-scale comparison. Proc R Soc Lond B 276: 401–405.
14. Catania KC, Remple MS (2002) Somatosensory cortex dominated by the representation of teeth in the naked mole-rat brain. Proc Natl Acad Sci U S A 99: 5692–5697.

15. Iwaniuk AN, Dean KM, Nelson JE (2004) A mosaic pattern characterizes the evolution of the avian brain. Proc R Soc Lond B 271: S148–S151.
16. Iwaniuk AN, Nelson JE (2001) A comparative analysis of relative brain size in waterfowl (Anseriformes). Brain Behav Evol 57: 87–97.
17. Kelber C, Rössler W, Kleineidam CJ (2010) Phenotypic plasticity in number of glomeruli and sensory innervation of the antennal lobe in leaf-cutting ant workers (*A. vollenweideri*). Dev Neurobiol 70: 222–234.
18. Kuebler LS, Kelber C, Kleineidam CJ (2010) Distinct antennal lobe phenotypes in the leaf-cutting ant (*Atta vollenweideri*). J Comp Neurol 518: 352–365.
19. Hölldobler B, Wilson EO (1990) The ants. Cambridge: Harvard University Press.
20. Darwin C (1871) The Descent Of Man, and Selection in Relation to Sex. London: John Murray.
21. Molina Y, Harris RM, O'Donnell S (2009) Brain organization mirrors caste differences, colony founding and nest architecture in paper wasps (Hymenoptera: Vespidae). Proc R Soc Lond B 276: 3345–3351.
22. Molina Y, O'Donnell S (2008) Age, sex, and dominance-related mushroom body plasticity in the paperwasp *Mischocyttarus mastigophorus*. Dev Neurobiol 68: 950–959.
23. Smith AR, Seid MA, Jiménez LC, Wcislo WT (2010) Socially induced brain development in a facultatively eusocial sweat bee *Megalopta genalis* (Halictidae). Proc R Soc Lond B 277: 2157–2163.
24. Gronenberg W, Riveros AJ (2009) Social brains and behavior - past and present. In: Gadau J, Fewell J, eds. Organization of insect societies: from genome to sociocomplexity. Cambridge: Harvard University Press. pp 377–401.
25. Wilson EO (2003) *Pheidole* in the New World: a dominant, hyperdiverse ant genus. Cambridge: Harvard University Press.
26. Seid MA, Traniello JFA (2006) Age-related repertoire expansion and division of labor in *Pheidole dentata* (Hymenoptera: Formicidae): a new perspective on temporal polyethism and behavioral plasticity in ants. Behav Ecol Sociobiol 60: 631–644.
27. Muscedere ML, Willey TA, Traniello JFA (2009) Age and task efficiency in the ant *Pheidole dentata*: young minor workers are not specialist nurses. Anim Behav 77: 911–918.
28. Muscedere ML, Traniello JFA, Gronenberg W (2011) Coming of age in an ant colony: cephalic muscle maturation accompanies behavioral development in *Pheidole dentata*. Naturwissenschaften 98: 783–793.
29. Seid MA, Harris KM, Traniello JFA (2005) Age-related changes in the number and structure of synapses in the lip region of the mushroom bodies in the ant *Pheidole dentata*. J Comp Neurol 488: 269–277.
30. Seid MA, Traniello JFA (2005) Age-related changes in biogenic amines in individual brains of the ant *Pheidole dentata*. Naturwissenschaften 92: 198–201.
31. Seid MA, Goode K, Li C, Traniello JFA (2008) Age- and subcaste-related patterns of serotonergic immunoreactivity in the optic lobes of the ant *Pheidole dentata*. Dev Neurobiol 68: 1325–1333.
32. Mertl AL, Traniello JFA (2009) Behavioral evolution in the major worker subcaste of twig-nesting *Pheidole* (Hymenoptera: Formicidae): does morphological specialization influence task plasticity? Behav Ecol Sociobiol 63: 1411–1426.
33. Mertl AL, Sorenson MD, Traniello JFA (2010) Community-level interactions and functional ecology of major workers in the hyperdiverse ground-foraging *Pheidole* (Hymenoptera, Formicidae) of Amazonian Ecuador. Insectes Soc 57: 441–452.
34. Strausfeld NJ, Sinakevitch I, Brown SM, Farris SM (2009) Ground plan of the insect mushroom body: functional and evolutionary implications. J Comp Neurol 513: 265–291.
35. Heisenberg M (1998) What do the mushroom bodies do for the insect brain? An introduction. Learn & Mem 5: 1–10.
36. Farris SM (2011) Are mushroom bodies cerebellum-like structures? Arthropod Struct Dev 40: 368–379.
37. Fahrbach SE (2006) Structure of the mushroom bodies of the insect brain. Annu Rev Entomol 51: 209–232.
38. Durst C, Eichmuller S, Menzel R (1994) Development and experience lead to increased volume of subcompartments of the honeybee mushroom body. Behav Neural Biol 62: 259–263.
39. Withers GS, Fahrbach SE, Robinson GE (1993) Selective neuroanatomical plasticity and division of labour in the honeybee. Nature 364: 238–240.
40. Gronenberg W, Heeren S, Hölldobler B (1996) Age-dependent and task-related morphological changes in the brain and the mushroom bodies of the ant *Camponotus floridanus*. J Exp Biol 199: 2011–2019.
41. O'Donnell S, Donlan NA, Jones TA (2004) Mushroom body structural change is associated with division of labor in eusocial wasp workers (*Polybia aequatorialis*, Hymenoptera: Vespidae). Neurosci Lett 356: 159–162.
42. Maleszka J, Barron AB, Helliwell PG, Maleszka R (2009) Effect of age, behaviour and social environment on honey bee brain plasticity. J Comp Physiol A 195: 733–740.
43. Fahrbach SE, Moore D, Capaldi EA, Farris SM, Robinson GE (1998) Experience-expectant plasticity in the mushroom bodies of the honeybee. Learn & Mem 5: 115–123.
44. Kühn-Bühlmann S, Wehner R (2006) Age-dependent and task-related volume changes in the mushroom bodies of visually guided desert ants, *Cataglyphis bicolor*. J Neurobiol 66: 511–521.
45. Fahrbach SE, Farris SM, Sullivan JP, Robinson GE (2003) Limits on volume changes in the mushroom bodies of the honey bee brain. J Neurobiol 57: 141–151.
46. Farris SM, Robinson GE, Fahrbach SE (2001) Experience- and age-related outgrowth of intrinsic neurons in the mushroom bodies of the adult worker honeybee. J Neurosci 21: 6395–6404.
47. Ehmer B, Gronenberg W (2004) Mushroom body volumes and visual interneurons in ants: comparison between sexes and castes. J Comp Neurol 469: 198–213.
48. Wilson EO (1976) The organization of colony defense in the ant *Pheidole dentata* Mayr (Hymenoptera: Formicidae). Behav Ecol Sociobiol 1: 63–81.
49. Patel AD (1990) An unusually broad behavioral repertory for a major worker in a dimorphic ant species: *Pheidole morrisi* (Hymenoptera: Formicidae). Psyche 97: 181–191.
50. Bucher D, Scholz M, Stetter M, Obermayer K, Pfluger HJ (2000) Correction methods for three-dimensional reconstructions from confocal images: I. tissue shrinking and axial scaling. J Neurosci Methods 100: 135–143.
51. Warton DI, Wright IJ, Falster DS, Westoby M (2006) Bivariate line-fitting methods for allometry. Biol Rev 81: 259–291.
52. Barton RA, Harvey PH (2000) Mosaic evolution of brain structure in mammals. Nature 405: 1055–1058.
53. Acebes A, Ferrus A (2001) Increasing the number of synapses modifies olfactory perception in *Drosophila*. J Neurosci 21: 6264–6273.
54. Winnington AP, Napper RM, Mercer AR (1996) Structural plasticity of identified glomeruli in the antennal lobes of the adult worker honey bee. J Comp Neurol 365: 479–490.
55. Ishikawa Y, Aonuma H, Miura T (2008) Soldier-specific modification of the mandibular motor neurons in termites. PLoS ONE 3: e2617.
56. Groh C, Rössler W (2011) Comparison of microglomerular structures in the mushroom body calyx of neopteran insects. Arthropod Struct Dev 40: 358–367.
57. Stieb SM, Muenz TS, Wehner R, Rössler W (2010) Visual experience and age affect synaptic organization in the mushroom bodies of the desert ant *Cataglyphis fortis*. Dev Neurobiol 70: 408–423.
58. Chittka L, Niven J (2009) Are bigger brains better? Curr Biol 19: R995–R1008.
59. Smith CR, Toth AL, Suarez AV, Robinson GE (2008) Genetic and genomic analyses of the division of labour in insect societies. Nat Rev Genet 9: 735–748.
60. Robinson EJH (2009) Physiology as a caste-defining feature. Insectes Soc 56: 1–6.
61. Bonabeau E, Theraulaz G, Deneubourg J-L, Aron S, Camazine S (1997) Self-organization in social insects. Trends Ecol Evol 12: 188–193.
62. Camazine S, Deneubourg JL, Franks NR, Sneyd J, Theraulaz G, et al. (2001) Self-organization in biological systems. Princeton: Princeton University Press.
63. Delgado J, Solé RV (1997) Collective-induced computation. Phys Rev E 55: 2338–2344.
64. Solé RV, Bonabeau E, Delgado J, Fernández P, Marín J (2000) Pattern formation and optimization in army ant raids. Artif Life 6: 219–226.
65. Garnier S, Gautrais J, Theraulaz G (2007) The biological principles of swarm intelligence. Swarm Intell 1: 3–31.
66. Detrain C, Pasteels JM (1992) Caste polyethism and collective defense in the ant *Pheidole pallidula*: the outcome of quantitative differences in recruitment. Behav Ecol Sociobiol 29: 405–412.
67. Nagy M, Akos Z, Biro D, Vicsek T (2010) Hierarchical group dynamics in pigeon flocks. Nature 464: 890–U899.

2

Different Cranial Ontogeny in Europeans and Southern Africans

Marina L. Sardi[1]*, Fernando V. Ramírez Rozzi[2]

1 División Antropología, Facultad de Ciencias Naturales y Museo, Universidad Nacional de La Plata - CONICET, La Plata, Buenos Aires, Argentina, **2** UPR 2147 CNRS, Paris, France

Abstract

Modern human populations differ in developmental processes and in several phenotypic traits. However, the link between ontogenetic variation and human diversification has not been frequently addressed. Here, we analysed craniofacial ontogenies by means of geometric-morphometrics of Europeans and Southern Africans, according to dental and chronological ages. Results suggest that different adult cranial morphologies between Southern Africans and Europeans arise by a combination of processes that involve traits modified during the prenatal life and others that diverge during early postnatal ontogeny. Main craniofacial changes indicate that Europeans differ from Southern Africans by increasing facial developmental rates and extending the attainment of adult size and shape. Since other studies have suggested that native subsaharan populations attain adulthood earlier than Europeans, it is probable that facial ontogeny is linked with other developmental mechanisms that control the timing of maturation in other variables. Southern Africans appear as retaining young features in adulthood. Facial ontogeny in Europeans produces taller and narrower noses, which seems as an adaptation to colder environments. The lack of these morphological traits in Neanderthals, who lived in cold environments, seems a paradox, but it is probably the consequence of a warm-adapted faces together with precocious maturation. When modern *Homo sapiens* migrated into Asia and Europe, colder environments might establish pressures that constrained facial growth and development in order to depart from the warm-adapted morphology. Our results provide some answers about how cranial growth and development occur in two human populations and when developmental shifts take place providing a better adaptation to environmental constraints.

Editor: Carles Lalueza-Fox, Institut de Biologia Evolutiva - Universitat Pompeu Fabra, Spain

Funding: Financial support was provided by: PICT 1822 Agencia Nacional de Promoción Científica y Técnica, Argentina [http://www.agencia.gov.ar/] and PI N531 Universidad Nacional de La Plata, Argentina [www.unlp.edu.ar]. The funders had no role in study design, data collection and analysis, decision to publish, or preparation of the manuscript.

Competing Interests: The authors have declared that no competing interests exist.

* E-mail: msardi@fcnym.unlp.edu.ar

Introduction

The variation of growth and development among modern humans has been studied since decades. Frequently these studies focused on nutritional and epidemiological aspects that influence life-history variables, such as the age of attainment of adult size, the age at menarche, age at first reproduction, etc., whereas some other studies suggest that differences in growth and development would be genetically programmed [1].

Populations of Sub-Saharan African ancestry, for instance, differ in body size and shape with respect to populations of European ancestry at similar ages and similar socioeconomic levels [1]. The former develop ossification centres and present skeletal maturation and sexual maturation at more advanced ages than the latter; however, these results have been contested [2]. Dental studies also suggest that Southern African populations are characterised by a more advanced development when they are compared with populations of European ancestry [3–5]. Comparing adult individuals, craniometric differences were observed in the jaw, midface and cranial base. On average, the upper nasal region is relatively more projected in Europeans, together with more retracted jaws; Southern Africans, in contrast, present low noses in low faces, some degree of prognatism, narrower midfaces and cranial bases and frontal flatness [6–9].

Similarities in phenotypes among individuals are produced by regularities in developmental systems but it remains unknown which developmental mechanisms does differ in order to produce variation of specific cranial structures between populations. The study of Strand Vidarsdottir et al. [10] carried on with ontogenetic series dealing with between-populations variation suggests that part of facial shape differentiation is already present in early stages of postnatal ontogeny and that postnatal development contribute to adult differentiation. Even if this study [10] included 10 human groups, some of them were represented by small sample sizes and most of the study focused on the relationship of shape versus size.

All changes produced by growth and development constitute an ontogenetic trajectory. Growth results by changes in size while development by changes in shape [11–13] according with biological and/or chronological ages. The parameters that determine an ontogenetic trajectory are: the onset (α) and the offset (β) of growth and development, the rate of change (k) and the initial value of the trait (y_0), which resulted from growth and development previous to the observation [12].

The link between developmental changes and diversification among species or populations is the heterochronic approach.

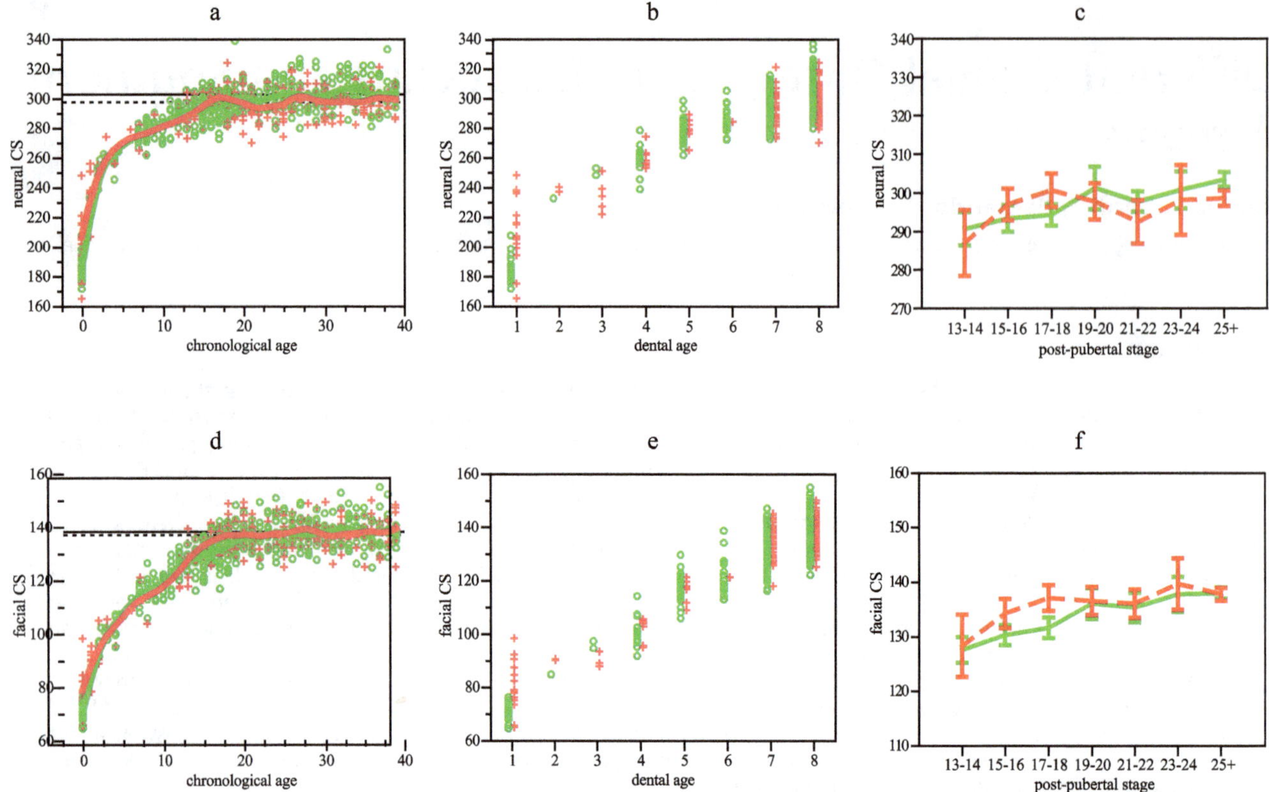

Figure 1. Neurocranial and facial size variation. (a) Neurocranial centroid size values vs chronological age. Smoothing splines accounted for 91.7 and 84.8% of variation for Europeans and Southern-Africans, respectively. (b) Neurocranial centroid size values vs dental age. (c) Mean and 95% standard error of neurocranial centroid size vs post-pubertal stages. (d) Plot of facial centroid size values vs chronological age. Smoothing splines accounted for 92 and 90.4% of variation for Europeans and Southern-Africans, respectively. (e) Facial centroid size values vs dental age. (f) Mean and 95% standard error of facial centroid size vs post-pubertal stages. Green: Europeans. Red: Southern Africans. Horizontal lines in a and b represent adult means: Europeans, solid line; Southern Africans, dotted line.

Heterochrony refers to evolutionary changes in rates and timing of developmental events, which modify ontogenetic trajectories of morphological units. Heterochrony has been described by formalisms of Gould [11] and Alberch et al. [12]. Any modification in α, k and β of a given trait, traditionally measured by a single variable, from one species to other [11–13] or from one population to other [14] indicates a heterochronic change. This concept as well as analytical approaches involved underwent several reformulations [13,15,16] leading to some confusions.

In the last decades, most of the studies of biological form are based on landmark configuration and shape is quantified by Principal Component Analysis (PCA) after a Procrustes superimposition. Some scholars have suggested that in multivariate comparisons Alberch's et al. terminology [12] cannot be used and some controversy has arisen because there is no consensus about how to interpret ontogenetic trajectories and the dissociation between size, shape and time from multivariate data. On the one hand, Mitteroecker et al. [17,18] evaluate ontogenetic changes in a shape space between species. These authors state that a change can only be interpreted as heterochrony when their trajectories are identical in the shape space, but differ just in the extension, which indicates that the offset of the development occurs at different time or size. One requisite is that the shape space encompasses all PCs since, according with Mitteroecker et al. [17,18], individual PCs are statistical constructions and they cannot be directly interpreted. On the other hand, Lieberman et al. [19] consider that an individual PC derived from geometric-morphometric data is an adequate measure of shape because each PC is statistically independent, being useful to derive testable hypotheses about developmental covariation among characters [19]. Lieberman et al. [19] interpret heterochronies from the analysis of single PCs following the method proposed by Alberch's et al. [12] and reinterpreted by Alba [20]. Lieberman

Table 1. ANOVA results for testing differences between adult means.

	F	P
neurocranium		
CS	**10.44**	**0.001**
PC1	0.04	0.844
PC2	**5.15**	**0.024**
PC3	**13.86**	**0.000**
face		
CS	0.05	0.824
PC1	**298.10**	**0.000**
PC2	**377.90**	**0.000**
PC3	0.56	0.454

Table 2. Regression equations and ANCOVA results.

	Regression equations				ANCOVA		ANCOVA without individuals of age 0	
	Europeans		S. Africans		F	F	F	F
neurocranium	**constant**	**slope**	**constant**	**slope**	**intercept**	**slope**	**intercept**	**slope**
logCAge vs logCS	**5.29**	**0.13**	**5.38**	**0.10**	**41.14**	**49.68**	**11.20**	0.96
logCAge vs PC1	**0.12**	**−0.04**	**0.13**	**−0.04**	1.25	0.44		
logCAge vs PC2	−0.00	0.00	**−0.01**	**0.01**	0.99	**4.37**	**11.49**	0.18
logCAge vs PC3	**−0.03**	**0.01**	**−0.01**	**0.01**	**45.09**	0.11		
logCS vs PC1	**1.78**	**−0.31**	**2.15**	**−0.38**	**8.12**	**8.08**	0.19	0.72
logCS vs PC2	0.03	−0.01	**−0.31**	**0.05**	**5.54**	**5.84**	**13.53**	0.41
logCS vs PC3	**−0.32**	**0.05**	**−0.40**	**0.07**	**45.39**	0.57		
face								
logCAge vs logCS	**4.31**	**0.18**	**4.38**	**0.16**	**21.33**	**17.92**	0.65	2.67
logCAge vs PC1	**−0.12**	**0.05**	**−0.12**	**0.03**	0.03	**49.56**	3.61	**7.31**
logCAge vs PC2	**0.05**	**−0.01**	**0.01**	**−0.01**	**1113.54**	2.27		
logCAge vs PC3	0.00	0.00	**−0.03**	**0.01**	**23.98**	**24.01**	0.38	1.84
logCS vs PC1	**−1.19**	**0.25**	**−0.97**	**0.19**	**17.84**	**25.36**	**9.83**	**13.32**
logCS vs PC2	**0.36**	**−0.07**	**0.44**	**−0.09**	3.34	**9.02**	**1094.94**	1.64
logCS vs PC3	**0.07**	**−0.01**	**−0.28**	**0.05**	**36.80**	**36.50**	0.26	0.10

Numbers in bold indicate probability under 0.05.

et al. [19] state that the requisite for indentifying heterochronies proposed by Mitteroecker et al. [17,18] is too stringent since the multivariate analysis will almost always result in divergence of one or more PCs, even in two closely related species. Furthermore, the approach of Mitteroecker et al. [17,18] does not include any measure of ontogenetic time (biological or chronological age), rendering difficult the assessment of heterochronies. Indeed, allometries (size-related shape changes) are sometimes taken as a surrogate of time, but this not always produces similar results because changes in the association between size and shape may be independent of that between shape and age [13,19]. When age is not available, it has been usual to compare ontogenetic series to explain morphologic divergence, avoiding inferences about heterochronies [21–23]. Different approaches can lead to very contradictory interpretations, as occurred in the evaluation of heterochronies between bonobos and chimpanzees. Whereas Mitteroecker et al. [17,18] explained variation between both species as result of non heterochronic transformations, Lieberman et al. [19], who used biological age as reference for size and shape modifications, suggested that bonobo is paedomorphic relative to chimpanzee due to initial shape underdevelopment.

In this work, we assess craniofacial changes throughout ontogeny in two human populations -Western Europeans and Southern Africans- by means of geometric-morphometric methods. Since we agree with Lieberman's et al. [19], we follow their approach in order to examine main patterns of variation in ontogenetic data. Two null hypotheses are stated: a) Southern Africans and Europeans present similar rates of cranial growth and development, and b) they undergo the offset of growth and development at similar age.

Results

Neurocranium

Size variation can be observed in Figure 1. Neurocranial ontogenetic changes in size were quite similar in both populations showing an important inflection point at ages 3–5 (Fig. 1A). At age 15 most of adult size is attained, however, Europeans achieve greater size (Table 1). According with dental age, changes are gradual being close to adult size around stage 7 (Fig. 1B). Size changes against chronological age (Table 2) on log-transformed data indicate that growth trajectories diverge. Southern Africans present greater size than Europeans at age 0, but it is probably because most of European individuals belong to the first trimester of postnatal life [24]. When individuals of age 0 were removed, slopes do not differ (Table 2). The ANOVA and Dunnet test among post-pubertal stages (Table 3, Fig. 1C) indicate that Europeans show highly significant differences, being those individuals between 13 and 18 years old smaller than adults. Significant difference in Southern Africans is only observed when adults are compared with the class of 13–14 years old. Dental stages 7 and 8 differ in both populations (Table 4).

From the GPA/PCA for neurocranial landmarks, the first three PCs obtained explain more than 57% of variation. Changes across PC1 (31% of variation) (Fig. 2A–B) show overlapped trajectories for both groups across all ontogeny (the divergence around age 10 may be a consequence of sample bias, see Methods), being adult means non-significant (Table 1). ANCOVA indicate that both trajectories are identical, considering age (Table 2). Allometric trajectories seem quite overlapped (Fig. 2C), but they diverge (Table 2). Nevertheless, since individuals of age 0 may modify slopes, analyses were done extracting them and divergence disappeared (Table 2). Transformation grids (Fig. 2D–E) show an expansion of the neurocranium in lateral view with the relative

Table 3. ANOVA and Dunnet's one-tailed test among post-pubertal stages and adults.

neurocranium				
Europeans	**CS: F = 10.20**	**PC1: F = 3.32**	**PC2: F = 0.55**	**PC3: F = 5.47**
25–39 vs 13–14	p = **0.000**	p = 0.144		p = **0.000**
25–39 vs 15–16	p = **0.000**	p = **0.024**		p = **0.003**
25–39 vs 17–18	p = **0.000**	p = 0.380		p = **0.002**
25–39 vs 19–20	p = 0.612	p = 0.114		p = 0.987
25–39 vs 21–22	p = **0.046**	p = 0.960		p = 0.617
25–39 vs 23–24	p = 0.509	p = 0.999		p = 0.142
S. Africans	**CS: F = 2.48**	**PC1: F = 1.82**	**PC2: F = 2.17**	**PC3: F = 1.79**
25–39 vs 13–14	p = **0.011**			
25–39 vs 15–16	p = 0.803			
25–39 vs 17–18	p = 0.999			
25–39 vs 19–20	p = 0.888			
25–39 vs 21–22	p = 0.068			
25–39 vs 23–24	p = 0.941			
face				
Europeans	**CS: F = 16.93**	**PC1: F = 3.56**	**PC2: F = 11.39**	**PC3: F = 2.46**
25–39 vs 13–14	p = **0.000**	p = **0.001**	p = **0.000**	p = **0.003**
25–39 vs 15–16	p = **0.000**	p = 0.054	p = **0.000**	p = 0.123
25–39 vs 17–18	p = **0.000**	p = 0.108	p = **0.000**	p = 0.144
25–39 vs 19–20	p = 0.379	p = 0.205	p = **0.019**	p = 0.338
25–39 vs 21–22	p = 0.165	p = 0.942	p = 0.085	p = 0.505
25–39 vs 23–24	p = 0.929	p = 0.998	p = 0.929	p = 0.373
S. Africans	**CS: F = 4.14**	**PC1: F = 1.74**	**PC2: F = 0.51**	**PC3: F = 0.86**
25–39 vs 13–14	p = **0.000**			
25–39 vs 15–16	p = 0.073			
25–39 vs 17–18	p = 0.843			
25–39 vs 19–20	p = 0.630			
25–39 vs 21–22	p = 0.543			
25–39 vs 23–24	p = 0.999			

Numbers in bold indicate probability under 0.05.

increment of the anterior (frontal) component as individuals increase in age and size. The ANOVA and Dunnet's test among post-pubertal stages (Table 3, Fig. 2F) indicate that Europeans differ, but only slightly due to the sub-sample of age 15–16; Southern Africans, in contrast, do not differ. Dental stages 7 and 8 differ among Europeans but not among Southern Africans (Table 4).

Table 4. ANOVA for dental stages 7 and 8.

	CS	**PC1**	**PC2**	**PC3**
neurocranium	F	F	F	F
Europeans	**59.78**	**7.47**	0.70	**8.06**
S. Africans	**10.04**	3.85	0.01	**4.26**
face				
Europeans	**76.66**	**10.93**	**75.35**	**5.68**
S. Africans	**19.93**	0.41	3.87	1.40

Numbers in bold indicate probability under 0.05.

Between-populations variation across PC2 (14% of variation) according with age and size seems overlapped (Fig. 3A–C), but adults present significant differences (Table 1). According with age, Southern Africans show significant changes, whereas Europeans do not change (Table 2). Trajectories on log-transformed data diverge, but divergence became non-significant removing individuals of age 0. Variation described by PC2 (Fig. 3D) indicates that Southern Africans, as increase in age and size, they develop a less rounded vault with frontal flatness in lateral view. Post-pubertal stages do not differ among Europeans, but they differ among Southern Africans, however, no group is smaller than adults (Table 3, Fig. 3E). Dental stages 7 and 8 do not differ in either population (Table 4).

According with PC3 (12.5% of variation), Europeans and Southern Africans show similar changes linked to age and size (Fig. 4A–C), showing adults highly significant differences (Table 1).

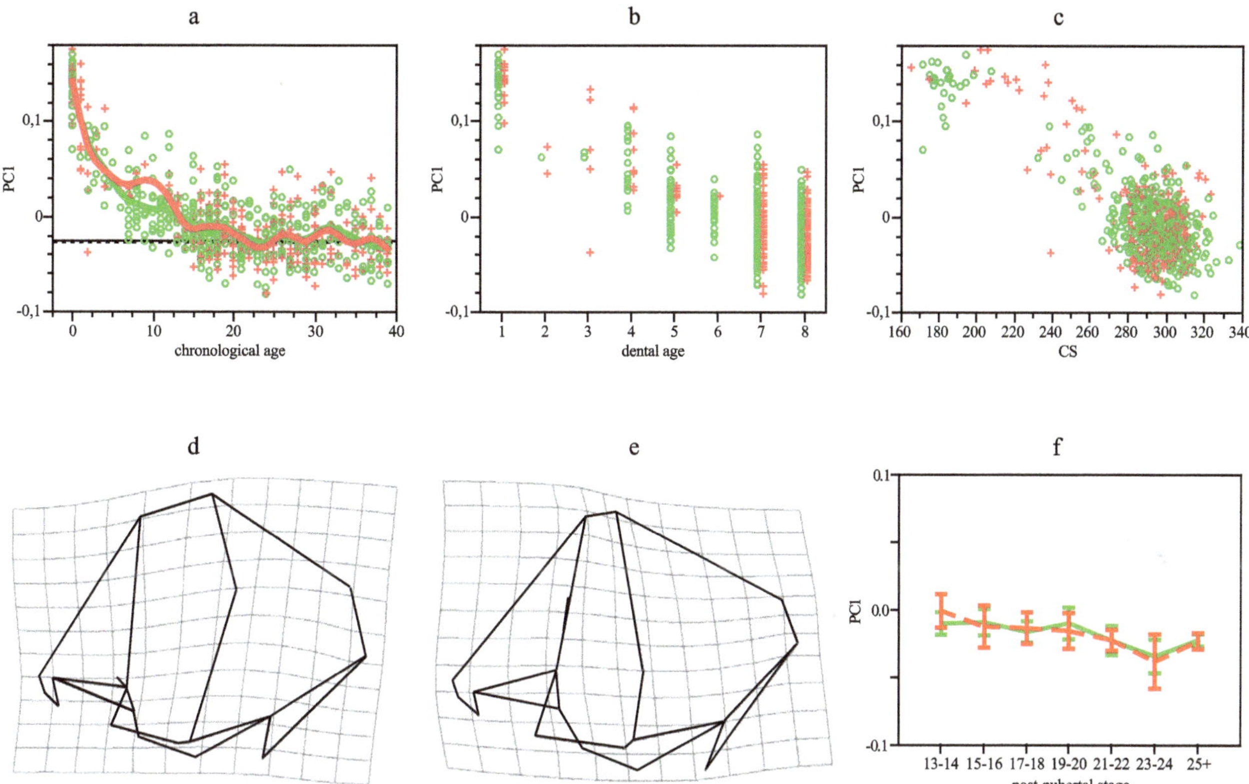

Figure 2. GPA/PCA results for neurocranial PC1. (a) PC1 scores vs chronological age. Smoothing splines accounted for 74% of variation in both distributions. (b) PC1 scores vs dental age. (c) PC1 socres vs CS. (d) Lateral view of neurocranial shape in extreme positive values (newborns = target), considering extreme negative values as the reference. (e) Lateral view of neurocranial shape in extreme negative values (adults = target), considering extreme positive values as the reference. (f) Mean and 95% standard error for PC1 scores vs post-pubertal stages. Green: Europeans. Red: Southern Africans. Horizontal lines in a represent adult means: Europeans, solid line; Southern Africans, dotted line.

Trajectories are parallel, but intercepts diverge (Table 2). Variation described by PC3 indicates that, since birth, Southern Africans present taller vaults in frontal view (Fig. 4D). In the European sample, individuals under 19 years old differ from adults, but post-pubertal stages do not differ among Southern Africans (Table 3, Fig. 4E). Dental stages 7 and 8 differ in both populations (Table 4).

Face

Size changes in facial ontogenies do not differ between Europeans and Southern Africans (Fig. 1D–E), neither adult size (Table 1). Slopes diverge as well as intercepts with log-transformed data, however this difference disappears when individuals of age 0 are removed (Table 2). The comparison of post-pubertal stages (Table 3, Fig. 1F) indicated that the offset of facial growth was more advanced in Southern Africans (13–14 years old) than in Europeans (up to 17–18 years old) (Table 3). Variation between dental stages 7 and 8 is highly significant for both populations (Table 4).

From the GPA/PCA for facial landmarks, the first three PCs obtained explain more than 47% of variation. Trajectories across PC1 (28% of variation), in contrast, show both groups overlapped during the first two years of life, but from this age, trajectories diverge progressively; divergence increases after 5 years old (Fig. 5A) and after dental stage 5 (Fig. 5B), resulting in an important difference in average adult shape among these populations (Table 1). ANCOVA indicates that slopes diverge, but not intercepts (Table 2), which means that there is no important variation among newborns. Adult differentiation is achieved because Southern Africans show lesser shape changes than Europeans. Differentiation between slopes is maintained after removing individuals of age 0 (Table 2). Size-related shape changes on PC1 are also lower in Africans than Europeans (Fig. 5C), being slopes significantly divergent (Table 2). Transformation grids indicate that main postnatal changes occur at the sagital plane; the nasal cavity became taller and narrower, being the nose and the palate more distally located in relation with the orbit (Fig. 5D–E). These changes are more pronounced in Europeans. According with the Dunnet's test, shape modifications are similar in both populations but the non-significant variation of Southern Africans may depend on their greater standard error (Fig. 5F), although differences between dental stages 7 and 8 were significant only for Europeans.

The PC2 (12.5% of variation) expressed variation associated with chronological and dental ages (Fig. 6A–B) and size (Fig. 6C). Adult differentiation is highly significant (Table 1), although both populations show parallel developmental changes since slopes of changes in shape according with log-transformed age and size do not differ (Table 2); the highly significant differences in intercepts reveals that differentiation has begun before birth (Table 2, Fig. 6A–C). Transformation grids indicate that shape changes are related with midfacial morphology (Fig. 6D–E). Southern Africans showed greater prognatism and wider nasal cavity than Europeans during all postnatal ontogeny, but these characteristics are accentuated up to adulthood. Post-pubertal stages do not reveal

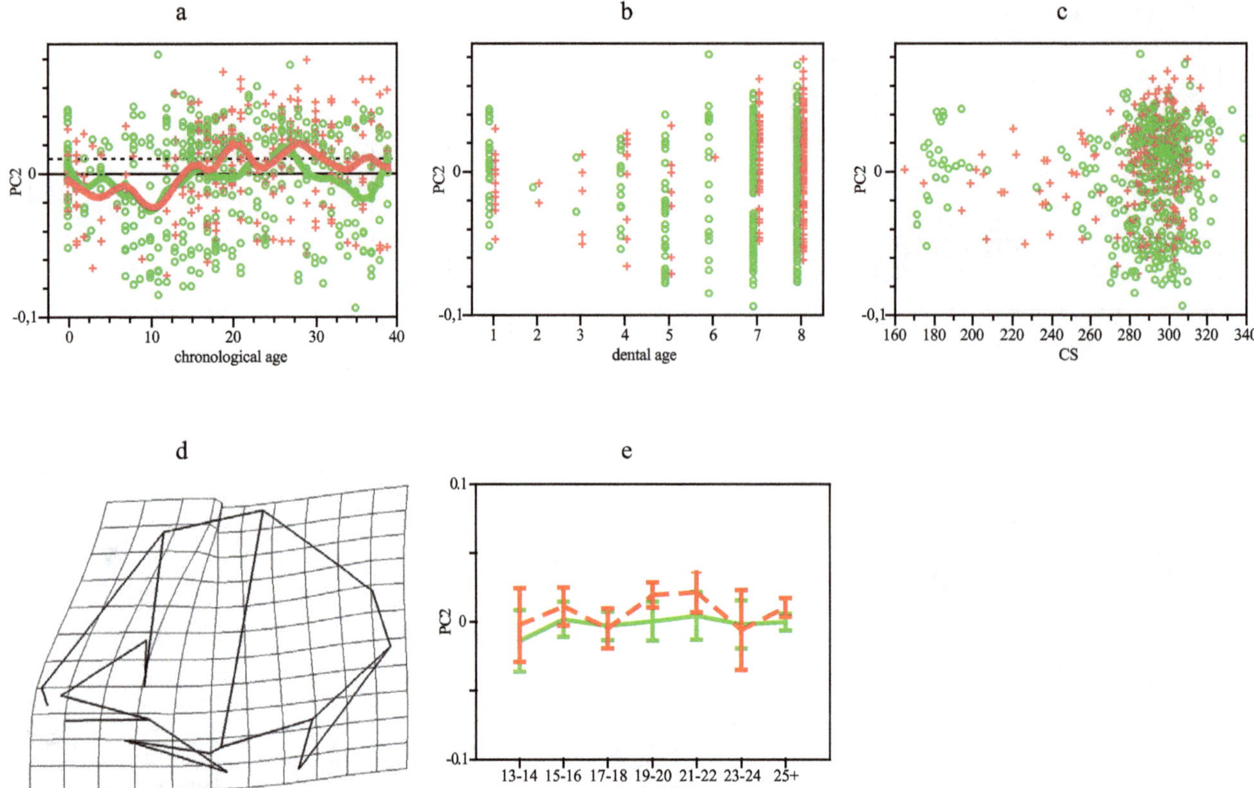

Figure 3. GPA/PCA results for neurocranial PC2. (a) PC2 scores vs chronological age. Smoothing splines accounted for 10.5 and 13% of variation in Europeans and Southern Africans, respectively. (b) PC2 scores vs dental age. (c) PC2 scores vs CS. (d) Lateral view of neurocranial shape, considering extreme negative values as the reference and extreme positive values as the target (other views do not show deformation). (e) Mean and 95% standard error for PC2 scores vs post-pubertal stages. Green: Europeans. Red: Southern Africans. Horizontal lines in a represent adult means: Europeans, solid line; Southern Africans, dotted line.

differences among Southern Africans, neither among latter dental stages, but individuals younger than 19 years old differ from adults among Europeans, as well as individuals of dental stage 7 with respect to those of stage 8 (Tables 3 and 4, Fig. 6D).

According with PC3 (7% of variation), both populations are quite overlapped across age and size, excepting among the youngest individuals (Fig. 7A–C). Adult shape does not differ (Table 1). Only Southern Africans present significant changes according with age and size (Table 2), diverging from Europeans in slopes and intercepts. Divergence becomes non-significant when individuals of age 0 are removed. Transformation grids indicate that this PC represents ontogenetic increases in prognatism (Fig. 7D). Differentiation among post-pubertal stages and among dental stages 7 and 8 are significant only for Europeans (Tables 3 and 4, Fig. 7E).

Discussion

Results of this study are somewhat coincident with previous research carried on with adults [6–9], regarding neurocranial and facial features, such as nasal morphology, alveolar projection, frontal flatness, among others (Figs. 1, 2, 3, 4, 5, 6 and 7). We demonstrated how those morphologic characters that make people look different are shaped during ontogeny, how some traits covary, and how morphologic traits differ among humans by small scale shifts in developmental rates and timing.

Both the neurocranium and the face express combinations of characters that differ since birth between Europeans and Southern Africans whereas other characters follow postnatal divergent patterns. Diversification produced during postnatal ontogeny is less evident in the neurocranium than in the face because neurocranial growth and development show parallel trajectories for PC1 (Fig. 2). Divergence expressed along other shape components was mainly produced by divergence among individuals belonging to age 0 (Table 2). This is probably because brain growth, which influences on neurocranial size and shape [25], show significant changes in rates during the first year of life [26]. Frontal flatness, a character that distinguishes Southern Africans [8], seems developed after birth (Fig. 3). Increments in the height of the external neurocranial structures, in contrast, are common for both groups and this morphology probably diverge from prenatal ontogeny. In this shape component (PC3), as well as in centroid size, Europeans take more time to achieve adult morphology (Tables 3 and 4; Figs. 1, 4).

Although main postnatal changes are shared between Europeans and Southern Africans, a small proportion of variation, linked to frontal flatness, is divergent due to greater developmental rates in the second group. Thus, both hypotheses proposed are rejected. In heterochronic terms, changes observed in neurocranium have to be interpreted very cautiously. Similar shape at similar age is attained in both populations following PC1, it suggests that there is no heterochrony between populations. In PC2, Europeans do not show change through age whereas Southern Africans do; PC2

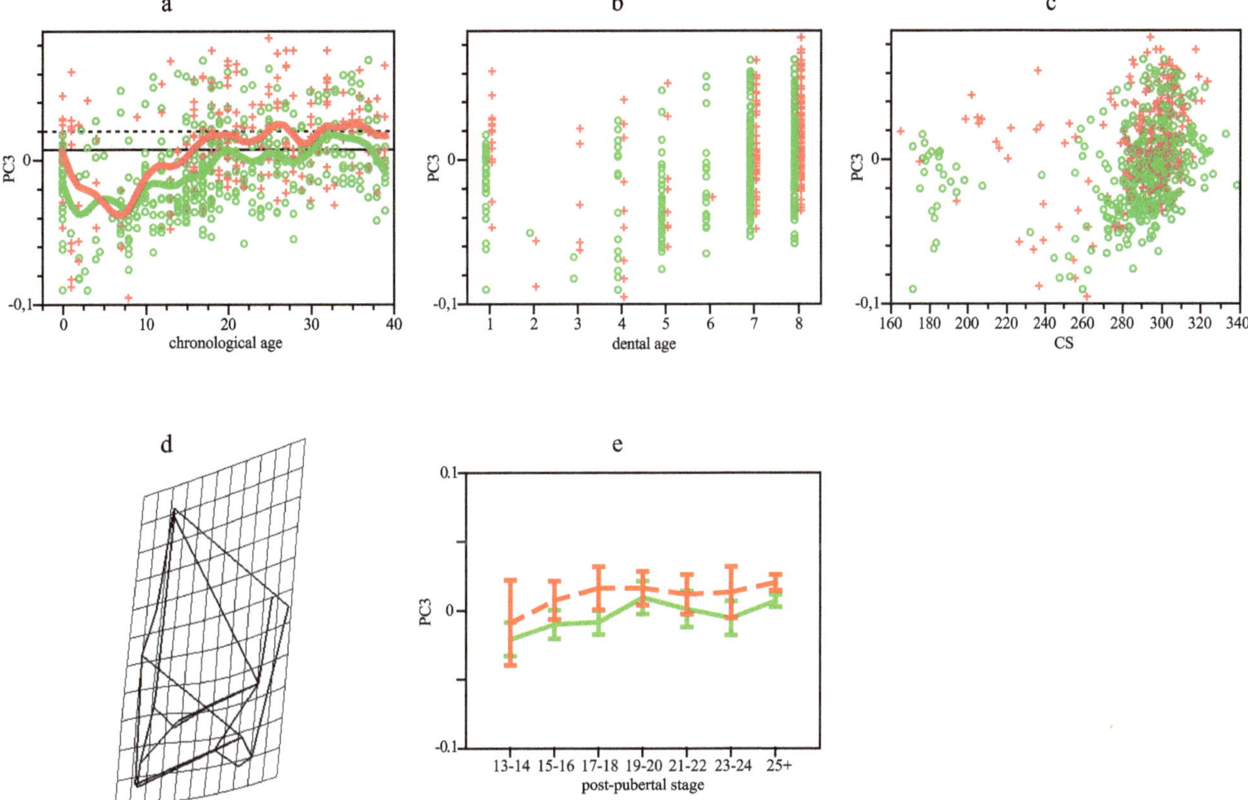

Figure 4. GPA/PCA results for neurocranial PC3. (a) PC3 scores vs chronological age. Smoothing splines accounted for 23 and 19% of variation in Europeans and Southern Africans, respectively. (b) PC3 scores vs dental age. (c) PC3 scores vs CS. (d) Frontal view of neurocranial shape, considering extreme negative values as the reference and extreme positive values as the target. (e) Mean and 95% standard error for PC3 scores vs post-pubertal stages. Green: Europeans. Red: Southern Africans. Horizontal lines in a represent adult means: Europeans, solid line; Southern Africans, dotted line.

represents most probably changes that cannot be interpreted from heterochrony. Differently, PC3 indicates a displacement, aspects represented by this PC are observed in Southern Africans at an earlier age that they occur in Europeans.

Facial ontogenetic trajectories are more divergent between populations (Tables 2, 3 and 4; Figs. 5, 6 and 7). The main axe of shape variation does not show differences at birth but progressively diverges, showing Europeans higher developmental rates than Southern Africans (Fig. 5) and attaining taller faces. During adulthood, Europeans attain highly significant shape differences with respect to Southern Africans, but with similar size (Table 1).

Other facial traits, linked to nasal width and prognatism, show parallel ontogenetic changes and are already different at birth (Fig. 6). Size-related variation is divergent because allometries are more pronounced in Southern Africans (Table 2). These ones present all over postnatal ontogeny wider noses and the superior alveolar arch more projected than Europeans. This differentiation may result from prenatal divergence and accentuated during postnatal ontogeny, along with changes in nasal height (Fig. 5).

Facial size and those morphologic characters represented by main PCs attain their adult state during pre-pubertal stages in Southern Africans and later in Europeans (Tables 3 and 4). Even if the smaller sample size among the former may produce statistical type II error for shape components, when dental age was used Southern Africans do not change after the second permanent molar is at the occlusal plane, contrarily to what occurs among Europeans. This indicates that Southern Africans undergo the offset of facial growth and development earlier than Europeans (Tables 3 and 4).

Considering facial ontogeny, thus, both hypotheses were rejected. Overall differentiation between both populations arises by a combination of processes that involve changes in rates and time of offset of facial growth and development. Facial development involves increases in facial and nasal height, being more accentuated in Europeans than in Southern Africans. These ones seem to retain younger traits during adulthood. In heterochronic terms, face in Europeans followed acceleration [20]; it undergoes greater changes in shape, with respect to age and size, with a similar final size (Table 2, Fig. 2). This set of characters does not differ at birth but progressively diverge.

Chronological age in Southern Africans might be biased by aging methods (see Methods). However, the use of a biological age as dental eruption produced similar results to those obtained with chronological age (Fig. 2). Dental maturation is supposed to be more advanced in Southern Africans [3–5], thus dental categories may encompass African individuals that are younger than Europeans, all of which expresses that ontogenetic differences would be more pronounced than observed here.

Although morphologic variation is continuous, significant differences for some body and cranial characters have been observed among worldwide modern human populations, especially for those that are geographically distant [6–9]. The pattern of

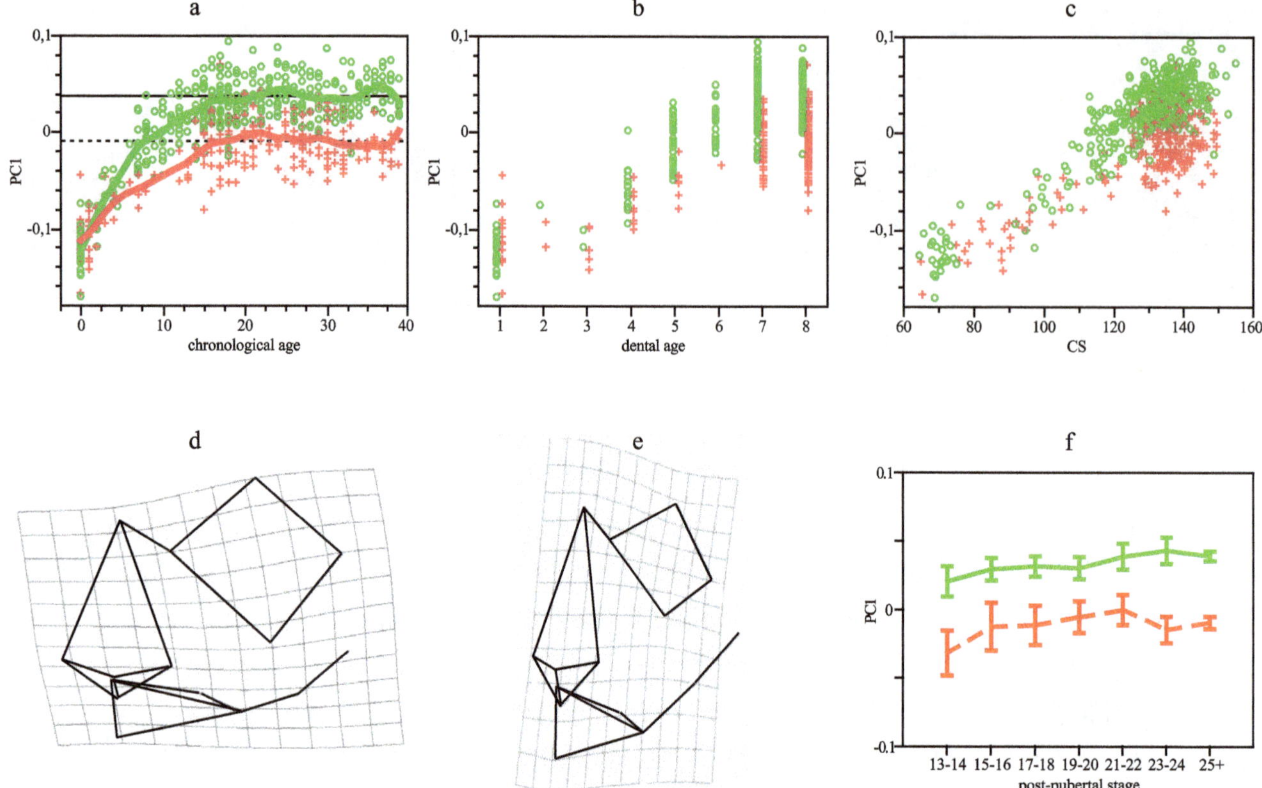

Figure 5. GPA/PCA results for facial PC1. (a) PC1 scores vs chronological age. Smoothing splines accounted for 83.4 and 68% of variation in the European and Southern African distributions, respectively. (b) PC1 scores vs dental age. (c) PC1 scores vs CS. (f) Frontal view of facial shape in extreme negative values (newborns = target), considering extreme positive values as the reference. (g) Frontal view of facial shape in extreme positive values (adults = target), considering extreme negative values as the reference. (g) Mean and 95% standard error for PC1 scores vs post-pubertal stages. Green: Europeans. Red: Southern Africans. Horizontal lines in a represent adult means: Europeans, solid line; Southern Africans, dotted line.

variation has been explained mainly as a consequence of population history. Environmental factors seem to have a minor influence in overall morphology [27,28]; however, some characters are likely to differ, as a consequence of selection or plasticity [29–33].

Lower limbs, whose morphology expresses climatic adaptation, also differ among these same populations, being Southern Africans taller than Europeans. The ontogenetic study of Frelat and Mitteroecker [34] indicates that this pattern results from postnatal divergence (in lower limb length) and prenatal divergence (relative length of tibia and femur). However, whereas our results of facial morphology suggest greater developmental rates for Europeans than Southern Africans, developmental rates of femur and tibia are greater among the latter. Postnatal ontogeny would reinforce the body climatic adaptation [34].

Similarly, nasal variation has been regarded as environmentally shaped. Nasal morphology varies across ecogeographic regions probably as a consequence of climatic adaptation [29,31,32]. Native populations of colder climates present taller and narrower noses with respect to warm-adapted populations (e.g. Subsaharan Africans), providing greater surface for warming and humidifying inspired air through the contact with the nasal mucosa, which enables a better thermoregulation [29,31,32]. Since this morphology is established at birth and accentuated early in postnatal life, it may express adaptative pressures involved.

These results may present evolutionary implications, providing clues for Neanderthal characters. Neanderthals have been considered as hominids adapted to cold climate, in part due to their supposed large paranasal sinuses. Recently, Rae et al. [35] demonstrated that Neanderthals were not characterized by relatively large paranasal sinuses, neither are they relatively smaller, as would be expected according with experimental studies of cold adaptation [33]. Neanderthals present relatively wide nasal apertures -associated to prognatism-, which is a character related to warm climate [36]. This paradox was explained by Holton and Franciscus [36], who suggested that a relative wide nasal aperture in Neanderthals is the retention of a plesiomorphic character. Facial morphology of Neanderthals differ from modern humans since very early in ontogeny [37,38]. The similar pattern of midfacial variation in Neanderthals and Southern Africans may suggest that facial ontogenies were also similar, along with more advanced maturation in the former [37,39] as well as the latter [1,3–5]. At the end of the Pleistocene, when modern *Homo sapiens* migrated into Asia and Europe, colder environments established pressures that might constrain facial growth and development for subsequently attain adapted morphology. This morphology was attained by increasing facial developmental rates and extending the attainment of adult size and shape but without affecting the reproductive output of the population.

The offset of cranial growth and development, especially in facial structures, differs between Europeans and Southern Africans, which clearly fits with developmental timing in other traits [1,3–5]. It is probable that genes and several substances that act on development are involved. Some circulating hormones,

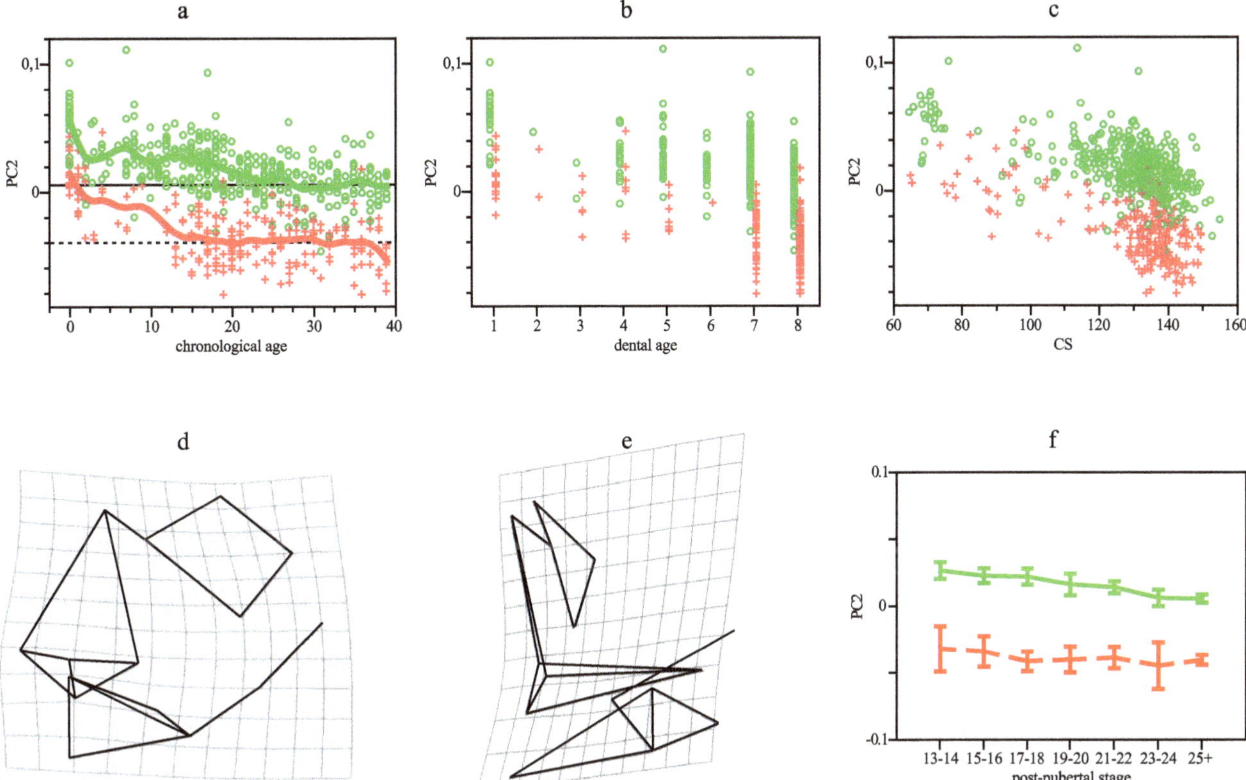

Figure 6. GPA/PCA results for facial PC2. (a) PC2 scores vs chronological age. Smoothing splines accounted for 42 and 46% of variation in the European and Southern African distributions, respectively. (b) PC2 scores vs dental age. (c) PC2 scores vs CS. (d) Frontal view of facial shape in extreme negative values (adult Southern Africans = target), considering extreme positive values (European newborns) as the reference. (e) Lateral view of facial shape in extreme negative values (adult Southern Africans = target), considering extreme positive values (European newborns) as the reference. (g) Mean and 95% standard error for PC2 scores vs post-pubertal stages. Green: Europeans. Red: Southern Africans. Horizontal lines in a represent adult means: Europeans, solid line; Southern Africans, dotted line.

such as growth hormone and IGF I, have effects on overall growth, promoting also growth on particular tissues or stimulating the local production of other growth factors [40]. Growth hormone and IGFI regulate systemically developmental rates and times of maturation, as is evident during the adolescent growth spurt. They may influence on ontogenetic allometries –e.g. the face- together with –or as a result of- other major developmental events, which occur quite late in ontogeny, and also probably associated to other variables of life-history, such as sexual maturation. Unfortunately, to the moment, there is no actual evidence suggesting differences in growth-promoting substances between native sub-Saharan populations and Europeans.

In sum, this study highlights the importance of examining the intra-specific variation in phenotypes and development for understanding evolutionary origins of interspecific diversification [41]. The adult differentiation between Europeans and Southern Africans arises by a combination of processes that involve traits modified during prenatal life and also others that diverge during postnatal ontogeny. If evolutionary developmental paleoanthropology is better defined by the questions it asks: how, when and why [42], our results provide some answers about how cranial growth and development occur in two human populations and when developmental shifts take place across individual's life. This enables to infer why variation does occurred, probably as a by product of the integration with other biological variables providing a better adaptation to environmental constraints. A further concern is that when anatomically modern humans are compared with other hominids, inferences about the differentiation must explicitly consider which human population is being compared.

Methods

Two cranial ontogenetic series derived from individuals whose age at death is between 0–39 years old were studied (Table 5). The West European sample encompasses, for the main part, Portuguese cemetery-derived individuals, which are housed at the Museo Antropologico, of the University of Coimbra (Portugal). A smaller proportion of this sample is composed of cadaver-derived skulls from French individuals, which are housed at Musée de L'Homme (France). Sex and age at death is known through cemeteries archives and direct observation of cadavers in the case of sex.

The second sample encompasses South African individuals of Bantu origins. The cranial material belongs to the Dart collection housed at the University of Witwatersrand (Johannesburg, South Africa). This collection comprises skulls of cadaver-derived origins. Sex was assessed by direct observation, whereas age at death was estimated by unknown methods [43].

Since chronological age may be biased in the Dart collection, dental maturation was recorded according with a ranking (Table 6) in both collections, which is a good proxy of biological development. Each dental class was established when some teeth are fully emerged. Thus, morphometric analyses were carried out

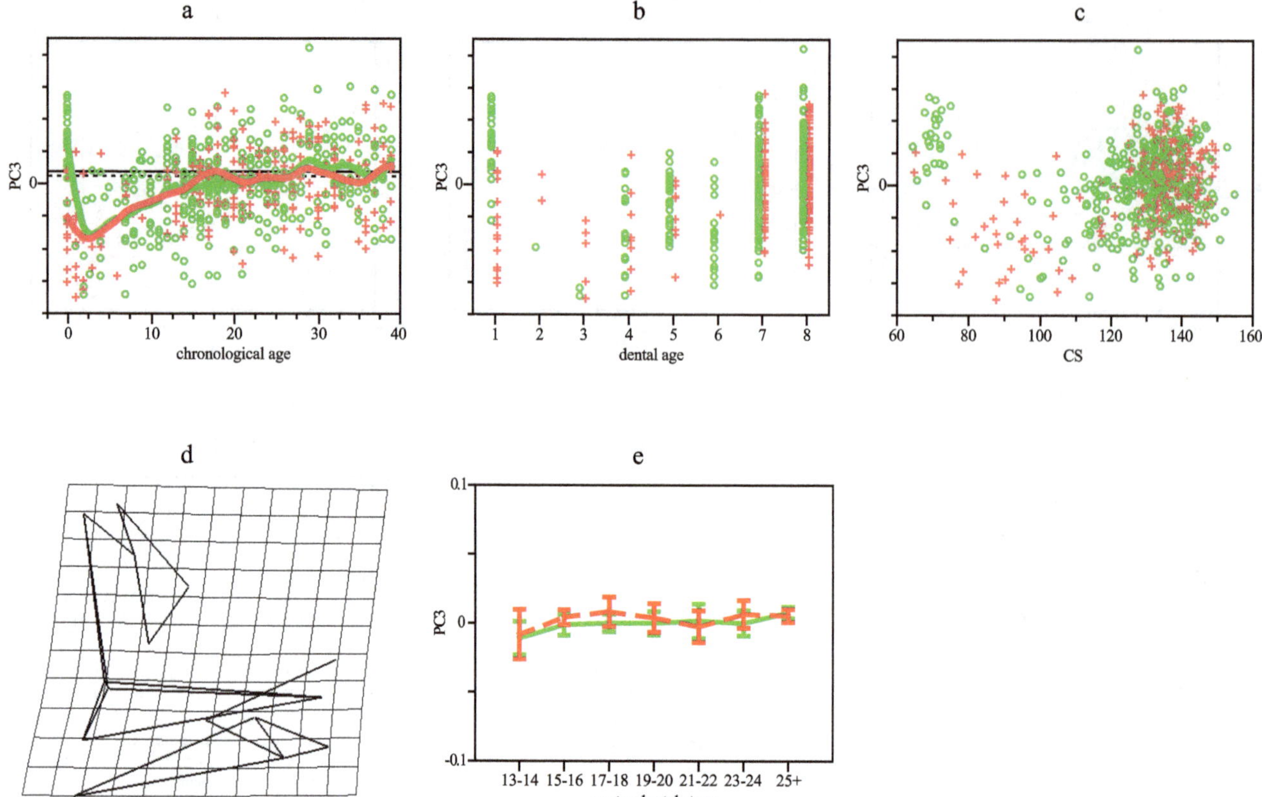

Figure 7. GPA/PCA results for facial PC3. (a) PC3 scores vs chronological age. Smoothing splines accounted for 24% of variation in both distributions. (b) PC3 scores vs dental age. (c) PC3 scores vs CS. (d) Lateral view of facial shape in PC3, considering extreme negative values as the reference and extreme positive values as the target (other views do not show deformation). (e) Mean and 95% standard error for PC3 scores vs post-pubertal stages. Green: Europeans. Red: Southern Africans. Horizontal lines in a represent adult means: Europeans, solid line; Southern Africans, dotted line.

considering as reference both chronological and biological (dental) ages.

Thirty three-dimensional (3D) landmarks, located in the vault, basicranium, and face (Table 7) were registered with Microscribe on the left side of the skull by one of the authors (M.L.S.). Wire-frames were built with landmarks located either on the face or neurocranium (Table 7).

All 3D coordinates of landmarks were analysed by geometric-morphometric methods. Geometric-morphometrics suit well with the analysis and representation of the relationships among size and shape because it enables the evaluation of heterochronies since the Procrustes superimposition provides measures of shape once all information due to scale, location and rotation was removed; and it provides a measure of size –the centroid size- that is uncorrelated with shape.

Cuadratic distances between homologous landmarks were minimised by means of Generalised Procrustes Analysis (GPA) with *Morphologika*. After Procrustes transformation, landmark configurations were analysed by means of a Principal Component Analysis (PCA). GPA and PCA enabled to obtain scores of shape variation and the centroid size (CS). Transformation grids were built to visualize morphologic changes.

Neurocranium and face were analysed separately because both morphological units present different embryological origins and they present different developmental rates [44,45]. The neurocranium encompasses two main skeletal structures –the vault and the basicranium- of different embryological origins, however growth rates are similar and associated to brain growth. From a phylogenetic perspective, some developmental shifts in both the neurocranium and the face can explain morphologic diversity among mammals [46] and, more specifically, among primates [19,47].

Individuals were plotted according with size (CS) and shape (PCs) variables against chronological and dental ages. In order to visualize trajectories according with chronological age, the smoothing spline was adjusted with Jump 5 (SAS Institute Inc.). This method requires the definition of the smoothing parameter λ, which establishes the trade-off between the bias and the variance along a trajectory. Some λ were explored but 10 were chosen by visual inspection. Greater detail is provided in those PCs that account for greatest percentage of variance.

Statistical analyses were done with an alpha level of 0.05 using with Systat 10.2 (Systat Software Inc.) and Statistica (Statsoft Inc.) softwares. Differences among adults were tested with ANOVA. In order to test for change of size against age, shape against age and shape against size, within-populations regression lines were adjusted after the transformation of chronological age and centroid size into natural logarithms to get linear distributions. Equality in trajectories was evaluated by means of ANCOVA. Firstly, ANCOVA for testing the homogeneity of slopes was performed introducing the interaction term between the covariate and the grouping variable. Population was the grouping variable, log-chronological age and log-centroid size were covariates and shape variables (PCs) were the dependent variables; log-centroid size was also a dependent variable using log-chronological age as a

Table 5. Sample distribution according with chronological ages.

	Europeans				Southern Africans		
chron. age	females	males	unknown	Total	females	males	Total
0	1	1	24	26	3	8	11
1	0	0	1	1	6	4	10
2	2	0	5	7	2	2	4
3	1	1	0	2	0	2	2
4	0	0	6	6	2	1	3
5	0	0	1	1	0	0	0
6	0	1	0	1	0	1	1
7	6	2	2	10	1	2	3
8	5	5	2	12	0	1	1
9	2	2	1	5	0	1	1
10	6	3	2	11	0	0	0
11	4	4	0	8	0	0	0
12	9	3	1	13	3	0	3
13	2	5	0	7	2	4	6
14	5	2	4	11	2	1	3
15	8	10	2	20	1	4	5
16	6	7	0	13	2	8	10
17	14	8	1	23	1	8	9
18	11	9	1	21	2	9	11
19	6	5	0	11	6	2	8
20	3	6	0	9	6	8	14
21	5	4	0	9	4	7	11
22	8	3	0	11	2	2	4
23	2	6	0	8	2	1	3
24	6	5	0	11	2	3	5
25–39	60	65	0	125	47	46	93
Total	**172**	**157**	**53**	**382**	**96**	**125**	**221**

Table 7. Landmarks registered with Microscribe on the left side of the skull.

neurocranium	face
Nasion	Nasion
Glabella	Subspinale
Bregma	Prosthion
Vertex	Palatine-maxillare suture
Lambda	Posterior nasal spine
Opisthocranion	Right alare
Opisthion	Left alare
Basion	Zygomaxillare
Hormion	Maximum alveolar width
Pterion	Maxillary tuberosity
Eurion	Inferior zygo-temporal suture
Asterion	Dacryon
Porion	Ectoconchion
Stephanion	Superior rim of the orbit
Posterior mandibular fossa	Orbital
Sphenotemporal crest	
Optic foramen	
Dacryon	
Superior rim of the orbit	

covariate. A non significant interaction between the grouping variable and the covariate indicates that the relation between the covariate and the response variable Y does not differ between groups. When slopes did not differ, a second ANCOVA pooling the regression slopes (removing the interaction term) was performed. This enabled to test for differences in y-adjusted values for any x-value, which is also a test or equality of populations intercepts [48]. When slopes and intercepts do not differ, trajectories are identical and potential adult variation may result from the extension or truncation of trajectories; but if intercepts differ, it is probably due to differentiation generated during prenatal life. A significant interaction, in contrast, indicates that slopes differ. When slopes differ significantly, certain values of X (i.e. age) were chosen and both ANCOVA methods were repeated with and without the interaction term in order to determine the regions of

Table 6. Ranking of dental maturation and sample distribution.

		Europeans				Southern Africans		
stage	maturation	females	males	unknown	total	females	males	total
1	no teeth at the occlusal plane	1	1	24	26	5	10	15
2	dec. incisors at the occlusal plane	0	0	1	1	1	1	2
3	dec. canines at the occlusal plane	0	0	2	2	4	1	5
4	dec. dentition completely erupted	3	2	10	15	3	6	9
5	1st permanent molar at the occlusal plane	16	12	6	34	3	4	7
6	3rd premolar is at the occlusal plane	9	8	2	19	1	0	1
7	2nd permanent molar at the occlusal plane	77	60	8	145	25	24	49
8	3rd molar fully erupted	66	74	0	140	54	79	133
	Total	**172**	**157**	**53**	**382**	**96**	**125**	**221**

significance [48]. Significant slopes can be associated with non-significant intercepts, which may indicate that both groups are not different during first stages of postnatal ontogeny and they diverge later. If the intercept also differs, no assumption about ontogeny can be done because the differentiation between intercepts is not maintained for other values of X [48].

In order to evaluate the offset of growth and development, individuals of different chronological ages in each population were compared with ANOVA and Dunnett's test (one tail). The Dunnett's t test is a method for comparing several group means to a control mean, which is useful to look for significant differences of those individuals that are older than 12 years old with respect to the adult reference. Adults encompass individuals aged between 25 and 39. Those individuals aged from 13 to 24 were grouped into 13–14, 15–16, 17–18, 19–20, 21–22, 23–24 classes in order to get greater sample sizes. When dental age was used, adults are those individuals belonging to dental class 8. These ones were compared only with those of dental class 7 which approximately corresponds to post-pubertal stage, given that M2 is fully emerged around 12.5–13.5 years [49]. Differences between means were compared with ANOVA.

Acknowledgments

We are indebted to P. Mennecier, M.E. Cunha, S. Wasterlain, N. Pather, P. Mamiane who provided access to collections and to anonymous reviewers who made corrections and suggestions that improved significantly this paper. Much of this work was improved by comments and help from Marisol Anzelmo, Amandine Blin, Pablo Sardi, Jimena Barbeito-Andrés, Fernando Ventrice, Yves Le Bouc and Evelia Oyhenart.

Author Contributions

Conceived and designed the experiments: MLS FVRR. Performed the experiments: MLS. Analyzed the data: MLS. Wrote the paper: MLS FVRR.

References

1. Eveleth PB (1986) Population differences in growth. Environmental and genetic factors. In: Falkner F, Tanner JM, eds. Human Growth: a Comprehensive Treatise. New York: Plenum. pp 221–239.
2. Pasquet P, Froment A, Pondi-Njiki O (1996) Le développement pubertaire des écoliers camerounais. Etude en milieu périurbain (banlieue de Yaoundé). Cahiers d'Anthropol Biom Hum 14: 223–236.
3. Reid D, Dean MC (2006) Variation in modern human enamel formation times. J Hum Evol 50: 329–346.
4. Olze A, van Niekerk P, Ishikawa T, Zhu BL, Schulz R, et al. (2007) Comparative study on the effect of ethnicity on wisdom tooth eruption. Int J Legal Med 121: 445–448.
5. Liversidge H (2008) Dental age revisited. In: Irish JD, Nelson GC, eds. Technique and application in dental anthropology. Cambridge: Cambridge University Press. pp 234–265.
6. Howells WW (1989) Skull shapes and the map. Cambridge: Harvard University Press.
7. Lahr MM (1996) The evolution of modern human diversity. Cambridge: Cambridge University Press.
8. Hanihara T (2000) Frontal and facial flatness of major human populations. Am J Phys Anthropol 111: 105–134.
9. Hennessy RH, Stringer CB (2002) Geometric morphometric study of the regional variation of modern human craniofacial form. Am J Phys Anthropol 117: 37–48.
10. Viðarsdóttir US, O'Higgins P, Stringer CA (2001) Geometric morphometric study of regional differences in the ontogeny of the modern human facial skeleton. J Anat 201: 211–229.
11. Gould SJ (1977) Ontogeny and phylogeny. Cambridge: Harvard University Press.
12. Alberch P, Gould SJ, Oster GF, Wake DB (1979) Size and shape in ontogeny and phylogeny. Paleobiology 5: 296–317.
13. Godfrey LR, Sutherland MR (1995) What's growth got to do with it? Process and product in the evolution of ontogeny. J Hum Evol 29: 405–431.
14. Ramírez Rozzi FV, González-José R, Pucciarelli HM (2005) Cranial growth in normal and low-protein-fed *Saimiri*. An environmental heterochrony. J Hum Evol 49: 515–535.
15. McKinney ML, McNamara KJ (1991) Heterochorony: The evolution of ontogeny. New York: Plenum Press.
16. Klingenberg CP (1998) Heterochrony and allometry: the analysis of evolutionary change in ontogeny. Biol Rev Camb Philos Soc 73: 79–123.
17. Mitteroecker P, Gunz P, Bookstein FL (2004) Regional dissociated heterochrony in multivariate analysis. Ann Anat 186: 463–470.
18. Mitteroecker P, Gunz P, Bookstein FL (2005) Heterochrony and geometric morphometrics: a comparison of cranial growth in *Pan paniscus* versus *Pan troglodytes*. Evol & Devel 7: 244–258.
19. Lieberman DE, Carlo J, Ponce de León M, Zollikofer CPE (2007) A geometric morphometric analysis of heterochrony in the cranium of chimpanzees and bonobos. J Hum Evol 52: 647–662.
20. Alba DM (2002) Shape and stage in htererochronic models. In: Minough-Purvis N, McNamara KJ, eds. Human evolution through developmental change. Baltimore: Johns Hopkins University Press. pp 296–317.
21. Cobb SN, O'Higgins P (2004) Hominins do not share a common postnatal facial ontogenetic shape trajectory. J Exper Zool 302B: 302–321.
22. Cobb SN, O'Higgins P (2007) The ontogeny of sexual dimorphism in the facial skeleton of the African apes. J Hum Evol 53: 176–190.
23. Gerber S, Neige P, Eble GJ (2007) Combining ontogenetic and evolutionary scales of morphological disparity: a study of early Jurassic ammonites. Evol Dev 9: 472–482.
24. Guihard-Costa AM (1988) Estimation de l'âge fœtal à partir des dimensions cranio-faciales. Bull Assoc Anat (Nancy) 72: 15–19.
25. Moss ML, Young RW (1960) A functional approach to craniology. Am J Phys Anthropol 18: 281–292.
26. Sgouros S, Goldin JH, Hockley AD, Wake MJ, Natarajan K (1999) Intracranial volume change in childhood. J Neurosurg 91: 610–616.
27. Relethford JH (2002) Apportionment of global human genetic diversity based on craniometrics and skin color. Am J Phys Anthropol 118: 393–398.
28. Relethford JH (2004) Boas and beyond: migration and craniometric variation. Am J Hum Biol 16: 379–386.
29. Franciscus RG, Long JC (1991) Variation in human nasal height and breadth. Am J Phys Anthropol 85: 419–427.
30. Roseman CC, Weaver TD (2004) Multivariate apportionment of global human craniometric diversity. Am J Phys Anthropol 125: 257–263.
31. Hubbe M, Hanihara T, Harvati K (2009) Climate signatures in the morphological differentiation of worldwide modern human populations. Anat Rec 292: 1720–1733.
32. Noback ML, Harvati K, Spoor F (2011) Climate-related variation of the human nasal cavity. Am J Phys Anthropol 145: 599–614.
33. Rae TC, Viðarsdóttir US, Jeffery N, Steegmann T (2006) Developmental response to cold stress in cranial morphology of *Rattus*: implications for the interpretation of climatic adaptation in fossil hominins. Proc R Soc B 273: 2606–2610.
34. Frelat MA, Mitteroecker P (2011) Postnatal ontogeny of tibia andfemur in two human populations: a multivariate morphometric analysis. Am J Hum Biol 23: 796–804.
35. Rae TC, Koppe H, Stringer CB (2011) The Neanderthal face is not cold adapted. J Hum Evol 60: 234–239.
36. Holton NE, Franciscus RG (2008) The paradox of a wide nasal aperture in cold-adapted Neandertals: a causal assessment. J Hum Evol 55: 942–951.
37. Ponce de León MS, Zollikofer CPE (2001) Neandertal cranial ontogeny and its implications for late hominid diversity. Nature 412: 534–538.
38. Krovitz GE (2003) Shape and growth differences between Neandertals and modern humans: grounds for a species-level distinction? In: Thompson JL, Krovitz GE, Nelson AJ, eds. Patterns of growth and development in the Genus *Homo*. Cambridge: Cambridge University Press. pp 320–342.
39. Ramírez Rozzi FV, Bermúdez de Castro JM (2004) Surprisingly rapid growth in Neanderthals. Nature 428: 936–939.
40. Raff RA (1996) The shape of life. Chicago: The University of Chicago Press.
41. Arthur W (2002) The emerging conceptual framework of evolutionary developmental biology. Nature 415: 757–764.
42. Zollikofer CPE, Ponce de León M (2010) The evolution of hominin ontogenies. Sem Cell Devel Biol 21: 441–452.
43. Dayal MR, Kegley AD, Strkalj G, Bidmos MA, Kuykendall KL (2009) The history and composition of the Raymond A. Dart Collection of Human Skeletons at the University of the Witwatersrand, Johannesburg, South Africa. Am J Phys Anthropol 140: 324–35.
44. Sperber GH (2001) Craniofacial development. Ontario: BC Decker Inc.
45. Enlow DH, Hans MG (1996) Crecimiento facial. México DF: McGraw-Hill Interamericana.
46. Smith KK (1996) Integration of craniofacial structures during development in Mammals. Amer Zool 36: 70–79.
47. Lieberman DE, Krovitz GE, McBratney-Owen B (2004) Testing hypotheses about tinkering in the fossil record: the case of the human skull. J Exper Zool 302b: 284–301.
48. Quinn G, Keough M (2002) Experimental design and data analysis for biologist. Cambridge: Cambridge University Press.

3

Shared Human-Chimpanzee Pattern of Perinatal Femoral Shaft Morphology and Its Implications for the Evolution of Hominin Locomotor Adaptations

Naoki Morimoto*, Christoph P. E. Zollikofer, Marcia S. Ponce de León

Anthropological Institute, University of Zurich, Zurich, Switzerland

Abstract

Background: Acquisition of bipedality is a hallmark of human evolution. How bipedality evolved from great ape-like locomotor behaviors, however, is still highly debated. This is mainly because it is difficult to infer locomotor function, and even more so locomotor kinematics, from fossil hominin long bones. Structure-function relationships are complex, as long bone morphology reflects phyletic history, developmental programs, and loading history during an individual's lifetime. Here we discriminate between these factors by investigating the morphology of long bones in fetal and neonate great apes and humans, before the onset of locomotion.

Methodology/Principal Findings: Comparative morphometric analysis of the femoral diaphysis indicates that its morphology reflects phyletic relationships between hominoid taxa to a greater extent than taxon-specific locomotor adaptations. Diaphyseal morphology in humans and chimpanzees exhibits several shared-derived features, despite substantial differences in locomotor adaptations. Orangutan and gorilla morphologies are largely similar, and likely represent the primitive hominoid state.

Conclusions/Significance: These findings are compatible with two possible evolutionary scenarios. Diaphyseal morphology may reflect retained adaptive traits of ancestral taxa, hence human-chimpanzee shared-derived features may be indicative of the locomotor behavior of our last common ancestor. Alternatively, diaphyseal morphology might reflect evolution by genetic drift (neutral evolution) rather than selection, and might thus be more informative about phyletic relationships between taxa than about locomotor adaptations. Both scenarios are consistent with the hypothesis that knuckle-walking in chimpanzees and gorillas resulted from convergent evolution, and that the evolution of human bipedality is unrelated to extant great ape locomotor specializations.

Editor: David Caramelli, University of Florence, Italy

Funding: Grant sponsor: Swiss National Science Foundation; Grant number: 3100A0-109344/1. The funders had no role in study design, data collection and analysis, decision to publish, or preparation of the manuscript.

Competing Interests: The authors have declared that no competing interests exist.

* E-mail: morimoto@aim.uzh.ch

Introduction

Humans and extant great apes exhibit a pattern of locomotor diversification [1,2,3,4], which stands in contrast with their phyletic relationships. While humans are obligate terrestrial bipeds, our closest living relatives, the chimpanzees, exhibit a wide range of arboreal locomotor behaviors [5,6], and their peculiar mode of terrestrial quadrupedal locomotion – knuckle-walking – differs substantially from human bipedal locomotion [7,8]. The more distantly-related gorillas also exhibit various arboreal locomotor behaviors, as well as terrestrial knuckle-walking [9,10,11]. Because knuckle-walking occurs in chimpanzees and gorillas, it has been proposed as an ancestral mode of locomotion from which human bipedality evolved [12]. This hypothesis has been challenged on anatomical, developmental and behavioral grounds [13,14,15], and the orangutan has been proposed, instead, as a model for the evolution of bipedality from a generalized bipedal/quadrupedal arboreal repertoire of locomotion [14]. In contrast to both hypotheses, the phyletic and functional analysis of the skeleton of *Ardipithecus ramidus* [16,17,18,19,20] provided evidence that hominin bipedality might have evolved from a locomotor mode no longer present in extant great apes.

During reconstruction of the evolutionary history of hominin bipedalism, fossil evidence from hind limb elements, especially from the femur, has played a central role. The surface topography of the proximal femoral diaphysis of *Ardipithecus ramidus* [16] and *Australopithecus afarensis* [21] has provided evidence for reorganization of the femoropelvic musculature toward bipedal locomotor behaviors [22,23]. Likewise, the proximal femoral morphology of *Orrorin tugenensis* indicates bipedal locomotor adaptations [24]. Form-function relationships of the femur are complex, however, as femoral morphology results from both long-term processes of selection and adaptation, and short-term processes of bone remodeling during an individual's lifetime (Wolff's Law [25] or bone functional adaptation [26,27]). Femoral morphology thus typically reflects a combination of (a) the impact of an individual's locomotor history on its musculoskeletal system, (b) taxon-specific adaptation of the musculoskeletal system to specialized locomotor behaviors, and (c) phyletic history not directly related to a taxon's actual locomotor adaptations (phyletic inertia) [26,27,28,29,30,31]. Discrimination between these factors is especially difficult in fossil specimens, for which *in-vivo* patterns of locomotion and species-specific locomotor behavior are unknown, and taxon affiliation is often uncertain.

Here we address these questions by studying femoral morphology in fetuses and neonates of extant great apes and humans. Phyletic relationships and locomotor behaviors of these taxa are well known. Great ape taxa show a remarkable variety of arboreal and terrestrial, quadrupedal and bipedal locomotor behaviors [3,5,6,11,32,33], the frequencies of which depend on taxon-specific, environmental and life-history factors [3,34,35]. While various modes of terrestrial locomotion are an important component of the locomotor repertoire of chimpanzees and gorillas [11,33], orangutans are highly restricted to arboreal habitats and are unique among great apes in showing pronograde suspensory behaviors and fist-walking [2,3].

Studying long bone morphology in fetuses and neonates permits analysis of the effects of the developmental program before the onset of locomotion, that is, before the skeletal morphology is modified by taxon-specific and/or individual mechanical loading regimes, and by environmental factors. Because epiphyses are not yet ossified around the time of birth, we focus on diaphyseal morphology. We ask whether perinatal femoral diaphyseal morphology reflects phyletic relationships independent of an extant taxon's locomotor adaptation (H0), or whether it reflects adaptation to taxon-specific locomotor behaviors (H1). According to hypothesis H0, humans and chimpanzees should exhibit similar femoral morphologies, to the exclusion of gorillas; according to H1, chimpanzees and gorillas are expected to exhibit largely similar diaphyseal morphologies, while modern human femoral diaphyses should be clearly distinct.

Long bone morphology is brought about by growth in longitudinal and radial directions. During this process, bone is deposited at diaphyseal growth plates and subperiosteal surfaces, respectively, and resorbed at endosteal surfaces [36,37,38,39,40,41]. Young et al. [42,43] have shown that hominoid long bone longitudinal relative to radial growth is more variable than in other primate taxa, and reflects taxon-specific locomotor adaptations. In hominoids, taxon-specific limb proportions are almost fully established at birth [44], indicating distinct taxon-specific longitudinal growth characteristics already before birth. Longitudinal diaphyseal growth characteristics and morphology thus provide support for hypothesis H1.

Here we complement this study by investigating radial diaphyseal morphology. Variability in radial growth results in variability in external (subperiosteal) surface morphology and cortical bone thickness. These features are correlated with musculoskeletal topography [21,45] and cross-sectional biomechanical properties [46,47,48], respectively. Specifically, we ask whether prenatal subperiosteal morphology of the hominoid femoral diaphysis reflects phyletic history (H0) or taxon-specific locomotor adaptations (H1). In the first case (H0), human and chimpanzee morphologies should exhibit several shared-derived features compared to gorilla and orangutan morphologies. In the second case (H1), the fetal/neonate diaphyseal surface morphology of humans should be distinct from that of all great ape taxa, while chimpanzees and gorillas should be more similar to each other than to orangutans.

Three-dimensional data of femoral diaphyses were acquired with computed tomography (CT) from a sample of late fetal to neonate humans, chimpanzees, gorillas, and orangutans (see Materials and Methods). Data were analyzed with methods of morphometric mapping (MM), which are well suited to quantify the morphology of relatively featureless cylindroid structures such as long bone diaphyses [29]. In contrast to standard geometric-morphometric techniques, MM does not require pre-defined morphological features such as anatomical landmarks. Rather, morphological features characterizing the sample as a whole, or subsamples, are identified by means of the MM analysis. Here, the shape of the external diaphyseal surface is quantified by its transverse curvature (= curvature around the shaft), which closely reflects the topography of muscular attachment sites [29,45] (during the fetal period the internal (endosteal) surface is not yet fully ossified and hence cannot be quantified reliably [37,49]). Hereafter we use *diaphyseal surface morphology* to denote the resulting MMs (see Materials and Methods, Fig. S1). MMs of all specimens of the sample were aligned so as to minimize differences in rotation around the diaphyseal longitudinal axis. The aligned MMs were then submitted to 2D Fourier Analysis. Principal Components Analysis (PCA) was used to reduce the high dimensionality of the data in Fourier space. This procedure permits to characterize principal patterns of shape variation in the sample (Fig. 1A), to quantify phenetic similarity between taxa [50] (Fig. 1B), and to visualize commonalities and differences between taxon-specific diaphyseal morphologies (Fig. 1C).

Results and Discussion

MM-based analysis shows that taxon-specific femoral diaphyseal surface morphologies are already present before birth (Fig. 1). Graphing the first two shape components (SC1 and SC2), which account for 23.9% and 15.1% of the total shape variation in the sample indicates that diaphyseal surface morphologies of gorillas (G) and orangutans (O) are more similar to each other than to any other taxon, while diaphyseal morphologies of chimpanzees (C) and humans (H) are approximately equally distant from GO morphologies (Table 1). Differences between taxa along SC1 partly reflect differences in neonatal body mass (Fig. S2A), while differences along shape component 2 are independent of body mass (Fig. S2B). Furthermore, taxon-specific differences in diaphyseal shape are not due to differences in diaphyseal length and cross-sectional area (Fig. S3). Also, sex-specific shape differences could not be found at this early stage of development.

Using orangutans as an outgroup, a phyletic tree evaluated from the data of Fig. 1A clearly groups humans with chimpanzees (HC), versus gorillas (Fig. 1B). Tree topology is well supported by bootstrapping (999 replications of the given tree out of 1000 resamplings). This phene-based tree is consistent with molecular trees of human and great ape phyletic divergence [51,52,53], supporting hypothesis H0 that femoral diaphyseal surface morphology in the fetal/neonatal period reflects hominoid phylogeny.

Taxon-specific perinatal femoral diaphyseal surface morphologies are visualized in Fig. 1C. The proximal femoral diaphysis of G and O is characterized by the presence of a prominent lateral spiral pilaster (*lp*) [16,21], which is delimited by fossae on its inferolateral and superomedial sides (*ilf* and *smf*) [45]. Also, GO femora are characterized by a marked lateral ridge (*lr*) on the distal diaphysis. H and C femoral diaphyses also exhibit a *lp*, but it is only weakly expressed compared to GO. Most notably, the HC femur is characterized by the presence of a linea aspera (*la*) along the posterolateral diaphysis. This feature has a similar position and

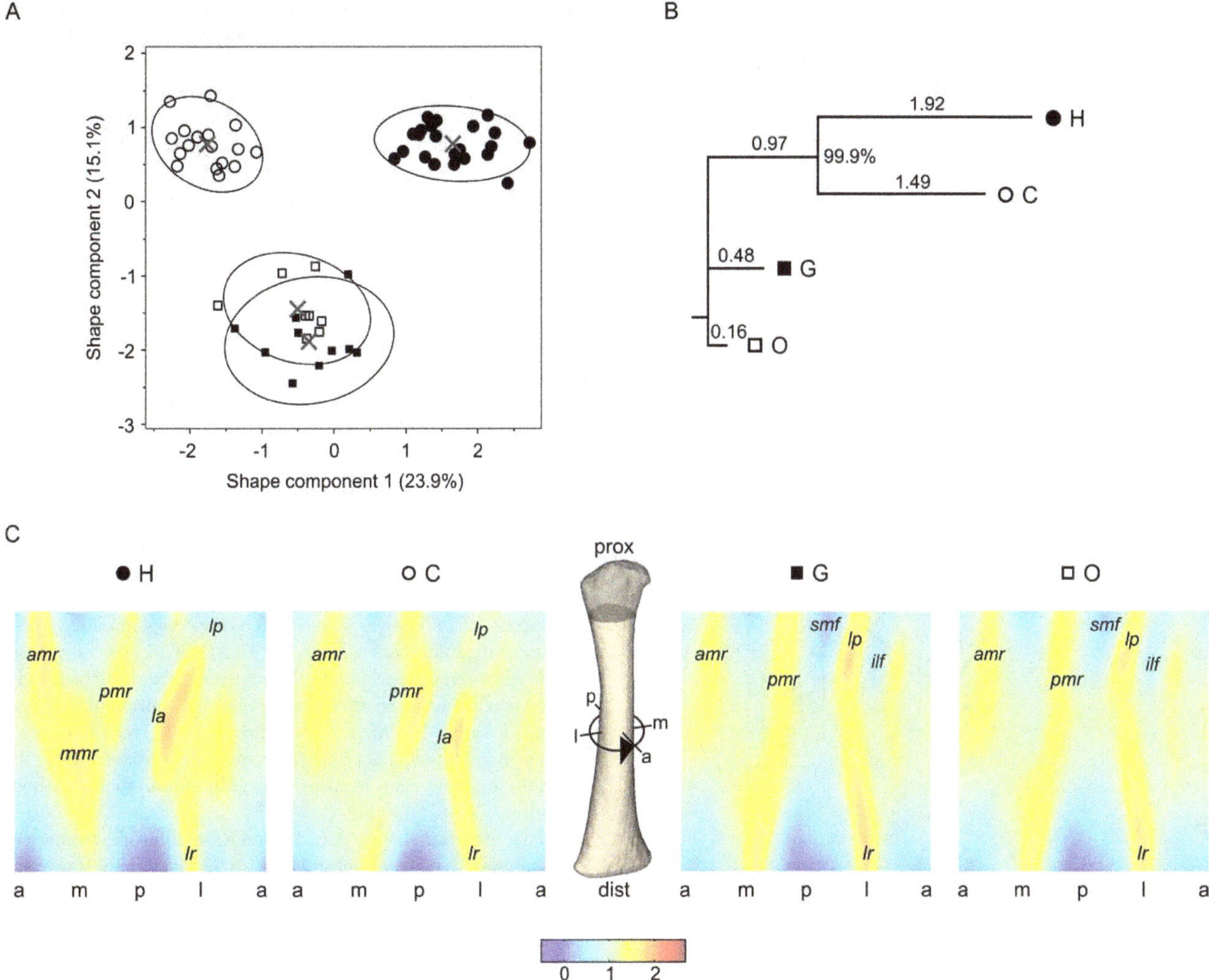

Figure 1. Femoral diaphyseal shape variation in hominoids. A, variation along shape components 1 and 2 of morphospace (humans: filled circles, chimpanzees: open circles, gorillas: filled squares, orangutans: open squares; crosses/ellipses indicate taxon-specific means/90%-density ellipses). B, neighbor-joining tree based on between-taxon distances (see Table 1); numbers above branches indicate branch lengths; number at the branch node indicates bootstrap support (999 of 1000 replications); H: humans, C: chimpanzees, G: gorillas, O: orangutans. C, morphometric maps [false-color images of external surface curvature (relative units)] visualizing taxon-specific mean morphologies (a-m-p-l: anterior-medial-posterior-lateral); *la*: linea aspera, *lp*: lateral pilaster, *ilf*: inferolateral fossa, *smf*: superomedial fossa, *lr*: lateral ridge, *amr*: anteromedial ridge, *pmr*: posteromedial ridge, *mmr*: midshaft medial ridge.

orientation in humans and chimpanzees, and is not present on GO femora (Fig. 1C).

Which evolutionary processes gave rise to this pattern of morphological similarity and dissimilarity between taxa? Before this question can be addressed, the potential influence of environmental factors and associated loading regimes on fetal long bone development has to be considered. In the uterus, the effects of gravitation are neutralized by buoyancy, but the fetal skeleton experiences loads through spontaneous fetal limb movements, as well as reactive and inertial forces elicited by maternal movements. Clinical evidence shows that spontaneous fetal limb movements are important for normal limb development [54]. These movements are mediated by central pattern generators [55], i.e., genetically programmed neural networks. Fetal movements thus reflect the developmental state of the neuromotor system rather than environmental factors [56]. Also, our results make it unlikely that taxon-specific maternal locomotor/postural behaviors influence fetal long bone morphology. For example, chimpanzee and gorilla neonatal femora have a clearly distinct morphology (Fig. 1) despite largely similar neonatal body size (Table S1) [57,58,59] and maternal locomotor behaviors, while gorilla and orangutan neonates have similar femoral diaphyseal

Table 1. Morphometric distances between taxon-specific mean shapes.

	H (*Homo*)	C (*Pan*)	G (*Gorilla*)
C (*Pan*)	3.41*	–	–
G (*Gorilla*)	3.32*	3.00*	–
O (*Pongo*)	3.10*	2.57*	0.64 ($p = 0.41$)

*$p < 0.001$.

morphology, despite significant differences in maternal locomotor behaviors. Overall, it appears unlikely that differences in intrauterine loading regimes contribute substantially to taxon-specific differences in femoral diaphyseal morphology.

The following evolutionary scenarios bringing about the observed differences between taxa may thus be considered (Fig. 2): (a) H and C similarities in femoral diaphyseal morphology represent shared-derived features, which go back to the last common ancestor (HC-LCA) (Fig. 2A), (b) H and C morphologies evolved independently from an African great ape ancestor (Fig. 2B), and (c) G and O morphologies represent derived states, while the HC-LCA represents the primitive state (Fig. 2C).

Scenarios (b) and (c) imply that similar morphologies result from parallel and convergent evolution, respectively. This is unlikely, given the substantial differences between H and C with respect to locomotor behaviors and associated selective pressures (obligate bipedalism versus predominant quadrupedalism), and between G and O (mostly terrestrial versus predominantly arboreal locomotion).

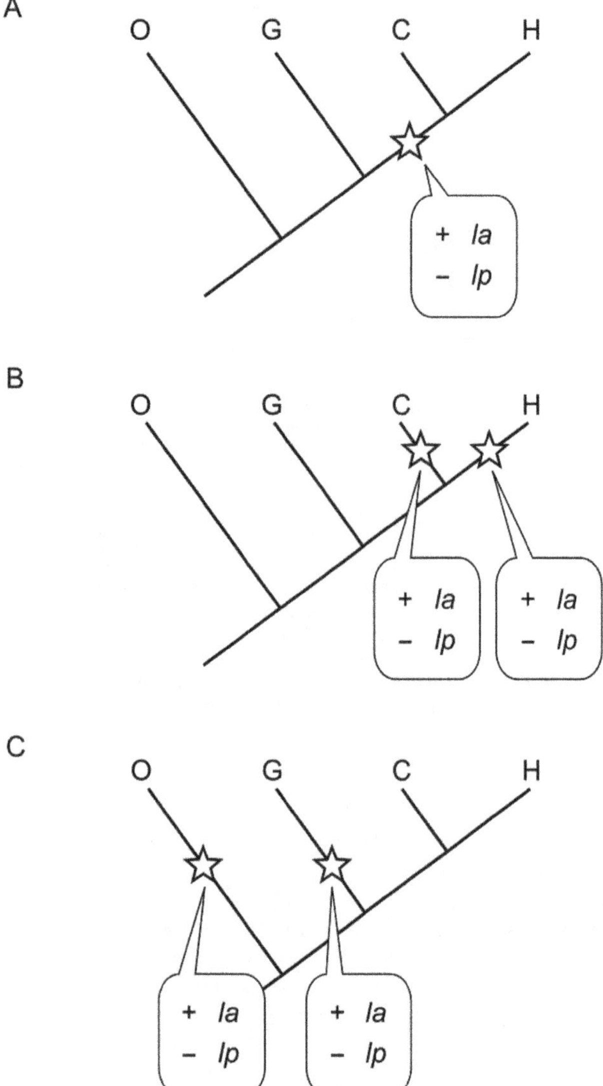

Figure 2. Hypothetical scenarios of femoral diaphyseal shape evolution. Scenario A: shared-derived formation of linea aspera and reduction of lateral pilaster in humans and chimpanzees. Scenario B: parallel evolution of *la* and reduction of *lp*. Scenario C: convergent evolution of similar orangutan/gorilla features (see Fig. 1C for feature codes).

Scenario (a) is more parsimonious. Adopting this scenario as the most likely one, we may thus infer that, in HC, prenatal femoral diaphyseal ontogeny follows a derived mode, while GO represent the primitive mode. It has been suggested that chimpanzee and gorilla femoral diaphyseal morphologies reflect a shared femoropelvic musculoskeletal organization [16,21]. In contrast, our results indicate that chimpanzee and gorilla femoral morphologies are distinct already during early development. Together with evidence from musculoskeletal anatomy of ref. [45], this adds to the growing evidence that HC phenetic similarities reflect their close phylogenetic relationship [13,45,50,60]. This is consistent with the hypothesis that knuckle-walking and associated skeletal adaptations of chimpanzees and gorillas evolved independently [15].

It remains to be clarified to which extent the inferred derived HC-LCA diaphyseal surface morphology resulted from neutral evolution (i.e., evolution by drift [61,62]), and/or from adaptation to taxon-specific locomotor behaviors, respectively. Since close links exist between femoral diaphyseal surface morphology and muscle topography [45], we hypothesize that the HC-LCA underwent an adaptive shift in femoropelvic musculoskeletal organization. Inferences on possible HC-LCA locomotor specialization must remain speculative. If we assume that the posteriorly-located *la* of H and C neonate femora (Fig. 1C) represents a shared-derived feature, its inferred presence in the HC-LCA might indicate a modified function of the muscles inserting along this structure (e.g. the *gluteus maximus*) during hind limb-mediated body propulsion [45].

While our data imply that H and C exhibit shared-derived femoral diaphyseal features relative to G, they also show that morphologies of both H and C diverged from the HC-LCA morphology, probably to a greater extent in H than in C (Figs. 1A,B). This is in concordance with fossil evidence from *Ardipithecus* indicating taxon-specific evolution of femoral morphology not only in hominins but also in panins since their split from the HC-LCA [16,17,19,20].

Human and chimpanzee femoral diaphyseal features unique to each taxon (Figs. 1A,C) most likely reflect taxon-specific locomotor adaptations. For example, humans differ from chimpanzees in exhibiting a prominent anteromedial ridge (*amr*) and a ridge along the medial diaphysis (*mmr*; Fig. 1C) while chimpanzees show a more prominent posteromedial ridge (*pmr*). These morphological differences might reflect differences in the relative size and attachment areas of locomotor muscles around the femur (e.g. large vastus muscles relative to adductor/hamstring muscles in humans compared to chimpanzees [21]). In addition to phyletic divergence, diaphyseal morphologies of H and C also diverge during postnatal development, with the effect that the morphology of the proximal femoral diaphysis of C becomes more similar to G, e.g. regarding the expression of the lateral spiral pilaster (*lsp*) [21,29]. It remains to be elucidated in greater detail to which extent each of the diaphyseal features identified in Fig. 1C reflects taxon-specific locomotor function, and to which extent they reflect homology versus homoplasy.

Our data provide evidence that the surface morphology of the perinatal hominoid femoral diaphysis reflects phylogenetic affinities (hypothesis H0) to a greater extent than locomotor adaptation (hypothesis H1). The underlying processes of prenatal radial diaphyseal ontogeny appear to be evolutionarily more conservative than those of longitudinal ontogeny. The latter have been shown to reflect taxon-specific locomotor adaptations in terms of limb segment lengths and proportions [42,43,44]. While the elongation of the hind limb – which is a key feature of human bipedality [43] – could have been effected by a relatively minor modification of the developmental program [41], radial ontogeny and associated femoral diaphyseal surface morphology seem to be constrained by muscular topography, which has been reported to reflect phyletic relationships in the hominoids [13,45,60].

These hypotheses require further testing, especially through detailed comparisons of H, C and G locomotor musculoskeletal development and topography, biomechanics, kinematics, and kinetics. Overall, our results suggest a two-stage approach to investigate the origins of human bipedal locomotion with actualistic data: first identify and analyze the shared-derived features of humans and chimpanzees compared to gorillas, then identify and analyze the uniquely derived features of humans and chimpanzees relative to the inferred HC-LCA, respectively.

Materials and Methods

Sample Structure

The sample consists of femora of *Homo sapiens* ($N=22$; femoral diaphyseal length: 41.6–63.4 mm), *Pan troglodytes* ($N=17$; 32.2–55.1 mm), *Gorilla gorilla* ($N=10$; 20.0–59.9 mm) and *Pongo pygmaeus* ($N=8$; 30.8–46.8 mm) from late fetal stages (3 months pre-term) to neonate stages (before the eruption of the first deciduous molar; <2 months). Since femoral shape does not exhibit significant sex-specific differences at this early stage of development, we used taxon-specific pooled-sex samples. All specimens are from the Collections of the Anthropological Institute and Museum of the University of Zurich.

Volumetric Data Acquisition

Femora of wet (formalin-preserved, frozen or fresh cadaver) specimens were scanned using a Siemens 64-detector-array CT device (beam collimation 1.0 mm; standard/bone kernels [B30/B60]; serial cross-sections reconstructed at 0.2 mm intervals). Small specimens were scanned using a micro-CT scanner (μCT80, Scanco Medical, Switzerland; volume data reconstructed at an isotropic voxel resolution of 75 μm). Cross sections orthogonal to the principal axis of the femoral shaft were obtained by resampling the original volumetric data using the software Amira 4.1 (Mercury Systems).

Morphometric Data Acquisition

In immature specimens, unfused epiphyses are often missing, or their position relative to the diaphysis cannot be reconstructed reliably. We thus focus here on diaphyseal morphology. The femoral diaphysis was extracted from the CT volume data using epiphyseal lines as proximal and distal delimiters. Femoral diaphyseal length was measured as the distance between proximal and distal epiphyseal lines. Subperiosteal (external) outlines of each cross section were parameterized with elliptical Fourier analysis (EFA) [63]. EFA was used to reduce noise, and to define parametric outline functions. The curvature of the external diaphyseal surface (k_{ext}) was calculated analytically using the parametric functions of EFA. Resulting positive/negative values of the curvature k_{ext} denote convex/concave regions, respectively (see ref. [29] for details).

Morphometric Analysis

For each specimen, measurements of k_{ext} were sampled around each cross-sectional outline, and along the entire diaphyseal shaft. These data were normalized to their respective median values, and mapped onto a cylindrical coordinate system (ρ, θ, z), where $\rho = 1/(2\pi) = \text{constant}$ denotes the radius of the cylinder. Specimens were prealigned manually such that angle θ denotes the anatomical direction ($\theta = 0°\rightarrow 360°$: anterior → medial → posterior → lateral → anterior), and z denotes the normalized position along the diaphysis ($z = 0 \rightarrow 1$: distal → proximal) [64,65]. Since $\rho = \text{constant}$, data can be visualized as two-dimensional morphometric maps $\mathbf{M}(\theta, z)$, and distributions $k_{ext}(\theta, z)$ (Fig. S1) can be represented as $K \times L$ matrices, where K and L denote the number of elements along z and θ, respectively ($K = L = 300$).

For the comparative analysis of the morphometric maps $\mathbf{M}_i$ of all specimens $i = 1 \ldots N$, differences between specimens in orientation around the diaphyseal long axis (θ) had to be minimized. This procedure is analogous to the Procrustes superposition used in anatomical landmark-based geometric morphometric analyses. However, because the morphometric maps of the femoral diaphysis do not contain predefined anatomical features, the alignment was performed in Fourier space. To this end, 2D-Fourier transforms $F(\mathbf{M}_i)$ of all $\mathbf{M}_i$ were calculated ($\mathbf{M}$ has a natural periodicity in θ), yielding $K \times L$ sets of Fourier coefficients, which define a specimen's diaphyseal shape as a point in multidimensional Fourier space. Specimens were aligned to each other by minimizing inter-specimen distances in Fourier space through rotation around θ (diaphyseal axis).

To reduce the high dimensionality of the data in Fourier space, and to identify principal patterns of shape variability in the sample, Fourier coefficient sets were submitted to Principal Components Analysis (PCA). To facilitate visual inspection and anatomical interpretation of the results of PCA, real-space morphometric maps were reconstructed by transforming a given point $\mathbf{P}^*$ in PC space into its corresponding set of Fourier coefficients $F(\mathbf{M}^*)$, and applying an inverse Fourier transform to obtain a morphometric map $\mathbf{M}^*$. This method was used to produce the MMs of Fig. 1C. Morphometric maps were false-color coded. All calculations were performed with MATLAB7.7 (MathWorks) (see ref. [29] for details).

Similarity Analysis

Dissimilarity matrices $\mathbf{D}$ were evaluated to represent all between-taxon distances D (quantified as Euclidean distances between taxon mean points; see Table 1) in shape space. Phenetic trees were evaluated for $\mathbf{D}$ with PHYLIP 3.69 [66], using the neighbor-joining method.

Supporting Information

Figure S1 Principle of morphometric mapping. A, 3D representation of the right femur. B, principle of cylindrical projection (anterior [0°] → medial [90°] → posterior [180°] → lateral [270°] → anterior [0°]).

Figure S2 Correlation between taxon-specific means of shape component scores and means of neonatal body mass (data summarized in Table S1; humans: filled circles, chimpanzees: open circles, gorillas: filled squares, orangutans: open squares). SC1 is weakly correlated with neonatal body mass ($p = 0.06$, $R^2 = 0.88$) (A). SC2, which distinguishes between human-chimpanzee and gorilla-orangutan, is not correlated with neonatal body mass (B).

Figure S3 Correlation between femoral diaphyseal shape component scores (SC1, SC2) and femoral size (humans: filled circles, chimpanzees: open circles, gorillas: filled squares, orangutans: open squares). Shape component scores are plotted against femoral diaphyseal length (A), and median femoral diaphyseal cross-sectional area (B). Each cross-sectional area was calculated as the total area of bone marrow-filled cross-section. Overall, taxon-specific differences in femoral diaphyseal length are not correlated with femoral diaphyseal morphology. Humans exhibit a weak correlation of SC1 with femoral diaphyseal length ($p<0.05$, $R^2=0.20$); chimpanzees exhibit a weak correlation of SC1 with femoral diaphyseal cross-sectional area ($p<0.05$, $R^2=0.28$).

Table S1 Neonatal body mass of hominoids.

Acknowledgments

We thank P. Jans for help with sample preparation and CT scanning. The comments of C. Finlayson and M. Walker are greatly acknowledged. We are also grateful to the two anonymous reviewers for their valuable comments and suggestions.

Author Contributions

Conceived and designed the experiments: NM MSPdL CPEZ. Performed the experiments: NM. Analyzed the data: NM. Contributed reagents/materials/analysis tools: NM MSPdL CPEZ. Wrote the paper: NM MSPdL CPEZ.

References

1. Hunt K (1991) Positional behavior in the Hominoidea. International Journal of Primatology 12: 95–118.
2. Hunt K, Cant J, Gebo D, Rose M, Walker S, et al. (1996) Standardized descriptions of primate locomotor and postural modes. Primates 37: 363–387.
3. Thorpe SK, Crompton RH (2006) Orangutan positional behavior and the nature of arboreal locomotion in Hominoidea. American Journal of Physical Anthropology 131: 384–401.
4. Crompton RH, Vereecke EE, Thorpe SKS (2008) Locomotion and posture from the common hominoid ancestor to fully modern hominins, with special reference to the last common panin/hominin ancestor. Journal of Anatomy 212: 501–543.
5. Doran DM (1992) The ontogeny of chimpanzee and pygmy chimpanzee locomotor behavior - a case-study of paedomorphism and its behavioral-correlates. Journal of Human Evolution 23: 139–157.
6. Doran DM (1993) Comparative locomotor behavior of chimpanzees and bonobos - the influence of morphology on locomotion. American Journal of Physical Anthropology 91: 83–98.
7. Gebo DL (1992) Plantigrady and foot adaptation in African apes: Implications for hominid origins. American Journal of Physical Anthropology 89: 29–58.
8. Richmond BG, Begun DR, Strait DS (2001) Origin of human bipedalism: The knuckle-walking hypothesis revisited. In: Ruff C, editor. Yearbook of Physical Anthropology, Vol 44. 70–105.
9. Tuttle RH, Watts DP (1985) The positional behavior and adaptive complexes of *Pan gorilla*. In: Kondo S, Ishida S, Okada M, Kimura T, Yamazaki M, editors. Primate Morphophysiology, Locomotor Analyses and Human Bipedalism. Tokyo: University of Tokyo Press. 261–288.
10. Remis M (1995) Effects of body size and social context on the arboreal activities of lowland gorillas in the Central African Republic. American Journal of Physical Anthropology 97: 413–433.
11. Doran DM (1997) Ontogeny of locomotion in mountain gorillas and chimpanzees. J Hum Evol 32: 323–344.
12. Richmond BG, Strait DS (2000) Evidence that humans evolved from a knuckle-walking ancestor. Nature 404: 382–385.
13. Gibbs S, Collard M, Wood B (2002) Soft-tissue anatomy of the extant hominoids: a review and phylogenetic analysis. Journal of Anatomy 200: 3–49.
14. Thorpe SKS, Holder RL, Crompton RH (2007) Origin of human bipedalism as an adaptation for locomotion on flexible branches. Science 316: 1328–1331.
15. Kivell TL, Schmitt D (2009) Independent evolution of knuckle-walking in African apes shows that humans did not evolve from a knuckle-walking ancestor. Proceedings of the National Academy of Sciences of the United States of America 106: 14241–14246.
16. Lovejoy CO, Suwa G, Spurlock L, Asfaw B, White TD (2009) The pelvis and femur of *Ardipithecus ramidus*: The emergence of upright walking. Science 326: 71.
17. Lovejoy CO, Simpson SW, White TD, Asfaw B, Suwa G (2009) Careful climbing in the Miocene: The forelimbs of *Ardipithecus ramidus* and humans are primitive. Science 326: 70.
18. Lovejoy CO, Latimer B, Suwa G, Asfaw B, White TD (2009) Combining prehension and propulsion: The foot of *Ardipithecus ramidus*. Science 326: 72.
19. Lovejoy CO, Suwa G, Simpson SW, Matternes JH, White TD (2009) The great divides: *Ardipithecus ramidus* reveals the postcrania of our last common ancestors with African apes. Science 326: 100–106.
20. White TD, Asfaw B, Beyene Y, Haile-Selassie Y, Lovejoy CO, et al. (2009) *Ardipithecus ramidus* and the paleobiology of early hominids. Science 326: 75–86.
21. Lovejoy CO, Meindl RS, Ohman JC, Heiple KG, White TD (2002) The Maka femur and its bearing on the antiquity of human walking: Applying contemporary concepts of morphogenesis to the human fossil record. American Journal of Physical Anthropology 119: 97–133.
22. Lieberman DE, Raichlen DA, Pontzer H, Bramble DM, Cutright-Smith E (2006) The human gluteus maximus and its role in running. Journal of Experimental Biology 209: 2143–2155.
23. Pontzer H, Raichlen DA, Sockol MD (2009) The metabolic cost of walking in humans, chimpanzees, and early hominins. Journal of Human Evolution 56: 43–54.
24. Richmond BG, Jungers WL (2008) *Orrorin tugenensis* femoral morphology and the evolution of hominin bipedalism. Science 319: 1662–1665.
25. Wolff J (1892) Das Gesetz der Transformation der Knochen. Berlin: A. Hirschwald.
26. Pearson OM, Lieberman DE (2004) The aging of Wolff's "Law": Ontogeny and responses to mechanical loading in cortical bone. Yearbook of Physical Anthropology 47: 63–99.
27. Ruff CB, Holt B, Trinkaus E (2006) Who's afraid of the big bad wolff? "Wolff is law" and bone functional adaptation. American Journal of Physical Anthropology 129: 484–498.
28. Lovejoy CO, McCollum MA, Reno PL, Rosenman BA (2003) Developmental biology and human evolution. Annual Review of Anthropology 32: 85–109.
29. Morimoto N, Zollikofer CPE, Ponce de León MS (2011) Exploring femoral diaphyseal shape variation in wild and captive chimpanzees by means of morphometric mapping: a test of Wolff's Law. Anatomical Record 294: 589–609.
30. Wallace IJ, Middleton KM, Lublinsky S, Kelly SA, Judex S, et al. (2010) Functional significance of genetic variation underlying limb bone diaphyseal structure. American Journal of Physical Anthropology 143: 21–30.
31. Wallace IJ, Tommasini SM, Judex S, Garland T, Demes B (2012) Genetic variations and physical activity as determinants of limb bone morphology: An experimental approach using a mouse model. American Journal of Physical Anthropology 148: 24–35.
32. Remis MJ (1998) The effects of body size and habitat on the positional behavior of lowland and mountain gorillas. In: Strasser E, Fleagle J, Rosenberger A, McHenry H, editors. Primate Locomotion -Recent Advances. New York: Plenum Press. 95–106.
33. Doran DM (1996) Comparative positional behabior of the African apes. In: McGrew MC, Marchant LF, Nishida T, editors. Great Ape Societies. Cambridge: Cambridge University Press.
34. Doran DM, Jungers WL, Sugiyama Y, Fleagle J, Heesy C (2002) Multivariate and phylogenetic approaches to understanding chimpanzee and bonobo behavioral diversity. In: Boesch C, Hohmann G, Marchant LF, editors. Behavioural diversity in Chimpanzees and Bonobos. Cambridge: Cambridge University Press. 14–34.
35. Thorpe SK, Crompton RH (2005) Locomotor ecology of wild orangutans (*Pongo pygmaeus abelii*) in the Gunung Leuser Ecosystem, Sumatra, Indonesia: a multivariate analysis using log-linear modelling. American Journal of Physical Anthropology 127: 58–78.
36. Schwartz JH (1995) Skeleton Keys: An Introduction to Human Skeletal Morphology, Development, and Analysis. Oxford: Oxford University Press.
37. Scheuer L, Black S, Christie A (2000) Developmental Juvenile Osteology. San Diego, San Francisco, New York, Boston, London, Sydney, Tokyo: Academic Press.
38. Van der Eerden BCJ, Karperien M, Gevers EF, Löwik CWGM, Wit JM (2000) Expression of indian hedgehog, parathyroid hormone-related protein, and their receptors in the postnatal growth plate of the rat: evidence for a locally acting growth restraining feedback loop after birth. Journal of Bone and Mineral Research 15: 1045–1055.
39. Rauch F (2005) Bone growth in length and width: the Yin and Yang of bone stability. J Musculoskelet Neuronal Interact 5: 194–201.
40. Kronenberg HM (2006) PTHrP and skeletal development. Annals of the New York Academy of Sciences 1068: 1–13.
41. Serrat MA, Lovejoy CO, King D (2007) Age- and site-specific decline in insulin-like growth factor-I receptor expression is correlated with differential growth plate activity in the mouse hindlimb. The Anatomical Record: Advances in Integrative Anatomy and Evolutionary Biology 290: 375–381.
42. Young NM, Hallgrimsson B (2005) Serial homology and the evolution of mammalian limb covariation structure. Evolution 59: 2691–2704.
43. Young NM, Wagner GP, Hallgrimsson B (2010) Development and the evolvability of human limbs. Proceedings of the National Academy of Sciences of the United States of America 107: 3400–3405.
44. Schultz AH (1973) Age changes, variability and generic differences in body proportions of recent hominoids. Folia Primatologica 19: 338–359.

45. Morimoto N, Zollikofer CPE, Ponce de León MS (2011) Femoral morphology and femoropelvic musculoskeletal anatomy of humans and great apes: a comparative virtopsy study. Anatomical Record 294: 1433–1445.
46. Ruff CB (2003) Ontogenetic adaptation to bipedalism: age changes in femoral to humeral length and strength proportions in humans, with a comparison to baboons. J Hum Evol 45: 317–349.
47. Ruff CB (2003) Growth in bone strength, body size, and muscle size in a juvenile longitudinal sample. Bone 33: 317–329.
48. Högler W, Blimkie CJR, Cowell CT, Inglis D, Rauch F, et al. (2008) Sex-specific developmental changes in muscle size and bone geometry at the femoral shaft. Bone 42: 982–989.
49. Standring S, editor (2004) Gray's Anatomy. 39 ed. Edinburgh/London/New York: Churchill Livingstone.
50. Lockwood CA, Kimbel WH, Lynch JM (2004) Morphometrics and hominoid phylogeny: Support for a chimpanzee-human clade and differentiation among great ape subspecies. Proceedings of the National Academy of Sciences of the United States of America 101: 4356–4360.
51. Kumar S, Filipski A, Swarna V, Walker A, Hedges SB (2005) Placing confidence limits on the molecular age of the human-chimpanzee divergence. Proceedings of the National Academy of Sciences of the United States of America 102: 18842–18847.
52. Patterson N, Richter DJ, Gnerre S, Lander ES, Reich D (2006) Genetic evidence for complex speciation of humans and chimpanzees. Nature 441: 1103–1108.
53. Hobolth A, Christensen OF, Mailund T, Schierup MH (2007) Genomic relationships and speciation times of human, chimpanzee, and gorilla inferred from a coalescent hidden Markov model. PLoS Genetics 3: 294–304.
54. Kuwata T, Matsubara S, Ohkusa T, Yada Y, Suzuki M (2011) Decreased fetal movement prompts investigation of prenatal/neonatal nemaline myopathy: The possible merit of fetal movement count. Journal of Obstetrics and Gynaecology Research 37: 921–925.
55. MacKay-Lyons M (2002) Central pattern generation of locomotion: a review of the evidence. Phys Ther 82: 69–83.
56. Lacquaniti F, Ivanenko YP, Zago M (2012) Development of human locomotion. Current Opinion in Neurobiology 22: 1–7.
57. Leigh SR, Shea BT (1996) Ontogeny of body size variation in African apes. Am J Phys Anthropol 99: 43–65.
58. Smith RJ, Leigh SR (1998) Sexual dimorphism in primate neonatal body mass. Journal of Human Evolution 34: 173–201.
59. DeSilva JM (2011) A shift toward birthing relatively large infants early in human evolution. Proceedings of the National Academy of Sciences of the United States of America 108: 1022–1027.
60. Gibbs S, Collard M, Wood B (2000) Soft-tissue characters in higher primate phylogenetics. Proceedings of the National Academy of Sciences of the United States of America 97: 11130–11132.
61. Kimura M (1968) Evolutionary rate at the molecular level. Nature 217: 624–626.
62. Nei M (2007) The new mutation theory of phenotypic evolution. Proc Natl Acad Sci U S A 104: 12235–12242.
63. Kuhl F, Giardina C (1982) Elliptic Fourier features of a closed contour. Computer graphics and image processing 18: 236–258.
64. Zollikofer CPE, Ponce de León MS (2001) Computer-assisted morphometry of hominoid fossils: the role of morphometric maps. In: De Bonis L, Koufos G, Andrews P, editors. Phylogeny of the Neogene Hominoid Primates of Eurasia. Cambridge: Cambridge University Press. 50–59.
65. Bondioli L, Bayle P, Dean C, Mazurier A, Puymerail L, et al. (2010) Technical note: Morphometric maps of long bone shafts and dental roots for imaging topographic thickness variation. American Journal of Physical Anthropology 142: 328–334.
66. Felsenstein J (1989) PHYLIP - Phylogeny Inference Package (Version 3.2). Cladistics 5: 164–166.

4

Don't Fall Off the Adaptation Cliff: When Asymmetrical Fitness Selects for Suboptimal Traits

Elodie Vercken[1]*, Maren Wellenreuther[2], Erik I. Svensson[2], Benjamin Mauroy[3]

1 Institut Sophia Agrobiotech, UMR 1355 ISA, Institut National de la Recherche Agronomique, Sophia-Antipolis, France, **2** Department of Biology, Lund University, Lund, Sweden, **3** Laboratoire J.A. Dieudonné, UMR CNRS 7351, Université de Nice-Sophia Antipolis, Nice, France

Abstract

The cliff-edge hypothesis introduces the counterintuitive idea that the trait value associated with the maximum of an asymmetrical fitness function is not necessarily the value that is selected for if the trait shows variability in its phenotypic expression. We develop a model of population dynamics to show that, in such a system, the evolutionary stable strategy depends on both the shape of the fitness function around its maximum and the amount of phenotypic variance. The model provides quantitative predictions of the expected trait value distribution and provides an alternative quantity that should be maximized ("genotype fitness") instead of the classical fitness function ("phenotype fitness"). We test the model's predictions on three examples: (1) litter size in guinea pigs, (2) sexual selection in damselflies, and (3) the geometry of the human lung. In all three cases, the model's predictions give a closer match to empirical data than traditional optimization theory models. Our model can be extended to most ecological situations, and the evolutionary conditions for its application are expected to be common in nature.

Editor: Martin Heil, Centro de Investigación y de Estudios Avanzados, Mexico

Funding: This work was supported by the Institut National de la Recherche Agronomique and the Centre National de la Recherche Scientifique. The funders had no role in study design, data collection and analysis, decision to publish, or preparation of the manuscript.

Competing Interests: The authors have declared that no competing interests exist.

* E-mail: elodie.vercken@sophia.inra.fr

Introduction

Evolutionary theory predicts that trait means in populations should evolve towards the value that maximizes fitness [1], which is also a central assumption in most optimality analyses [2]. However, in many cases the evolution of fitness-related traits might be constrained by genetic or physiological trade-offs that cause negative genetic correlations between traits [3]. Fitness in these situations will be maximized in a way that depends on the balance between the counteracting selective forces on traits, and the net fitness functions are expected to be bell-shaped [4]. For instance, increased annual reproductive effort is expected to negatively affect adult survival in long-lived species, leading to a fitness optimum where lifetime reproductive success will be maximized by intermediate reproductive effort [5]. Optimality theory predicts that natural selection will drive the population towards this 'optimal' trait value that maximizes fitness [6], which has a close connection to the concept of 'adaptive peaks' in population genetics [7]. This classical optimization approach has been successfully applied in evolutionary ecology to predict the population mean of many phenotypic traits.

In particular, the evolution of reproductive traits, such as offspring number, has received considerable interest and has provided evolutionary ecologists with a solid conceptual foundation for optimality theory in life-history evolution [8–10]. However, in many species of birds and mammals, the number of offspring most commonly observed is often less than the maximum [11–16]. Several alternative theories have been advanced to explain this pattern [15,17]: (1) costs of reproduction due to trade-offs with parental survival or future reproduction [18]; (2) inter-annual variation in juvenile survival related to variation in environmental quality [12]; (3) individual optimization in relation to individual condition and local resource availability [19]; and (4) the interaction between asymmetrical fitness costs and individual variance in brood size ('cliff-edge hypothesis') [12,20].

Among these hypotheses concerning the evolution of litter size, cliff-edge effects have the potential to provide a unifying framework for understanding the optimization of phenotypic traits. This theory predicts that when juvenile survival is asymmetrically low in large broods, moderate variance around the optimal brood size will result in large differences in survival between clutches slightly smaller or larger than the optimal. As a consequence of these asymmetric costs, females producing larger than the most productive broods will leave fewer descendants than females producing smaller than the most productive broods, and the evolutionary optimal should be smaller than the most productive brood size.

In what follows, the relationship between phenotypic value and reproductive value will be referred to as 'phenotype fitness' (which defines the single most productive brood size). In contrast, the relationship between genotypic value and the reproductive value averaged over the phenotype range for each genotype will be referred to as 'genotype fitness' (which defines the evolutionary optimum). The difference between these two definitions of fitness is illustrated on Figure 1. In the absence of any phenotypic variance, these two definitions merge.

On a more general note, cliff-edge effects are related to the properties of convex functions known as Jensen's inequality.

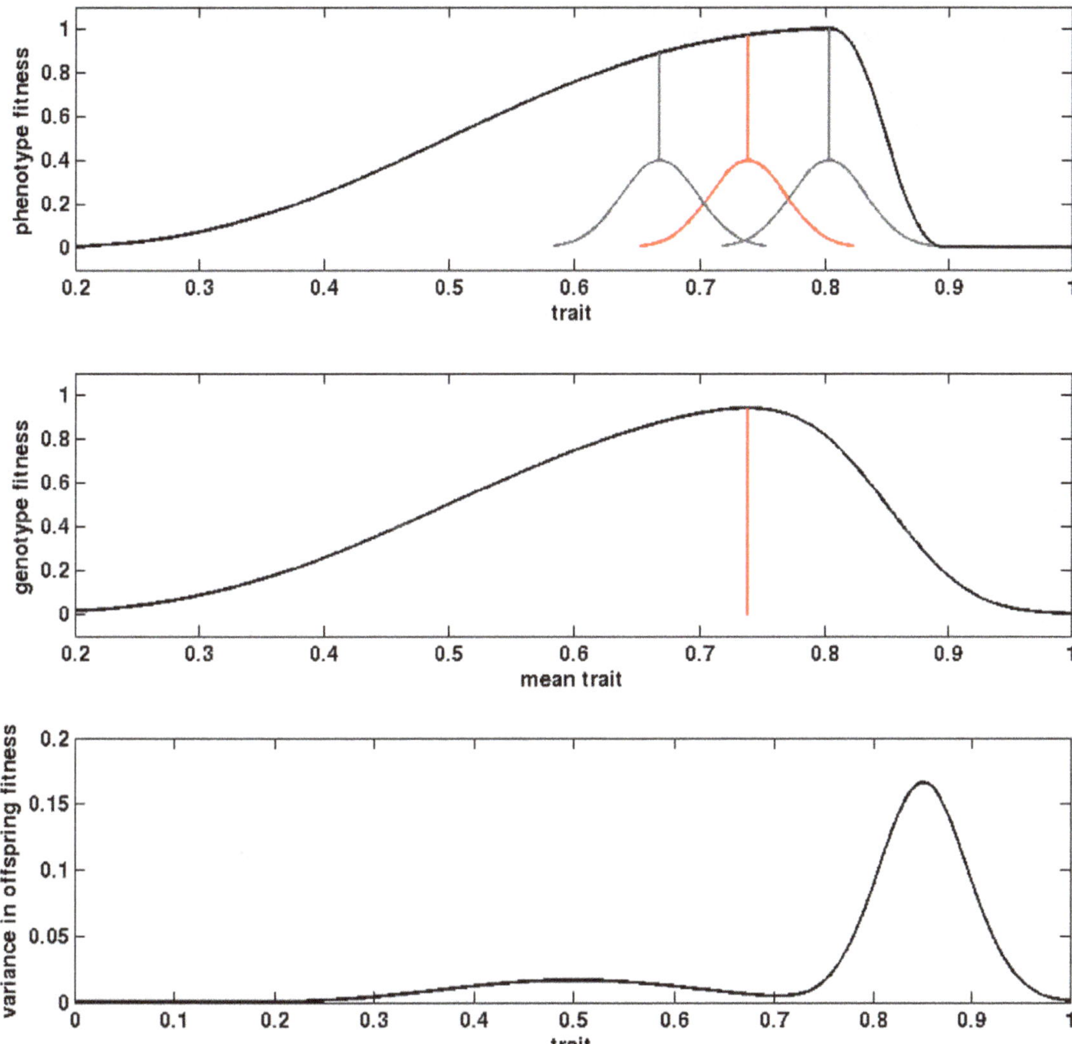

Figure 1. Comparison between (a) the phenotype fitness function, and (b) genotype fitness function relative to the trait variation (after Martin and Huey [27]). Because of the steep part (cliff) of function (a), the optimal trait value (red phenotypic distribution) is shifted downwards from the trait value that maximizes phenotype fitness. Function (c) represents the variance in offspring phenotype fitness for different mean values of the trait.

Jensen's inequality states that, for a convex function f and a set of values $x = (x_1,\ldots,x_n)$ with a mean of $\overline{x}$ and non-zero variance, the average image of x, $\overline{f(x)}$ does not equal the image of the average x, $f(\overline{x})$. In other words, the trait value that gives the highest genotype fitness in the presence of phenotypic variance is not the value that gives the highest phenotype fitness. These analytical principles have been successfully applied to a wide range of evolutionary questions, including the evolution of reproductive systems and life-history strategies, individual behaviour or population dynamics in variable environments [21–28]. In this article, we propose a general mathematical formalization of the cliff-edge problem that can be applied to any fitness-related trait that exhibits phenotypic variability and has asymmetric fitness costs. These conditions are expected to be quite common in nature; hence the generality of a model incorporating these effects is likely to be high.

We develop an analytical model that describes the evolution of a population with random variation in the expression of a fitness-related trait and an asymmetrical fitness function. We demonstrate that these conditions select for apparent sub-optimal genotypes with regard to phenotype fitness, and we show that the optimal genotypic value depends on both the amount of variance of the trait and on the skewness of the fitness function. The model provides quantitative predictions of the position of the optimum and the distribution of phenotypic variance. To illustrate our method, we apply the model to three different evolutionary systems for which we were able to estimate realistic fitness functions from empirical data. Two of these examples (evolution of litter size in guinea pigs, and the evolution of male sexual ornaments in a damselfly) are in line with the classical framework of life history evolution. The third example (evolution of human respiratory tract geometry) stems from evolutionary medicine and demonstrates that cliff-edge effects can act on any trait that is targeted by natural selection.

Results

1- Model of the evolution of a population in presence of phenotypic variation

We consider a population with a continuous trait (e.g. reproductive effort, physiological parameters). We assume that there is no genetic variation for this trait in the population: all genotypes take the value g, but the phenotypic expression can vary randomly from φ_{min} to φ_{max}. This phenotypic variance can arise from several different processes (e.g. developmental instability, environmental variability, maternal effects, epistasis) as long as it is a random process such that any individual of genotype g can experience any phenotype between φ_{min} and φ_{max}. The simple evolutionary processes involved in the population model were chosen to isolate the effect of phenotypic variability on trait evolution and to increase the generality of model predictions.

The function $a(\varphi,t)$ denotes the frequency of individuals in the population having the phenotype value φ at time t. We assume that instantaneous fitness depends on the phenotype, so these individuals have a reproductive rate $b(\varphi)\geq 0$ and a mortality rate of $m(\varphi)>0$. Therefore, at each instant t and per unit of time, $m(\varphi)a(\varphi,t)$ individuals of phenotype φ die, while $b(\varphi)a(\varphi,t)$ descendants are produced by individuals of phenotype φ.

For any individual with genotype g, the phenotype φ of its offspring is randomly distributed around the value g following the distribution function $G(\varphi,g,\sigma)$. We consider that $G(\varphi,g,\sigma)$ is a Gaussian function (default hypothesis for a quantitative characters, see [29]) centred around g with a variance σ^2.

Therefore, offspring of phenotype φ are produced by parents of genotype g in the proportion $G(\varphi,g,\sigma)$, regardless of the parents' own phenotype, i.e. there are no cross-generational effects. Therefore, at each time a total quantity $\int_{\vartheta_{min}}^{\vartheta_{max}} b(l)\,a(l,t)dl$ of new individuals are produced in the population among which only a fraction $G(\varphi,g,\sigma)$ will have the phenotype φ.

The variation of the distribution of individuals of parameter φ over time is then given by the differential equation:

$$\frac{\partial a}{\partial t}(\vartheta,t)=\underbrace{-m(\vartheta)\,a(\vartheta,t)}_{mortality}\text{\%hbrace} \\ \underbrace{+\,G(\vartheta,g,\sigma)\times\int_{\vartheta_{min}}^{\vartheta_{max}} b(l)\,a(l,t)dl}_{reproduction}\text{\%hbrace} \quad (1)$$

We show in Supplementary Text S1 that such a population does not go extinct as long as

$$F(g)=\int_{\vartheta_{min}}^{\vartheta_{max}}\frac{G(l,g,\sigma)\,b(l)}{m(l)}dl\ >\ 1 \quad (2)$$

The term in the integral represents the per capita growth rate $w(l)=b(l)/m(l)$ (phenotype fitness) of individuals with phenotype l multiplied by their frequency in the population. Thus, the function $F(g)$ represents the sum of phenotype fitness of all phenotypes weighted by their respective frequency, i.e. the weighted mean of phenotype fitness in the population. The growth rate function $F(g)$ is therefore the genotype fitness of genotype g. This result holds for populations with limited resources (Supplementary Text S1).

Then, we consider two populations with limited growth, one with a genotype g_1 and the other with a genotype $g_2\neq g_1$. They are represented by their respective distribution $a_1(\varphi,t)$ and $a_2(\varphi,t)$. They are interacting with each other due to mutually shared resources that are limited. In order to have true competition, we assume that each population does not go extinct if it is alone, which is equivalent to $F(g_1)>1$ and $F(g_2)>1$.

The equations that describe the evolution of these populations and their distributions along time are:

$$\begin{cases} \frac{\partial a_1}{\partial t}(\vartheta,t)=-m(\vartheta)\,a_1(\vartheta,t)+\left(1-\int_{\vartheta_{min}}^{\vartheta_{max}}(a_1(l,t)+a_2(l,t))dl\right) \\ \times\,G(\vartheta,g_1,\sigma)\times\int_{\vartheta_{min}}^{\vartheta_{max}} b(l)\,a_1(l,t)dl \\ \frac{\partial a_2}{\partial t}(\vartheta,t)=-m(\vartheta)\,a_2(\vartheta,t)+\left(1-\int_{\vartheta_{min}}^{\vartheta_{max}}(a_1(l,t)+a_2(l,t))dl\right) \\ \times\,G(\vartheta,g_2,\sigma)\times\int_{\vartheta_{min}}^{\vartheta_{max}} b(l)\,a_2(l,t)dl \end{cases} \quad (4)$$

There are four equilibrium points: coexistence of both populations; extinction of both populations; only one population survives while the other goes extinct (two combinations). Coexistence is possible only if $g_1=g_2$, which is excluded by hypothesis. Moreover since $F(g_1)$ and $F(g_2)$ are assumed to be strictly greater than 1 for each population, it can be shown that the extinction of both populations is not possible. Thus, under these hypotheses one population must invade the other. Then, a successful invasion of population g_1 into population g_2 (i.e. equilibrium $a_1\neq 0$ and $a_2=0$ stable) is possible if and only if:

$$F(g_1)=\int_{\vartheta_{min}}^{\vartheta_{max}}\frac{G(l,g_1,\sigma)\,b(l)}{m(l)}dl\ >\ F(g_2)=\int_{\vartheta_{min}}^{\vartheta_{max}}\frac{G(l,g_2,\sigma)\,b(l)}{m(l)}dl \quad (5)$$

Hence, the evolutionary stable strategy (ESS) corresponds to the genotype g^* that maximises the growth rate function $F(g)$. The population of parameter g^*, also called the super-mutant population, will invade any population of parameter $g\neq g^*$, while it cannot be invaded by other populations with a parameter $g\neq g^*$. Thus, the genotype g^* is an ESS and should be observed in population at the equilibrium state, although the phenotype varies in the population.

To determine g^*, it is necessary to calculate the maximum of the growth rate function $F(g)$. Hence, the most efficient genotype is the one that maximises the success of the whole population by cumulating the relative success of each phenotypic trait weighted by their frequency.

When the fitness function is symmetric or if there is no phenotypic variance at all, the genotypic value g^* associated with the maximum of the function F is equal to the value that maximizes the phenotype fitness w. However, when the fitness function is asymmetric and the phenotypic variance is non-zero, these two values do not match anymore, as predicted by Jensen's inequality. In this case, the optimum genotype g^* value is systematically shifted from the maximum of the phenotype fitness in the direction of the least slope (Supplementary Text S2).

2- Application of the model using three biological examples

Most heritable traits are expected to include some degree of non-additive genetic, environmental or developmental variability that affects their optimal expression. Our first example focuses on optimal litter size in laboratory strains of Guinea pigs (*Cavia porcellus*). For this, we re-analyzed Mountford's original data set [20] that was used in his development of the cliff-edge hypothesis. Our second example deals with a secondary sexual trait in the damselfly *Calopteryx splendens*, using survival and mate choice data obtained in the field ([30], M. Wellenreuther, E. Vercken and E. Svensson, unpublished data). The third example is based on modelling work about the impact of lung geometry on respiratory performance in humans [31]. In all these examples, fitness functions are not symmetrically shaped around their maximum value, and we show consistent matching between empirical data and model predictions.

2-1 Example 1: Optimal litter size in Guinea pigs. In his seminal paper, Mountford [20] showed that the litter size that leads to the maximum number of surviving offspring in Guinea pigs was not the most frequent one. He suggested that the phenotypic variability associated with high asymmetric fitness costs for large litter sizes could explain this observation. To prove the validity of his theory, he produced a theoretical example based on simple numeric calculations. In what follows, we show that the frequency distribution of litter size in Guinea pigs could be predicted by the general model described above, and that this model can also provide information about both the optimal genotype as well as the variance in the expressed phenotype.

In our model, the litter size is called L and extents from $L_{min} = 0$ to $L_{max} = 9$. Reproductive rate $b(L)$ corresponds to offspring survival in relation to the litter size (data reproduced from Mountford [20]). Life-history theory predicts that, as a consequence of trade-offs between present and future reproduction, female survival should decrease when litter size increases ([32–36] but see [15,37,38] for counter-examples). In the absence of any data on the precise relationship between litter size and female mortality rate in Guinea pigs, we assumed a simple linear model $m(L) = m_r + \alpha \times L$, where m_r is the mortality rate for non-reproductive individuals and α is a constant estimated by least-squares method. The intercept value m_r does not affect the position of the genotype fitness maximum as it plays the role of a scaling factor once α is chosen. The phenotype fitness function is asymmetric around a maximum plateau for litter sizes between 2 and 3 (Figure 2a).

The general method to compare the predictions of our model to the measured data is similar in the three different examples used in this paper and is described in detail in Supplementary Text S3. First, we determine the optimal genotype $L_o(\sigma)$ for each of the possible values of phenotypic variance σ^2. Second, we find the value of (σ, $L_o(\sigma)$) that best fits empirical data using the mean-square method, here in the first example, the distribution of litter size in Guinea pigs.

Figure 2b represents the relationship between the standard deviation σ and optimal genotype $L_o(\sigma)$ (first step). Because the phenotype fitness function has a maximum plateau, this relationship is not monotonic. For small values of σ, most offspring will fall within the phenotype fitness plateau. However, because of the long tails of the Gaussian distribution, a small proportion will be outside of it. Therefore, the optimal genotype fitness is initially shifted away from the steepest slope (left side).

When σ increases, at first a significant proportion of offspring phenotypes reaches values of $L>3$, while fewer phenotypes reach values of $L<2$ (because the optimum is initially right-shifted). Therefore, the optimum genotype value decreases as it shifts away from the closest fitness slope, accounting for the initial negative relationship in Figure 2b.

For higher values of σ, a more significant proportion of offspring phenotype reaches values of $L<2$ and thus 'fall off the cliff'. Because the phenotype fitness loss is higher for phenotypes that reach values of $L<2$ than for those with values of $L>3$, the optimum genotype will shift away from the steepest slope and tends to obtain higher values of L as σ increases (positive relationship in Figure 2b).

In Guinea pigs, the optimal phenotypic trait predicted was *3.05* with a phenotypic variance of *1.30* ($\sigma = 1.14$). The difference between the distribution of the trait in the population predicted by

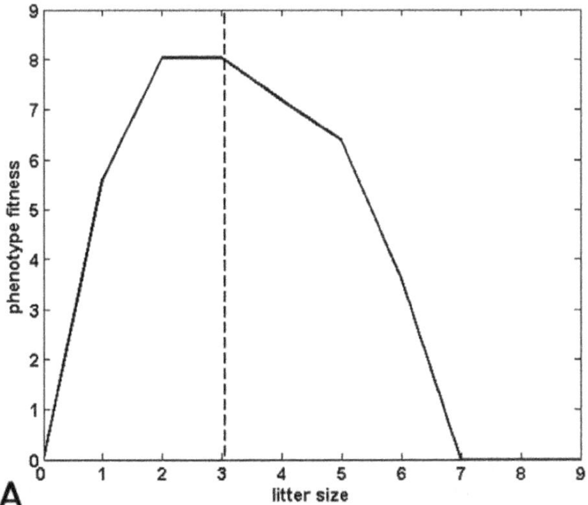

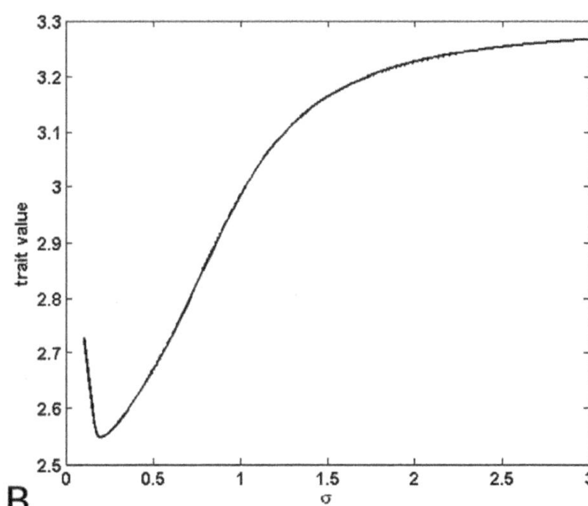

Figure 2. Effect of phenotypic variance on optimal litter size in Guinea pigs. (a) Asymmetric fitness function $b(L)/m(L)$. The curve reaches its maximum value on the plateau between $L = 2$ and $L = 3$. The dashed line corresponds to the optimal genotype $L = 3.05$ ($\sigma = 1.14$) that best fits the empirical data from Mountford [20]. (b) Position of the optimal genotype relative to the standard deviation σ.

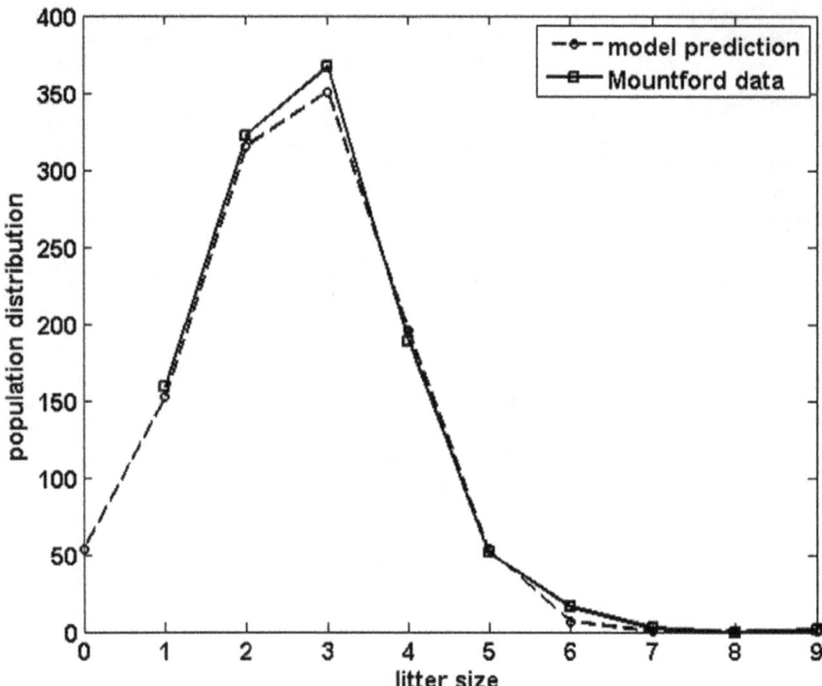

Figure 3. Comparison between model predictions (dashed line) and empirical data (Mountford [20], solid line) for the distribution of litter sizes in the population. The model predicts an optimal genotypic value for litter size at *3.05* with a variance σ^2 of *1.30*.

the model and the distribution observed by Mountford [20] was less than *3.5%* (Figure 3).

2-2 Example 2: Selection on male wing patch size in *Calopteryx splendens*. In the damselfly *Calopteryx splendens*, males have a dark melanized wing patch that covers approximately 50% of the wing [39]. Wing patches in this species function as secondary sexual traits and are only carried by males [40–42]. Wing melanization affects male predation risk, as males with larger wing patches suffer higher mortality from avian predators [30]. This trait is also under sexual selection in this population, where female mating response increases with male wing patch size (Supplementary Text S4).

The phenotype fitness function is asymmetric (Figure 4) around a maximum at wing patch length $x = 17.52$ *mm*. The fitness decrease is steeper for larger patches; hence the optimal genotype is expected to be smaller than the phenotype fitness maximum.

The best fit was obtained with an optimal trait x_o of *16.93 mm* and a standard deviation σ of *2 mm*. The difference between the values predicted by the model and those measured in the field was less than *17%* (Figure 5).

2-3 Example 3: Estimating optimal lung geometry. Mauroy et al. [31] developed a model of the human bronchial tree to study the relationship between the geometry of the tree and its hydrodynamical resistance. They modelled the distal part of the lungs as a dichotomical tree branching in a homothetical way: at each bifurcation: each branch divides in two identical smaller branches, whose length and diameter are reduced by a constant factor *h*, the homothetical factor (Figure 6, for $h_l = h_d = h$).

They showed that the mean phenotypic parameter *h*, observed from empirical data [43], is around *0.8470*, while the optimum value predicted by their model was *0.7937*. Although small, this difference is expected to have major effects on the resistance and volume of the lung because of the multiplicative nature of the homothetical transformation (i.e. if the tree bifurcates 10 times, the deepest branches will be h^{10} smaller than the first generation branch). Large lung hydrodynamic resistance (small *h*) requires more energy for lung ventilation, while large lung volume (large *h*) results in a reduced exchange surface (less volume is available for alveoli). We consider that the fitness of an individual is a function of respiratory efficiency and thus depends on the value of the parameter *h*.

We further extended the model originally formulated by Mauroy et al. [31] to estimate a more realistic fitness function for *h*. Based on morphometric data [44] we assumed that lengths and diameters of the tree branches are not reduced by the same factor at each bifurcation (Figure 6, $h_l \neq h_d$). Under this hypothesis, the resulting phenotype fitness function (Figure 7) is asymmetrical around the optimum (Supplementary Text S4). The steepest decrease in fitness occurs for values below the optimum, thus we expect the optimal value of the genotype fitness to be higher than the most efficient phenotype.

Empirical data indicate a mean phenotypic value of $h = 0.8470$ [36]. The model predicts that this mean phenotype is reached for an optimal genotype $h = 0.8504$ and a standard deviation $\sigma = 0.2$. Compared with the value maximizing the phenotype fitness function $h_{min} \sim 0.7937$, the optimal genotype corresponds to a resistance that is *3.3* times smaller and a volume that is *2.4* times larger. This result quantitatively confirms the hypothesis stated by Mauroy et al. [31] that this shift from the phenotypic optimum acts as a security margin to protect the bronchial tree from phenotypic variations. This analysis is representative of an optimality problem related to the geometry of a transport tree, and as such it can be extended to many other well-known theoretical contexts, such as the Metabolic Theory of Ecology [44,45].

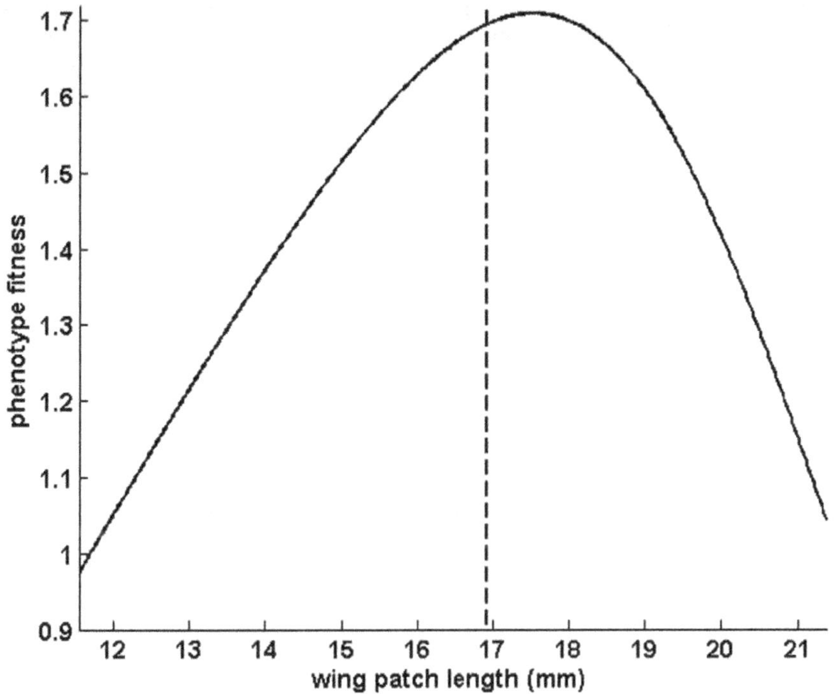

Figure 4. Estimate of the asymmetric phenotype fitness function *z* for *Calopteryx splendens* in relation with wing patch length *x*. The maximum is reached for a wing patch length of *17.52 mm*. The dashed line represents the optimal genotype predicted by our model that best fits the observed population distribution, i.e. a patch length of *16.93 mm*.

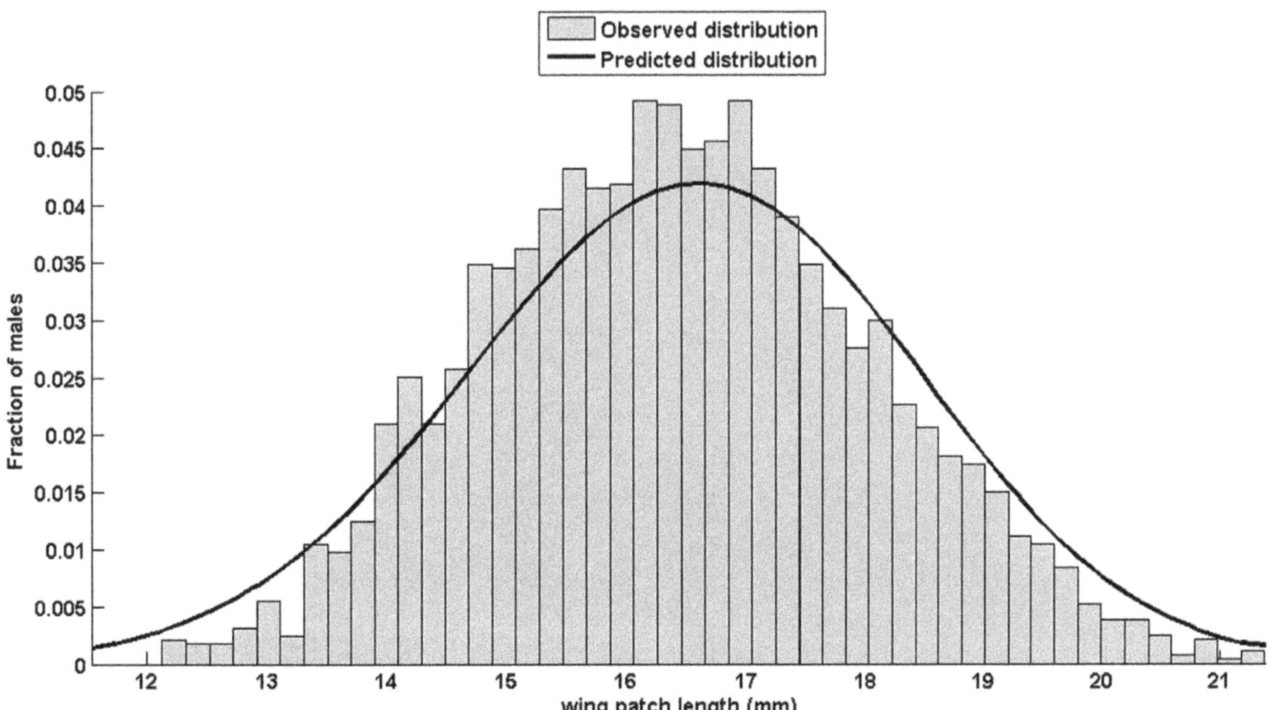

Figure 5. Distribution of wing patch length of *Calopteryx splendens* at the population Naturreservat Klingavälsån (55.6384, 13.54142) in southern Sweden. The bars correspond to males caught in the field. The line shows the distribution predicted by the model with an optimal patch length of *16.93* mm and a standard deviation of 2 mm.

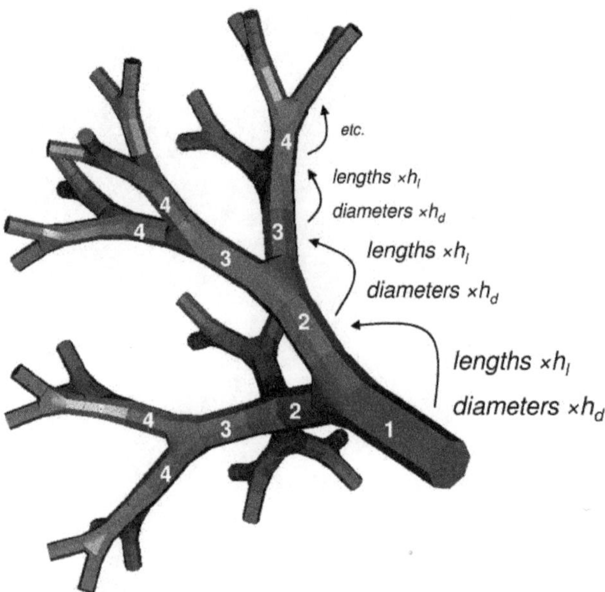

Figure 6. Model of the distal part of the bronchial tree used in this study. After each bifurcation the generation index is incremented by one (white numbers). The full model consists of *11* generations.

Discussion

Environmental variation is ubiquitous in nature and can generate substantial levels of phenotypic variation in fitness-related traits [20,23,46]. Such unpredictable variation (e. g. developmental plasticity) can have profound effects on optimal trait values [22,47]. Our model explicitly incorporates such unpredictable variation through its effects on phenotypic variance and shows that the position of the genotype fitness optimum will ultimately depend on both the amount of phenotypic variance and the shape of the fitness function. For symmetrical fitness and variance functions, the optimal value for the fitness-related trait matches the value that maximizes phenotype fitness. However, this classical optimization scenario does not hold when fitness functions are asymmetrical and when environmental variance leads to a variable expression of genetic traits. Then, the genotype fitness optimum is instead expected to shift from the phenotype optimum value in the direction of the least slope. Such qualitative predictions can be driven directly from the shape of the fitness function. Furthermore, as illustrated in our three examples, quantitative predictions of random phenotypic variance and genotype fitness optima can be derived from empirical data.

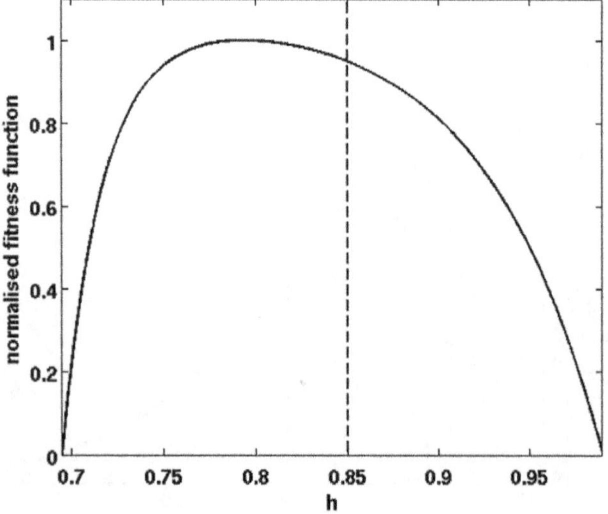

Figure 7. Relationship between homothetical factor *h* and fitness (trade-off between lung volume and hydrodynamical resistance). The vertical dashed line corresponds to the optimal genotype $h = 0.8504$ with $\sigma = 0.2$, which best fits empirical data.

Model applications

In the first two examples, we used empirical data on the frequency distribution of traits and their relationship with fitness components (reproductive rate and mortality) to estimate σ, the amount of random phenotypic variance in the population, and to validate the predictions from the model. If the theoretical distribution predicted by the value of σ closely matches the observed distribution of the trait, then the cliff-edge hypothesis is a sufficient condition to explain the shift in the distribution away from the phenotype fitness maximum. In this case, the most frequent phenotype in the population is located at the predicted value of the genotype fitness optimum. Furthermore, independent of the intrinsic quality of the prediction, the difference between the theoretical and observed descriptors provides information about the importance of processes other than cliff-edge effects in the evolution of the trait. In these two examples, we obtained less than 3.5% deviation for the Guinea pig data and less than 17% for the damselfly data. Data on litter size in Guinea pigs are more likely to meet the model's assumptions (e.g. laboratory strains with low amount of additive genetic variance, controlled environment with few external selective pressures). In contrast, natural damselfly populations should contain more genetic variability, and many selective pressures in addition to predation and female mate choice are expected to affect the evolution of male wing patch size [40]. These processes are partly responsible for the variability that is not incorporated in the model. Yet, the model gives a closer prediction of the actual fitness optimum than the phenotype fitness maximum, thus supporting the claim that cliff-edge effects are likely to play a strong role in the evolution of natural populations.

Alternatively, the model can be used as an *a priori* hypothesis to predict the value of key parameters when empirical data is not available, which is a classical approach in physics and biomechanics modelling. In the lung example, the frequency distribution of the trait h is unknown, but its mean value in the population has been estimated. Mauroy et al. [31] suggested that the value of h is shifted from the phenotype fitness optimum in order to confer higher robustness to the lung geometry in response to developmental variation. In this context, the cliff-edge hypothesis provides a formal framework to calculate the expected value of σ, which can then be used to implement other models and derive further predictions that can be empirically tested.

Generality of the conditions of the model

The keystone hypotheses of the model are the existence of an asymmetrical fitness function and a certain amount of phenotypic variance, and the qualitative model predictions appear robust to the precise shape of these functions (Supplementary Text S2). Although the exact geometry of trade-offs will differ between different ecological situations, asymmetrical trade-offs are likely to be the rule rather than the exception, especially for traits under stabilizing selection [48]. Indeed, if a trade-off results from the interaction between two unrelated traits, there is no reason why their respective effects on phenotype fitness should be exactly opposing each other, i.e. completely symmetric. Similarly, there are many processes that can generate random phenotypic variation. For instance, condition-dependence, phenotypic plasticity, developmental effects, environmental fluctuations and/or non-additive genetic effects (e. g. epistasis) are all common and well-known processes that are likely to increase phenotypic variance [49,50]. These processes are expected to have a significant influence on traits with low heritability, which is a common characteristic of traits strongly related to fitness [29]. Therefore, sources of stochastic phenotypic variation have been suggested to be key factors in the evolutionary ecology of populations [51–54].

Finally, we modelled the effects of phenotypic variance during development of individual phenotypes. These phenotypes are then assumed to be stable during life, i.e. we explicitly considered inter-generational phenotypic variance. However, the model can be generalized and the same predictions can also be made in the case of intra-individual variance, for example, when the value of a fitness-related trait changes during an individual's life. Such an example was recently documented by Martin and Huey [27] in the context of thermoregulation in reptiles. The authors showed that the optimal range of body temperatures for an individual should not be centred at the temperature for which the instantaneous fitness is maximised, but should be shifted towards a lower temperature and that the magnitude of the shift increased with the asymmetry of the fitness function.

Phenotypic variability and species adaptation

In the context of optimization problems, stochastic effects can influence the predictions of theoretical models (e.g. bet-hedging strategies, optimization of the geometric mean fitness, [23,55,56], this study). However, these effects depend quite strongly on the relationship between an individual's genotype and the variability of its phenotype.

First, if the variability in phenotype expression is independent from the individual's genotype (as is the case in our model), its association with an asymmetrical fitness functions can be a significant limit to adaptation. Our example of bronchial tree geometry illustrates this situation and provides an adaptive explanation relating to the fact that the human lung is probably not as efficient as it could be. Several other examples in evolutionary medicine appear to be consistent with the existence of cliff-edge effects [57–60]. Strong directional selection for traits that are globally advantageous would sometimes drive their mean too close to the 'fitness cliff', which could set the stage for counter-selection of extreme phenotypes. Such mechanisms would limit the long-term directional evolution of heightened physiological, mental and immune capacities in humans, and the average performance of individuals would be lower than their maximum potential. These examples tentatively suggest that similar processes might have operated in many different species to constrain the evolution of phenotypic traits within a smaller range than their full physiological potential.

Alternatively, when phenotype variability is related to the individual's genotype, the evolutionary consequences are likely to be quite different. In certain conditions, selection can favour the ability of phenotypes to resist random developmental or environmental perturbations, a process known as canalization [61]. Canalization is expected to be favoured in spatially or temporally variable environments, or in environments connected by high levels of gene flow, because it allows the persistence of high genetic variation and evolutionary potential [62]. In contrast, different genotypes might display different reaction norms in response to environmental variations, i.e. the amount of phenotypic variance and the shape of its distribution can differ between individuals. In this case, asymmetrical variance functions can be selected as a way to compensate for an asymmetrical fitness function by avoiding most detrimental phenotypes (Supplementary Text S2). However, such a strategy could be selected only if environmental variations do not affect the fitness function itself, and if they are restricted within a limited range so that extreme phenotypes will be rare.

Conclusion

In this study, we propose a simple formalization, validated by three empirical examples, of an evolutionary process known as the 'cliff-edge' effect. Our predictions stand for any trait (i) associated with an asymmetrical fitness function and (ii) when phenotypic expression is subjected to random variation, which are conditions expected to be common in nature [63,64]. In this framework, future studies should aim at analysing the optimization of genotype fitness instead of phenotype fitness. If only cliff-edge effects are shaping the evolution of the trait, then the most frequent value of the trait should match the genotype fitness optimum, i.e. the genotype fitness optimum is the null hypothesis for trait optimization. On the contrary, if the trait is non-optimal with regard to the genotype fitness, other evolutionary processes should be considered. For instance, unmeasured fitness components might cause undetected trade-offs that constrain the adaptation of the trait.

Author Contributions

Conceived and designed the experiments: EV BM MW ES. Performed the experiments: EV BM MW ES. Analyzed the data: EV BM. Wrote the paper: EV BM MW ES. Wrote and analyzed the population model: EV BM.

References

1. Fisher RA (1930) The genetical theory of natural selection. Oxford: Clarendon Press. 272 p.
2. Maynard Smith J (1978) Optimization theory in evolution. Ann Rev Ecol Syst 9: 31–56.
3. Stearns SC (1992) The evolution of life histories. Oxford: Oxford University Press. 262 p.
4. Fairbairn DJ, Reeve JP (2001) Natural selection. In: Fox CW, Roff DA, Fairbairn DJ, eds. Evolutionary Ecology, Concepts and Case Studies. Oxford: Oxford University Press. pp 29–43.
5. Roff DA (1992) The evolution of life histories: theory and analysis. New York: Chapman and Hall. 535 p.
6. Parker GA, Maynard Smith J (1990) Optimality theory in evolutionary biology. Nature 348: 27–33.

7. Wright S (1932) The roles of mutation, inbreeding, crossbreeding and selection in evolution. Proceedings of the 6th International Congress on Genetics 1: 355–366.
8. Lack D (1947) The significance of clutch size. Ibis 89: 302–352.
9. Charnov EL, Skinner SW (1984) Evolution of host selection and clutch size in parasitoid wasps. Flor Entomol 67: 5–21.
10. Mock DW, Parker GA (1998) The evolution of sibling rivalry. Oxford: Oxford University Press. 480 p.
11. Lessells CM (1986) Brood size in canada geese - a manipulation experiment. J Anim Ecol 55: 669–689.
12. Boyce MS, Perrins CM (1987) Optimizing great tit clutch size in a fluctuating environment. Ecology 68: 142–153.
13. Moller AP (1991) Ectoparasite loads affect optimal clutch size in swallows. Funct Ecol 5: 351–359.
14. Boutin S, Moses RA, Caley MJ (1988) The relationship between juvenile survival and litter size in wild muskrats (*Ondatra zibethicus*). J Anim Ecol 57: 455–462.
15. Morris DW (1992) Optimum brood size: tests of alternative hypotheses. Evolution 46: 1848–1861.
16. Risch TS, Dobson FS, Murie JO (1995) Is mean litter size the most productive? A test in Columbian ground squirrels. Ecology 76: 1643–1654.
17. Risch TS, Michener GR, Dobson FS (2007) Variation in litter size: a test of hypotheses in Richardson's ground squirrels. Ecology 88: 306–314.
18. Williams GC (1966) Natural selection, the costs of reproduction, and a refinement of Lack's principle. Am Nat 100: 687–690.
19. Smith CC, Fretwell SD (1974) The optimal balance between size and number of offspring. Am Nat 108: 499–506.
20. Mountford MD (1968) The significance of litter size. J Anim Ecol 37: 363–367.
21. Smallwood PD (1996) An introduction to risk sensitivity: the use of Jensen's inequality to clarify evolutionary arguments of adaptation and constraint. Am Zool 36: 392–401.
22. Ruel JJ, Ayres MP (1999) Jensen's inequality predicts effects of environmental variation. Trends Ecol Evol 14: 361–366.
23. Pasztor L, Kisdi E, Meszéna G (2000) Jensen's inequality and optimal life history strategies in stochastic environments. Trends Ecol Evol 15: 117–118.
24. Wilson WG, Harder LD (2003) Reproductive uncertainty and the relative competitiveness of simultaneous hermaphroditism versus dioecy. Am Nat 162: 220–241.
25. Drake JM (2005) Population effects of increased climate variation. Proc Roy Soc B Biol Sci 272: 1823–1827.
26. Freedberg S, Taylor DR (2007) Sex ratio variance and the maintenance of environmental sex determination. J Evol Biol 20: 213–220.
27. Martin TL, Huey RB (2008) Why "suboptimal" is optimal: Jensen's inequality and ectotherm thermal preferences. Am Nat 171: 102–118.
28. Turnbull LA, Rees M, Purves DW (2008) Why equalising trade-offs aren't always neutral. Ecol Lett 11: 1037–1046.
29. Falconer DS (1981) Introduction to Quantitative Genetics, 2nd Edition. New-York: Longman. 340 p.
30. Svensson EI, Friberg M (2007) Selective predation on wing morphology in sympatric damselflies. Am Nat 170: 101–112.
31. Mauroy B, Filoche M, Weibel ER, Sapoval B (2004) An optimal bronchial tree may be dangerous. Nature 427: 633–636.
32. Millar JS (1977) Adaptive features of mammalian reproduction. Evolution 31: 370–386.
33. Stearns SC (1989) Trade-offs in Life History Evolution. Funct Ecol 3: 259–268.
34. Boyd IL, Croxall JP, Lunn NJ, Reid K (1995) Population demography of Antarctic fur seals - the costs of reproduction and implications for life-histories. J Anim Ecol 64: 505–518.
35. Ruusila V, Ermala A, Hyvarinen H (2000) Costs of reproduction in introduced female Canadian beavers (*Castor canadiensis*). J Zool 252: 79–82.
36. Koivula M, Koskela E, Mappes T, Oksanen TA (2003) Cost of reproduction in the wild: Manipulation of reproductive effort in the bank vole. Ecology 84: 398–405.
37. Hare JF, Murie JO (1992) Manipulation of Litter Size Reveals No Cost of Reproduction in Columbian Ground Squirrels. J Mammal 73: 449–454.
38. Humphries MM, Boutin S (2000) The determinants of optimal offspring number in free-ranging red squirrels. Ecology 81: 2867–2877.
39. Rüppel G, Hilfert-Rüppel D, Rehfeldt G, Schütte C (2005) Die Prachtlibellen Europas. Hoehnwarsleben: Westarp Wissenschaften. 255 p.
40. Siva-Jothy M (1999) Male wing pigmentation may affect reproductive success via female choice in a Calopterygid damselfly (Zygoptera). Behaviour 136: 1365–1377.
41. Svensson E I, Kristoffersen L, Oskarsson K, Bensch S (2004) Molecular population divergence and sexual selection on morphology in the banded demoiselle (*Calopteryx splendens*). Heredity 93: 423–433.
42. Svensson EI, Eroukhmanoff F, Friberg M (2006) Effects of natural and sexual selection on adaptative population divergence and premating isolation in a damselfly. Evolution 60: 1242–1253.
43. Weibel ER (1963) Morphometry of the human lung. Berlin: Springer. 151 p.
44. West GB, Brown JH, Enquist BJ (1997) A general model for the origin of allometric scaling laws in biology. Science 276: 122–126.
45. Enquist BJ, West GB, Charnov EL, Brown JH (1999) Allometric scaling of production and life-history variation in vascular plants. Nature 401: 907–911.
46. Morris DW (1996) State-dependent life-histories, Mountford's hypothesis, and the evolution of brood size. J Anim Ecol 65: 43–51.
47. Yoshimura J, Shields WM (1992) Components of uncertainty in clutch-size optimization. Bull Math Biol 54: 445–464.
48. De Witt TJ (1997) Optimizing clutch size in birds. Trends Ecol Evol 12: 443.
49. van Noordwijk AJ, de Jong G (1986) Acquisition and allocation of resources: their influence on variation in life-history tactics. Am Nat 128: 137–142.
50. Mazer SJ, Damuth J (2001) Evolutionary significance of variation. Fow CW, Roff DA, Fairbairn DJ, eds. Evolutionary Ecology, Concepts and Case Studies. Oxford: Oxford University Press. pp 16–28.
51. Real LA, Ellner S (1992) Life history evolution in stochastic environments: a graphical mean-variance approach. Ecology 73: 1227–1236.
52. Fox GA (1993) Life history evolution and demographic stochasticity. Evol Ecol 7: 1–14.
53. Whitlock MC (1995) Variance-induced peak shifts. Evolution 49: 252–259.
54. Wilson AJ, Pemberton JM, Pilkington JG, Coltman DW, Mifsud DV, Clutton-Brock TH, Kruuk LEB (2006) Environmental Coupling of Selection and Heritability Limits Evolution. PLoS Biol 4: e216.
55. Philippi T, Seger J (1989) Hedging one's evolutionary bets, revisited. Trends Ecol Evol 4: 41–44.
56. Yoshimura J, Jensen VAA (1996) Evolution and population dynamics in stochastic environments. Res Pop Ecol 38: 165–182.
57. Nesse RM, Williams GC (1996) Why we get sick: the new science of Darwinian medicine. New York: Vintage Books. 290 p.
58. Burns JK (2004) Elaborating the social brain hypothesis of schizophrenia. Behav Brain Sci 27: 868–885.
59. Nesse RM (2004) Cliff-edged fitness functions and the persistence of schizophrenia. Behav Brain Sci 27: 862–863.
60. Stearns SC, Koella JK (2008) Evolution in health and disease (2nd edition). Oxford: Oxford University Press. 400 p.
61. Flatt T (2005) The evolutionary genetics of canalization. Q Rev Biol 80: 287–316.
62. Stearns SC (2002) Progress on canalization. Proc Natl Acad Sci USA 99: 10229–10230.
63. Lajus DL, Graham JH, Kozhara AV (2003) Developmental instability and the stochastic component of total phenotypic variance. In: Polak M, ed. Developmental Instability: causes and consequences. Oxford: Oxford University Press. pp 343–366.
64. Hansen TF, Carter AJR, Pélabon C (2006) On adaptive accuracy and precision in natural populations. Am Nat 168: 168–181.

5

Mechanistic Explanations for Restricted Evolutionary Paths That Emerge from Gene Regulatory Networks

James Cotterell[1,2]*, James Sharpe[1,3]*

1 EMBL-CRG Systems Biology Research Unit, Centre for Genomic Regulation (CRG), Barcelona, Spain, 2 Developmental Biology Section, MRC Human Genetics Unit, Edinburgh, United Kingdom, 3 Institució Catalana de Recerca i Estudis Avançats (ICREA) Professor, Centre for Genomic Regulation (CRG), Barcelona, Spain

Abstract

The extent and the nature of the constraints to evolutionary trajectories are central issues in biology. Constraints can be the result of systems dynamics causing a non-linear mapping between genotype and phenotype. How prevalent are these developmental constraints and what is their mechanistic basis? Although this has been extensively explored at the level of epistatic interactions between nucleotides within a gene, or amino acids within a protein, selection acts at the level of the whole organism, and therefore epistasis between disparate genes in the genome is expected due to their functional interactions within gene regulatory networks (GRNs) which are responsible for many aspects of organismal phenotype. Here we explore epistasis within GRNs capable of performing a common developmental function – converting a continuous morphogen input into discrete spatial domains. By exploring the full complement of GRN wiring designs that are able to perform this function, we analyzed all possible mutational routes between functional GRNs. Through this study we demonstrate that mechanistic constraints are common for GRNs that perform even a simple function. We demonstrate a common mechanistic cause for such a constraint involving complementation between counter-balanced gene-gene interactions. Furthermore we show how such constraints can be bypassed by means of "permissive" mutations that buffer changes in a direct route between two GRN topologies that would normally be unviable. We show that such bypasses are common and thus we suggest that unlike what was observed in protein sequence-function relationships, the "tape of life" is less reproducible when one considers higher levels of biological organization.

Editor: Arnar Palsson, University of Iceland, Iceland

Funding: The authors have no support or funding to report.

Competing Interests: The authors have declared that no competing interests exist.

* E-mail: james.cotterell@crg.es (JC); james.sharpe@crg.es (JS)

Introduction

It remains unclear how restricted evolving populations are to move through any route in genotype space. This is due to our lack of understanding of how genotype maps to phenotype, since mutations occur at the level of genotype yet selection acts at the level of phenotype. In particular it has been suggested that certain combinations of genetic interaction are not viable or less fit, meaning that evolution cannot tinker through any form; there are restrictions on trajectories. Evolutionary trajectories can be viewed through the guise of a neutral network which derives from Maynard-Smith's original concept of protein space [1]. A neutral network assumes that genotypes fall into two classes; those that are viable and those that are non-viable. Only those that are viable are included in the neutral network, and structurally similar genotypes are connected based on particular criteria. The shape of a neutral network can lie between two extremes: from regular and smooth to irregular with concavities. Only on irregular neutral networks are evolutionary trajectories restricted since the path to another viable genotype may involve a step to a non-viable genotype. Indeed neutral networks were first utilized to explore evolutionary routes in RNA sequence-shape space [2–4].

Irregular neutral networks are caused by the phenomenon of epistasis (Figure 1a and b). Epistasis essentially means that the fitness of certain genetic combinations does not have a simple linear relationship to the component parts [5–8]. A particular form of epistasis, *reciprocal sign epistasis* (RSE), describes a 2-loci situation where a wildtype and a double mutant are viable or fit yet the corresponding single mutants are unviable or significantly less fit. In particular this type of epistasis captures the basic idea of restricted evolutionary paths since no direct route is possible between the two viable/fit genotypes. The study of reciprocal sign epistasis is thus important to understand how evolution is constrained. Important questions include: How prevalent is RSE in nature? What is the underlying mechanistic basis? And can it be bypassed?

Previous studies exploring how epistasis affects evolutionary trajectories have mainly focused on sequence-function relationships in enzymes [9–16]. However since selection generally acts at the level of whole organisms, and multicellular phenotypes are largely controlled by gene regulatory networks (GRNs) it therefore remains an important challenge to go beyond interactions within single genes/proteins, and to consider possible epistasis within the context of dynamical GRNs. We therefore chose a biologically-validated model of how GRNs can create multicellular spatial patterns as a new paradigm within which to explore the extent and the mechanistic basis of RSE in developmental processes. We previously employed such a biologically-validated model of gene regulation to explore how cells can arrange themselves into organized domains of expression state, by interpreting a morphogen gradient to generate a stripe of gene expression (File S1 and

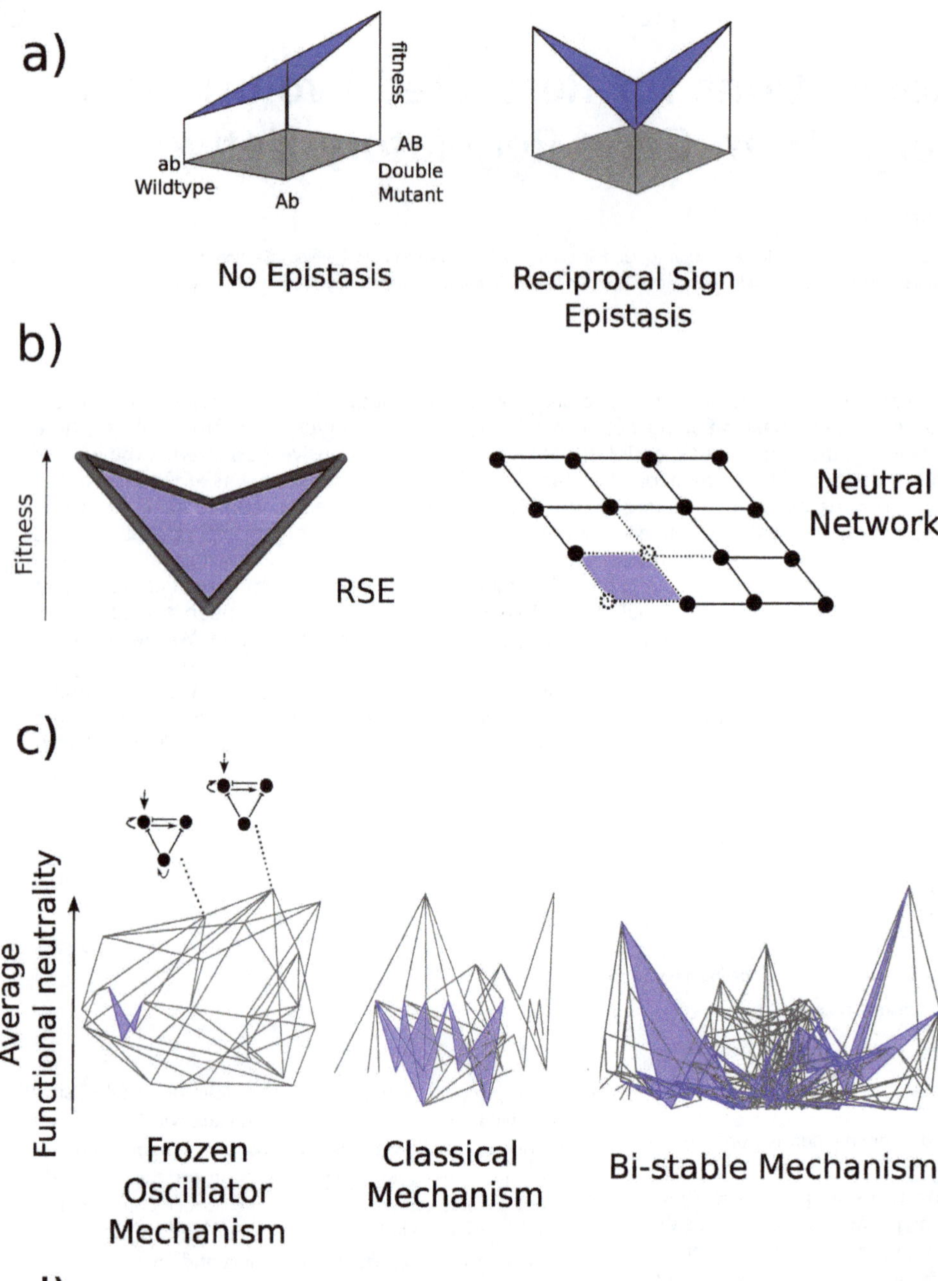

	Incoherent Feed-forward-1	Bi-stable	Classical	Frozen Oscillator	Mutual Inhibition	Overlapping Domains
Number of Topologies	97	109	39	27	44	56
Number of RSE	24	19	4	1	12	3
RSE/ Topology	0.25	0.17	0.10	0.04	0.32	0.05

Figure 1. Epistasis and neutral networks. a) Illustrating reciprocal sign epistasis. When there is no epistasis the combined effects of two mutations are the result of the addition of the fitness effect of each individual mutation. There is reciprocal sign epistasis when the two individual mutations negatively affect fitness yet the double mutant is fitter than the combination of individual mutations. Figure adapted from [15]) b) Illustrating how reciprocal sign epistasis causes the irregularity in the shape of a neutral network. Dots are viable genotypes and edges connect genotypes equal except for one mutation. We assume the two unfit genotypes in an RSE geometry are unviable (dashed dots) and the two fit genotypes are viable. Therefore RSE will be responsible for the gaps in neutral networks that lead to the irregular shape (loss of the dashed edges and dots in the neutral network shown). c) Examples of average functional neutrality landscapes for 3 gene regulatory network mechanisms capable of interpreting a morphogen gradient. Topologies are vertices and single mutant neighbors are connected via edges (as illustrated by the two topologies in the Frozen Oscillator average functional neutrality landscape). Topologies are spaced in the y-axis according to their average functional neutrality (fraction of viable parameter space capable of performing the morphogen interpretation function) and in the x-axis to reduce edge-crossing. Mutant neighborhoods where statistically significant RSE exists are colored blue. d) The number of topologies and the incidence of RSE for each of the different mechanisms. The amount of RSE normalized to the number of topologies in the landscape can be found on the bottom row.

Figure S1) [17]. We chose the most basic functional measure; simply whether the GRN (genotype) produces a stripe of gene expression or not giving a simple binary fitness score of 0 or one for each genotype which defines whether it is included in the neutral network or not. Because of the continuous model of gene regulation used here, the resulting stripes of gene expression can have different sizes and shapes (and still be considered functional; See Methods). As such the functional genotypes are "nearly" neutral though we will refer to them as neutral throughout this manuscript for purposes of clarity.

Gene regulatory network (genotype) space is not naturally discrete like protein or RNA space. Protein or RNAs have discrete amino acids or bases at each sequence position unlike gene regulatory networks that have a continuous parameter range at each gene-gene interaction position. One way to make genotype space discrete is to use a representation known as topology space, which is based on the gene network structure or wiring design (See File S1 and Figure S2). A Step within topology space then is a discrete addition/removal of a single gene-gene interaction. All genotypes with a given topology, regardless of their underlying parameter values map to that point in the corresponding topology space. We devised a "topology atlas" – a metagraph (a graph of graphs) which represents all possible topologies, and directly links together those which have just a single topological difference. Such metagraphs have been shown to be useful for exploring the relationship between innovation and robustness in GRNs [18,19].

A topology represents a population of individual GRNs with a particular wiring structure. If one samples a number of random parameter sets for each topology (each parameter set is an individual genotype) and sums the resulting binary fitness scores then an average functional neutrality score is generated for that region of genotype space (which is also a measure of robustness with respect to parameter sets for the corresponding topology). These average functional neutrality landscapes are thus a form of abstract neutral network that contain information about the number of functional individuals within that region of genotype space. We postulated that this average functional neutrality should correlate with the number of functional paths through that region of genotype space represented by the corresponding topology. Hence we used the average functional neutrality landscape of discrete topologies as a tool to measure the frequency of RSE and restricted evolutionary paths within the underlying neutral network of continuous genotypes.

Results

To measure the prevalence of RSE in this system, we first simulated every topology from the topology atlas with 30,000 random parameter sets. Functional topologies are those that could produce a stripe of gene expression for at least one tested parameter set (See Methods for the definition of functionality). In the previous study we found that the vast majority of topologies were working by one of 6 different dynamical mechanisms. We split the functional topologies into the different mechanism groups and analyzed each group separately (See Methods and Supplementary Methods). This was required since the different mechanisms are using distinctly different regions of parameter space [17]. As such it is not appropriate to analyze RSE with all topologies included in one set since it is probably not possible to mutate between genotypes of differing mechanism and maintain function at the level of genotype (topology with a specific parameter set). In other words the different mechanisms probably represent disconnected neutral networks at the level of genotype and thus it is appropriate to analyze them separately.

We then defined the average functional neutrality of a topology as the fraction of parameter space that is functional (that can produce a stripe of gene expression). We systematically searched through the list of successful GRNs for pairs (reference topologies) that are 2 steps apart in the atlas, but whose 2 direct intermediate topologies both display significantly lower average functional neutrality (i.e. we searched for groups of 4 connected topologies whose surface in the average functional neutrality landscape displays the V-shaped fitness geometry typical of RSE in Figure 1a). Statistically significant lower average functional neutrality was defined as a threshold of one standard deviation of the average functional neutrality of the reference topologies assuming a bootstrapped binomial distribution (see Methods). This analysis revealed that RSE is widely-distributed across average functional neutrality landscapes for all 6 mechanisms (see Figure 1c for 3 examples) however its prevalence appears to vary between the different mechanisms (Figure 1d). This variation is not simply the result of different sized landscape – indeed the proportion of topologies involved in RSE changes between the different mechanisms. This suggests that to achieve the same biological function a choice exists between different dynamical mechanisms, which may possess average functional neutrality landscapes with intrinsically smoother or more irregular distributions.

What then is the mechanistic cause of RSE within our spatial patterning GRNs? To address this question, we studied in detail a local region of the average functional neutrality landscape responsible for the Bi-stable mechanism (Figure 2a). This local region displays the V-shaped viability/fitness valley due to RSE – the direct routes between two functional GRN topologies show extremely low average functional neutrality (Figure 2b). All versions of this mechanism display a common underlying core topology (black regulatory interactions in Figure 2b). One of the two viable versions of the topology (topology 1) contains only this core design, while the other viable version (topology 4) contains two extra regulatory links (green arrows). Importantly these two extra regulatory links are counter-balanced; that is they are opposite in sign and affect the same gene. By analyzing the pattern-forming dynamics of the networks, we discovered that these counter-balanced regulatory inputs explain the observed

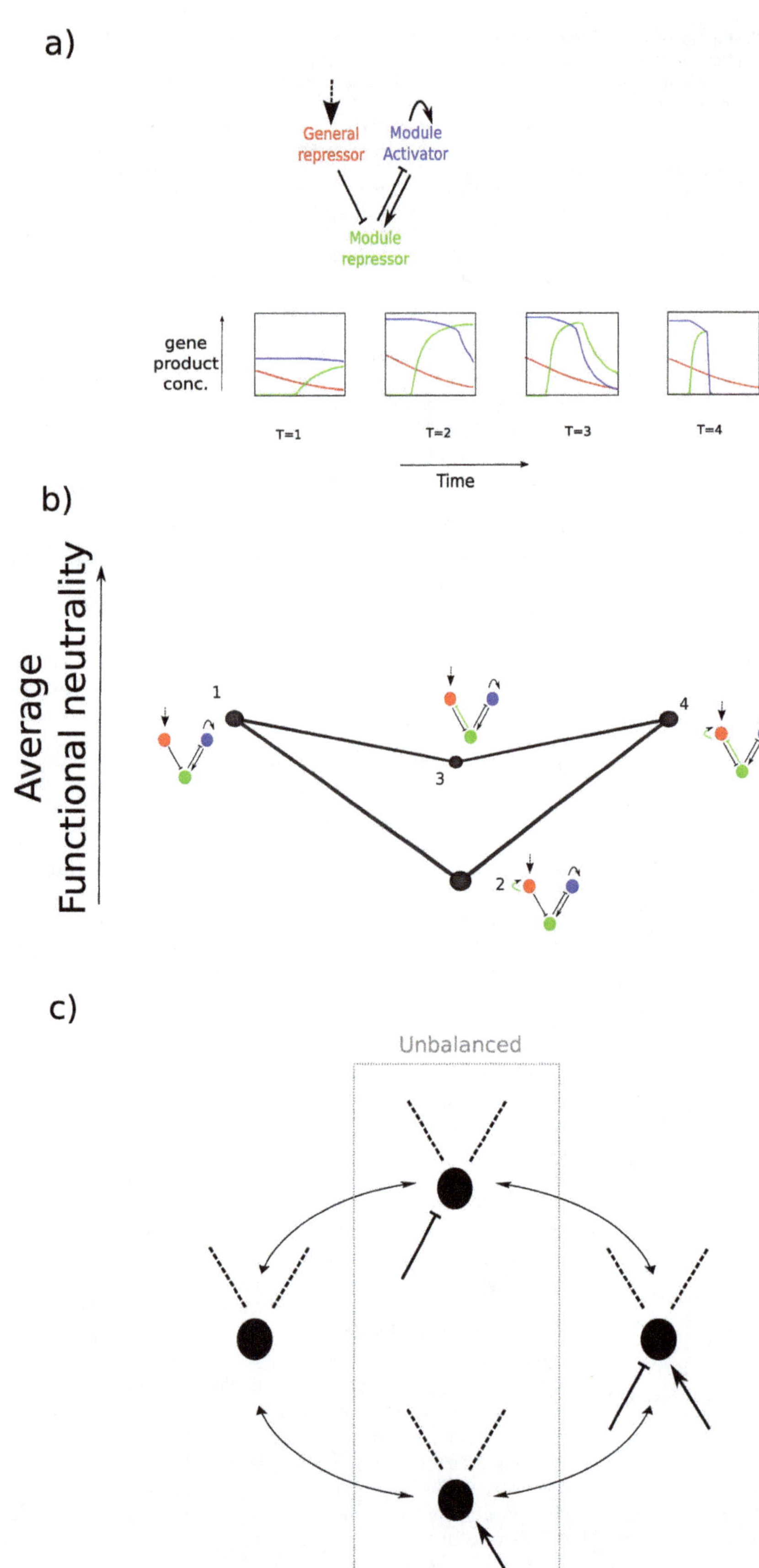
a)
General repressor
Module Activator
Module repressor
gene product conc.
T=1
T=2
T=3
T=4
Time
b)
Average Functional neutrality
1
2
3
4
c)
Unbalanced

Figure 2. Counter-balanced gene regulatory inputs as a cause of RSE. a) A description of the bi-stable mechanism. (Top) The core topology. The genes are named based on their role in the mechanism. The mechanism involves a "general repressor" which regulates the activity of a bi-stable module. The bi-stable module consists of an auto-activating gene that activates a gene that represses itself. (Bottom) The space time behavior of the core topology with a typical parameter set. The x-dimension represents space and the y-dimension represents the gene product concentration. The color corresponds to the gene in the topology above. T indicates the representative stage of the mechanism. The morphogen feeds into the general repressor that correspondingly forms a similar gradient. The "module activator" starts to be expressed everywhere due to positive auto-regulation. The "module repressor" starts to be expressed on the right hand side due to activation everywhere by the module activator and repression on the left hand side by the general repressor. *Middle (T = 2):* At the very right hand side, the module repressor reaches a high enough concentration to start to force the activator off. *Late (T = 3):* The module repressor concentration also starts to drop due to lack of activation from the module activator. *Final (T = 4).* The result is a stripe of expression of the module repressor gene. b) An example of where counter-balanced gene-gene regulatory inputs cause RSE. The core topology of the Bi-stable mechanism (black gene-gene interactions) is viable/fit (1) along with the core topology with two additional interactions of opposing sign (green gene-gene interactions) that feed into the same gene (4). However either of the individual mutations alone has significantly less average functional neutrality (2 and 3). c) An illustration of the general concept of counter-balanced gene regulatory inputs causing RSE. The inputs must feed into the same gene though the interaction could come from the same gene or another gene.

RSE. With only the activating input (topology 2), or only the repressive input (topology 3), the mechanism is unbalanced and is unable to correctly interpret the morphogen gradient (Figure 2c). However, when both regulatory inputs are present, the system is balanced, and able to form a stable stripe of expression.

Are counter-balanced regulatory inputs a general cause of RSE in dynamical networks, or is our example a specific one-off? To address this question, we explored how many of the RSE examples found for our 6 mechanisms were explained in the same way. For each case of RSE described earlier, we analyzed what mutational changes there are in the two intermediates. If one of those mutational changes is the addition of a positive interaction feeding into a particular gene, and the other is a repression feeding into the same gene, then we classified this as an example of counter-balanced regulatory inputs. We find that counter-balanced regulatory inputs underlie RSE in most of the 6 mechanisms studied. Interestingly however, it is mechanism dependent, for example it was never observed for the Classical mechanism, but explains 36% of RSE in the Bi-stable average functional neutrality landscape (when the topologies of the Bi-stable average functional neutrality landscape are re-sampled with a million parameter sets). We have thus uncovered a likely common cause of RSE underlying biological systems, but which is influenced by the specific dynamical design of the GRN.

The widespread appearance of RSE suggests that the underlying neutral networks at the level of continuous genotype space are irregular and that evolutionary paths will be restricted. Irregular neutral networks can be considered an abstract version of rugged fitness landscapes. However it has been suggested in that the appearance of rugged fitness landscapes is simply the result of not taking into account all of the appropriate genetic loci that describe a particular biological system [20]. When the appropriate number of genetic loci are taken into account so that there are more than 3 dimensions required to describe a fitness landscape, fitness ridges exist between peaks in the higher dimensions. Gavrilets (2004) has described those paths through unaccounted for dimensions that connect fitness peaks as "extra-dimensional bypasses". Such bypasses have been shown to exist in the contexts of inflorescence architectures, flower color and enzyme sequence-function relationships [14,21,22]. Ortlund *et al.*, 2007 in the context of enzyme sequence-function relationships have demonstrated how a "permissive" mutation allowed an unfavorable mutation to occur without loss of function. The presence or absence of the permissive interaction can be thought of as an extra dimension in genotype space [16,20]. Returning to our example of RSE from the Bi-stable mechanism, we too see that a permissive mutation is possible: Adding a repressive interaction to the network (red link in Figure 3a) "buffers" the system, allowing one of the mutations that previously resulted in a low average functional neutrality to be incorporated (green repressive link). From this design, the second of the epistatic mutations is also viable, and from the resulting topology (7) the buffering link can be removed. Hence higher-dimensional fitness ridges exist in both enzyme sequence-function relationships and GRNs responsible for pattern formation. The underlying similarity of the landscape structure can be seen by comparing Figures 3a and 3b.

We next investigated exactly how the permissive interaction maintains the functionality of topology 1 when the direct interactions are added. The key to understand how the permissive mutation allows for a bypass is to consider how the stripe of gene expression moves in the spatial domain when the two inhibitory mutations are added from topology 1 to topology 3 and 5. To test the affect on stripe position of the topology 1>3 and 1>5 transitions we mutated all functional parameter sets of topology 1 into topology 3 and 5 by giving them the appropriate interaction. We gave all parameter sets a small strength interaction (0.1) so that we could see the typical movement of the position of the stripe without losing function. We simulated the un-mutated and mutated parameter sets and asked how much the peak (highest value) of the stripe moves. The results are striking and show that on average mutating to topology 5 moves the peak 1.1 cells to the right, whereas mutating to topology 3 on average moves the peak 8.3 cells to the left for an inhibitory interaction of 0.1. This means that typically a mutation to topology 3 will destroy the function since most mutations involve a change far greater than 0.1 which will shift the stripe out of the spatial configuration of the model (off the left hand boundary as depicted in Figure 3c second panel). However a mutation to topology 5 will shift the stripe several cells to the right and most probably maintain functionality (Figure 3c third panel). That means that typically parameter sets from topology 5 will have their stripes further to the right of the spatial domain. Such a configuration gives much more scope for functionality when one adds the inhibition of topology 3 to generate topology 6 as depicted in Figure 3c (bottom panel). There is now more space to shift the stripe of expression and maintain functionality. Hence the permissive mutation functions by having a small counteracting affect to the detrimental mutation thus improving the likelihood that the system will stay within the spatial configuration of the model when the permissive mutation occurs before the detrimental mutation.

The observation of higher dimensional fitness ridges in multiple contexts suggests that the concept of 2-loci epistasis may not be appropriate for the study of many biological systems. This point depends on whether these extra-dimensional bypasses are a common feature of the neutral networks that underlie biological systems? To explore how common extra dimensional bypasses are in these situations where there is RSE, we analyzed the extent of bypasses between the RSE topologies of the Bi-stable mechanism. A bypass must only contain topologies with significant average functional neutrality (using the same significance measure de-

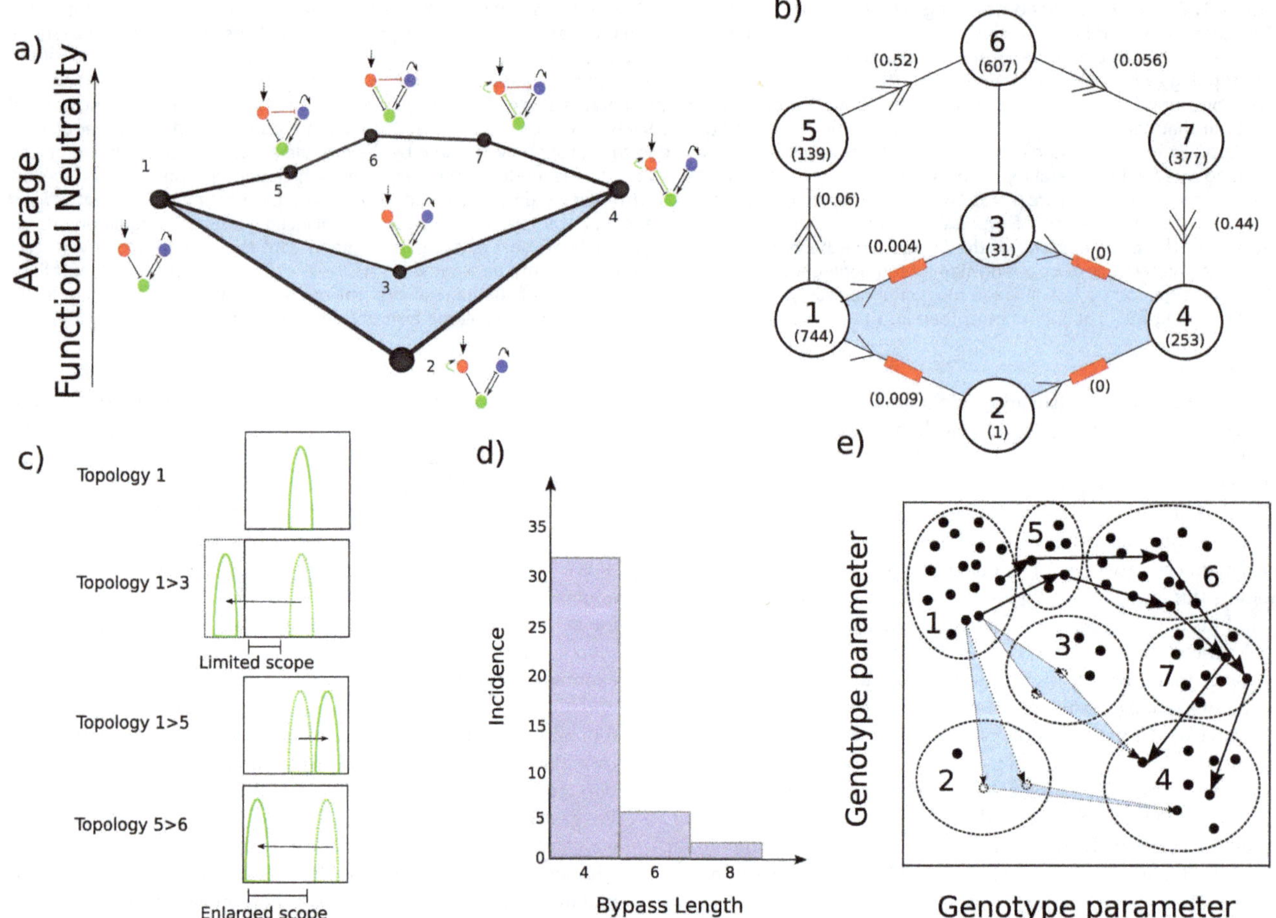

Figure 3. Permissive mutations allow RSE to be bypassed. a) The same example of RSE from figure 2 with an extra-dimensional bypass. The addition of a permissive mutation (red interaction in topology 5) before the original individual mutations are added (green interactions in topologies 6 and 7) allows the topology to retain a high average functional neutrality. The permissive mutation can then be removed (7 to 4) resulting in a bypass of the RSE. b) A scheme adapted from Ortlund et al., (2007) highlighting the similarity of permissive mutations in gene regulatory networks and in enzyme sequence-function relationships [14]. Topologies are shown in the circles. Large numbers in the circles represent the topology number and the number in the bracket is the number of functional parameter sets for that topology (i.e. the average functional neutrality; note that this was measured after the 7 topologies were re-simulated with a million parameter sets). The numbers on the edges represent the probability of a mutation of one of the successful parameter sets being functional when changed to the next topology when stepping through parameter space (in the direction of the arrows; see Methods). Non-viable steps are shown as red blocks (defined by a chance of less that 1%) and viable steps are shown by double headed arrows. c) The permissive interaction is caused by opposing effectors of differing strength. A typical mutation from a genotype with topology 1 to a genotype with topology 3 will significantly shift the stripe of gene expression to the left (second panel). The majority of these mutations leave the stripe outside the spatial field as defined by the morphogen gradient and therefore they are not functional. The mutation of a genotype with topology 1 to a genotype with topology 5 on the other hand typically shifts the stripe to the right and by a much smaller amount than the 1>3 shift (see main text). Therefore many more of the 1>5 transitions are viable. Because many more of the genotypes that correspond to topology 5 have their stripes at the extreme of the spatial domain, there is now more scope to introduce the green inhibition of topology 3. There is more space to shift the stripe by introducing this interaction and still result in a functional stripe. d) A histogram of the extra-dimensional bypass lengths between pairs of topologies in RSE. e) An illustration of how an average functional neutrality landscape can be envisaged as a density map of functionally neutral genotypes and how our measure of RSE and bypasses using topology is an underestimate. The Axes are genotype parameters. In reality there are 3 dimensions for the 3 interactions added between topology 1 and 7 but here we have reduced to 2 for simplicity. Black dots represent functional genotypes (topology with specific parameter set) and dashed dots represent non-functional genotypes. Large dashed circles represent the topologies (numbers) to which those genotypes map. Multiple different RSE and extra dimensional bypasses between topology 1 and 4 are shown by the light blue RSE geometry and the paths of solid black arrows respectively.

scribed earlier) and can be of any length. The results are shown as a histogram in Figure 3d. In 71% (40 out of 56) of the cases of RSE a high fitness ridge exists. Hence extra dimensional bypasses are almost as common as RSE since in most cases where we observe RSE there is also a route between the two topologies that does not involve a significant average functional neutrality decrease. Those situations of RSE where there is a bypass of length 4 have a direct permissive mutation between them. In other words like the example in Figure 3a, the two direct routes result in loss of fitness but the addition and removal of a single gene-gene interaction buffers these changes allowing a transition without loss of average functional neutrality. The longer (>4) bypasses between the two RSE topologies involve more complex buffering than simple single permissive mutations such as multiple subtle effects of individual mutations.

As mentioned in the introduction we postulated that the average functional neutrality of a topology should correlate with the number of functional paths through that underlying region of genotype space. To directly test this postulation we attempted to mutate each functional genotype (topology with specific parameter set) belonging to one topology class directly into the neighboring functional topology class by the addition or removal of a single gene-gene interaction with the rest of the parameter set equal. For each functional parameter set of each topology we added/removed the appropriate interaction which was given a random strength (See Methods). The resulting parameter set was simulated after each mutation and the stripe criterion was used to ask whether the mutant genotype was functional. We performed 100 mutations for each functional genotype and scored the fraction of mutations that were functional for each topology (labeled on the edges of Figure 3b). The result demonstrates that it is far more probable (approximately 10 fold comparing transition 1>5 versus 1>2 or 1>3) to generate successful mutants moving through the bypass than through the direct routes. This result shows that there is a larger fraction of functional paths through the topologies that have larger average functional neutrality. The probability that a mutation is functional and the fraction of functional parameter space of the resulting topology have a Pearson's correlation coefficient of 0.74 (one-tailed probability of 0.046). Furthermore we can perform walks in parameter space that change topology 1 to 4 all the way through the bypass that maintain functionality yet none of an equal number of walks through the direct routes maintain functionality (See File S1 and Figure S3). Together these results validate our postulation and the use of topology space as a meaningful way of measuring RSE and restricted evolutionary paths in GRNs.

Discussion

In summary, we have taken the analysis of RSE into the realm of dynamical GRNs. We demonstrate that most mechanisms for performing even a simple developmental function display some degree of RSE in their underlying functionality landscapes, suggesting that this is a common intrinsic property for evolving networks. This analysis demonstrated that the irregularity of the neutral network in genotype space is related to the underlying mechanism. Since the shape of a neutral network has been suggested to influence evolutionary innovation it is possible that certain mechanisms are favorable from an evolutionary standpoint [19]. This analysis also revealed that counter-balanced regulatory inputs may often be responsible for RSE, and that RSE can be bypassed by means of permissive mutations which buffer the changes that occur in the direct route between two GRN topologies (Figure 3b).

Validity of the Model

A key assumption for the validity of the model used in this study is that the core networks described do actually control variation in phenotype in natural systems. Intriguingly both the classical and the mutual inhibition type networks can be found as sub-networks of the *Drosophila* gap gene network [17]. The gap gene network of the *Drosophila* blastoderm has been shown to reduce variation in gene expression from Bicoid to the downstream GAP genes by the process of canalization [23,24]. Indeed knock out of Kruppel, Knirps or Tailless increases this variation [25,26]. Hence it can be demonstrated that for at least 2 core GRNs evaluated in this study there exists evidence that these networks control variation in gene expression and the resulting phenotype since the Gap genes control the segmentation of the embryo. The model is thus valid to explore evolutionary trajectories in developmental GRNs.

One criticism of the model we have used is that it does not utilize a typical fitness measure. Typical fitness measures involve some aspect of the phenotype that can be considered a selectable trait. For example the sharpness of the stripe of gene expression would be one such measure that could potentially give a selective advantage. These features have important influences on the likelihood that different routes are taken in genotype space. However the most important contribution to fitness is whether the genotype can achieve the function or not irrespective of how "well" it performs the function. Our average functional neutrality is based on the likelihood of a genotype that corresponds to that topology being functional. We have shown average functional neutrality of topologies correlates with the fraction of functional paths through them (Figure 3b). As such it is a valid statistical measure of the amount of RSE and restriction in evolutionary paths within a genotype-phenotype structure.

Furthermore this idea can be illustrated when we envisage our sampling in topology space as a density map of functional genotypes (topology with specific parameters) in underlying continuous genotype space (Figure 3e). Here the regions of space corresponding to the topologies with high average functional neutrality are dense, while those corresponding to the topologies with low average functional neutrality are sparse. For one to travel from a genotype in one dense region to a genotype in the other dense region directly through topologies 2 and 3, statistically many of the routes will involve a non functional intermediate as illustrated by the 2 RSE geometries in figure 3e (this point is confirmed by our mutational walks through parameter space; values on the edges in figure 3b). Hence though we have only scored this topological geometry as a single RSE, in the underlying continuous genotype space it may represent multiple RSEs. In the same way, though we have scored just a single extra dimensional bypass from topology 1 through 5, 6 and 7 to 4, in underlying continuous genotype space there maybe multiple routes as illustrated by the bold arrows in figure 3e. Taken together our measures of RSE and bypass frequency are likely to be underestimates of the actual amounts of RSE and bypasses in these genotype-phenotype maps. This further strengthens our conclusions that RSE and bypasses are frequent in genotype-phenotype maps for higher levels of biological function.

A General Theory of Permissive Mutations

Our analysis suggests one general theory for permissive mutations in gene regulatory networks and possibly other genotype-phenotype systems. There is an order by which mutations with opposing affect on functionality must occur. A mutation with a minor affect (the permissive mutation) must occur before a mutation with an opposite larger effect to maintain the system in the dynamic range as defined by the system configuration. For example in this work one aspect of the system configuration is the maximum and the minimum of the morphogen gradient which defines a dynamic range between 1 and 0.1. The minor permissive mutation shifts the functionality (stripe of gene expression) to the edge of the configuration limits (edge of the morphogen gradient or spatial boundary in our case). The probability of maintaining function with this mutation is greater than the probability of maintaining function with the large opposing mutation since the large mutation is likely to shift the functionality out of the dynamic range of the system (shift the stripe off the end of the spatial boundary so that the stripe only occurs between two morphogen thresholds both above 1– Figure 3c second panel). However once the permissive mutation has occurred, the scope of the dynamic range of the system is now much larger for the second larger mutation since its affect on functionality (the stripe) will push the system to the opposite side of the configuration limits. Hence the larger mutation can be a viable mutation, but only if the permissive mutation has already occurred. Note such situations would not arise if the morphogen gradient would not have configuration limits and could range from infinity to 0. However such boundaries or limits to the system like the concept of saturation for example are physically unavoidable aspects of GRN systems. As such the concept of smaller permissive mutations followed by larger mutations with opposing effect are probably a common explanation for extra-dimensional bypasses in both genotype-phenotype maps of GRNs and other genotype-phenotype systems.

Testable Hypothesis

How then can we experimentally test out hypothesis that mutually balanced interactions are responsible for RSE and interactions of opposing action yet differing strength are responsible for permissive mutations? Testing such hypothesis will require going beyond the identification of essential genes (for which data is currently available in multiple species) to a situation where essential individual gene-gene interactions are identified. A complementary experiment for validating our hypothesis would involve constructing our GRNs synthetically. If one could synthetically construct such networks then one could directly explore the affects of adding complementary positive and negative interactions of differing strength. We would predict for example that adding either alone (of similar strength) would diminish function, but in combination should have minimal effect. We would also predict that in general mutations of opposing effect but of different strength maintain function only if added in a particular order (minor permissive mutation first). At least 3 of our core networks have already been built in multiple species and others are currently being constructed [27–30]. The power of such constructions is that they only contain the basic core interactions needed for the function rather than other natural networks that have become baroque in nature (contain a more complicated architecture than seems necessary for the function [31]) probably through drift. Indeed our observation that extra dimensional bypasses are common in genotype-phenotype maps for GRNs suggests one explanation for why GRN networks have a baroque structure in the first place since there are many possibilities for adding functionally neutral gene-gene interactions which seemingly are not important for core function.

Replaying the Tape of Life at Higher Levels of Biological Organization?

How can the results of this study be compared to those results exploring similar phenomena in single proteins? A related phenomenon to permissive interactions with restricted evolutionary paths due to RSE has also been observed for specific examples in protein sequence-function relationships [14]. The existence of such a phenomenon in these diverse contexts suggests that permissive routes through higher genotypic dimensions are a general feature of the evolution of biological systems. However, one of the key findings of studies of evolutionary constraints in single proteins was that many intramolecular combinations are non-viable and therefore the trajectories open to evolution are limited and often the "tape of life" can be replayed such that that evolution often takes the same mutational path [13,32,33]. Here we have studied a higher level of biological organization and instead find that there are many more routes between functional genotypes (almost all incidences of 2 topologies linked via RSE have at least one bypass for example). This discrepancy between these conclusions could simply result from the fact that we have used a binary fitness function. Although many routes produce a functional stripe of gene expression, some routes may result in the production of a "better" stripe than others such that those routes are more likely. If a less abstract fitness function was used then involving some aspect of the quality of the stripe this may biased the likelihood of some specific paths over others. Alternatively the discrepancy may result from the fact that when considering a single protein the ways to increase fitness are greatly limited due to pleiotropy and conformational epistasis. Protein sequence space is discrete and any change can have a great effect on multiple aspects of the protein potentially destroying the function. Genotype space by contrast is continuous meaning that changes can be finer allowing for more oblique traverses without loss of function. The tape of life for the evolution of development at the level of GRNs then, may not be as predictable as that for protein function.

Methods

Enumerating All GRN Topologies

A topology can be represented in the form of a matrix w_{ij} where i and j represent the position in those matrices and values 1, -1 and 0 represent activation, repression and no interaction respectively. We generated all possible matrices that correspond to unlabelled topologies and then removed isometric equivalents by comparing them in all possible permutations. There are 19,683 gene network matrices before non-isometric topologies have been removed and this is reduced to 3,284 topologies in the fully enumerated set.

The morphogen gene is a gene that activates one of the genes of the GRN but is not affected by the GRN. Each GRN topology is represented multiple times with the morphogen feeding into the different genes (exact number depends on the amount of symmetry in the GRN topology). The morphogen is taken account of in the topology generation by extending the GRN matrix ($i=i+1$) to include the input from the morphogen (which is permutated independently). When the morphogen is included the number of isometric topologies increases from 3,284 to 9,710.

Creating an Atlas of GRNs by Including Explicit Neighbour Definitions

Two GRN topologies are considered neighbours in the atlas if the two GRN topologies are one Hamming distance apart (a single gene-gene interaction change). The Hamming distance can be measured by the following equation where

$$D(w,w')=\sum_{i,j}|\mathrm{sgn}(w_{ij})-\mathrm{sgn}(w'_{ij})|. \quad (1)$$

D is the Hamming distance between the matrices of two GRN topologies w and w' whilst i and j represent the position in those matrices. The matrices are compared in every permutation and the lowest D of those permutations is taken as the Hamming distance. Hence two GRN topologies are neighbours if the gain or removal of any one interaction can transform one of the GRN topologies into the other.

The Gene Regulation Model

We employed a biologically-verified model of gene regulation for this problem, and therefore adapted the continuous mathematical model developed over the last 20 years by Reinitz et al [34] which quantitatively captures the spatio-temporal dynamics of gap gene patterning in response to the Bicoid morphogen gradient during *Drosophila* embryogenesis. The model is described by

$$\frac{dg_{ij}}{dt} = \chi\left[\phi\left[\sum_{l=1}^{Ng} W^{li}g_{lj} + M\right]\right] + D_i\nabla^2 g_{ij} - \lambda g_{ij} + \eta(t)g_{ij}, \quad (2)$$

where g_{ij} is the concentration of the *ith* gene in the *jth* cell, $\phi(x)$ is a function defining the interaction amongst genes (which can take the form of a Michaelis-Menten, sigmoid or other non-linear input function), W^{li} is a matrix containing the strength of gene-to-gene regulation parameters, M is the morphogen input described in more detail in the section "configuration of the spatial domain" below, $\chi(x)$ is the Heaviside function (to prevent negative gene product production rates), D_i is the diffusion constant for the *ith* gene which we use to represent local cell-cell signaling, λ is the decay rate (set to 0.05), and $\eta(t)$ is a noise term which adds uniformly-distributed fluctuations (+/−1%) to the concentration of every gene in every cell at every time step. There is zero auto-correlation in the noise term. The parameters that could vary in the model were regulation W^{li}, and diffusion D_i. The input function describes the relationship between the activation and inhibition of a gene and its actual expression. The input function used in this work took the form of a Michaelis-Menten function which is defined by

$$O = \frac{I}{(1+I)}, \quad (3)$$

where I is the total input into the gene and O is the output of the function.

The Discretized Form of the Equations

How the concentration x of a gene i will change in any given cell j at time t is described by

$$\begin{aligned} x_{t+1}^{i,j} &= x_t^{i,j} + \chi\left[\phi\left[\sum_{l=1}^{Ng} W^{li}x_t^{l,j}\right]\right] \\ &+ D_i\left[x_t^{i,j-1} + x_t^{i,j+1} - 2x_t^{i,j}\right] - \lambda x_t^{i,j} + \eta(t)x_t^{i,j}. \end{aligned} \quad (4)$$

Here W is the interaction matrix described earlier and ϕ is the input function. D is the diffusion coefficient for each gene i and λ is the decay parameter which is the same for every gene. The noise term $\eta(t)$ selects a random number within a given range.$\chi(x)$ is the Heaviside function where $\chi(x)=x$ for $\forall x>0$ and $\chi(x)$ otherwise. Its purpose is to make sure the regulation term can only take positive values.

Parameter Range Distributions

For each GRN topology 30,000 different parameter sets were tested (a GRN topology with a specific set of parameters we called a genotype). There are up to 12 variable parameters for a 3 gene network; diffusion for each individual gene and then the strengths of the interaction values between the genes. The parameters are chosen randomly though biased towards lower numbers through a logarithmic probability distribution. The logarithmic probability distribution was implemented in order to take account of the fact that a small change in a small parameter value will have a greater effect on a network's behaviour than a small change in a larger parameter value. The logarithmic probability distribution is described by

$$V = 0.9995^i R, \quad (5)$$

Where i is a random number between 0 and 10,000 and R is the parameter range and V is the resulting parameter value. Parameter ranges are as follows; regulation 0–10 and diffusion 0–0.05.

Configuration of the Spatial Domain

The simulations take place on a theoretical one dimensional row of 32 cells. Zero-flux boundary conditions are used throughout this work. The simulation starts with every gene in every cell set to have a concentration of 0.1. This was necessary because the noise term used is a percentage noise term and thus if the concentration were always 0 at the start of the simulation then the products of any genes with positive feedbacks without any other input would remain at 0. The simulation is also initiated by the positive input from the morphogen gradient that does not change throughout the simulation.

The morphogen strength was chosen to give an approximate input range to the receiving gene of 10–50% of the maximal activation. The morphogen input is defined by

$$M = Id^c, \quad (6)$$

Where M is the morphogen input, I is the morphogen concentration in the left-most cell of the field, d is the reduction of morphogen concentration in each subsequent cell of the morphogen gradient and c is the cell position. For the 10–50% input range, $I=1$ and $d=0.93$ was used for the Michaelis-Menten function.

Stripe Forming Functional Definition

For a genotype (GRN topology with a specific parameter set) to be considered functional it had to reach an equilibrium (described in Supplementary Methods) and it had to produce a stripe of gene expression for at least one of the genes. For each gene we measured an abstraction of its gene expression over the one-dimensional field where each cell was defined as low or high. We defined a cell as low if the gene expression was below 10% of the maximum possible allowed by the model. We defined a cell as high if the gene expression was above 10% of the maximum gene expression allowed by the model. A gene was considered to have a stripe pattern if it had a single region of low for 2 consecutive cells followed by a single region of high for a maximum of 16 consecutive cells followed by a single region of low for at least 2 consecutive cells. The two low regions must occur at the extremities of the field. The definition is intentionally loose in the sense that the single stripe can be of any width up to 16 cells and be in any position in the spatial domain. This is because we are interested in the basic design principles of the system, not the details of how to control a specific width. Functional parameter sets that can produce the single stripe of gene expression we term "solutions". Hence a single topology has multiple genotypes and can have multiple solutions. The number of solutions that each topology has is a measure of its mutational robustness. A GRN topology must have at least one solution to be considered functional.

Splitting the Functional Topologies into Subsets each Responsible for a Different Mechanism and Creating Average Functional Neutrality Landscapes

In order to explore whether changes in the mechanism by which a GRN is functioning is responsible for the prevalence of RSE in an average functional neutrality landscape, we split the functional GRN topologies into 6 categories, each corresponding to the 6 different mechanisms we identified in our previous study [17]. If a GRN topology could perform a mechanism for at least one parameter set then it was included in that particular subset. Therefore a GRN topology can be present in multiple subsets. GRN topologies were assigned to mechanistic classes using the method for mapping mechanisms to the complexity atlas described in Supplementary Methods. Each subset is then used to build an average functional neutrality landscape for each particular mechanism. The total number of functional GRN topologies was 471. The number of GRN topologies in each of the average functional neutrality landscapes was as follows; Incoherent Feed-Forward type 1 (97), Mutual Inhibition (44), Frozen Oscillator (27), Overlapping Domains (56), Bi-stable (109) and Classical (39). Topologies are assigned an average functional neutrality score based on their parameter robustness (Number of parameter sets that successfully produced the stripe of gene expression). Topologies are vertices in the landscapes and they are connected by edges based on the neighbor definitions of the atlas described earlier.

Calculating the Extent of RSE

In order to calculate the extent of RSE in the individual average functional neutrality landscapes we analyzed all pair-wise combinations of topologies in the average functional neutrality landscape to see if they conform to the V-shaped viability/fitness geometry. The V-shaped viability/fitness geometry of RSE is defined as two GRN topologies (topologies A and B) that are two hamming distances apart that have exactly two direct intermediate topologies. These intermediate topologies both must have significantly lower average functional neutrality than topologies A and B. Significantly low average functional neutrality is defined by

$$F = F_A - S_A \quad and \; F = F_B - S_B. \tag{7}$$

Where F_A is the average functional neutrality of topology A (which is defined as the number of functional parameter sets for that topology) and S_A is the standard deviation of the average functional neutrality of topology A, assuming a binomial distribution. The standard deviation of a binomial distribution is described by

$$S = \sqrt{np(1-p)}. \tag{8}$$

Where n is the number of trials and p is the probability of functionality (the true average functional neutrality). Since we do not know the true average functional neutrality, this measure is bootstrapped using the measured average functional neutrality F_A as a proportion of the total number of parameter sets tested (number of trials, n) giving

$$S = \sqrt{F_A\left(1 - \frac{F_A}{n}\right)}. \tag{9}$$

Calculating the Extent of Permissive Mutations/extra-dimensional Bypasses

In order to calculate the extent of extra-dimensional bypasses between pairs of GRN topologies that have the RSE V-shaped viability/fitness geometry we analyzed all mutational routes between these pairs of topologies. For a route to be considered an extra-dimensional bypass every GRN topology within the route must have average functional neutrality within one standard deviation of topologies A and B as described in equation 7. For each pair-wise combination of GRN topology showing the RSE geometry we record the length of each extra-dimensional bypass (measured by the number of gene-gene interaction changes). The direct permissive mutations are those with a bypass length of 4 since they involve the 2 original mutations required to change topology A into B (and vice versa) and the two other mutations that involve the addition and removal of the permissive mutation.

Performing Mutational Walks between Topologies

We measured the fraction of mutations from a functional parameter set of one topology to a functional parameter set of another topology (specifically in the direction of the arrows in figure 3b). We start with a functional parameter set of our reference topology and remove or add the appropriate interaction. If we add an interaction, we give the new interaction the appropriate sign and with a random value using the criteria that we describe in parameter range distributions above. If we remove an interaction the value is simply set to 0. We then re-simulate this mutated genotype and ask if it is functional using our functional stripe definition described above. We do this 100 times for every specific parameter set and calculate the proportion of tests that are functional. For example we perform 74,400 tests of mutating topology 1 to topology 5. The proportions of functional tests are found in brackets on the edges of figure 3b.

Supporting Information

Figure S1 Summarizing the key results of the previous study [17]. A complexity landscape identifies core mechanisms responsible for generating a stripe of gene expression. Vertices are topologies and edges connect topologies one hamming distance apart (one gene-gene interaction change). Topologies are spaced manually in the x-axis to reduce edge crossing and in the y-axis by their complexity (number of gene-gene interactions). Stalactites of complexity emerge out of the bottom of the landscape converging to minimal core topologies that represent distinct mechanisms. Mechanisms were mapped to the complexity landscape by coloring topologies according to their mechanism class (See Supplementary Methods). Topologies were colored Light green (A: Incoherent feed-forward type 1 mechanism), Light Blue (B: Mutual inhibition mechanism), light red (C: Frozen oscillator mechanism), Dark green (D: Overlapping domains mechanism), Dark blue (E: Bi-stable mechanism) or Dark red (F: Classical mechanism). Topologies were colored yellow if they were capable of acting via multiple mechanisms depending upon their exact parameter set. The corresponding core topologies of these mechanisms (those at the very bottom of the stalactite) are shown below the complexity landscape. 3 examples space-time behaviors of each mechanism are also shown. Here the spatial dimensions represent time and space and the intensity of red, green or blue based on the expression value of that gene at that time in that cell.

The final gene expression graphs at equilibrium are shown below (corresponding to the bottom space time plot).

Figure S2 Topology space is a discrete representation of an underlying continuous genotype (parameter) space. The concept of a topology space (adapted from reference 17). ***(a)*** A GRN topology where two of the gene-gene interactions α and β correspond to the parameter space in *(b)*. ***(b)*** A parameter space of the two parameters α and β. Dots are random parameter sets from this space. ***(c)*** A topology space is created if all values of α and β that are positive are considered gene-gene activations, those values of α and β that are negative are considered gene-gene repressions and those values of α and β that are 0 are considered to generate no gene-gene interaction. Regions of parameter space corresponding to the different topologies are indicated by the different colored circles surrounding the topologies and the different colored dots in *(d)*. Where topologies differ by a single gene-gene interaction (one Hamming distance) they are linked by a blue line. Such links connect regions of close parameter space.

Figure S3 Not all topology changes represent viable steps in an evolutionary walk on a neutral network at the level of the underlying genotype space. Two examples are shown. Boxes represent a simplified version of the parameter spaces of 3 topologies called A, B and C. Black circles represent the functional parameter volumes. Thick arrows represent viable mutational steps and dashed arrows represent unviable steps. (Top) The two topologies (A and B) each work by a single but different mechanism. Changing A into B or vice versa then leads to a non-viable region of parameter space. (Bottom) The topology (B) can generate the gene expression pattern with two mechanisms. Two other neighboring topologies (A and C) can function but using either one of these two mechanisms. Changing from either A or C to B is viable. However changing from A to B to C or vice versa is only possible if the change between the viable regions of topology B involves a change in only a single parameter.

Acknowledgments

We thank Fyodor Kondrashov, Ben Lehner and Johannes Jaeger for critical discussions and reading of the manuscript.

Author Contributions

Conceived and designed the experiments: JC. Performed the experiments: JC. Analyzed the data: JC. Contributed reagents/materials/analysis tools: JC. Wrote the paper: JC JS.

References

1. Maynard Smith J (1970) Natural selection and the concept of protein space. Nature 225: 563–564.
2. Schuster P, Fontana W, Stadler PF, Hofacker IL (1994) From Sequences to Shapes and Back: A Case Study in RNA Secondary Structures. Proc. Roy. Soc. Lond 255: 279–284.
3. van Nimwegen E, Crutchfield JP, Huynen M (1999) Neutral Evolution of Mutational Robustness. Proc Natl Acad of Sci USA 96: 9716–9720.
4. van Nimwegen E, Crutchfield JP (2000) Metastable Evolutionary Dynamics: Crossing Fitness Barriers or Escaping via Neutral Paths? Bulletin of Mathematical Biology 62: 799–848.
5. Phillips PC (2008) Epistasis–the essential role of gene interactions in the structure and evolution of genetic systems. Nat Rev Genet 9: 855–67.
6. Phillips PC (1998) The language of gene interaction. Genetics 149: 1167–71.
7. Beerenwinkel N, Pachter L, Sturmfels B (2007) Epistasis and shapes of fitness landscapes. Statistica Sinica 17: 1317–1342.
8. Beerenwinkel N, Pachter L, Sturmfels B, Elena SF, Lenski RE (2007) Analysis of epistatic interactions and fitness landscapes using a new geometric approach. BMC Evol Biol 7: 60.
9. Kondrashov AS, Sunyaev S, Kondrashov FA (2002) Dobzhansky-Muller incompatibilities in protein evolution. Proc Natl Acad Sci U S A 99: 14878–83.
10. Lunzer M, Miller SP, Felsheim R, Dean AM (2005) The biochemical architecture of an ancient adaptive landscape. Science 310: 499–501.
11. Miller SP, Lunzer M, Dean AM (2006) Direct demonstration of an adaptive constraint. Science 314: 458–61.
12. Bridgham JT, Carroll SM, Thornton JW (2006) Evolution of hormone-receptor complexity by molecular exploitation. Science 312: 97–101.
13. Weinreich DM, Delaney NF, Depristo MA, Hartl DL (2006) Darwinian evolution can follow only very few mutational paths to fitter proteins. Science 312: 111–114.
14. Ortlund EA, Bridgham JT, Redinbo MR, Thornton JW (2007) Crystal structure of an ancient protein: evolution by conformational epistasis. Science 317: 1544–8.
15. Poelwijk FJ, Kiviet DJ, Weinreich DM, Tans SJ (2007) Empirical fitness landscapes reveal accessible evolutionary paths. Nature 445: 383–6.
16. Bridgham JT, Ortlund EA, Thornton JW (2009) An epistatic ratchet constrains the direction of glucocorticoid receptor evolution. Nature 461: 515–9.
17. Cotterell J, Sharpe J (2010) An atlas of gene regulatory networks reveals multiple three-gene mechanisms for interpreting morphogen gradients. Molecular Systems Biology 6: 425.
18. Ciliberti S, Martin OC, Wagner A (2007) Robustness can evolve gradually in complex regulatory gene networks with varying topology. PLoS Comput Biol 3: 0164–0173.
19. Ciliberti S, Martin OC, Wagner A (2007) Innovation and robustness in complex regulatory gene networks. Proc Natl Acad Sci U S A 104: 13591–6.
20. Gavrilets S (2004) Fitness landscapes and the origin of species. Princeton: Princeton University Press
21. Whibley AC, Langlade NB, Andalo C, Hanna AI, Bangham A, et al. (2006) Evolutionary paths underlying flower color variation in Antirrhinum. Science 313: 963–6.
22. Prusinkiewicz P, Erasmus Y, Lane B, Harder LD, Coen E (2007) Evolution and development of inflorescence architectures. Science 316: 1452–6.
23. Manu, Surkova S, Spirov AV, Gursky VV, Janssens H, et al. (2009) Canalization of gene expression in the Drosophila blastoderm by gap gene cross regulation. PLoS Biol 7: 3.
24. Manu, Surkova S, Spirov AV, Gursky VV, Janssens H, et al. (2009) Canalization of gene expression and domain shifts in the Drosophila blastoderm by dynamical attractors. PLoS Comput Biol 5: 3.
25. Surkova S, Golubkova E, Manu, Panok L, Mamon L, et al. (2013) Quantitative Dynamics and Increased Variability of Segmentation Gene Expression in the Drosophila Kruppel and knirps mutants. Developmental Biology 13: 00013–4.
26. Janssens H, Crombach A, Wotton KR, Cicin-Sain D, Surkova S, et al. (2013) Lack of tailless Leads to an Increase in Expression Variability in Drosophila Embryos. Developmental Biology 13: 00015–8.
27. Basu S, Gerchman Y, Collins CH, Arnold FH, Weiss R (2005) A synthetic multicellular system for programmed pattern formation. Nature 434: 1130–4.
28. Entus R, Aufderheide B, Sauro HM (2007) Design and implementation of three incoherent feed-forward motif based biological concentration sensors. Syst Synth Biol 1: 119–28.
29. Sohka T, Heins RA, Phelan RM, Greisler JM, Townsend CA, et al. (2009) An externally tunable bacterial band-pass filter. Proc Natl Acad Sci U S A 106: 10135–40.
30. Greber D, Fussenegger M (2010) An engineered mammalian band-pass network. Nucleic Acids Res 38: e174.
31. Lynch M (2007) The evolution of genetic networks by non-adaptive processes. Nat Rev Genet 10: 803–13.
32. Orr HA (2005) The genetic theory of adaptation: a brief history. Nat. Rev. Genet 6: 119.
33. Gould SJ (1989) Wonderful Life: The Burgess Shale and the Nature of History. New York: W. W. Norton and Company.
34. Reinitz J, Sharp DH (1995) Mechanism of eve stripe formation. Mech Dev 49: 133–58.

6

Contrasting Evolutionary Dynamics of the Developmental Regulator PAX9, among Bats, with Evidence for a Novel Post-Transcriptional Regulatory Mechanism

Caleb D. Phillips[1]*, Boyd Butler[1], John W. Fondon III[2], Hugo Mantilla-Meluk[1], Robert J. Baker[1]

1 Department of Biological Sciences, Texas Tech University, Lubbock, Texas, United States of America, **2** Department of Biology, University of Texas at Arlington, Arlington, Texas, United States of America

Abstract

Morphological evolution can be the result of natural selection favoring modification of developmental signaling pathways. However, little is known about the genetic basis of such phenotypic diversity. Understanding these mechanisms is difficult for numerous reasons, yet studies in model organisms often provide clues about the major developmental pathways involved. The paired-domain gene, PAX9, is known to be a key regulator of development, particularly of the face and teeth. In this study, using a comparative genetics approach, we investigate PAX9 molecular evolution among mammals, focusing on craniofacially diversified (Phyllostomidae) and conserved (Vespertilionidae) bat families, and extend our comparison to other orders of mammal. Open-reading frame analysis disclosed signatures of selection, in which a small percentage of residues vary, and lineages acquire different combinations of variation through recurrent substitution and lineage specific changes. A few instances of convergence for specific residues were observed between morphologically convergent bat lineages. Bioinformatic analysis for unknown PAX9 regulatory motifs indicated a novel post-transcriptional regulatory mechanism involving a Musashi protein. This regulation was assessed through fluorescent reporter assays and gene knockdowns. Results are compatible with the hypothesis that the number of Musashi binding-elements in PAX9 mRNA proportionally regulates protein translation rate. Although a connection between morphology and binding element frequency was not apparent, results indicate this regulation would vary among craniofacially divergent bat species, but be static among conserved species. Under this model, Musashi's regulatory control of alternative human PAX9 isoforms would also vary. The presence of Musashi-binding elements within PAX9 of all mammals examined, chicken, zebrafish, and the fly homolog of PAX9, indicates this regulatory mechanism is ancient, originating basal to much of the animal phylogeny.

Editor: Michael Schubert, Laboratoire de Biologie du Développement de Villefranche-sur-Mer, France

Funding: Funding was provided by the Texas Biological Database program. The funders had no role in study design, data collection and analysis, decision to publish, or preparation of the manuscript.

Competing Interests: The authors have declared that no competing interests exist.

* E-mail: caleb.phillips@ttu.edu

Introduction

The genetic mechanisms underlying human development as well as the morphological variation observed in natural populations remain poorly understood. Yet, it is clear that proper development is the result of complex regulation of developmental signaling pathways. This has been partly manifested in the anomalous human phenotypes often resulting from single, or a few, mutational events [1]. Furthermore, experiments using model organisms have shown that a subset of genes of regulatory function have major effect on craniofacial and dental development [2]. Although such findings derived from model organisms are highly relevant, identifying salient genomic changes underlying morphological adaptation is complicated by the general inability to produce test crosses, the difficulty of separating neutral and functional genetic variation, and understanding the relative importance of focusing on regulatory versus open-reading frame variation. In light of this, the study of interspecific morphological differences often relies on candidate genes identified in model systems. The benefit of candidate gene studies is that they exploit *a priori* established links between genotype and phenotype to enable the characterization of evolutionarily significant molecular divergences in genetically intractable non-model organisms.

The paired-box gene family of proteins (PAX) is an evolutionarily conserved and ancient gene family, having originated during the pre-Cambrian [3–4]. One member of this family, PAX9, is considered a candidate gene for craniofacial and dental evolution. PAX9 is known to play an important role in development as both a transcription factor and through protein-protein interactions [5–8]. PAX9 expression is necessary for proper development of the craniofacial and dental regions, and is also expressed during development of the vertebral column, tongue, limbs, and other organs [9–13]. Mutations in coding and putative regulatory regions of PAX9 are often associated with phenotypic abnormalities. Human phenotypes associated with PAX9 mutations are most commonly various forms of tooth agenesis [14–18]. PAX9 knock-outs are non-viable owing to defective development of the pharyngeal pouches resulting in malformed palatal and neck structures, and also exhibit tooth agenesis and polydactyly [9]. Clearly, improving models for PAX9 sequence evolution and how PAX9 is regulated are needed to better understand complex PAX9 involved developmental signaling pathways [9–10]. For example, it is reasonable that expression levels of Pax9 during the developmental process influence downstream effector expression and hence, the timing of cellular differentiation. Variance in this process would have implications to both the evolution of phenotypes in nature, as well as candidate mechanisms for anomalous human phenotypes.

Highly variable intra-ordinal and intra-familial patterns of craniofacial and dental evolution are observed within the order Chiroptera. Currently, 19 families of bats are recognized [19–20]. Although the family Phyllostomidae is thought to be among the youngest families, this group displays the most craniofacial, dental, and dietary diversity [21–22]. It is hypothesized that the morphological variability observed within Phyllostomidae is the result of niche adaptation as multiple dietary strategies are represented in the family, including insectivory, frugivory, nectarivory, sanguivory, omnivory, and carnivory [21–22]. Comparison of lineages illustrates often dramatic craniofacial morphological differences (Figure 1) for structures in which Pax9 is expressed during development. For example, there is extensive variation in rostral length, cranial morphology, dental formula, tooth shape, as well as extreme variation in tongue length among nectarivorous lineages. By comparison, the family Vespertilionidae, which is perhaps only slightly older than the Phyllostomidae [19], is very speciose, mostly represented by insectivorous taxa, and all maintain similarly conserved craniofacial and dental morphologies. The difference in occurrence of morphological diversification observed across these families enabled the development of a case-control type experimental design to understand open-reading frame and regulatory sequence evolution of PAX9. We investigated evolutionary rate, substitution patterns, and convergence of PAX9 in phyllostomids, vespertilionids, and other mammalian orders to understand if patterns were compatible with the hypothesis that PAX9 open-reading frame evolution has contributed to morphological diversification. Through comparison of regulatory regions, we developed a hypothesis for a novel translational regulatory mechanism, for which we provided experimental validation, and discussed potential developmental and evolutionary ramifications.

Materials and Methods

Sampling and Sequence Collection

Approximately 50 milligrams of tissue (heart, kidney, or liver) from 12 phyllostomids, 13 vespertilionids, one miniopterid, and two pteropodid species were loaned from the Genetic Resources Collection of the Natural Science Research Laboratory at Texas Tech University. Sampling included representatives of each of the major morphological diversifications associated with dietary diversifications within family Phyllostomidae (morphologies characteristic of insectivorous, sanguivorous, frugivorous, and nectarivorous lineages) as well as species of family Vespertilionidae (and one Miniopteridae, a sister family) selected to include a similar amount of evolutionary time as that sampled in the phyllostomids. Deeper evolutionary comparisons were also made to another divergent family, the family Pteropodidae (Old World megabats), as well as other mammalian orders and super-orders (outlined below). The two pteropodid species included are nectarivorous and exhibit some of the cranial and dental morphologies characteristic of this life history (elongated rostrum, increased spacing between molars, and reduced molar morphological diversity) that are shared with the four genera of nectarivorous phyllostomids also examined in this study (although derived from independent insectivorous ancestors). Museum identifications, taxonomic and dietary classifications are available in Table S1.

Genomic DNAs were isolated by phenol/chloroform extraction and precipitated isolations were rehydrated in TE buffer to a concentration of 125 ng/µl, determined by agarose electrophoresis and spectrophotometry. Two-hundred fifty ng of template genomic DNA was used in 50 µl PCR reaction volumes with 12.5 µmoles each deoxynucleoside, 1 µl Phire polymerase (New England Biolabs, Ipswich, MA), manufacturer buffer, and 25 pmoles each primer (all primers are described in Table S2 and Table S3). Amplicons were excised from 0.8% agarose gels and purified using silica membrane columns (Qiagen, Valencia, CA). Between 10 and 150 ng of amplicon was dehydrated depending on fragment size and cycle sequenced with 15 pmoles sequencing primer using Applied Biosystems BigDye Terminator Cycle Sequencing Kit version 3.1 (Applied Biosystems Inc., Foster City, CA). Sequencing products were purified by centrifugation through sephadex columns (Princeton Separations, New Jersey, NJ), dehydrated, and resuspended in formamide. Sequencing was performed on an ABI 3100-Avant Genetic Analyzer. Sequences were aligned in Sequencher version 4.9 (Gene Codes Corporation, Ann Arbor, MI) and manually edited.

Additional sequences for PAX9 were retrieved from GenBank to provide broader taxonomic comparison. The order Primates was represented by 15 species using data previously reported by [23] (GenBank accession numbers DQ067505–DQ067556). Additional PAX9 sequences were retrieved from genome sequence data from *Myotis lucifugus*, *Canis familiaris*, *Felis catus*, *Bos taurus*, *Mus musculus*, *Rattus norvegicus*, *Loxodonta africana*, and *Homo sapiens*. These sequences enabled a comparison across basal nodes of the Eutherian phylogeny. In addition, PAX9 UTR sequences were also retrieved for *Gallus gallus* and *Danio rerio*. Sequences for these bird and fish species were included to provide a general understanding about the timing of origin of identified putative regulatory motifs.

Open-reading Frame Sequence Evolution

Multiple tests and summary statistics were calculated to describe PAX9 evolutionary patterns (Table 1). Best fit models of molecular evolution for the nucleotide and predicted amino acid data were determined using the Bayesian Information Criterion and using neighbor-joining trees generated from the alignment. A Maximum Likelihood test of the molecular clock was performed to test if mutational distances from the root of the tree to the tips were equidistant. To quantify the overall pattern of selection a codon-based test for strict neutrality by averaging across sequence-pairs was performed with variance of the difference obtained through 1000 bootstrap replicates [24]. Codon-by-codon estimated dN-dS ratios (based on joint Maximum Likelihood reconstructions of ancestral states) were calculated to identify codons exhibiting an excess of non-synonymous substitutions (P values reflect the probability of rejecting the null hypothesis of neutral evolution and are the exact binomial probability of obtaining the observed or more biased numbers of synonymous and nonsynonymous changes for each codon site [25]). These analyses were performed using MEGA 5 [26] and codon-by-codon selection tests were performed using HyPhy [27] within MEGA 5.

Saturation plots were calculated to provide perspective about how synonymous and nonsynonymous mutations have accumulated among lineages at different evolutionary timescales. Measures of pairwise number of nucleotide and predicted amino acid

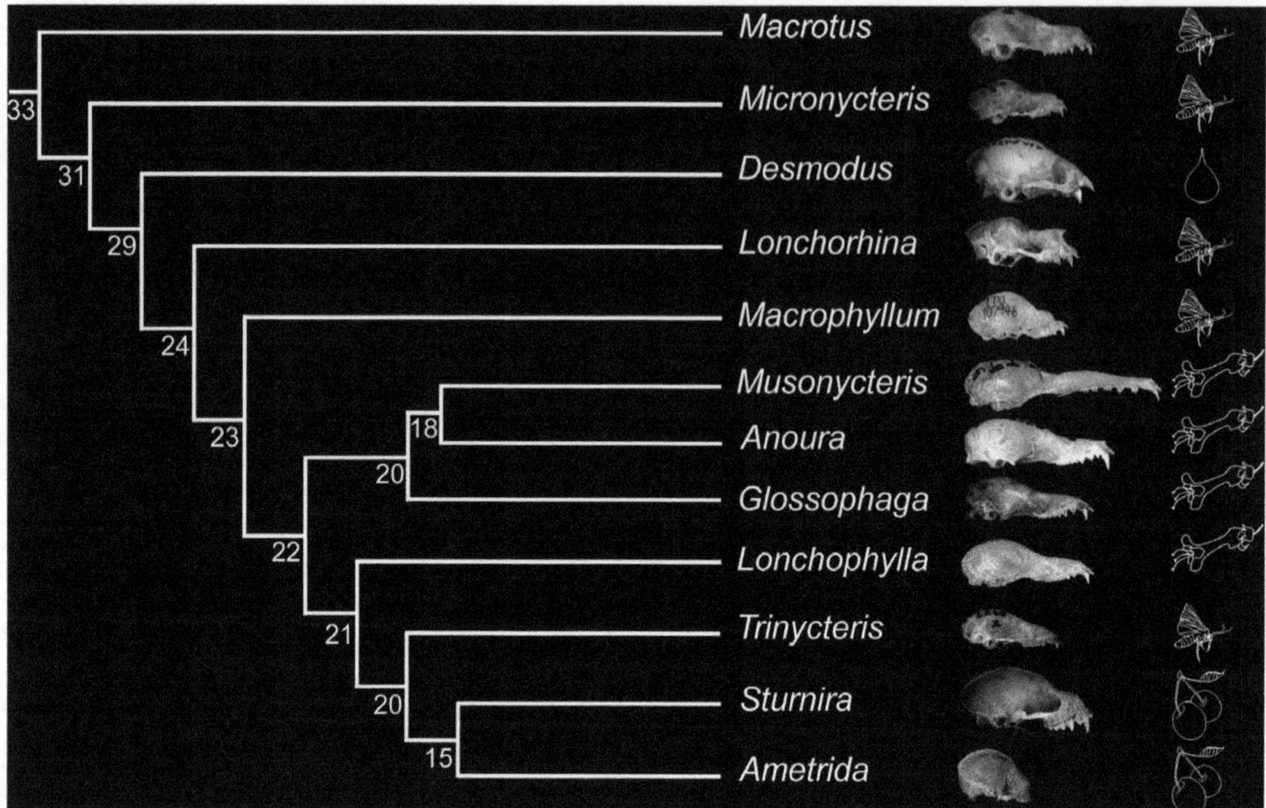

Figure 1. Cladogram reconstructed from [22,28–29] for phyllostomid bats included in this study. Node values are estimated divergence times taken from [22,28–29]. Each leaf of the cladogram includes genus, lateral image of skulls, and symbols of insect, blood, flower, or fruit to indicate dietary strategy of that genus.

differences were made among species of bat within families, and among species of primates. In addition, calculations of average pairwise distance were also made between orders. The orders represented by the above described sampling included Primates and Rodentia (super-order Euarchontoglires), Chiroptera, Carnivora, and Artiodactyla (super-order Laurasiaplacentalia), as well as Proboscidea (super-order Afrotheria). These measures of distance were regressed against time to most recent common ancestor (t_{mrca}). Divergence times for the data set were synthesized from several studies including [22,28–29] for the family Phyllos-

Table 1. Major analyses implemented in this study with short descriptions for the information obtained from each analysis.

Analysis	Information
Open-Reading Frame	
Molecular clock test	Detects substitution rate heterogeneity across phylogeny
Neutrality tests	Detection of selection
Saturation plots	Characterizes recurrent substitution over time
Ancestral state reconstruction	Estimates mutational pathways along given phylogeny
Structural prediction	Deduces polypeptide secondary structure
Binomial probability	Random probability of observed substitution frequencies by exon
Reticulation network	Phylogenetic inference, uncertainty displayed as reticulations
Untranslated Regions	
Phylogenetic shadowing	Defines statistically quantifiable conserved motifs
Regulatory element scanning	Identifies presence of any known regulatory motifs
Reporter expression assay	Quantification of translational regulatory control
RNAi	Gene knockdown to implicate target in proposed function

Analyses are grouped into open-reading frame or regulatory subheadings, indicating to which major hypothesis each analysis was relevant.

tomidae, [20,30] for the families Vespertilionidae and Miniopteridae, [31–35] for divergences within order Primates, and [36] for divergence times among mammalian orders.

Following the established evolutionary relationships among taxa, ancestral character states of the predicted amino acid variation were reconstructed following the rules of parsimony implemented in the Mesquite software package [37]. Variable positions and the frequency of recurrent substitution were plotted along a secondary structural prediction of PAX9 calculated using PSIPRED [38]. The location of the paired-binding domain reported by [7] was referenced here for additional structural relation of the distribution of mutations observed across the class Mammalia. Using the information on total detectable substitutions, cumulative binomial probabilities for the observed number of substitutions per exon were also calculated (the random probability of obtaining the same or greater number of amino acid replacements by chance, given the total number of replacements observed and the length of the exon). To examine the extent to which patterns of non-synonymous sequence evolution have resulted in phylogenetic convergence among lineages a bootstrapped neighbor-joining reticulation network based on uncorrected distances was constructed using SplitsTree4 version 4.12.3 [39]. This method of phylogenetic reconstruction was chosen because analysis provides information about the uncertainty of evolutionary relationships depicted as reticulations. Visualized reticulations (closed-loops in an otherwise bifurcating phylogenetic reconstruction) can provide greater clarification about sequence evolution among taxa than does nodal support values generally associated with bifurcating phylogenies. For comparison however, bootstrapped (1000 replicates) neighbor-joining trees were also estimated from nucleotide and predicted amino acid alignments incorporating the best-fit models of molecular evolution (determined above).

Regulatory Motif Analyses and Reporter Assays

The 5′ and 3′ UTR of PAX9 from all bat species, *Homo sapiens*, *Canis familiaris*, and *Mus musculus* (genome reads with full coverage of the UTRs) were screened for the presence of conserved islands by phylogenetic shadowing. Conserved islands are regions of nucleotide conservation that are identified by sequence comparison among taxa and among nucleotide positions within an alignment. Identifying conserved islands is important because they disclose nucleotide regions that may serve regulatory functions. For this analysis the hidden Markov model (HMM) approach available in the eShadow program was implemented [40]. The analysis consisted of the construction of a multiple sequence alignment, describing the alignment as a percent variation plot, and statistical evaluation to detect conserved islands [40]. For the statistical analysis, open-reading frames were defined for rate calibration to train the Hidden Markov Model parameters used to signify conserved islands. Following conserved island identification, nucleotide sequences of both open-reading frames and UTRs were surveyed for the presence of known regulatory elements. Regulatory elements observed in mRNAs are often sites of mRNA-protein interactions and can function to promote regulation of mRNAs at the post-transcriptional level. To determine the presence of any such elements within PAX9, sequences from each bat species was screened for 46 known UTR embedded regulatory motifs using UTRscan [41]. For this analysis complete PAX9 sequences from *Homo sapiens*, *Canis familiaris*, *Mus musculus*, *Danio rerio*, and *Gallus gallus* were included, as well as open-reading frame sequences from the available primates. The purpose of including these divergent species in addition to bat taxa was to describe the evolutionary conservation of identified regulatory elements contained within PAX9.

The analyses of PAX9 disclosed the presence of Musashi binding-elements (MBEs) that varied in frequency among lineages (specific findings reported in Results). Musashi is involved in maintenance of stem cell state, cellular differentiation, and tumorigenesis and has been shown to accomplish this regulation through repression of target mRNAs [42]. To investigate the functionality of these identified regulatory motifs a fluorescent reporter assay was developed. A tetracycline responsive bicistronic expression vector (pTRE-Tight-Bi-AcGFP1; Clontech Laboratories, Mountain View, CA) was constructed to contain mOrange fluorescent protein followed by the 3′ 130 bp of the open-reading frame and the contiguous 3′ UTRs (Figure 2). The purpose of including the 3′ end of the open-reading frame as part of the heterologous 3′ UTR was to include the MBE identified in this region, as well as those found in 3′ UTRs. Multiple constructs were created that differed only in the specific bat lineage from which the PAX9 sequence was amplified. Bat lineages were selected to include the major phylogenetic and morphological divergence among nectarivorous taxa, the highly derived vampire bat lineage, and the observed frequency variation of Musashi binding elements among phyllostomids. Final vectors contained two MBEs (TK19556, *Musonycteris harrisoni*), two different spatial combinations of three MBEs (TK101009, *Desmodus rotundus*; TK104582, *Lonchophylla concava*), four MBEs (TK101008, *Glossophaga soricina*), or only the vector's native simian virus 40 (SV40) polyadenylation signal containing no MBE, resulting in five unique expression constructs. In all constructs acGFP1 served as an internal control, being expressed in the opposite direction of the bicistronic promoter region from mOrange. Vectors were transfected (Nucleofector Reagent R; Lonza, Basel, CH) into Tet-Off Advanced HeLa cells (Clontech Laboratories, Mountain View, CA). HeLa was selected as the cell line to be used for these experiments because Western blotting confirmed the expression of both Musashi-2 and Musashi-1 (data not shown). Forty-eight hours post-transfection, relative fluorescence of both acGFP1 and mOrange was measured using a SpectraMax Gemini XPS plate reader (Molecular Devices, Sunnyvale, CA). Transfection and measurement for each cell line was replicated four times and the ratio of acGFP1 to mOrange for each measurement was taken as the comparable measure of reporter protein expression across cell lines and transfections. Expression levels across cell lines were compared using a Kruskal-Wallis test after a Shapiro-Wilk test indicated the data were not normally distributed ($W = 0.9$, $P < 0.01$).

For each cell line transient Musashi knockdowns were created using RNAi. These experiments were carried out to confirm the role that Musashi played in any observed differences among lines during the experiments described above. Sets of three siRNAs for both Musashi-1 and Musashi-2 (Origene, Rockville, MD) were used at a concentration of 3 nM to silence Musashi expression. Knockdown curves indicated 95% silencing at 48 hours post-knockdown and this time point was used for subsequent measurements. Each cell line was transfected with siRNAs using lipofectamine 2000 (Invitrogen, Grand Island, NY) and fluorescence was measured as described above. The knockdown experiment was replicated three times for each cell line.

Results

Open-reading Frame Sequence Evolution

The open-reading frame of chiropteran PAX9 was found to be 1026 bp. Sequence coverage among bats was represented by only

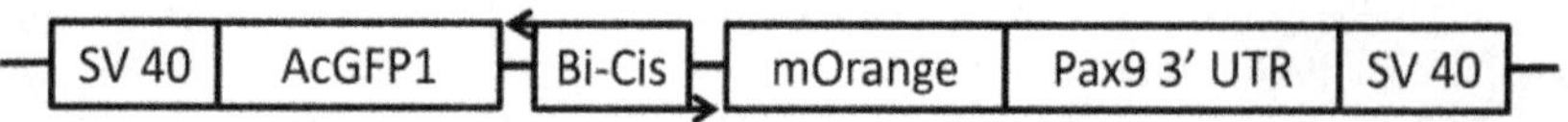

Figure 2. Diagram of reporter construct design. SV 40 = Simian virus 40 polyadenylation signal, AcGFP1 = green fluorescent protein, Bi-Cis = bicistronic promoter, mOrange = orange fluorescent protein, Pax9 3′ UTR = 3′ sequences of PAX9 from bat species described in the text. Experimental constructs differed only in the species from which PAX9 sequence was amplified. The control construct did not include a PAX9 3′ sequence.

1.1% missing data, and exons were sequenced in forward and reverse directions. Sequences are available from GenBank under accession numbers [KC549918–KC549944]. In order to characterize patterns of PAX9 open-reading frame variation several molecular analyses were conducted (Table 1). Using sequences of bats species generated for this study and others retrieved from GenBank (including several mammalian orders and primate taxa, see Materials and Methods) the best fit models of molecular evolution at the nucleotide and amino levels were T92+ G (0.9)+I (0.6) and JTT+G (0.05), respectively. The molecular clock was rejected at the nucleotide (P = 0.0001) and predicted amino levels (P = 0.0001). A history of purifying selection was signified through the overall test of strict neutrality ($d_N - d_S = -16.99$, P = 0.0001). Similarly, codon-by-codon analysis disclosed no codon positions that were significant for positive selection (i.e., no positive d_N-d_S were observed).

In order to characterize the distribution genetic variation over evolutionary time, pairwise nucleotide and predicted amino acids differences at different taxonomic levels were regressed against t_{mrca} (Figure 3a–d). The regression plots indicated saturation at the nucleotide level among orders (manifested as a weak linear relationship between pairwise differences and divergence time), but not within comparisons among bat families and among primates. Conversely, amino acid evolution was not heavily saturated in comparisons between orders, or through comparisons within bats and within primates.

Subsequent analyses were directed toward understanding the distribution of substitution variation and recurrent substitution across different polypeptide domains and exons. The plot of amino acid changes along the secondary structural prediction (Figure 3e) confirmed previous studies [23] indicating conservation of the paired-binding domain, although a single species of primate, *Callithrix jacchus*, has acquired a single change (Ala135Thr) one residue from the end of the paired-binding domain following the length reported by [7]. The octapeptide motif (residues 196–203; with a putative role in transcriptional regulation) was also found to be entirely conserved across mammals examined. However, parsimony reconstructions of ancestral character states described a history of recurrent substitution in areas downstream of the paired-binding domain and outside of the octapeptide motif. Among 38 variable positions, 18 were inferred to have undergone multiple substitutions among mammals, with a total of 87 substitutions inferred to have occurred over these 38 sites (an average of 2.3 changes per variable site). The specific number of changes inferred per site is included in the caption of Figure 3. For exons 2, 3, and 4 the cumulative probability for the observed number of nonsynonymous mutations given each exons length was found to be 0.61, 1.2×10^{-11}, and 0.02, respectively.

Phylogenetic analysis by neighbor-joining reticulation network of the predicted amino acid translations (Figure 3f) illustrated recurrent substitution identified in the above analyses occurred both among orders and among families. The centrality of the network consisted entirely of reticulations. Similarly, low nodal support was observed for the predicted polypeptide neighbor-joining phylogeny (Figure S1). Generally, the primates, vespertilionids, and phyllostomids clustered together in separate parts of the network, although relationships within these taxonomic groups were confounded by clade-specific recurrent substitution. Several instances of misplacement of specific lineages were also observed. For example, *Loxodonta* was placed within a clade of phyllostomid bats, and rodents were positioned amongst primates. Among misplacements within bats, independently derived nectar feeders clustered on two separate occasions. This occurred once when the nectarivorous pteropodids *Macroglossus* and *Eonycteris* were positioned on a clade sister to the nectarivorous phyllostomid *Glossophaga* (the result of Ala236Ser in the pteropodids and Ala236Thr in Glossophaga). The second misplacement of nectar feeders occurred in the grouping of *Musonycteris* and *Lonchophylla*, two independently derived phyllostomid nectarivorous genera. This placement was the result of convergence to Glu252Asp. Of additional interest within phyllostomids was the observation that the morphologically similar [22], yet evolutionarily divergent, insectivorous genera *Micronycteris* and *Trinycteris* were identical at the amino acid level.

Cis-regulatory Evolution

Outlier areas of nucleotide conservation within non-coding regions indicate potential regions of regulatory function. To identify any such motifs UTRs were analyzed for the presence of conserved islands. Both UTRs were found to contain conserved islands when comparisons included all non-bat mammalian taxa and phyllostomids or vespertilionids. When all samples were included the same conserved nucleotide domains were retained, although not statistically classifiable as conserved islands. Within the 5′ UTR a 146 bp conserved island 113 bp from the start codon was identified, and further analysis determined that this island is directly upstream from a putative internal ribosomal entry site. Within the 3′ UTR a 204 bp conserved island was identified 196 bp downstream of the stop codon. Next, nucleotide sequences of UTRs were surveyed for the presence of known regulatory elements identified in other studies using UTRscan [41]. This analysis revealed the presence of four Musashi-binding elements (MBEs; RT_nAGT (n = 1 to 3)) within conserved regions of the 3′ UTR. However, among bats there was variation in the number of MBEs present in a given species 3′ UTR. Two MBEs were found to occur within the upstream conserved island, while another two were identified within a 64 bp region 588 bp downstream of the stop codon that is also highly conserved (Figure 4a). Among bats, there was variation in the number and combination of MBEs present in given species 3′ UTR; however this variation was almost exclusively observed in phyllostomids (Figure 4b). Within phyllostomids the four identified MBEs were observed in 92%, 0%, 25%, and 75% of species, respectively. Vespertilionids were fixed for presence of the first and fourth MBE, with the exception of *Murina*, which only had the first MBE. The first, second, and fourth MBEs were observed in both pteropodids examined. Analysis of the 3′ UTRs of human, dog, mouse, chicken, and zebrafish confirmed the presence of a single MBE within each of these species 3′ UTRs, with the exception of a proposed alternative transcript in human which contained seven MBEs.

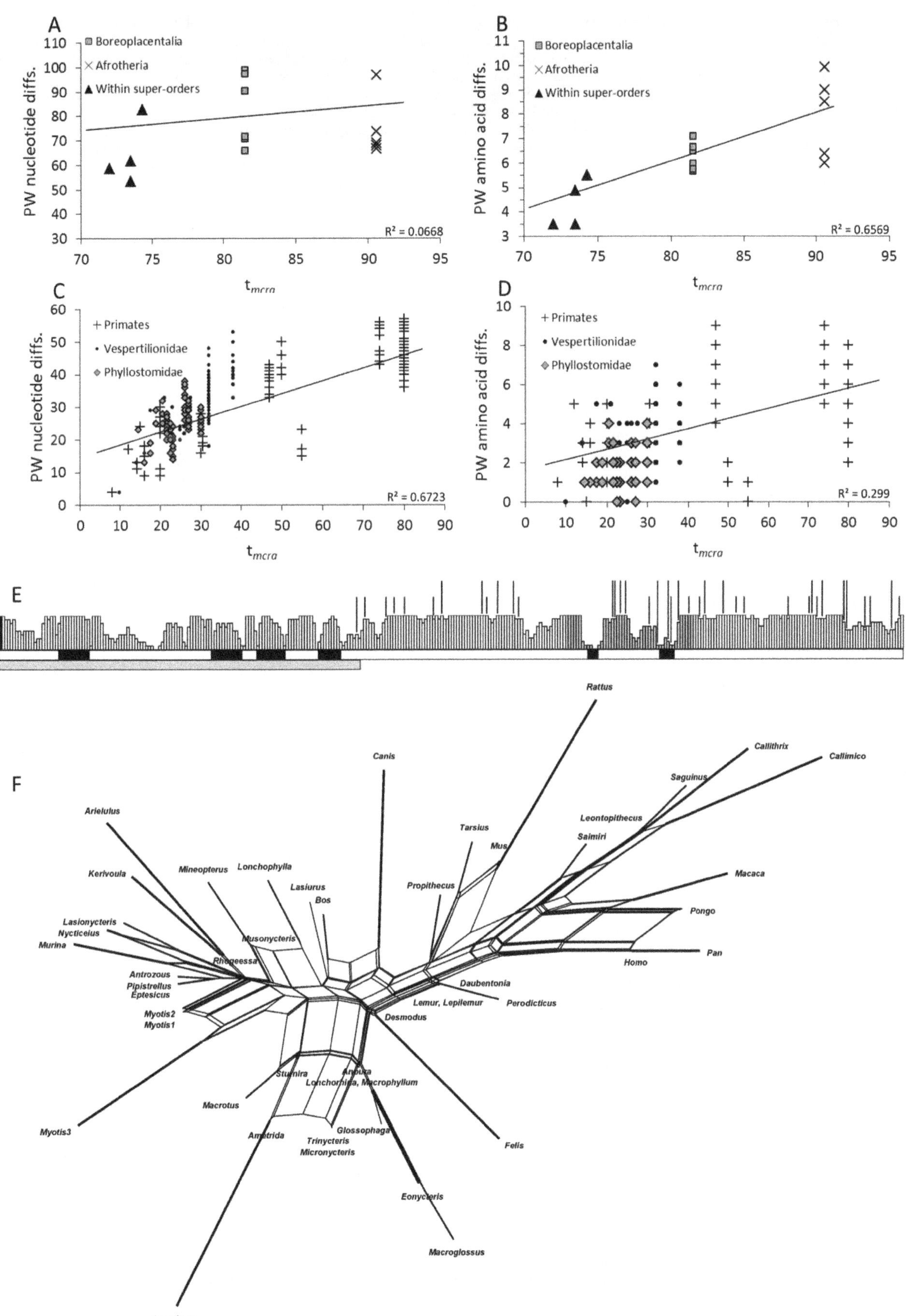
A
PW nucleotide diffs.
Boreoplacentalia
Afrotheria
Within super-orders
R² = 0.0668
t_mcra
B
PW amino acid diffs.
Boreoplacentalia
Afrotheria
Within super-orders
R² = 0.6569
t_mcra
C
PW nucleotide diffs.
Primates
Vespertilionidae
Phyllostomidae
R² = 0.6723
t_mcra
D
PW amino acid diffs.
Primates
Vespertilionidae
Phyllostomidae
R² = 0.299
t_mcra
E
F
Rattus
Canis
Callithrix
Callimico
Saguinus
Arielulus
Leontopithecus
Tarsius
Saimiri
Mus
Kerivoula
Mineopterus
Lonchophylla
Macaca
Lasiurus
Bos
Propithecus
Lasionycteris
Nycticeius
Musonycteris
Pongo
Murina
Rhogeessa
Pan
Homo
Antrozous
Pipistrellus
Eptesicus
Daubentonia
Lemur, Lepilemur
Perodicticus
Myotis2
Myotis1
Desmodus
Sturnira
Artibeus
Lonchorhina, Macrophyllum
Macrotus
Myotis3
Glossophaga
Ametrida
Trinycteris
Micronycteris
Felis
Eonycteris
Macroglossus
Loxodonta

Figure 3. Patterns of PAX9 open-reading frame evolution. Although under tight purifying selection, PAX9 exhibits patterns of saturation and recurrent substitution contingent on the level of comparison (amino acid versus nucleotide) as well as the amount of evolutionary time considered in the data. A) and B) display the pairwise nucleotide and predicted amino acid differences among orders regressed against t_{mrca}, respectively. C) and D) show similar plots, but among species from the families Phyllostomidae, Vespertilionidae and Miniopteridae, and the order Primates. E) Confidence in PSIPRED secondary structure prediction (greater confidence represented by larger bars) for each of the 341 residues of Pax9, and shading demarks exons. Directly below this histogram is the structural prediction in which white bars represent coiled structures and black represent helical regions. The solid grey bar at the bottom defines the limits of the paired-binding domain described by [7]. Vertical lines pointing to each codon position in the histogram indicate positions that vary across the mammalian taxa examined, and taller lines demark sights inferred to have accumulated recurrent substitutions. The number of inferred substitutions at these sights from left to right are as follows: 3, 2, 3, 2, 7, 4, 3, 4, 8, 6, 3, 2, 4, 4, 2, 3, 2, and 5. F) Reticulation network based on predicted amino acid translations. Closed loops in the network indicate homoplasies, and edge thickness is in proportion to bootstrap support.

The discovery of MBEs within UTR regions led to an additional survey of open-reading frames for regulatory motifs. This analysis disclosed the presence of an additional MBE 32 bp from the stop codon in all chiropteran lineages except *Murina*.

These results, along with additional evidence gathered through literature and bioinformatic surveys (see Discussion), led to the hypothesis that variation in the number of MBEs in PAX9 could contribute to evolutionary divergence in Pax9 expression. In order to test this hypothesis, we developed a fluorescence reporter assay (see Materials & Methods). Results of Kruskal-Wallis testing based on four independent transfections derived a non-significant test statistic ($H = 9.08$, $P = 0.06$). However, additional comparisons revealed that statistical significance was heavily influenced by the removal of either cell line containing three MBEs (removing 3 MBEs cell line resulted in $H = 8.45$, $P = 0.04$, and removing 3b MBEs cell lines resulted in $H = 9.41$, $P = 0.02$). Comparison of expression across cell lines indicated that level of expression of reporter protein corresponded to number of MBEs in the 3′ UTR of reporter protein (Figure 4c). Specifically, reporter expression under the control of two MBEs was the lowest; while reporter expression under the control of four MBEs was the highest. The cell lines with two different spatial combinations of three MBEs were intermediate.

In order to determine if the differences between UTRs from different species was due to Musashi activity, we knocked down endogenous Musashi expression using RNAi and repeated the assays. RNAi knock-down of Musashi completely abrogated reporter expression for all experimental lines ($H = 13.17$, $P < 0.001$), while having no effect on reporter controls possessing only SV40 3′ UTRs containing no MBEs ($H = 0.05$, $P = 0.83$; Figure 4d).

Discussion

Open-reading Frame Sequence Evolution

Understanding the functional significance of changes observed among sequence alignments are, as a rule, challenging. Determining the functional modification of substitutions observed in genes that are highly conserved over long evolutionary timescales, such as PAX9, is a special case of this difficulty. Within open-reading frames such genes are typified by continual purifying selection and emergence of rare variants that are often lineage specific. In this study, the PAX9 open-reading frame was found to adhere to this pattern of variation in which most genera have developed diagnostic amino acid sequences, yet a relatively small fraction of total residues were observed to vary across the broad mammalian evolutionary timescales considered. This overall pattern of variation was driven by recurrent substitution at a handful of sites, and invariance at many others. Understanding the selective basis for observed convergences is confounded by the fact that the signatures left by selection for functional change and those left by selection for functional redundancy are often the same. In genes that exhibit this type of selective history, many tests of positive selection are not empowered to detect selection. Because of this, instances of convergent sequence evolution between independent lineages convergent for phenotypic characters can be extremely informative and are often pursued by the evo-devo community [43–45]. Such convergences at the sequence and phenotypic levels are used as an indicator for convergent directional selection for protein function modifications. Subsequently in this section, we discuss general aspects of PAX9 open-reading frame evolution, specific instances of convergence, and the experimental possibilities for understanding observed variation.

Although PAX9 open-reading frame comparisons have previously been limited to comparisons among a narrow range of taxa, PAX9 has been shown to be highly conserved over those evolutionary time scales [7,23,46]. By far, exon 2, coding for the paired-binding domain, is the most highly conserved gene region, and this same pattern held true in the broader comparisons of this study. However, downstream from this region residues do vary. All but one of the 87 estimated residue substitutions among mammals occurred downstream from the paired-binding domain even though the downstream region accounts for less than 40% of the total polypeptide length. Although this region is highly variable as compared to the paired-binding domain, especially exon 3 (considering the total number of changes compared to exon length), the fact that it would be considered highly conserved in comparison to many other genes denotes an overall history of purifying selection. Pax9 is thought to physically interact with other proteins, although the specifics of these interactions are not well understood [8]. Because it is known that the C terminus of PAX9 functions in protein-DNA interactions during transcriptional activation, the function of the more variable N terminus could be specifically directed toward protein-protein interactions. Following this line of reasoning, adaptive roles of combinations of recurrent substitutions and lineage specific substitutions would relate to modifications of protein-protein interactions.

One could expect that open-reading frame variation would be greater among morphologically diversified phyllostomids relative to the craniofacially and dentally conserved vespertilionids. No indication of an increased rate in phyllostomids was observed, and there was a strong linear relationship between pairwise distance and t_{mrca}. In fact, total number of synonymous and nonsynonymous sites was higher in vespertilionids. Given the overall evidence for stabilizing selection of the PAX9 sequence, a more reasonable expectation would be convergences between specific taxa. Phylogenetic clustering as a result of polypeptide convergence was observed both between the independently derived nectarivorous genera *Lonchophylla* and *Musonycteris* [22,28–29] as well as between the pteropodids, *Eonycteris* and *Macroglossus*, and the phyllostomid *Glossophaga* (although the later relationship could be an artifact of taxon sampling and long-branch attraction). Similarly, *Micronycteris* and *Trinycteris* were found to be identical at the amino acid level and convergent for at least Thr337Ala. These

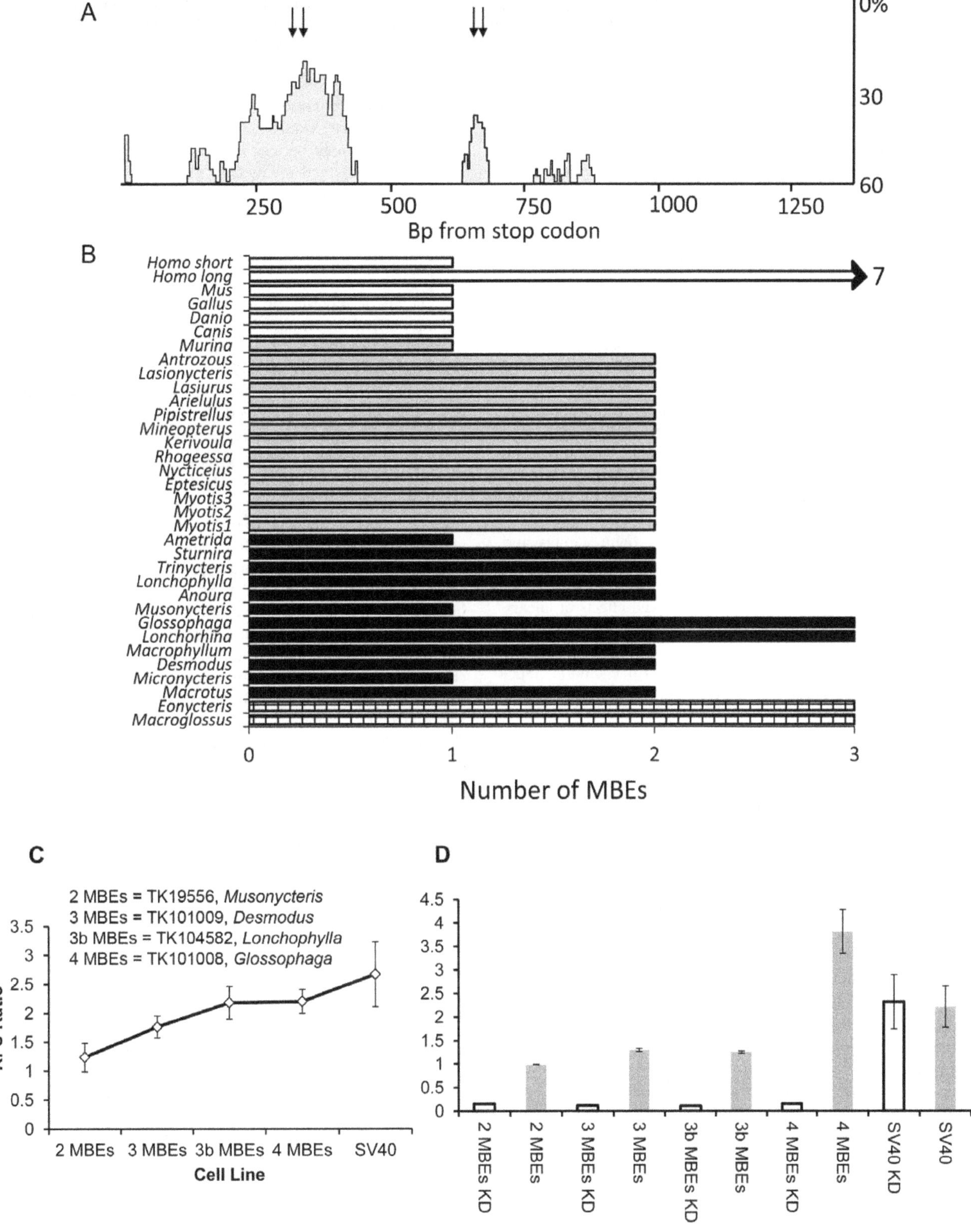

Figure 4. Comparative genetic and post-transcriptional reporter analysis. A) Patterns of nucleotide conservation across the 3′ UTR of PAX9 based on an alignment of *Homo*, *Canis*, *Mus*, and all bat species of this study, in which the vertical axis represents sequence divergence and the horizontal axis represents base pairs from the stop codon. Conserved domains are represented by grey areas and the locations of MBEs are demarked with arrows. The location of all seven MBEs occurring in a human alternative transcript are not show to conserve space (length >3Kb). B) Histogram summarizing the frequency of MBEs across taxa surveyed (checkered = pteropodids, black = phyllostomids, grey = vespertilionids, and white = non-bat taxa). *Homo* long and *Homo* short refer to the two alternative transcripts of PAX9 observed in humans. C) Results of the fluorescent reporter assay, with error bars representing standard mean error. Cell lines labeled as '3 MBEs' and '3b MBEs' are alternative spatial combination of this motif

frequency and, SV40 represents the construct with only SV40 polyadenylation signal containing no MBE. D) Result of the knockdown assay. Cell lines denoted with KD indicate knockdown lines. Error bars represent standard mean error, and are not visible in some instances due to the error being within histogram bar thickness.

two species share a similar insectivore dentition and cranial morphology, and craniometric assessments have classified these lineages as congeneric [47]. It is hypothesized that the craniofacial characteristics observed in these genera represent retention of the ancestral morphological condition [22]. However, multiple recent genetic studies have established these lineages diverged between 27–34 mya, being separated by many intermediate nodes exhibiting independent and extensive morphological divergences [22,28–29].

Summarizing the open-reading frame variation observed in this study, recurrent substitution was common, and would be expected to continue if sampling across all orders was increased to that available for Chiroptera and Primates. Specific instances of convergence within bats display patterns consistent with a functional role of open-reading frame evolution in morphological distinctiveness. Yet, it is important to note that other instances of convergence were observed with no obvious morphological connection between involved lineages. Understanding the biochemical effect of such changes would require experimental validation. However, the experimental designs to test these mutational effects, such as modifications to protein stability and protein-protein interactions, would be complicated by the effects of unknown PAX9 cellular interactions, difficulties of determining specific residues of functional modification, and effects of lineage specific mutations at other relevant loci. This scenario emphasizes the experimental difficulties of understanding the contribution of open-reading frame variation to morphological evolution, especially within evolutionarily conserved genes.

Cis-regulatory Evolution

Very little is known about the mechanisms controlling PAX9 expression. It has been postulated that PAX9 transcriptional activation is controlled by a Hoxa3 influence, as Hoxa3 mutants exhibit down regulated PAX9 expression [48]. Similarly, expression patterns of Pax9 during development are altered following full functional loss of Satb2 [49]. A cis-regulatory element has also been identified which is capable of inducing PAX9 expression in the ventromedial portion of the medial nasal process [50]. Several lines of evidence from the comparative genetic analysis of this study supported the hypothesis of a Musashi mediated post-transcriptional regulation of Pax9. The evolutionary conservation of MBEs within this gene indicates the proposed regulatory mechanism is ancient and evolutionarily conserved. However, the possibility of this regulation is contingent upon whether co-expression of Pax9 and Musashi in fact occurs. Important to the findings of the current study, it has been found that Msi-2 was among the top 100 most highly expressed novel gene discoveries in the developing mouse molar [51]. Additional survey of that microarray data set confirmed the co-expression of Msi-2 and Pax9. Further investigation of independent microarray data generated from mouse chondrocytes and osteoblasts [52], as well as human odontoblasts [53], establish the co-expression of Msi-2 and Pax9. These data are important because they confirm co-expression in developmentally relevant cell types.

Analysis of the regulatory capabilities of PAX9 3′ sequences containing different numbers of MBEs disclosed a statistically significant and directly proportional expressional effect according to MBE number. These findings were corroborated by knockdown analysis through which silencing of Musashi proteins resulted in near complete translational repression of MBE containing cell lines, but not the controls. This observation was particularly interesting, not only because it provided strong support for the hypothesized Musashi interaction, but also because it indicated interaction of regulatory protein(s) in addition to Musashi. In a regulatory scenario only involving a Musashi protein, the expressional effect of Musashi silencing on reporter proteins in MBE containing cell lines would be expected to be similar to that observed for the controls, yet this was not the case. The attenuation of reporter protein translation in MBE containing cell lines after Musashi knockdown describes a more complicated scenario, in which a combination of unidentified proteins and Musashi are required to achieve proper post-transcriptional regulation. It is possible that additional unknown regulatory motifs within the large conserved island of the 3′ UTR play a part in this putative multi-protein regulation.

An interesting finding to emerge from this study was the variation in MBE frequency within the 3′ UTR of morphologically variable phyllostomid bats, whereas the morphologically conserved vespertilionids, with the exception of one species, were invariant. Results pose the possibility that Pax9 is differentially expressed among evolutionary lineages in proportion to level of Musashi regulation. Variation of regulatory element frequency has previously been identified to affect genes involved in *Drosophila* developmental signaling pathways [54], emphasizing the role that regulatory copy number variation can have in modifying gene expression. Heterochronic effects of the Musashi-mediated regulation of PAX9 seem particularly plausible given Musashi's established role in maintenance of pluripotency and expression timing [50,55–59], and the role of PAX9 in cellular differentiation [5–8]. Furthermore, post-transcriptional control of expression in general is critical to proper developmental timing [60]. Yet, comparing the number of MBEs across lineages to gross morphological differences does not indicate major effects due to MBE frequency. Although evolution of Pax9 regulation is a candidate in the process of morphological diversification, changes at loci in addition to PAX9 certainly underlie observed morphologies.

In conclusion, the experimental design of this study involved comparative genetic analyses and reporter assays based on genetic variation that has evolved under the constraints of natural selection. Reporter assays yielded valuable insights into the evolutionary history of PAX9 and has supported a hypothesis for Pax9 post-transcriptional regulation. The discoveries resulting from this study were a consequence of sampling among natural variation, and these findings would not have been readily apparent through comparisons limited to model organisms. Subsequent experiments will be developed to understand how modulating Pax9 expression levels influences downstream effectors and the timing of cellular differentiation, a spatio-temporal perspective about the expression of Musashi during embryogenesis, and the specific mechanism by which Musashi-mediated post-transcriptional regulation is achieved.

Supporting Information

Figure S1 Neighbor-joining phylogenetic reconstructions. A) Previously reported evolutionary relationships among all taxa included. B) Relationships estimated from PAX9 nucleotide variation in which the established relationships among families, orders, and super-orders are largely recovered. C) Relationships estimated from among amino acid predictions. Bootstrap support values based on 1000 iterations are labeled adjacent to nodes.

Table S1 Sampling of bats included in this study categorized by diet and taxon.

Table S2 PCR primer combinations.

Table S3 Sequencing primers used in addition to PCR primers.omy.

Acknowledgments

Tissues were provided from the Genetics Resources Collection at the Museum at Texas Tech University, and we are grateful to Heath Garner and Kathy McDonald for curation and access to samples.

Author Contributions

Conceived and designed the experiments: CDP BB JWF HMM RJB. Performed the experiments: CDP BB. Analyzed the data: CDP. Contributed reagents/materials/analysis tools: BB JWF RJB. Wrote the paper: CDP BB JWF RJB.

References

1. Epstein CJ, Erickson RP, Wynshaw-Boris A (2008) Inborn errors in development. New York: Oxford Monographs on Medical Genetics, Oxford University Press.
2. Liu B, Rooker S, Helms J (2010a) Molecular control of facial morphology. Sem. Cell. Dev. Biol. 21: 309–313.
3. Vorobyov E, Horst J (2006) Getting the Proto-Pax by the Tail. J. Mol. Evol. 63: 153–64.
4. Wang W, Zhong J, Wang YQ (2010) Comparative genomic analysis reveals the evolutionary conservation of the Pax gene family. Genes Genet. Syst. 85: 193–206.
5. Ogawa T, Kapadia H, Wang B, D'Souza R (2005) Studies on PAX9-Msx1 protien interactions. Arch. Oral. Biol. 501 41–145.
6. Ogawa T, Kapadia H, Feng J, Raghow R, Peters H, et al. (2006) Functional consequences of interactions between PAX9 and Msx1 genes in normal and abnormal tooth development. J. Biol. Chem. 281: 18363–18369.
7. Wang Y, Groppe JC, Wu J, Ogawa T, Mues G, et al. (2009) Pathogenic mechanisms of tooth agenesis linked to paired domain mutations in human PAX9. Hum. Mol. Gen. 18: 2863–2874.
8. Kong H, Wang Y, Patel M, D'Souza RN (2011) Regulation of bmp4 expression in odontogenic mesenchyme: from simple to complex. Cell. Tiss. Org. 194: 156–160.
9. Peters H, Neubüser A, Kratochwil K, Balling R (1998) PAX9-deficient mice lack pharyngeal pouch derivatives and teeth and exhibit craniofacial and limb abnormalities. Genes Dev. 12: 2735–2747.
10. Peters H, Wilm B, Sakai N, Imai K, Maas R, et al. (1999) Pax1 and PAX9 synergistically regulate vertebral column development. Dev. 126: 5399–5408.
11. Jonker Z, Kist R, Aw A, Wappler I, Peters H (2004) PAX9 is required for filiform papilla development and suppresses skin-specific differentiation of the mammalian tongue epithelium. Mech. Dev. 121: 1313–1322.
12. Sasaki Y, O'kane S, Dixon J, Dixon M, Ferguson M (2007) Temporal and spatial expression of PAX9 and Sonic hedgehog during development of normal mouse palates and cleft palates in TGF-beta3 null embryos. Arch. Oral Biol. 52: 260–267.
13. Mammoto T, Mammoto A, Torisawa YS, Tat T, Gibbs A, et al. (2011) Mechanochemical control of mesenchymal condensation and embryonic tooth organ formation. Dev. Cell. 18: 758–769.
14. Bianch FJ, de Oliveira TF, Saito CB, Peres RC, Line SR (2007) Association between polymorphism in the promoter region (G/C-915) of PAX9 gene and third molar agenesis. J. Appl. Oral. Sci. 15: 382–386.
15. Londhe SM, Viswambaran M, Kumar P (2008) Multidisciplinary management of oligodontia. Med. J. Arm. F. Ind. 64: 67–69.
16. Mendoza-Fandino GA, Gee JM, Ben-Dor S, Gonzalez-Quevedo C, Lee K, et al. (2010) A novel g.-1258G>A mutation in a conserved putative regulatory element of PAX9 isassociated with autosomal dominant molar hypodontia. Clin. Genet. 80: 265–272.
17. Paixão-Côrtes VR, Braga T, Salzano FM, Mundstock K, Mundstock CA, et al. (2011) PAX9 and MSX1 transcription factor genes in non-syndromic dental agenesis. Arch. Oral. Biol. 56: 337–44.
18. Suda N, Ogawa T, Kojima T, Saito C, Moriyama K (2011) Non-syndromic oligodontia with a novel mutation of PAX9. J. Dental Res. 90: 382–386.
19. Teeling E, Springer M, Madsen O, Bates P, O'brien S, et al. (2005) A molecular phylogeny for bats illuminates biogeography and the fossil record. Sci. 307: 580–584.
20. Lack JB, Roehrs Z, Stanley C, Ruedi M, Van Den Bussche R (2010a) Molecular phylogenetics of Myotis indicate familial-level divergence for the genus Cistuo (Chiroptera). Journ. Mamm. 91: 976–992.
21. Dumont ER, Dávalos LM, Goldberg A, Santana SE, Rex K, et al. (2011) Morphological innovation, diversification and invasion of a new adaptive zone. Proc. Biol. Sci. 7: 1797–1805.
22. Baker R, Bininda-Emonds O, Manilla-Meluk H, Porter C, Van Den Busche (2012) Molecular timescale of diversification of feeding strategy and morphology in new world leaf-nosed bat (Phyllostomidae): a phylogenetic perspective. In: Gunnell G, Simmons N, editors. Evolutionary history of bats: fossils, molecules and morphology. Cambridge: Cambridge University Press. 385–409.
23. Perry G, Verrelli B, Stone A (2006) Molecular evolution of the primate developmental genes MSX1 and PAX9. Mol. Biol. Evol. 23: 644–654.
24. Nei M, Gojobori T (1986) Simple methods for estimating the number of synonymous and nonsynonymous nucleotide substitutions. Mol. Biol. Evol. 3: 418–426.
25. Suzuki Y, Gojobori T (1999) A method for detecting positive selection at single amino acid sites. Mol. Biol. Evol. 16: 1315–1328.
26. Tamura K, Peterson D, Peterson N, Stecher G, Nei M, et al. (2011) MEGA5: Molecular evolutionary genetics analysis using Maximum Likelihood, evolutionary distance, and Maximum Parsimony methods. Mol. Biol. Evol. 28: 2731–2739.
27. Kosakovsky Pond SL, Frost SDW, Muse SV (2005) HyPhy: hypothesis testing using phylogenies. Bioinfo. 21: 676–679.
28. Datzmann T, von Helversen O, Mayer F (2010) Evolution of nectarivory in phyllostomid bats (Phyllostomidae Gray, 1825, Chiroptera: Mammalia). BMC Evol. Biol. 4: 165–179.
29. Rojas D, Vale Á, Ferrero V, Navarro L (2011) When did plants become important to leaf-nosed bats? Diversification of feeding habits in the family Phyllostomidae. Mol. Ecol. 20: 2217–2228.
30. Lack JB, Van Den Bussche R (2010b) Identifying the confounding factors in resolving phylogenetic relationships in Vespertilionidae. Journ. Mamm. 91: 1435–14448.
31. Jameson NM, Hou ZC, Sterner KN, Weckle A, Goodman M, et al. (2011) Genomic data reject the hypothesis of a prosimian primate clade. J. Hum. Evol. 61: 295–305.
32. Matsui A, Rakotondraparany F, Munechika I, Hasegawa M, Horai S (2009) Molecular phylogeny and evolution of prosimians based on complete sequences of mitochondrial DNAs. Gene 441: 53–66.
33. Opazo JC, Wildman DE, Prychitko T, Johnson RM, Goodman M (2006) Phylogenetic relationships and divergence times among New World monkeys (Platyrrhini, Primates). Mol. Phyl. Evol. 40: 274–280.
34. Steiper M, Young N (2006) Primate molecular divergence dates. Mol. Phyl. Evol. 41: 384–394.
35. Wilkinson RD, Steiper ME, Soligo C, Martin RD, Yang Z, et al. (2011) Dating Primate Divergences through an Integrated Analysis of Palaeontological and Molecular Data. Syst. Biol. 60: 16–31.
36. Hallström B, Janke A (2010) Mammalian evolution may not be strictly bifurcating. Mol. Biol. Evol. 27: 2804–2816.
37. Maddison W, Maddison D (2010) Mesquite: a modular system for evolutionary analysis. Version 2.73 Mesquite website. Available: http://mesquiteproject.org. Accessed 2012 Jan 2.
38. Buchan DW, Ward SM, Lobley AE, Nugent TC, Bryson K, et al. (2010) Protein annotation and modeling servers at University College London. Nucl. Acids. Res. 38 Suppl, 38: W563–W568.
39. Huson DH, Bryant D (2006) Application of phylogenetic networks in evolutionary studies. Mol. Biol. Evol. 23: 254–267.
40. Ovcharenko I, Boffelli D, Loots G (2004) eShadow: a tool for comparing closely related sequences. Genome Res. 14: 1191–1198.
41. Grillo G, Turi A, Licciulli F, Mignone F, Liuni S, et al. (2010) UTRdb and UTRsite (RELEASE 2010): a collection of sequences and regulatory motifs of the untranslated regions of eukaryotic mRNAs. Nuc. Acid. Res. 38: 75–80.
42. Okano H, Imai T, Okabe M (2002) Musashi: a translational regulator of cell fate. J. Cell. Sci. 115: 1355–1359.
43. Li Y, Liu Z, Shi P, Zhang J (2010) The hearing gene Prestin unites echolocating bats and whales. Curr. Biol. 20: 55–56.
44. Liu Y, Cotton J, Shen B, Han X, Rossiter S, et al. (2010b) Convergent sequence evolution between echolocating bats and dolphins. Curr. Biol. 20: 53–54.
45. Liu Y, Han N, Franchini L, Xu H, Pisciottano F, et al. (2012) The voltage-gated potassium channel subfamily KQT member 4 (KCNQ4) displays parallel evolution in echolocating bats. Mol. Biol. Evol. 29: 1441–1450.

46. Pereira TV, Salzano FM, Mostowska A, Trzeciak WH, Ruiz-Linares A, et al. (2006) Natural selection and molecular evolution in primate PAX9 gene, a major determinant of tooth development. Proc. Natl. Acad. Sci. USA 103: 5676–5681.
47. Jones K, Purvis A, MacLarnon A, Bininda-Emonds O, Simmons N (2002) A phylogenetic supertree of the bats. Biol. Rev. 77: 223–259.
48. Su D, Ellis S, Napier A, Lee K, Manley N (2001) Hoxa3 and Pax1 regulate epithelial cell death and proliferation during thymis and parathyroid organogenesis. Dev. Bio. 236: 316–329.
49. Britanova O, Depew MJ, Schwark M, Thomas BL, Miletich I, et al. (2006) Stab2 haploinsufficiency phenocopies 2q32-q33 delections, whereas loss suggests a fundamental role in the coordination of jaw development. Amer. Journ. Hum. Gen. 79: 668–678.
50. Santagati F, Abe K, Schmidt V, Schmitt-John T, Suzuki M, et al. (2003) Identification of cis-regulatory elements in the mouse Pax9/Nkx2–9 genomic region: implication for evolutionary conserved synteny. Gen. Soc. Amer. 156: 235–242.
51. Pemberton T, Li F-Y, Oka S, Mendoza-Fandino G, Hsu Y, et al. (2007) Identification of novel genes expressed during mouse tooth development by microarray gene expression data. Dev. Dyn. 236: 2245–2257.
52. Fukada T, Civic N, Furuichi T, Shimoda S, Mishima K, et al. (2008) The zinc transporter SLC39A13/ZIP13 is required for connective tissue development; its involvement in BMP/TGF-β signaling pathways. PLoS One 3:e3642.
53. Lin D, Huang Y, He F, Gu S, Zhang G, et al. (2007) Expression survey of genes critical for tooth development in the human embryonic tooth germ. Dev. Dyn. 236: 1307–1312.
54. Lai EC, Posakony JW (1997) The bearded box, a novel 3′ UTR sequence motif, mediates negative post-transcriptional regulation of Bearded and Enhancer of split Complex gene expression. Devel. 124: 4847–4856.
55. Sakakibara S, Imai T, Hamaguchi K, Okabe M, Aruga J, et al. (1996) Mouse-musashi-1, a neural RNA binding protein highly enriched in the mammalian CNS stem cell. Dev. Biol. 176: 230–242.
56. Imai T, Tokunaga A, Yoshida T, Hashimoto M, Mikoshiba K, et al. (2001) The neural RNA-binding protein musashi1 translationally regulates mammalian numb gene expression by interacting with its mRNA. Mol. Cell. Biol. 21: 3888–3900.
57. Sakakibara S, Nakamura Y, Satoh H, Okano H (2001) RNA-binding protein musashi2: developmentally regulated expression in neural precursor cells and subpopulations of neurons in mammalian CNS. J. Neurosci. 21: 8091–8107.
58. Wuebben EL, Mallanna SK, Cox JL, Rizzino A (2012) Musashi2 is required for the self-renewal and pluripotency of embryonic stem cells. PLoS One. 7:e34827.
59. Charlesworth A, Wilczynska A, Thampi P, Cox LL, MacNicol AM (2006) Musashi regulates the temporal order of mRNA translation during *Xenopus* oocyte maturation. The EMBO Journ. 25: 2792–2801.
60. Kong J, Lasko P (2012) Translational control in cellular and developmental processes. Nat. Rev. Gen. 13: 383–394.

7

A Comparison of Inbreeding Depression in Tropical and Widespread *Drosophila* Species

Jesper S. Bechsgaard[1]*, Ary A. Hoffmann[2], Carla Sgró[3], Volker Loeschcke[1], Trine Bilde[1], Torsten N. Kristensen[1,4,5]

1 Department of Bioscience, Aarhus University, Aarhus, Denmark, **2** Department of Genetics and Bio21 Institute, Melbourne University, Melbourne, Australia, **3** School of Biological Sciences, Monash University, Melbourne, Australia, **4** Department of Molecular Biology and Genetics, Aarhus University, Tjele, Denmark, **5** NordGen - Nordic Genetic Resource Center, Ås, Norway

Abstract

The evolutionary history of widespread and specialized species is likely to cause a different genetic architecture of key ecological traits in the two species groups. This may affect how these two groups respond to inbreeding. Here we investigate inbreeding effects in traits related to performance in 5 widespread and 5 tropical restricted species of *Drosophila* with the aim of testing whether the two species groups suffered differently from inbreeding depression. The traits investigated were egg-to-adult viability, developmental time and resistance to heat, cold and desiccation. Our results showed that levels of inbreeding depression were species and trait specific and did not differ between the species groups for stress resistance traits. However, for the life history traits developmental time and egg-to adult viability, more inbreeding depression was observed in the tropical species. The results reported suggest that for life history traits tropical species of *Drosophila* will suffer more from inbreeding depression than widespread species in case of increases in the rate of inbreeding e.g. due to declines in population sizes.

Editor: Axel Imhof, Ludwig-Maximilians-Universität München, Germany

Funding: This work was funded by the Danish Natural Research Council with a Steno stipend to TNK and a frame grant to VL. AAH was supported via Australian Research Council Fellowship and Discovery grant schemes.

Competing Interests: The authors have declared that no competing interests exist.

* E-mail: torsten.nygaard@nordgen.org

Introduction

A species' genetic architecture, referring to the number of segregating functional variants, including dominance and epistatic interactions, underlying phenotypic traits, is formed by phylogenetic, demographic and ecological processes, and determines its biological performance, such as its basal resistance to environmental stressors and the ability to adapt evolutionary and through adaptive phenotypic plasticity to environmental changes [1]. The genetic architecture of a population or species also dictates how it will cope with inbreeding since the cost of inbreeding is under the influence of allele frequencies, amount of dominance variance and the level of inbreeding [2].

Some *Drosophila* species are restricted to narrow habitat ranges in the tropics ('tropical species'), whereas others occupy habitats ranging from tropic to temperate areas ('widespread species'). Widespread species are exposed to different, more variable and diverse ecological conditions compared to specialized restricted tropical species, which will likely lead to different selection pressures and subsequent genetic divergence of the two species groups through evolutionary adaptation [3–6]. In addition tropical species of *Drosophila* are on average more closely related to each other than to widespread species [6] adding a phylogenetic component to the differences among the two species groups. Widespread and tropical restricted species of *Drosophila* are for those reasons likely to have a different genetic architecture, especially in key ecological traits involved in environmental stress resistance.

Tropically restricted *Drosophila* species have recently been hypothesized to be more prone to experience reduced population sizes and extinction as compared to more widespread *Drosophila* species as the environment changes. This is based on evidence from a number of studies showing that widespread species have higher levels of basal resistance to climatic extremes (temperature and humidity) when compared to tropical climate specialists [6–9], and that some tropical restricted species of *Drosophila* have lower adaptive evolutionary potential in the ecological key traits desiccation and cold resistance [10,11]. Demographic effects like bottlenecks, inbreeding and/or genetic drift seem to be excluded as causing these differences since levels of neutral genetic variation seem to be similar among the two species groups [11,12] suggesting that historical (phylogenetic) and current (ecological) selection pressures are responsible [6,9].

Based on the hypothesis that tropical and widespread species of *Drosophila* have a different genetic architecture, it is of interest to investigate if the two species groups respond differently to inbreeding. If the cost of inbreeding differs between the species groups, this may influence the extinction probabilities of these species groups when they experience an increase in inbreeding rates in their natural habitats, due to for instance reductions in population size as may occur under future climate changes. Here, we study the consequences of inbreeding in 5 widespread and 5 tropical restricted species of *Drosophila* by investigating inbred and outbred lines for the traits egg-to-adult viability, developmental time, and resistance to cold, heat and desiccation stresses. Generally, we found strong line, species and trait specific effects of inbreeding. Tropical species on average suffered more from inbreeding depression compared to widespread species for life history traits, whereas no differences among species groups were observed in stress resistance traits.

Materials and Methods

Collection of flies

A population from each of 10 *Drosophila* species was collected during 2007 and 2008 (December to April) in North Queensland, along the Australian east coast. Five of the species are restricted to the tropics, mostly rainforest habitats (*D. birchii*, *D. bunnanda*, *D. bipectinata*, *D. sulfurigaster* and *D. pseudoananassae*), while the other 5 are cosmopolitan or widespread species whose habitat includes temperate regions as well (*D. melanogaster*, *D. simulans*, *D. hydei*, *D. repleta* and *D. serrata*). Average latitude of the tropical and widespread species investigated here range from 13.04° to 18.15° and 101.75° to 129.40°, respectively (based on registrations entered in the taxodros database available at http://www.taxodros.uzh.ch). For each species, 18 to 22 inseminated females from a single population were caught and brought to the laboratory to establish mass bred populations, which were maintained for 15 generations with at least 1000 individuals before outbred and inbred lines were established. The expected loss of heterozygosity during this process is calculated according to [13] and constitutes about 3%. Flies were reared on an oat-sugar-yeast-agar medium under a 12:12 h light/dark cycle at 20°C prior to performing the experiments. Details regarding sampling locations are described in [8]. No permits were required for the described field collections.

Breeding regimes

From each species, 3 outbred control lines and 20 inbred lines were established from the mass bred population. The outbred lines were started from about 750 individuals, whereas the inbred lines were started from single pairs of virgin females and males which were subsequently run through four generations of full sib mating, resulting in an expected inbreeding level of 0.59 ($F_t = 1/4(1+2F_{t-1}+F_{t-2})$). Thereafter inbred lines were expanded to about 500 individuals before performing the experiments. Between 3 and 10 inbred lines per species were used for experiments (Table S1). For species where more than 10 inbred lines were available 10 randomly collected lines were used (11, 13 and 14 lines were available to select from from *D. sulfurigaster*, *D. repleta* and *D. pseudoananassae*, respectively). The experiments were done in the second generation after the last inbreeding round (see [14] for more details).

Egg collection

From each inbred and control line 20 eggs were collected into each of 10 vials with 7 ml medium, and incubated at 20°C. Flies emerging from these eggs were used for assessing stress resistance traits (see below). From prior knowledge of generation times, collection of eggs was timed so that flies from all species emerged over as short a time span as possible. Flies used for experiments were all between 3 and 8 days old.

Developmental time and egg-to-adult viability

Ten vials (with 7 ml medium) per line each containing 20 eggs were set up as specified above. Emerging flies were counted twice a day (at 8:00 and 20:00 h). Egg-to-adult viability was scored as the number of males and females emerging from each vial and developmental time estimated as the time taken to develop from an egg to an adult female or male fly. Flies were all reared at 20°C degrees since this temperature is considered non-stressful for all species investigated. Emerging flies were used for the temperature and desiccation assays described below.

Temperature resistance assays

1) Heat knock down resistance. Ten flies per line and sex were individually put into 5 ml glass vials and submerged into a preheated 38°C water bath. Heat resistance was scored as the time until the flies were unable to move any of their body parts.
2) Chill coma recovery time. Ten flies per line and sex were individually put into 5 ml glass vials and submerged into a 0°C water bath and kept there for 3 hours. The vials were then put at room temperature, and chill coma recovery was scored as the time when the flies were able to stand on their legs.
3) Critical thermal minimum (CTmin). Ten flies per line and sex were individually put into 5 ml glass vials and submerged into a 20°C water bath. The temperature was then decreased by 0.1°C per minute. CTmin was scored as the temperature at which flies were in coma unable to move any body parts (see [14] and [15] for details regarding this assay).

Desiccation resistance assay

Ten flies per line and sex were individually put into 5 ml glass vials covered with gauze. Vials were then put into sealed aquariums. Prior to setting up the experiment, desiccant was added to the aquariums to reach air humidity of 0–5%. Desiccation resistance was scored as the time until the flies were unable to move any of their body parts. Vials were checked every hour.

Statistics

For each trait we used a nested ANOVA to test for effects of line (nested within species and breeding regime, as random effects), species, breeding regime (inbred or outbred), sex and the interactions between the fixed effects. Inbreeding depression (δ) was estimated for each trait using the formula $\delta = 1 -$ mean inbred/mean control [16]. For traits where individuals with lower values are expected to be more fit(develommental time, CTmin and chill coma recovery time) inbreeding depression was estimated using the formula $\delta = 1 -$ mean control/mean inbred. Mean control was the average of the three control lines and each inbred line was standardized to this value for all traits in all species. Positive values were indicative of inbreeding depression for all traits. . Effects of species, sex and their interaction on the level of inbreeding depression in each trait were tested with ANOVAs. Data were generally normally distributed although deviations from normality were observed in a few cases. Egg-to-adult viability data were arc sin square root transformed to improve normality of the data. For the other traits analyses were performed on non-transformed data. Differences between levels of inbreeding depression between widespread and tropical species were tested with a non-parametric Mann-Whitney test for life history traits (developmental time and egg-to-adult viability) and stress resistance traits (CTmin, cold recovery time, heat knock down time and desiccation resistance), respectively. For each species within the two species groups, estimates of inbreeding depression were averaged across males and females for each line. Thus for each trait and species we had an estimate of inbreeding depression for each line. These data were used to test for effects of species group using an ANOVA with species nested within distribution (tropical or widespread species group).

Results

Developmental time: Line (nested within species and breeding regime), species, breeding regime (inbred or outbred), sex and the species×sex interaction all significantly affected developmental time (line: $F_{79,1767} = 17.47$, $P<0.001$; species: $F_{9,1767} = 207.95$, $P<0.001$; breeding regime: $F_{1,1767} = 33.52$, $P<0.001$; sex: $F_{1,1767} = 231.53$, $P<0.001$; species×sex: $F_{9,1767} = 5.65$, $P<0.001$; Fig. 1a, Table S1). Remaining factors in the analysis were non-significant (results not shown). Inbreeding on average increased developmental time by 3.8 and 4.5% across species in males and females, respectively (Fig. 1a). The level of inbreeding depression (faster developmental time is interpreted as beneficial) differed between species ($F_{9,118} = 6.09$, $P<0.001$, Fig. 1), whereas it was similar for males and females of the same species ($F_{1,118} = 0.80$, NS). Levels of inbreeding depression were also not affected by the species×sex interaction ($F_{9,118} = 0.22$, NS).

Egg-to-adult viability: Line, species, breeding regime (inbred or outbred) and the species×breeding regime interaction all significantly affected egg-to-adult viability (line: $F_{80,1862} = 12.31$, $P<0.001$; species: $F_{9,1862} = 2.25$, $P<0.05$; breeding regime: $F_{1,1862} = 37.68$, $P<0.001$; species×breeding regime $F_{9,1862} = 3.46$, $P<0.01$; Fig. 1b, Table S1). Other factors in the model were non-significant (results not shown). Inbreeding on average decreased viability by 33.1 and 30.0% across species in males and females respectively (Fig. 1b). The level of inbreeding depression differed between species ($F_{9,120} = 21.29$, $P<0.001$, Fig. 1). Neither sex ($F_{1,120} = 0.26$, NS) nor the interaction species×sex impacted on the level of inbreeding depression ($F_{9,120} = 0.52$, NS).

Heat knock down resistance: Line (nested within species and breeding regime), species, and the breeding regime×sex and species×sex interactions significantly affected heat knock down resistance (line: $F_{69,1512} = 1.93$, $P<0.001$; species: $F_{9,1512} = 69.90$, $P<0.001$; breeding regime×sex $F_{9,1512} = 7.73$, $P<0.01$; species×sex: $F_{9,1512} = 1.96$, $P<0.05$; Fig. 1c, Table S1). Breeding regime, sex and the remaining interactions between the main effects did not affect heat knock down resistance (results not shown). Heat knock down time was on average 4.2% and 6.0% lower in inbred compared to outbred flies (Fig. 1c). The level of inbreeding depression differed between males and females ($F_{1,85} = 6.32$, $P<0.05$). The impact of inbreeding was not affected by species ($F_{9,85} = 0.67$, NS) nor by the interaction species×sexes ($F_{9,85} = 1.12$, NS) (Fig. 1c).

Chill coma recovery time: Line (nested within species and breeding regime), sex, species and the species×breeding regime and species×sex interactions all significantly affected egg-to-adult viability in chill coma recovery time (line: $F_{77,1704} = 1.97$, $P<0.001$; sex: $F_{1,1704} = 11.18$, $P<0.001$; species: $F_{9,1704} = 116.99$, $P<0.001$; species×breeding regime $F_{9,1704} = 3.38$, $P<0.01$; species×sex: $F_{9,1704} = 5.11$, $P<0.001$; Fig. 1d, Table S1). Remaining factors in the model were non-significant (results not shown). On average inbreeding increased chill coma recovery time (decreased cold resistance) by 6.1 and 4.6% in males and females, respectively. The impact of inbreeding differed between species ($F_{9,108} = 6.46$, $P<0.001$, Fig. 1d). Neither sex ($F_{1,108} = 0.36$, NS) nor the interaction species×sex impacted on the level of inbreeding depression ($F_{9,108} = 1.61$, NS).

CTmin: Line (nested within species and breeding regime), sex, species and breeding regime all significantly affected CTmin (line: $F_{81,2063} = 1.88$, $P<0.001$; sex: $F_{1,2063} = 6.66$, $P<0.01$; species: $F_{9,2063} = 149.60$, $P<0.001$; breeding regime: $F_{1,2063} = 6.84$, $P<0.05$; Fig. 1e, Table S1). Remaining factors in the model were non-significant (results not shown). Inbred females and males had on average a CTmin that was respectively 5.3 and 8% higher (less cold resistant) compared to outbred females and males. The level of inbreeding depression differed between species ($F_{9,116} = 3.00$, $P<0.01$), whereas it was the same in males and females of the same species ($F_{1,116} = 1.13$, NS). The interaction species×sex was significant ($F_{9,116} = 2.05$, $P<0.05$) (Fig. 1e).

Desiccation resistance: Line (nested within species and breeding regime), sex, species, breeding regime (inbred or outbred) and the species×sex and the species×sex×breeding regime interactions all significantly affected desiccation resistance (line: $F_{74,1708} = 6.06$, $P<0.001$; sex: $F_{1,1708} = 46.64$, $P<0.001$; species: $F_{9,1708} = 44.89$, $P<0.001$; breeding regime: $F_{1,1708} = 4.93$, $P<0.05$; species×sex: $F_{9,1708} = 34.30$, $P<0.001$; species×sex×breeding regime: $F_{9,1708} = 2.23$, $P<0.05$; Fig. 1f, Table S1). The remaining factors in the model were non-significant (result not shown). Inbreeding on average decreased desiccation resistance by 9.0 and 5.1% across species in males and females respectively (Fig. 1f). The level of inbreeding depression did not differ between species ($F_{9,101} = 1.73$, NS) nor sexes ($F_{1,101} = 1.17$, NS), but was affected by the interaction species×sex ($F_{9,101} = 2.03$, $P<0.05$) (Fig. 1f).

Inbreeding effects in widespread and tropical restricted species

For developmental time more inbreeding depression was observed in tropical compared to widespread species for both sexes (females: $F_{1,8} = 9.71$, $P<0.05$; males: $F_{1,8} = 14.86$, $P<0.01$, Fig. 1a). A non-significant trend in the same direction was observed for egg-to-adult viability (females: $F_{1,8} = 1.22$, $P<0.10$; males: $F_{1,8} = 1.07$, $P<0.20$, Fig. 1b). For the remaining traits, no suggestion of a difference in the impact of inbreeding on tropical and widespread species was evident (Fig. 1c–f).

Splitting data into life history (developmental time and egg-to adult viability) and stress resistance traits, we found that tropical species suffered on average more from inbreeding depression in life history traits compared to widespread species ($P<0.001$; Fig. 2a) whereas no difference was observed for stress resistance traits (NS, Fig. 2b). When considering all traits together, there was no effect of species group on the level of inbreeding depression (NS).

Discussion

In this study we investigated the effect of inbreeding in 5 tropical restricted and 5 widespread species of *Drosophila*. The two species groups have a different evolutionary history and they occupy different habitats and thus are likely to have experienced different selection pressures. We tested the hypothesis that, due to likely differences in the genetic architecture of the two species groups, the level of inbreeding depression differs between widespread and tropical species groups. For the stress resistance traits investigated we found no evidence to support this hypothesis (Fig. 2b), but for the life history traits we found that tropical species suffered more from inbreeding depression compared to widespread species, although large species differences were observed (Fig. 2a). The higher inbreeding depression in tropical species for life history traits is partly due to the fact that for two widespread species inbred flies have higher fitness compared to outbred flies (Fig. 1a,b and 2a). Purging of deleterious recessive alleles during the process

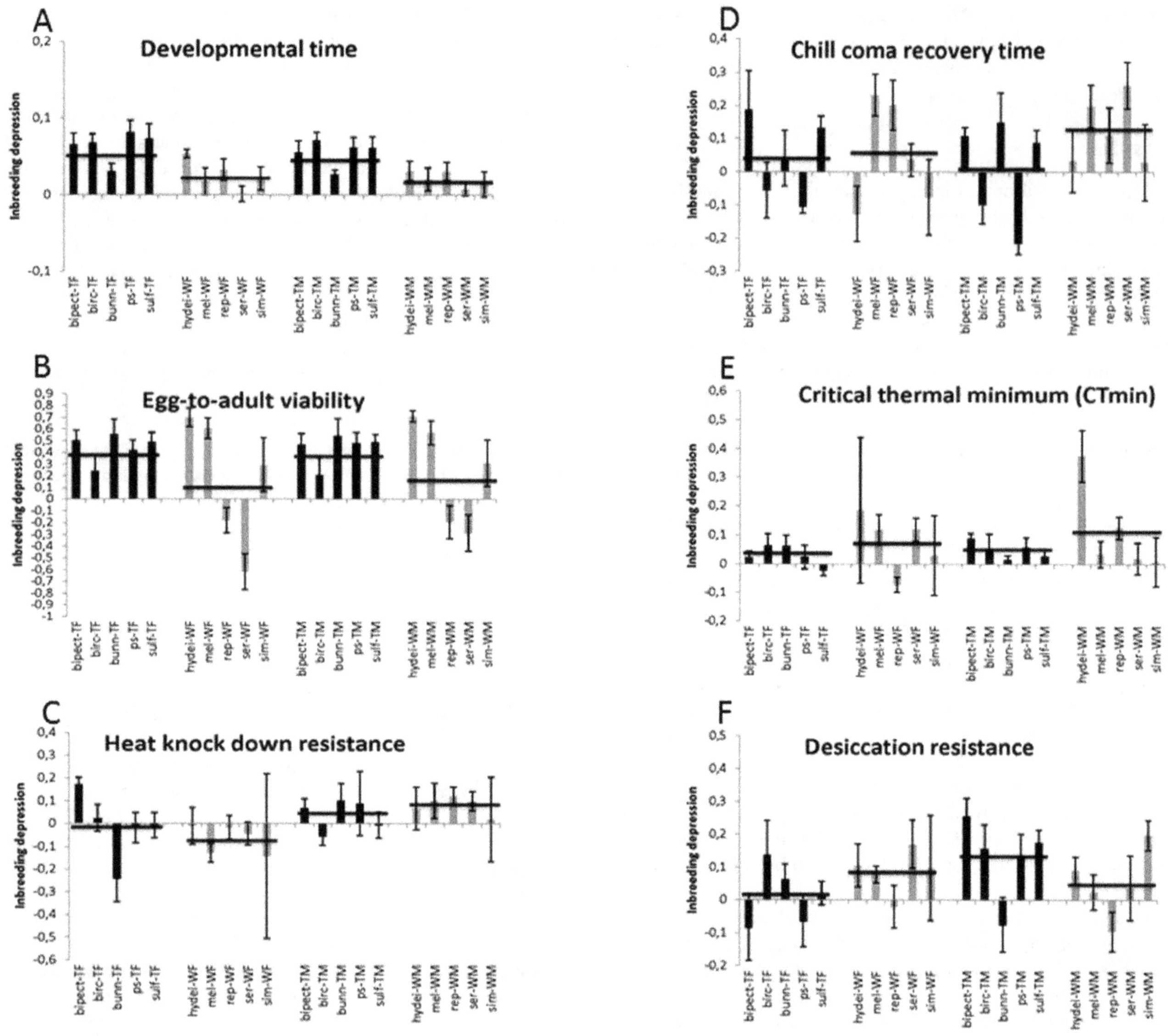

Figure 1. Inbreeding depression for each of 6 traits (± SE) (a: 'Developmental time', b: 'Egg-to-adult viability', c: 'Heat knock down resistance', d: 'Chill coma recovery time', e: 'Critical thermal minimum (CTmin)', f: 'Desiccation resistance'). Data are split into 4 groups: tropical females (TF), tropical males (TM), widespread females (WF) and widespread males (WM). Horisontal lines represent averages for each of the 4 groups (TF, TM, WF, WM). The species are: *D. bipectinata* (bipect), *D. birchii* (birc), *D. bunnanda* (bunn), *D. hydei* (hydei), *D. melanogaster* (mel), *D. pseudoananassae* (ps), *D. repleta* (rep), *D. serrata* (ser), *D. simulans* (sim) and *D. sulfurigaster* (sulf). See 'Materials and methods' for a description of each assay.

of inbreeding could explain this result if the efficiency of purging were species specific. However, given the fast rate of inbreeding (full sib mating) used here purging is not likely to be efficient [17,18]. Thus despite a large species effect and a relatively low number of species investigated the observation that inbreeding depression on average tend to be higher for life history traits in tropical compared to widespread species is likely caused by a distinct genetic architecture in the two species groups.

For cold and desiccation resistance, previous studies have provided evidence suggesting that the genetic architecture is indeed different between the two species groups for these traits [10,11]. Tropical species have, in contrast to widespread species, very low heritability for cold and desiccation resistance [10,11]. Two opposing hypotheses have been suggested to explain this result: 1) loci influencing those two traits have decayed (loss of function) in the tropical species [19], the reason being that they are never in use in a humid and warm habitat, and 2) loci influencing cold and desiccation resistance have been under strong directional selection in the near past so that alleles have been fixed in the tropical species. With the decay hypothesis, where loci involved in desiccation and cold resistance would be non-functional we would expect no inbreeding depression in tropical species for cold and desiccation resistance. Our results fail to support this hypothesis since no difference in levels of inbreeding depression between tropical and widespread species were observed for the two traits (Fig. 1d,e,f). According to the second hypothesis we would expect more inbreeding depression in tropical species for desiccation and cold resistance. This is because selection theory and empirical evidence suggests that inbreeding depression will be more severe for traits shaped by (past) directional selection [20–22]. Our data

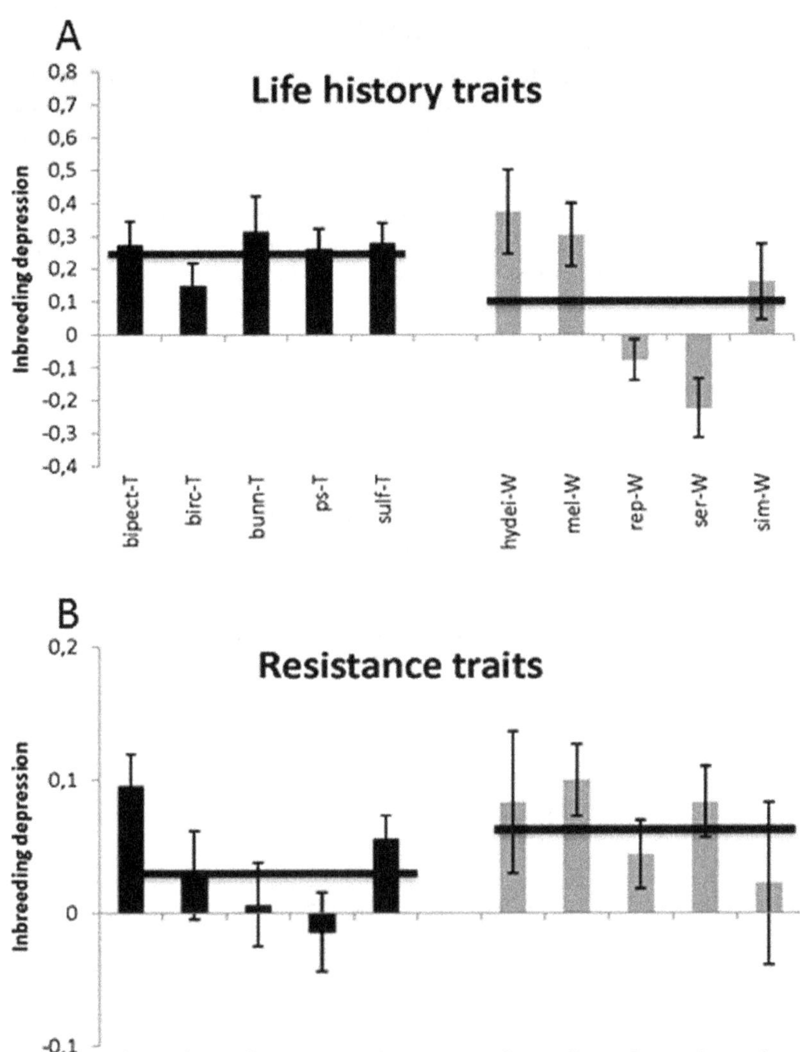

Figure 2. Average inbreeding depression for the 10 species based on a) average inbreeding depression of the two investigated life history traits for each species (± SE) and b) average inbreeding depression of the four investigated stress resistance traits for each species (± SE). Black bars are tropical species and grey bars are widespread species. Horizontal black lines represent averages for the tropical and widespread species. The species are: *D. bipectinata* (bipect), *D. birchii* (birc), *D. bunnanda* (bunn), *D. hydei* (hydei), *D. melanogaster* (mel), *D. pseudoananassae* (ps), *D. repleta* (rep), *D. serrata* (ser), *D. simulans* (sim) and *D. sulfurigaster* (sulf). Symbols: 'T' = 'tropical' and 'W' = 'widespread'.

do also not support the hypothesis that low additive genetic variance in tropical species can be explained by depletion of additive genetic variance induced by directional selection, since we do not see a difference in inbreeding depression in cold and desiccation resistance between tropical and widespread species (Fig. 1d,e,f).

Our results show that inbreeding depression was strongest for the trait egg-to-adult viability and that tropical species on average tended to suffer relatively more from inbreeding depression in egg-to-adult viability and developmental time. If we assume that the level of inbreeding depression is more severe in traits under directional selection for which there is strong empirical evidence [20–22], this may indicate that 1) among the six traits we investigated, egg-to-adult viability has been under strongest directional selection in the past, and 2) egg-to-adult viability on average is under stronger selection in tropically restricted species compared to widespread species.

Effective population sizes are expected to influence the effect of inbreeding, since at low effective population sizes more slightly deleterious variants may be segregating due to strong drift and weak selection. One might speculate that tropical species in general would have low effective population sizes due to small distribution ranges, and therefore predicted to show stronger effects of inbreeding. On the other hand historical inbreeding might also purge deleterious alleles and thus reduce deleterious consequences of inbreeding [23]. The net outcome of effects of effective population size on fitness consequences of consecutive full sib mating is therefore difficult to predict. We have no information on effective population sizes in the investigated species but their high fertility and observations done during sampling of the populations suggest that census sizes are very high for all species.

High and similar heritabilities for morphological traits in the investigated species provide indirect evidence for this claim [10,11].

Inbreeding depression is commonly observed to be more severe for traits closely associated with fitness. Thus, the observation that egg to adult viability is the traits mostly affected by inbreeding support previous results; our data provide evidence from multiple species that there is a fundamental difference between life history and stress response traits with respect to the underlying genetic variation that gives rise to inbreeding depression.

The environment is expected to change dramatically in the future. Temperatures will increase and be more variable and many other climatic variables will change [24]. A central question in regard to this is whether animals and plants are able to respond to these changes through evolutionary and plastic adaptations. As discussed above, tropical species have been hypothesized to be especially prone to population reductions and extinctions due to phylogenetic constraints, lower basal resistance levels and lower adaptive potential in ecological key traits [9–11,25]. For life history traits our results suggest that this will be reinforced by more inbreeding depression in tropical species. The variation in impact of inbreeding between species and lineages within and between species groups observed in this study however reveals the complexity of inbreeding effects; to fully understand the effect of species group on levels of inbreeding depression we suggest testing more species in multiple and ecologically relevant environments.

Supporting Information

Table S1 Mean trait values of control ('c') and inbred ('i') flies (with standard errors in parentheses) for each sex, species and trait observed in this study. 'N' equals the number of inbred lines tested (in parentheses the number of lines available after the 4 generations of full sub mating). N equals 3 for all traits in both males and females in the control lines. The species are: *D. bipectinata* (bipect), *D. birchii* (birc), *D. bunnanda* (bunn), *D. hydei* (hydei), *D. melanogaster* (mel), *D. pseudoananassae* (ps), *D. repleta* (rep), *D. serrata* (ser), *D. simulans* (sim) and *D. sulfurigaster* (sulf). Symbols: 'T' = 'tropical' and 'W' = 'widespread'.

Acknowledgments

We are grateful to D. Andersen, M. R. Hansen, K. Noreikienė, M. Ondrésik, V. Kellermann, J. Wit and A. B. Stentebjerg for laboratory assistance and to R. Goode and J. Shirriffs for field collections.

Author Contributions

Conceived and designed the experiments: JSB TNK. Performed the experiments: JSB TNK. Analyzed the data: JSB TNK. Contributed reagents/materials/analysis tools: AAH CS. Wrote the paper: JSB AAH CS VL TB TNK.

References

1. Hoffmann AA, Parsons PA (1997) Extreme environmental change and evolution. Cambridge: Cambridge University Press. xii,259p p.
2. Kristensen TN, Sorensen AC (2005) Inbreeding - lessons from animal breeding, evolutionary biology and conservation genetics. Animal Science 80: 121–133.
3. Angilletta MJ (2009) Thermal adaptation : a theoretical and empirical synthesis. Oxford: Oxford University Press. xii, 289 p., 282 p. of plates p.
4. Chown SL, Terblanche JS (2007) Physiological diversity in insects: Ecological and evolutionary contexts. In: Simpson SJ, editor. Advances in Insect Physiology, Vol 33. pp. 50–152.
5. Ghalambor CK, Huey RB, Martin PR, Tewksbury JJ, Wang G (2006) Are mountain passes higher in the tropics? Janzen's hypothesis revisited. Integrative and Comparative Biology 46: 5–17.
6. Kellermann V, Loeschcke V, Hoffmann AA, Kristensen TN, Fløjgaard C, et al. (2012) Phylogenetic constraints in key functional traits behind species' climate niches: pattersn of desiccation and cold tolerance across 95 Drosophila species. Evolution, In press.
7. Hoffmann AA, Watson M (1993) Geographical Variation in the Acclimation Responses of Drosophila to Temperature Extremes. American Naturalist 142: S93–S113.
8. Overgaard J, Kristensen TN, Mitchell KA, Hoffmann AA (2011) Thermal Tolerance in Widespread and Tropical Drosophila Species: Does Phenotypic Plasticity Increase with Latitude? American Naturalist 178: S80–S96.
9. Kellermann V, Overgaard J, Hoffmann AA, Flojgaard C, Svenning J-C, et al. (2012) Upper thermal limits of Drosophila are linked to species distributions and strongly constrained phylogenetically. Proceedings of the National Academy of Sciences of the United States of America 109: 16228–16233.
10. Hoffmann AA, Hallas RJ, Dean JA, Schiffer M (2003) Low potential for climatic stress adaptation in a rainforest Drosophila species. Science 301: 100–102.
11. Kellermann V, van Heerwaarden B, Sgro CM, Hoffmann AA (2009) Fundamental Evolutionary Limits in Ecological Traits Drive Drosophila Species Distributions. Science 325: 1244–1246.
12. van Heerwaarden B, Kellermann V, Schiffer M, Blacket M, Sgro CM, et al. (2009) Testing evolutionary hypotheses about species borders: patterns of genetic variation towards the southern borders of two rainforest Drosophila and a related habitat generalist. Proceedings of the Royal Society B-Biological Sciences 276: 1517–1526.
13. Nei M, Maruyama T, Chakraborty R (1975) BOTTLENECK EFFECT AND GENETIC-VARIABILITY IN POPULATIONS. Evolution 29: 1–10.
14. Kristensen TN, Loeschcke V, Bilde T, Hoffmann AA, Sgro C, et al. (2011) No Inbreeding Depression for Low Temperature Developmental Acclimation across Multiple Drosophila Species. Evolution 65: 3195–3201.
15. Overgaard J, Kristensen TN, Sorensen JG (2012) Validity of Thermal Ramping Assays Used to Assess Thermal Tolerance in Arthropods. Plos One 7.
16. Lande R, Schemske DW (1985) The Evolution of Self-Fertilization and Inbreeding Depression in Plants .1. Genetic Models. Evolution 39: 24–40.
17. Reed DH, Lowe EH, Briscoe DA, Frankham R (2003) Inbreeding and extinction: Effects of rate of inbreeding. Conservation Genetics 4: 405–410.
18. Wang JL, Hill WG, Charlesworth D, Charlesworth B (1999) Dynamics of inbreeding depression due to deleterious mutations in small populations: mutation parameters and inbreeding rate. Genetical Research 74: 165–178.
19. Hoffmann AA (2010) Physiological climatic limits in Drosophila: patterns and implications. Journal of Experimental Biology 213: 870–880.
20. DeRose MA, Roff DA (1999) A comparison of inbreeding depression in life-history and morphological traits in animals. Evolution 53: 1288–1292.
21. Falconer DS, Mackay TFC (1996) Introduction to quantitative genetics: Longman Group Limited. xv + 464 p.
22. Lynch M, Walsh B (1998) Genetics and analysis of quantitative traits: Sinauer Associates, Inc. {a}, 108 North Main Street, Sunderland, Massachusetts 01375, USA. xvi+980p p.
23. Swindell WR, Bouzat JL (2006) Reduced inbreeding depression due to historical inbreeding in Drosophila melanogaster: evidence for purging. Journal of Evolutionary Biology 19: 1257–1264.
24. Solomon S, Intergovernmental Panel on Climate Change., Intergovernmental Panel on Climate Change. Working Group I. (2007) Climate change 2007 : the physical science basis : contribution of Working Group I to the Fourth Assessment Report of the Intergovernmental Panel on Climate Change. Cambridge; New York: Cambridge University Press. viii, 996 p. p.
25. Deutsch CA, Tewksbury JJ, Huey RB, Sheldon KS, Ghalambor CK, et al. (2008) Impacts of climate warming on terrestrial ectotherms across latitude. Proceedings of the National Academy of Sciences of the United States of America 105: 6668–6672.

8

Phylogeny-Based Comparative Methods Question the Adaptive Nature of Sporophytic Specializations in Mosses

Sanna Huttunen[1]*, Sanna Olsson[2], Volker Buchbender[3], Johannes Enroth[4], Lars Hedenäs[5], Dietmar Quandt[3]

1 Department of Biology, University of Turku, Turku, Finland, **2** Department of Agricultural Sciences, University of Helsinki, Helsinki, Finland, **3** Nees-Institute for Biodiversity of Plants, University of Bonn, Bonn, Germany, **4** Department of Biosciences and Botanical Museum, University of Helsinki, Helsinki, Finland, **5** Department of Cryptogamic Botany, Swedish Museum of Natural History, Stockholm, Sweden

Abstract

Adaptive evolution has often been proposed to explain correlations between habitats and certain phenotypes. In mosses, a high frequency of species with specialized sporophytic traits in exposed or epiphytic habitats was, already 100 years ago, suggested as due to adaptation. We tested this hypothesis by contrasting phylogenetic and morphological data from two moss families, Neckeraceae and Lembophyllaceae, both of which show parallel shifts to a specialized morphology and to exposed epiphytic or epilithic habitats. Phylogeny-based tests for correlated evolution revealed that evolution of four sporophytic traits is correlated with a habitat shift. For three of them, evolutionary rates of dual character-state changes suggest that habitat shifts appear prior to changes in morphology. This suggests that they could have evolved as adaptations to new habitats. Regarding the fourth correlated trait the specialized morphology had already evolved before the habitat shift. In addition, several other specialized "epiphytic" traits show no correlation with a habitat shift. Besides adaptive diversification, other processes thus also affect the match between phenotype and environment. Several potential factors such as complex genetic and developmental pathways yielding the same phenotypes, differences in strength of selection, or constraints in phenotypic evolution may lead to an inability of phylogeny-based comparative methods to detect potential adaptations.

Editor: Pär K. Ingvarsson, University of Umeå, Sweden

Funding: This research was supported by a Marie Curie Intra-European Fellowship (MEIF-CT-2005-009452; years 2005–2008 to SH); a Marie Curie Reintegration Grant (PERG03-GA-2008-230953; years 2008–2011 to SH) within the 6th European Community Framework Program; a post doctoral researcher grant from the Academy of Finland (project no. 121373; years 2008–2011 to SH); researcher exchange grants by the Academy of Finland/German Academy Exchange Service (DAAD) to JE, DQ, VB, SO and DAAD/The Swedish Foundation for International Cooperation in Research and Higher education (STINT) to VB, LH, SH, SO, DQ; the Swedish Research Council (Vetenskapsrådet project no. 621 2003 3338; to LH); the Deutsche Forschungsgemeinschaft (DFG QU 153/3-1, 153/3-2 to DQ); and SYNTHESYS grants financed by the European Community Research Infrastructure Action under the FP6 'Structuring the European Research Area' Programme (http://www.synthesys.info) to VB, JE, SO). The funders had no role in study design, data collection and analysis, decision to publish, or preparation of the manuscript.

Competing Interests: The authors have declared that no competing interests exist.

* E-mail: shuttu@utu.fi

Introduction

Since Darwin's *Origin of Species* [1], correlations between ecology and morphological traits in organisms have led biologists to postulate that adaptive diversifications are the driving force for morphological evolution. Speculations on the adaptive evolution of morphological traits still tend to be common, especially in papers dealing with evolutionary history and morphological evolution. Based on field observations of covariation between phenotypes and environments, for example, xerophytic plants in a Mediterranean-type vegetation, succulent plants in arid environments, and many specialized morphological structures in aquatic plants are called classical examples of true adaptations in many botany text books, because these traits may aid survival in the respective environments [2,3]. Only rarely, however, evidence emerges on evolutionary processes that result in the phenotypes. For traits that have evolved in response to environmental selection pressure, the shift in ecology should take place before the shift in phenotype, but usually the evolutionary order between these shifts is unknown (see, however [4,5]). Thus, as long as it is unclear if natural selection by the habitat is the driving force behind the evolution of observed traits, it is questionable whether these are adaptations in the strictest sense [6].

Testing the hypothesis of adaptive diversifications across a wide taxonomic scale may be rather challenging. Studies dealing with the origin of the adaptations are most often restricted to showing adaptive evolution within or between populations or, sometimes, between two or very few species. Most genetic methods commonly used for detecting adaptive evolution are non-applicable or will require rather extensive research efforts if a group with potentially adaptive traits involves a large number of species scattered among taxonomically diverse groups. However, phylogenetic approaches utilizing molecular phylogenies and information on distribution of traits among terminals allow detection of correlated evolution between ecology and morphological traits [7–10]. Correlated evolution of ecological and morphological shifts often appears as evidence for adaptive evolution [11–13]. Correlated evolution alone, however, does not directly confirm natural selection or fitness differences between the phenotypes, key factors that are needed for adaptive evolution. Phylogenetic approaches can, however, reveal potential adaptations by showing correlated evolution and a relative order of evolutionary changes in ecology and phenotypes. They can thus serve to point out a potential evolutionary link between a change in environment and a shift in phenotype and to detect whether the change in environment was followed by a shift in morphology.

Mosses growing as epiphytes on other plants form a taxonomically diverse group including species from most major lineages among the division Bryophyta. Epiphytes are especially common and are scattered among almost all families of pleurocarpous mosses, the crown clade of the subphylum including about half of the total 10,000 moss species in the world [14,15]. The pleurocarpous mosses comprise typically perennial mosses with creeping stems and sporophyte-producing lateral branches; various pleurocarpous moss lineages have also repeatedly and independently conquered epiphytic habitats [16–20]. The major radiation among the group took place more than 165 to131 mya ago, during the late Jurassic and Cretaceous [21]. One hypothesis is that at least one of the driving forces behind the major pleurocarp radiation is the evolution of epiphytic life forms, especially on woody angiosperms [16,22,23]. This time-frame of pleurocarp evolution coincides with the radiation of angiosperms [24].

Despite their diverse origin, epiphytic mosses tend to share certain morphological characteristics. In particular, their sporophytes are often reduced to various degrees: the seta that carries the spore capsule is short, the capsule is orthotropous, making the capsule horizontal on vertical substrates, the peristome that regulates spore dispersal at the mouth of the capsule is reduced and is capable of only weak hygroscopic movement (Fig. 1). The sporophyte reductions seem to be linked with xerophytic habitats [25,26], and epiphytes grow in a special form of xerophytic habitat. Drought in epiphytic habitats has also been assumed to result in trade-offs in evolution of sexual systems and life history traits in epiphytic lineages of the liverwort genus *Radula* [27]. As early as 1908, Grout [28] observed in mosses an association between these specialized sporophytic characters and the epiphytic habitat and suggested this phenomenon to be an adaptation. The specializations appear indeed to be very common and easy to find among pleurocarps. For example, among the 439 pleurocarpous moss species studied by Hedenäs [26], some 10% of species with capsules have an erect capsule and a peristome with some reduced traits. As sporophyte structures are responsible for producing and dispersing spores, changes in a sporophyte may also impact strongly upon fitness. A plausible explanation for the repeated evolution of similar morphological traits under similar environmental conditions is therefore that they are beneficial for the survival of individuals in those environments.

We aim to test here whether evolution of morphological specialization in mosses is correlated with their shifts into exposed or epiphytic habitats. We use data from the pleurocarpous moss families Neckeraceae and Lembophyllaceae that both show several independent transitions to epiphytic habitats and a considerable degree of variation in their sporophytic traits [20]. Based on a connection that was noted in earlier studies [17,19,20,25,26,28], we selected for further analysis eight candidate traits. Their evolution seemed to be connected with a shift to epiphytic or exposed habitats, and we tested correlated evolution between the traits and a habitat shift using a Bayesian approach [10]. We tested the order of the character-state changes for morphology and habitat shifts by contrasting ancestral state reconstructions and by comparing the fit of two evolutionary models that had different transition rates for dual character states. Based on the results, we will distinguish morphological specializations that are potentially adaptations to epiphytic and exposed habitats, discuss other possible explanations for convergent evolution, and evaluate the utility of this method in detecting adaptive evolution in general.

Results

Ancestral State Reconstructions for Habitat Shifts and Morphological Traits

Ancestral state reconstructions favored with high probability a scenario that the ancestor of the Lembophyllaceae – Neckeraceae clade (node I, Fig. 2) as well as the ancestor of all Neckeraceae species (node II, Fig. 2) lived on soil or in an unexposed habitat (Fig. 2, Supporting information, Appendix S2). These ancestors had higher probabilities for plesiomorphic character states for seven morphological traits of the total eight studied. Only character 2 (c2), the operculum shape, showed a character-state shift between nodes I and II (Fig. 2, Supporting information, Appendix S2; see Supporting information, Appendix S1 for a list of all characters). The ancestor of the Lembophyllaceae – Neckeraceae clade thus had a conical to rostrate operculum, i.e a plesiomorphic state, whereas the ancestor of all Neckeraceae species had a derived state with a long-rostrate operculum (Fig. 2, Supporting information, Appendix S2).

Shifts to exposed epilithic or epiphytic habitats have occurred in five lineages (Fig. 2; Table 1, Supporting information, Appendix S2): 1) in the *Dolichomitriopsis diversifolia - Bryolawtonia vancouveriensis* clade (Lembophyllaceae, node A), 2) in the core Lembophyllaceae (node B), 3) in the *Neckera* clade (Neckeraceae, node C), 4) in the *Porotrichum frahmii* – *Porotrichodendron superbum* clade (Neckeraceae, node D), and 5) in the *Pinnatella* clade (Neckeraceae, node E). Despite higher probabilities for habitats on exposed or epiphytic substrates, Bayes Factors (BFs) lend positive support (BF >2) to a derived character state only at nodes B, C, D and two lineages within the *Pinnatella* clade (Fig. 2, Table 1). For none of the morphological traits studied did transitions to derived states appear always in the same nodes along with a shift in habitat (Fig. 2; Supporting information, Appendix S2).

Correlated Evolution between Habitat Shift and Morphology

For four morphological traits, Reversible Jumping Markov Chain Monte Carlo (RJ MCMC) applying a dependent model (D; a model where morphological and habitat evolution are dependent on each other), the harmonic mean of log-likelihood scores sampled during the chain was significantly higher than for the chain with an independent model (I; a model where morphology and habitat evolve independently) (Table 2). These four traits were peristome orientation in a dry condition (c3), height of endostomial basal membrane (c5), and endostomial cilia (c6), and seta length (c8; Table 2). BF strongly favored the D model of evolution for these (BF >5; Table 2). The D models were also visited during the chain more frequently than expected (Table 2), which also supported their better fit to the data. For two morphological traits: peristome orientation in a dry condition and endostomial cilia, I models were not visited at all during the best RJ MCMC, lending the strongest possible support for correlated evolution of morphology and habitat preferences (Table 2).

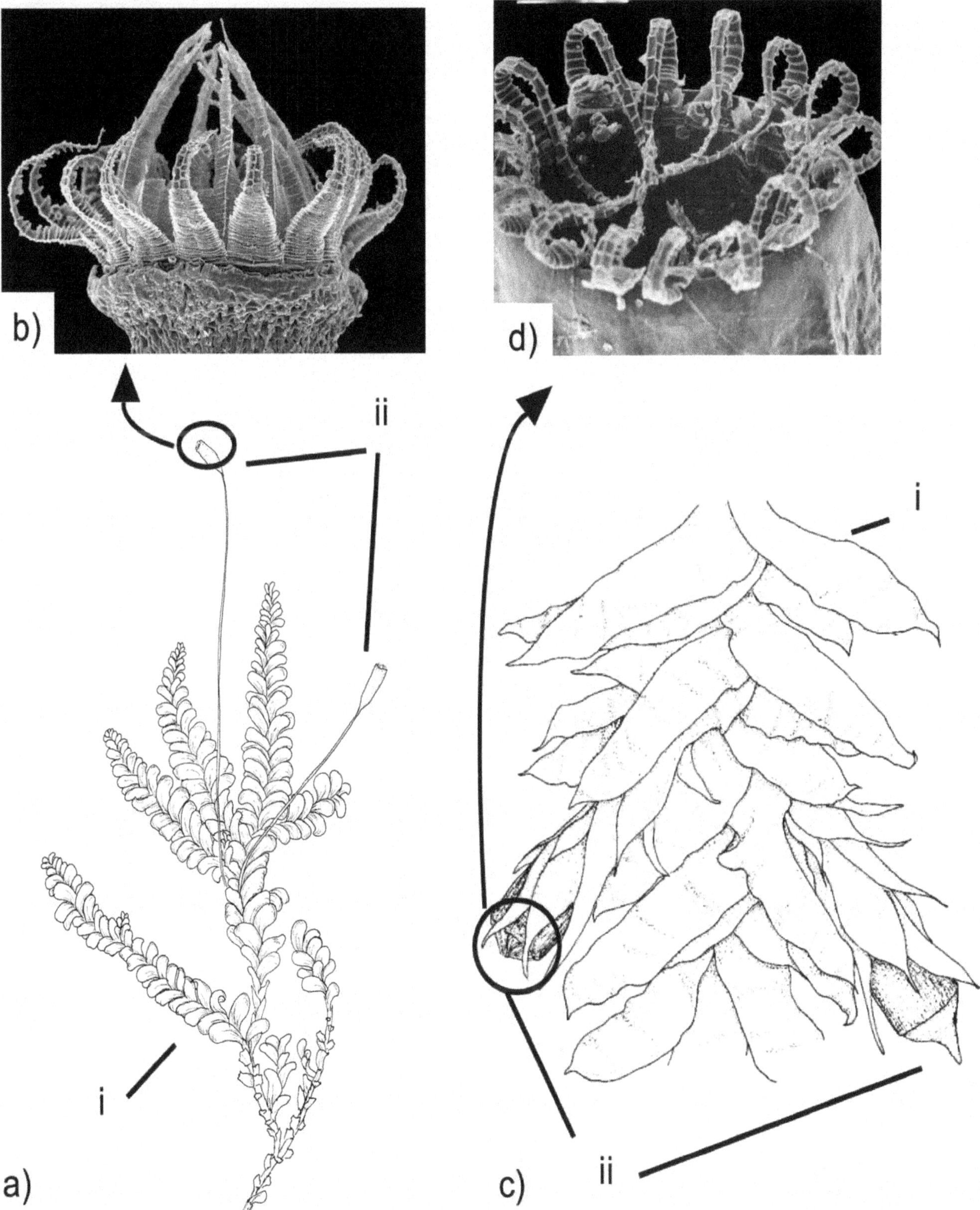

Figure 1. Sporophyte structure in Neckeraceae. Example of a perfect and a specialized sporophyte structure in Neckeraceae. a) *Homalia trichomanoides* gametophyte (i) and sporophytes (ii) with long setas and slightly inclined capsules; b) SEM view of well-developed hypnalean peristome in *H. trichomanoides*. c) *Neckera pennata* gametophyte (i) and sporophytes (ii) that have short setas immersed in perichaetial leaves and upright capsules; d) SEM view of reduced peristome in *N. pennata*. Pictures reprinted with permission of M. S. Ignatov and E. Ignatova.

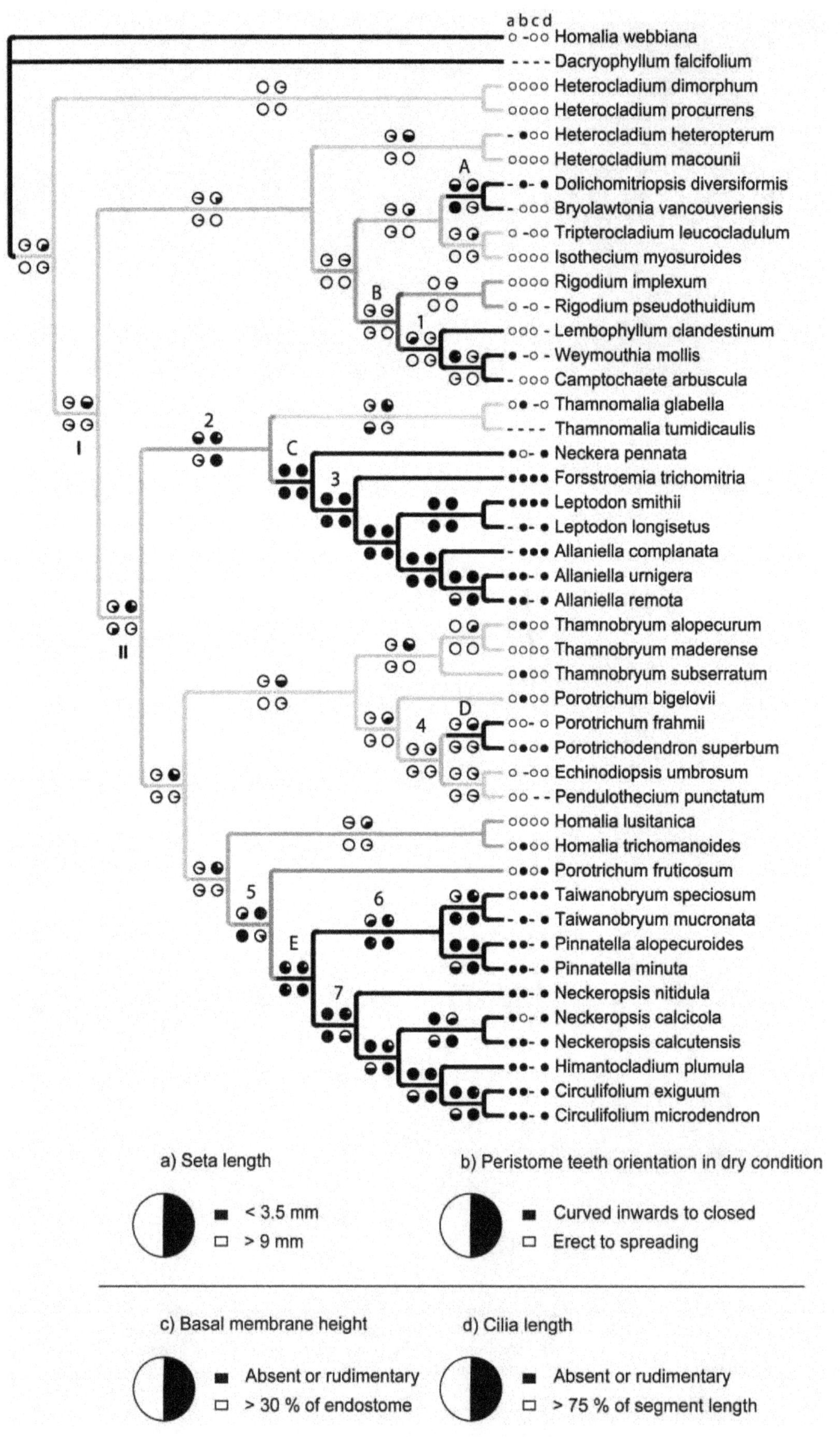
a b c d
Homalia webbiana
Dacryophyllum falcifolium
Heterocladium dimorphum
Heterocladium procurrens
Heterocladium heteropterum
Heterocladium macounii
Dolichomitriopsis diversiformis
Bryolawtonia vancouveriensis
Tripterocladium leucocladulum
Isothecium myosuroides
Rigodium implexum
Rigodium pseudothuidium
Lembophyllum clandestinum
Weymouthia mollis
Camptochaete arbuscula
Thamnomalia glabella
Thamnomalia tumidicaulis
Neckera pennata
Forsstroemia trichomitria
Leptodon smithii
Leptodon longisetus
Allaniella complanata
Allaniella urnigera
Allaniella remota
Thamnobryum alopecurum
Thamnobryum maderense
Thamnobryum subserratum
Porotrichum bigelovii
Porotrichum frahmii
Porotrichodendron superbum
Echinodiopsis umbrosum
Pendulothecium punctatum
Homalia lusitanica
Homalia trichomanoides
Porotrichum fruticosum
Taiwanobryum speciosum
Taiwanobryum mucronata
Pinnatella alopecuroides
Pinnatella minuta
Neckeropsis nitidula
Neckeropsis calcicola
Neckeropsis calcutensis
Himantocladium plumula
Circulifolium exiguum
Circulifolium microdendron
A
B
C
D
E
1
2
3
4
5
6
7
I
II
a) Seta length
< 3.5 mm
> 9 mm
b) Peristome teeth orientation in dry condition
Curved inwards to closed
Erect to spreading
c) Basal membrane height
Absent or rudimentary
> 30 % of endostome
d) Cilia length
Absent or rudimentary
> 75 % of segment length

Figure 2. Ancestral character state reconstruction for habitat preference and four morphological traits. Ancestral character state reconstruction for habitat preference and four morphological traits that evolution may correlate with the habitat shifts among Neckeraceae and Lembophyllaceae. The color of the branches in the inferred Bayesian topology represents two states of the habitat: on soil/unexposed (light gray) and epiphytic/exposed (for branches with probability >0.95 = black). Branches with probability >0.90 but <0.95 for epiphytic/exposed habitats are with dark gray color. Probabilities for morphological ancestral character state are shown as pie diagrams in the nodes. BayesFactor (BF) support for epiphytic/exposed habitat preference is shown below branches. For morphological traits BF for a derived character state is indicated with color of pie diagrams: BF <2 light gray, BF >2 dark gray; and BF >5 with black (see Table 1). Pie diagrams along branches are in the same order as in the legend showing their character states (a–d). Character states for terminals are stated before the taxon name. Dash (−) indicates missing or inapplicable data. Nodes A–E with show lineages with shifts to epiphytic or other exposed habitats.

Change First in Habitat or in Morphology?

In ancestral character state reconstructions, a shift to a derived morphological character state appears before the shift to exposed or epiphytic habitats at least in some lineages for all correlated traits (Fig. 2, Supporting information, Appendix S2). Derived morphology was frequently gained before habitat shift especially in nodes C and E. However, even if probabilities for derived states were already higher before reaching the nodes C and E, the BFs mostly lend support for shifts only at those nodes or even after (Fig. 2, Table 1; see e.g. c8). Derived character states were gained with at least positive BF support at the same node with shift in habitats or after the shift for the following traits: at node A, cilia length (c6); at nodes B and D, none of the correlated traits; at node C, dry peristome (c3), basal membrane height (c5), cilia length (c6), and seta length (c8); and at node E, dry peristome (c3), cilia length (c6), and seta length (c8) (Fig. 2, Table 1).

Rate coefficients for dual character state change indicated that the shift in habitats occurred before the change in morphology for three of the correlated traits, height of endostomial basal membrane (c5) and cilia (c6), and seta length (c8). For these, rate coefficients q12 (change in habitat preference but not in morphology) were significantly larger than q13 (change in morphology but not in habitat preference) (Table 3). For only one morphological trait, orientation of dry peristome (c3), q13 was smaller than q12 (Table 3). However, for all morphological traits, the fit of the evolutionary model where the rate coefficients were

Table 1. Bayes Factor (BF) support for four morphological traits and habitat preference.

node (Fig. 2)	node (Fig. S1)	Habitat preference			3) Dry peristome			5) Basal membrane			6) Cilia			8) Seta length		
		P(0)	P(1)	BF (1)	P(0)	P(1)	BF (1)	P(0)	P(1)	BF (1)	P(0)	P(1)	BF (1)	P(0)	P(1)	BF (1)
I	5	−25.44	−29.03	−7.17	−23.95	−25.57	−3.24	−8.53	−13.44	−9.81	−15.62	−21.93	−12.61	−15.44	−20.25	−9.61
II	16	−25.51	−28.45	−5.87	−24.61	−24.18	0.86	−8.34	−12.79	−8.91	−15.26	−20.96	−11.40	−15.36	−19.48	−8.23
A	10	−26.39	−26.04	0.71	–	–	–	–	–	–	−20.31	−15.79	**9.05**	−14.97	−18.05	−6.17
B	14	−28.37	−25.52	**5.70**	–	–	–	–	–	–	–	–	–	–	–	–
	15	−29.55	−25.37	**8.36**	–	–	–	–	–	–	–	–	–	−15.48	−16.63	−2.29
	17	−25.52	−25.76	−0.48	−25.12	−24.22	1.81	−9.11	−9.48	−0.73	–	–	–	−15.19	−17.73	−5.08
C	20	−27.67	−25.72	**3.89**	−26.84	−23.76	**6.16**	−12.05	−8.53	**7.03**	−18.23	−16.02	**4.43**	−18.04	−15.32	**5.44**
	21	−29.87	−25.56	**8.63**	−27.55	−23.77	**7.57**	−13.97	−8.80	**10.34**	−21.19	−15.59	**11.20**	−20.02	−15.09	**9.86**
	22	−30.30	−25.75	**9.11**	−28.35	−24.12	**8.45**	−14.58	−8.42	**12.32**	−22.07	−15.44	**13.26**	−21.01	−15.44	**11.14**
	23	−32.75	−25.47	**14.56**	−29.94	−23.91	**12.07**	−14.44	−8.55	**11.78**	−25.41	−15.80	**19.22**	−21.21	−15.01	**12.41**
	24	−31.41	−25.46	**11.91**	−29.19	−24.09	**10.21**	−14.05	−8.50	**11.09**	−23.91	−15.44	**16.94**	−21.42	−15.08	**12.67**
	25	−31.85	−25.48	**12.75**	−28.92	−24.22	**9.40**	−13.08	−8.22	**9.72**	−24.53	−15.46	**18.13**	−22.28	−15.08	**14.41**
D	32	−26.96	−25.61	**2.71**	–	–	–	–	–	–	–	–	–	–	–	–
	35	−25.48	−25.54	−0.12	−26.07	−23.99	**4.16**	−8.67	−12.96	−8.59	−17.91	−15.63	**4.57**	–	–	–
E	36	−25.64	−26.95	−2.62	−25.99	−23.62	−4.75	−8.62	−10.57	−3.91	−18.81	−15.39	**6.85**	−15.31	−17.18	−3.73
	37	−26.66	−25.78	1.75	−26.49	−24.16	**4.66**	−9.21	−9.25	−0.08	−20.33	−15.38	**9.89**	–	–	–
	38	−25.93	−26.56	−1.26	−26.30	−23.66	**5.27**	−8.93	−10.63	−3.40	−21.08	−15.60	**10.94**	−17.77	−15.31	**4.90**
	39	−27.32	−25.73	**3.18**	−26.13	−24.24	**3.79**	−9.61	−9.32	0.58	−20.87	−15.94	**9.84**	–	–	–
	40	−30.14	−25.44	**9.40**	−28.28	−24.13	**8.30**	−8.87	−9.29	−0.84	−23.52	−15.52	**15.99**	−19.67	−15.01	**9.32**
	41	−25.71	−26.47	−1.51	−26.18	−24.82	**2.72**	−8.78	−10.07	−2.58	−22.48	−15.38	**14.20**	−19.68	−15.12	**9.13**
	42	−25.71	−27.39	−3.37	−25.55	−23.69	**3.71**	−8.51	−10.42	−3.82	−22.69	−15.53	**14.32**	−20.41	−15.28	**10.27**
	43	−27.22	−25.57	**3.30**	−26.95	−23.97	**5.97**	−9.08	−9.81	−1.46	−23.39	−15.51	**15.76**	−21.03	−15.16	**11.74**
	44	−27.54	−25.65	**3.78**	−26.89	−23.86	**6.06**	−8.66	−9.95	−2.57	−23.36	−15.64	**15.44**	−21.26	−15.12	**12.26**

Bayes Factor (BF) support for ancestral states that earned higher probabilities at the nodes with probability >0.9 for a derived state and nodes I and II for four morphological traits and habitat preference. BFs are based on difference in harmonic means of likelihoods derived from two analyses, where character state at a given node is constrained to be either 0 or 1. BF >2 is considered as positive evidence and BF >5 as strong support for the character state's gaining the higher likelihood at the node. For character descriptions and coding of characters states and nodes see Fig. 2, Supporting information, Appendix S1, and Fig. S1. Probabilities for derived character states at each node are in the Supporting information Appendix S2 and in Fig. 2.

Table 2. Correlated evolution between change in morphology and shift to exposed epiphytic or epilithic habitat.

								Number of I visits					
Character	**D mean lnL**	**D st.dev. lnL**	**D max lnL**	**I mean lnL**	**I st.dev. lnL**	**I max lnL**	**BF**	**D1**	**D2**	**D3**	**posterior odds**	**posterior /prior odds**	**BF**
1) Post fertilization growth of perichaetial leaves	−43.94	0.33	−43.65	−43.76	0.06	−43.70	0.10	18	22	**25**	799.00	1.93	1.32
2) Operculum shape	−45.02	0.36	−44.61	−44.36	0.57	−43.97	−1.27	34	36	**30**	665.67	1.61	0.95
3) Dry peristome	−46.57	0.18	−46.43	−49.38	0.09	−49.31	**5.75**	**0**	3	6	NA	NA	**NA**
4) Spore size	−51.85	0.12	−51.72	−52.07	0.02	−52.06	0.67	**14**	15	10	1 427.57	3.45	2.48
5) Basal membrane	−32.14	0.20	−32.00	−34.79	0.20	−34.63	**5.26**	5	**1**	0	19 999.00	48.35	**7.76**
6) Cilia	−40.97	1.47	−40.02	−42.68	0.14	−42.55	**5.07**	**0**	1	3	NA	NA	**NA**
7) Peristome	−37.95	0.13	−37.80	−35.90	0.08	−35.81	3.98	1	**4**	4	4 999.00	12.09	4.98
8) Seta length	−38.42	0.14	−38.30	−41.58	0.13	−41.44	**6.28**	**1**	2	2	19 999.00	48.35	**7.76**

Results from test of correlated evolution between change in morphology and shift to exposed epiphytic or epilithic habitat. Test result based on i) comparisons of harmonic means of likelihoods (lnL) from reversible-jump Markov Chain Monte Carlo (RJ MCMC) runs with an independent (I) and a dependent (D) model of character evolution; and ii) numbers of visits in I models during RJ MCMC runs (D1, D2, D3). For i) mean of harmonic means (mean lnL), standard deviations (st.dev. lnL), and maximum harmonic mean of likelihood (max lnL) for three I and D runs are given. Bayes Factor values (BF) are calculated using the maximum harmonic mean of likelihood obtained from the best I and D run, i.e. the run yielding the highest likelihood after 200 000 000 iterations (I max lnL and D max lnL in the table). For ii) chains were run three times (D1, D2, D3), and for the best run, number of visits to I models was compared with the prior odds (see 10). BFs >5 based on prior and posterior odds give support for unexpectedly high number of visits to D models, and thus strongly support the evolutionary model assuming correlated evolution between morphological character change and habitat shift. When the support for the D model is the strongest and visits to I models are absent, zero values in the divisor yield non-applicable (NA) BF. BF >5 are considered strong evidence for correlated evolution [46,48] and are bolded.

restricted as equal (q13 = q12, i.e. a model with seven parameters for dual character state transitions; R in Table 3) was not significantly worse than the model where rates were allowed to vary freely (a full model with eight parameters; Table 3). Differences between the rate coefficients were thus so small that they did not lead to a significant difference in harmonic means of log likelihoods from MCMC chains with the restricted model and the full eight-parameter model.

Discussion

Adaptive Evolution in Explaining the Match between Habitat and the Presence of Specialized Traits

Adaptation alone is not able to explain the convergent evolution of sporophytic specializations in mosses that grow in epiphytic and other exposed habitats. Only for four specialized sporophytic traits, the short seta and three traits of the peristome, shifts in phenotype were correlated with a habitat shift (Fig. 2). In accordance with differences between transition rates, the shift to the derived morphological character state occurred after the shift in the habitats; three of these, seta length and two endostomial traits, may possibly be adaptations in the strict sense (Fig. 2) [6,29]. These three traits as well as the majority of other specialized traits are sporophytic reductions. As the function of alternative sporophytic phenotypes in different environments has not been explored, their effect on fitness and presumed role of natural selection will, however, remain untested. Besides direct impact of sporophytic reductions on dispersal and fitness, their evolution may be explained by an indirect increase in fitness via bionenergetics, because less biomass and energy need to be spent for sporophyte production. Habitats high above the ground can facilitate dispersal of spores, and thus eliminate, e.g., the need of a long seta as the whole plant body takes over its role. In contrast, strong stabilizing selection in sheltered low-elevation habitats in the forest-floor layer may favor retaining the long seta and capsules with a perfect peristome that actively disperses the spores by its hygroscopic movements [19,25]. In epiphytic and other exposed habitats ecological constraints may be relaxed, and some of the complex sporophytic traits that have become unnecessary are reduced [30]. In general, loss or reduction of structures is considered to have a simpler genetic basis than their gaining [31], a fact which may favor parallel evolution.

Adaptive evolution could not explain correlated evolution between the orientation of dry peristome and habitat shift (Table 3), since a peristome in which the teeth are curved inwards to close the capsule in the dry state had already evolved before the shift in habitats. It could be a pre-adaptation, i.e. of an evolutionary origin of which is not necessarily linked to selection or higher fitness in the current environment (exaptation [6]). Evolution of some other morphological traits that bryologists have traditionally linked with a shift to epiphytism, such as capsule orientation, does not correlate at all with the habitat shift, but these have appeared before the shift (see also [20]).

Other Processes Potentially Contributing to the Match between Morphology and Environment

Although a functional fit between organisms and their environment is often assumed to be due to adaptive evolution, ecological processes may also contribute significantly to the observed association between morphology and the environment [32]. Community assembly processes and habitat tracking, together with the higher fitness of the derived phenotype in exposed epiphytic and epilithic habitats, may explain the frequent occurrence of these traits in Lembophyllaceae and Neckeraceae species that grow in these habitats. Organisms with an already existing phenotype that fits better for a certain environment will thus be found more often in that habitat; this results in an observable fit between the specialized phenotypes and environments. After establishment in a new habitat, habitat selection and stabilizing selection will enable the creation and maintenance of the association [4].

If the evolutionary order of the shift in habitats versus phenotype goes unstudied, this may result in a false impression of the potential adaptive origin of the traits. For example, a low specific leaf area (SLA) and small leaves in flowering plants did not evolve as adaptations to dry Mediterranean climates. These features appeared in tropical forests that formed the ancestral vegetation in areas that nowadays host chaparral vegetation [4].

Table 3. Comparisons between rates of dual character state change in morphology and habitat.

Character	mean lnL D	max lnL D	mean lnL R	max lnL R	BF	q12	q13	P	Mann-Whitney U
3) Dry peristome	−46.96	−46.93	−47.13	−47.05	−0.18	55.23	**68.65**	**0.00**	133 000 000
stdev	0.03		0.09			23.13	21.23		
5) Basal membrane	−33.71	−33.60	−34.28	−33.72	−0.01	**65.39**	57.17	**0.00**	165 800 000
stdev	0.11		0.54			24.48	27.35		
6) Cilia	−40.63	−40.39	−41.36	−40.79	−0.79	**72.32**	45.66	**0.00**	92 495 436
stdev	0.27		0.78			20.81	27.89		
8) Seta length	−39.71	−39.63	−39.94	−39.87	−0.49	**66.29**	47.92	**0.00**	124 900 000
stdev	0.07		0.11			23.80	28.13		

Comparisons between rates of dual character state change in morphology and habitat. q12 is the rate coefficient for character change where morphology changes while habitat preference remains unchanged ([0,0]-> [0,1]), and q13 is the rate of the change where morphology remains unchanged while habitat changes ([0,0]-> [1,0]). Difference in rates was tested by running an MCMC chain applying the model of dependent evolution for morphological and habitat character state change (D; 8 parameters) and with the restricted model where q12 and q13 were forced to be the same (R; 7 parameters). Bayes Factors (BF) served to estimate whether the difference in likelihoods for R and D models was statistically significant. Both for R and D models, MCMC runs were repeated three times; means for D runs (mean lnL D) and for R runs (mean lnL R) are in the table. BF was calculated based on the best run, i.e. the one yielding the highest likelihood (max lnL D and max lnL R). BF >5 were regarded as strong support. Rate coefficients were also sampled during MCMC chains with a D model and used for testing the difference between q12 and q13. Means and standard deviations for the rate parameters (columns q12 and q13) from the run yielding the best likelihood are given and significance of differences between the rates is tested.

The parallel evolution of C_4 photosynthesis in some grass lineages was often considered an adaptation to arid environments or to changing atmospheric CO_2 concentrations, but the C_4 phenotype evolved before the shift to arid habitats [5]. It could thus represent a pre-adaptation to arid habitats, where the increased frequency of the phenotype as well as positive selection of the genes behind it are undoubtedly due to better fitness in arid environments [5,33]. The patterns observed in these two cases are thus analogous to some of the traits in epiphytic pleurocarpous mosses.

Phylogeny-Based Comparative Methods in Recognizing Shared Adaptations

The ability of phylogeny-based comparative methods to successfully detect correlated evolution between ecological and morphological traits, and thus potential adaptations, is based on the assumption that the same environmental selection pressure leads to similar phenotypic changes in different lineages. This implies that, first, the selective pressures do not vary significantly among populations due to highly similar ecological conditions. Second, in the different taxa and lineages the underlying genetic and other mechanisms for adapting to environmental change must be fairly similar. Third, the selection should lead to similar changes in phenotype despite differing combination of the original character states across taxa.

However, once the evolution of shared adaptation does not follow the assumptions made in phylogeny-based comparative methods they would have been undetected in our study group (see also [29]). In the few cases where we could not detect a significant correlation between morphological and habitat shifts, an analysis of genes that regulate the evolution of specialized sporophytic traits for epiphytic mosses might probably find signs of selection in some lineages.

Neither of the above mentioned assumptions are necessarily true when adaptations are studied on the present macroevolutionary scale. The strength and direction of environmental pressures may vary between taxa due to the world-wide sampling, because a wide geographical scale leads to a wider variation in microhabitat quality. In addition, morphological or physiological trait complexes can be acquired via differing underlying genetic pathways in different lineages [33–38]. Recent results suggest a surprising number of alternative genetic and developmental pathways behind similar trait complexes in different lineages; this may be explained by variation in evolutionary patterns among groups [37,39–41]. In mosses, slight differences in structure and trait combinations of specialized sporophytes between epiphytic lineages could indicate developmental or genetic differences in phenotype regulation. Due to the unique life cycle among land plants, with a dominant perennial gametophyte generation (see Fig. 1), gene expression and genetic regulation of sporophytic traits in mosses differ from those of derived land plants [42,43]. The limited information on their functional genetics hampers further evaluation along these lines in bryophytes.

Finally, phenotypic, genetic and developmental constraints may either prevent or enhance the shift to the adapted phenotype and favor convergent evolution [37,38,40,44]. Conflicting responses on selection in two traits in which genetic or developmental pathways are linked may constrain the changes in the phenotype [45]. Coevolution within character complexes and constraints that allow traits to shift to the adapted state only in combination with some other changes may explain why three of the four correlated traits in our study were endostomial traits.

Three Promising Candidate Traits for Further Studies on Adaptive Evolution of Epiphytic Mosses

The phylogenetic approach that we apply here provides a simple and cost-effective way to test hypotheses regarding the evolution of morphological specializations in relation to the habitat. Three traits that are correlated with habitat shift, seta length and two endostomial traits, may be adaptations to epiphytic or other exposed habitats (Fig. 2). Adaptive evolution is, however, not the only process that explains a high frequency of some derived traits in those habitats [25,26,28], since several specializations did not evolve as a response to the habitat shift. Additional studies are also needed to confirm selection due to differential fitness of reduced and perfect sporophyte morphology in epiphytic or other exposed habitats versus the forest floor. Research on the evolution of adaptations and adaptive diversification are mostly limited to populations or lower taxonomic levels and their methods are often difficult to apply to macroevolutionary studies such as the ones detecting selection in distantly related taxa. Although recent advances on the background of adaptive and convergent evolution [41] suggest the phylogenetic approach may in some cases have limitations in pointing out potential adaptations, any positive result will still be useful for sorting out the most promising candidate traits [29]. Our results provide information on processes that contribute to ecological specialization on a taxonomic scale that is rarely explored and allow valuable insights into the mechanisms of diversification and evolution of differences among organisms. Both are central questions in biological research.

Materials and Methods

Phylogenetic and Ancestral Character State Reconstruction

In order to test correlated evolution and to reconstruct ancestral states with BayesTraits [46], we scored character states for eight morphological characters and habitat preference with binary coding (Supporting information, Appendix S1). Morphological traits were selected among larger selection of traits that were studied in our earlier study for their utility to delimit taxonomic groups in Lembophyllaceae and Neckeraceae [20]. For habitat preference species were coded as occurring in the habitat where it is most typically found. Some moss species, however, can be found in variety of different habitats in rendering the assignment of habitat preference difficult. Especially in the case of predominantly epiphytic and epilithic species decision between these two character states may be difficult [47]. The basis for coding was our field experience of the species in different parts of their distribution areas, the information given in the literature and local floras, as well as habitat information on herbarium labels. We calculated support for preferred ancestral states at critical nodes with a shift in character state with Bayes Factors using the "fossil" command in BayesTraits [46].

The molecular data, methods of phylogeny reconstruction and ancestral states to reconstruction with BayesTraits [46] were described in our earlier study that aimed at clarifying taxonomy of the group [20].

Tests of Correlated Evolution

We performed tests of correlated evolution between habitat shifts and morphological traits using a Bayesian approach as implemented in BayesTraits [10]. The method utilizes a molecular phylogeny ([20]; Supporting information, Fig. S1) and distribution of morphological and habitat traits in terminals. It compares fit of two evolutionary models for two discrete characters, i.e. a model of correlated evolution (dependent evolution; D) employing up to eight rate parameters for dual character state transitions and an independent model of character evolution (I) with up to four rate parameters.

A reversible-jump Markov Chain Monte Carlo (RJ MCMC) served to sample trees and model parameters according to their posterior probabilities under the D and I models. Rate priors were set to vary within a uniform distribution between 0 and 100. We monitored acceptance rates and they were set to a rate deviation of approximately 20%. Each RJ MCMC was run for 1 000 000 000 generations and for all combinations of morphological character – habitat runs was repeated three times to check that log-likelihood values and harmonic means did not significantly differ between converted chains [46].

The fit of two competing models, I and D, was evaluated by two methods. First, RJ MCMC was run three times with both an independent (I) and a dependent (D) model of evolution. Support for either of the models was estimated by comparing harmonic means of likelihoods from I and D runs with logarithmic Bayes Factors (BFs). BF >5, based on one of the three D runs and the three I runs, was regarded as strong support for correlated evolution between a morphological trait and the habitat shift(s) [10]. The second approach was based on a property of the RJ MCMC that in the chain in which all eight dual character-state transitions can occur freely, the number of visits to the dependent or independent model is propositional to the posterior probability of the model [10]. Support for correlated evolution was thus also evaluated by comparing the ratio of prior and posterior odds for visits in I and D models during the chains [10]. Support for either model was estimated by use of BF.

Change First in Habitat or in Morphology?

We detected the order of character change in the phylogeny by three different methods. First, we compared ancestral state reconstructions for habitat preference and morphology. Second, rate coefficients were sampled from one out of three RJ chains with the dependent model of evolution. We tested for difference in the posterior distribution of rate coefficients for change in morphology but not in habitat preference (q13; from [state for habitat preference = 0, state for morphology = 0] to [state for habitat preference = 0, state for morphology = 1]) and for change in habitat preference but not in morphology (q12). Rate coefficients for character change where morphology changes while habitat preference remains unchanged (q12; [0,0]-> [0,1]) were compared with rates of the change where morphology remains unchanged while the habitat changes (q13; [0,0]-> [1,0]). The statistical significance of the difference was tested by the non-parametric Mann-Whitney t-test. Third, evolutionary significance of difference in rate coefficients was confirmed by running an MCMC chain with the dependent model of evolution (eight rate parameters) and with restricted dependent model where q12 and q13 were set to equal [8]. Settings and method for running MCMC with BayesTraits were the same as above. The fit of these models was compared with BFs based on harmonic means of the posterior probability of likelihoods.

Supporting Information

Figure S1 Bayesian tree for moss families Neckeraceae and Lembophyllaceae. Majority consensus of trees sampled after stationarity in the Bayesian analysis of the matrix including indels (for details, see [20] Olsson et al. 2009). Values along the branches indicate posterior probabilities (above the branches) and bootstrap support values from the parsimony analysis (below). The first value corresponds to the analyses with the matrix with insertion-deletion coding included in the analyses. Correlated evolution of habitat shift and morphological traits was tested for the subtree within a shaded box. Numbers indicate the nodes for which probabilities for derived ancestral character state are given in Supporting information Appendix S2.

Appendix S1 Coding for habitat preferences and morphological character states.

Appendix S2 Ancestral character state reconstructions for evolution of eight morphological characters and habitat preferences in the moss families Neckeraceae and Lembophyllaceae.

Author Contributions

Conceived and designed the experiments: SH LH DQ. Performed the experiments: SH SO. Analyzed the data: SO SH. Contributed reagents/materials/analysis tools: VB JE LH. Wrote the paper: SH SO.

References

1. Darwin C (1859) Origin of Species. On the Origin of Species by Means of Natural Selection, or the Preservation of Favoured Races in the Struggle for Life. London: John Murray, Albemarle Street.
2. Ridge I (2002) Water and transport in plants. In: Ridge I., editor. Plants. New York: Oxford University Press. 105–166.
3. Smith AM, Coupland G, Dolan L, Harberd N, Jones J, et al. (2010) Plant Biology. Garland Science, Taylor and Francis Group, LLC.
4. Ackerly DD (2004) Adaptation, niche conservatism, and convergence: comparative studies of leaf evolution in the California Chaparral. Am Nat 163: 654–671.
5. Osborne CP, Freckleton RP (2009) Ecological selection pressures for C-4 photosynthesis in the grasses. Proc Roy Soc London Ser B Biol Sci 276: 1753–1760.
6. Gould SJ, Vrba ES (1982) Exaptation–A missing term in science of form. Paleobiology 2: 4–15.
7. Maddison WP (1990) A method for testing the correlated evolution of two binary characters: Are gains and losses concentrated on certain branches of a phylogenetic tree? Evolution 44: 539–557.
8. Pagel M (1994) Detecting correlated evolution on phylogenies: A general method for the comparative analysis of discrete characters. Proc Roy Soc London Ser B Biol Sci 255: 37–45.
9. Huelsenbeck JP, Rannala B (2003) Detecting correlation between characters in a comparative analysis with uncertain phylogeny. Evolution 57: 1237–1247.
10. Pagel M, Meade A (2006) Bayesian analyses of correlated evolution of discrete characters by reversible-jump Markov Chain Monte Carlo. Am Nat 167: 808–825.
11. Patterson C (1988) Homology in classical and molecular biology. Mol Biol Evol 5: 603–625.
12. Schluter D, Nagel L (1995) Parallel speciation by natural selection. Am. Nat. 146: 292–301.
13. Nosil P, Crespi BJ, Sandoval CP (2002) Host-plant adaptation drives the parallel evolution of reproductive isolation. Nature 417: 440–443.
14. Buck WR, Goffinet B (2000) Morphology and classification of mosses. In: Shaw AJ, Goffinet B editors. Bryophyte biology. Cambridge: Cambridge Univ. Press. 71–123.
15. Bell NE, Quandt D, O'Brien TJ, Newton AE (2007) Taxonomy and phylogeny in the earliest diverging pleurocarps: square holes and bifurcating pegs. Bryologist 110: 533–560.
16. Buck WR, Goffinet B, Shaw AJ (2000) Testing morphological concepts of orders of pleurocarpous mosses (Bryophyta) using phylogenetic reconstructions based on *trnL-trnF* and *rps4* sequences. Mol Phylogen Evol 16: 180–198.
17. Vanderpoorten A, Hedenäs L, Cox CJ, Shaw AJ (2002) Phylogeny and morphological evolution of the Amblystegiaceae (Bryopsida). Molec Phylogenet Evol 23: 1–21.
18. Huttunen S, Ignatov M (2004) Phylogenetic analyses of Brachytheciaceae (Bryophyta) based on morphology, and sequence level data. Cladistics 20: 151–183.

19. Huttunen S, Ignatov MS, Müller K, Quandt D (2004) Phylogeny and evolution of epiphytism in the three moss families Meteoriaceae, Brachytheciaceae, and Lembophyllaceae. In: Goffinet B, Hollowell VC, Magill RE, editors. Molecular systematics of bryophytes. St. Louis: Missouri Botanical Garden Press. 328–355.
20. Olsson S, Buchbender V, Enroth J, Huttunen S, Hedenäs L, et al. (2009) Evolution of the Neckeraceae (Bryophyta): resolving the backbone phylogeny. Syst Biodiv 7: 419–432.
21. Newton AE, Wikström N, Bell N, Forrest LL, Ignatov MS (2007) Dating the diversification R. of the pleurocarpous mosses. In: Newton AE, Tangney RS editors. Pleurocarpous mosses: systematics and evolution. The Systematics Association Special Volume Series 71: 337–366.
22. Buck WR (1991) The basis for familial classification of pleurocarpous mosses. Adv Bryol 4: 169–185.
23. Shaw AJ, Cox CJ, Goffinet B, Buck WR, Boles SB (2003) Phylogenetic evidence of a rapid radiation of pleurocarpous mosses (Bryophyta). Evolution 57: 2226–2241.
24. Stuessy TF (2004) A transitional-combinational theory for the origin of angiosperms. Taxon 53: 3–16.
25. Vitt D (1981) Adaptive modes of the moss sporophyte. Bryologist 84: 166–186.
26. Hedenäs L (2001) Environmental factors potentially affecting character states in pleurocarpous mosses. Bryologist 104: 72–91.
27. Devos N, Renner MAM, Gradstein R, Shaw AJ, Laenen B, et al. (2011) Evolution of sexual systems, dispersal strategies and habitat selection in the liverwort genus *Radula*. New Phytologist 192: 225–236.
28. Grout AJ (1908) Some relations between the habitats of mosses and their structures. Bryologist 11: 97–100.
29. Losos JB (2011) Convergence, adaptation, and constraint. Evolution 65: 1827–1840.
30. Fong DW, Kane TC, Culver DC (1995) Vestigialization and loss of nonfunctional characters. Ann Rev Ecol Syst 26: 249–268.
31. Cresko WA, Amores A, Wilson C, Murphy J, Currey M, et al. (2004) Parallel genetic basis for repeated evolution of armor loss in Alaskan threespine stickleback populations. Proc Natl Acad Sci U S A 101: 6050–6055.
32. Ackeley DD (2003) Community assembly, niche conservatism, and adaptive evolution in changing environments. Int J Plant Sci 164: S165–S184.
33. Wang XY, Gowik U, Tang HB, Bowers JE, Westhoff P, et al. (2009) Comparative genomic analysis of C4 photosynthetic pathway evolution in grasses. Genome Biol 10: r68.
34. Crayn DM, Winter KJ, Smith AC (2004) Multiple origins of crassulacean acid metabolismand the epiphytic habit in the Neotropical family Bromeliaceae. Proc Natl Acad Sci U S A 101: 3703–3708.
35. Besnard G, Muasya AM, Russier F, Roalson EH, Salamin N, et al. (2009) Phylogenomics of C4 photosynthesis in sedges (Cyperaceae): multiple appearances and genetic convergence. Mol Biol Evol 26: 1909–1919.
36. Besnard G, Christin P-A (2010) Evolutionary genomics of C-4 photosynthesis in grasses requires a large species sampling. CR Biol 333: 577–581.
37. Christin P-A, Weinreich DM, Besnard G (2010) Causes and evolutionary significance of genetic convergence. Trends Gen 26: 400–405.
38. Yoon H-S, Baum DA (2004) Transgenic study of parallelism in plant morphological evolution. Proc Natl Acad Sci U S A 101: 6524–6529.
39. Orr HA (2001) The genetics of species differences. Trends Ecol Evol 16: 343–350.
40. Arendt J, Reznick D (2007) Convergence and parallelism reconsidered: what have we learned about the genetics of adaptation? Trends Ecol Evol 23: 26–32.
41. Elmer KR, Meyer A (2011) Adaptation in the age of ecological genomics: insights from parallelism and convergence. Trends Ecol Evol 26: 298–306.
42. Quatrano RS, McDaniel SF, Khandelwal A, Perroud P-F, Cove DJ (2007) *Physcomitrella patens*: mosses enter the genomic age. Curr Opin Plant Biol 10: 182–189.
43. Lang D, Zimmer AD, Rensing SA, Reski R (2008) Exploring plant biodiversity: the *Physcomitrella* genome and beyond. Trends Plant Sci 13: 542–549.
44. Wake DB (1991) Homoplasy the result of natural selection, or evidence of design limitations? Am Nat 138: 543–567.
45. Ogburn RM, Edwards EJ (2009) Anatomical variation in Cactaceae and relatives: trait lability and evolutionary innovation. Am J Bot 96: 391–408.
46. Pagel M, Meade A (2004) BayesTraits. Available: www.evolution.rdg.ac.uk. Accessed 2006 Nov 1.
47. Piippo S (1982) Epiphytic bryophytes as climatic indicators in Eastern Fennoscandia. Acta Bot Fenn 57: 1–39.
48. Kass RE, Raftery AE (1995) Bayes factors. J Am Stat Assoc 90: 773–795.

9

Asymmetrical Interference Effects between Two-Dimensional Geometric Shapes and Their Corresponding Shape Words

Bradley R. Sturz*, Joshua E. Edwards, Ty W. Boyer

Department of Psychology, Georgia Southern University, Statesboro, Georgia, United States of America

Abstract

Nativists have postulated fundamental geometric knowledge that predates linguistic and symbolic thought. Central to these claims is the proposal for an isolated cognitive system dedicated to processing geometric information. Testing such hypotheses presents challenges due to difficulties in eliminating the combination of geometric and non-geometric information through language. We present evidence using a modified matching interference paradigm that an incongruent shape word interferes with identifying a two-dimensional geometric shape, but an incongruent two-dimensional geometric shape does not interfere with identifying a shape word. This asymmetry in interference effects between two-dimensional geometric shapes and their corresponding shape words suggests that shape words activate spatial representations of shapes but shapes do not activate linguistic representations of shape words. These results appear consistent with hypotheses concerning a cognitive system dedicated to processing geometric information isolated from linguistic processing and provide evidence consistent with hypotheses concerning knowledge of geometric properties of space that predates linguistic and symbolic thought.

Editor: Philip Allen, University of Akron, United States of America

Funding: This research was supported by funds from the Office of the Vice President for Research and the Jack N. Averitt College of Graduate Studies at Georgia Southern University to BRS and TWB. The funders had no role in study design, data collection and analysis, decision to publish, or preparation of the manuscript.

Competing Interests: The authors have declared that no competing interests exist.

* E-mail: bradleysturz@georgiasouthern.edu

Introduction

For centuries nativists and empiricists have debated the extent to which spatial thinking is innate or learned [1]. Recent scientific inquiry has provided evidence in support of innate spatial mechanisms for navigation and orientation via geometric cues [2–7] as well as for facial recognition via spatial properties [8], and the strongest evidence in support of a nativist approach comes from cross-species experiments on geometric encoding of the environment [9–13], see also [14–15]. Experiments conducted on species from ants to adult humans have provided evidence that incidental learning of geometric properties of an environment is a fundamental and ubiquitous component of spatial cognition that occurs across phylogeny and ontogeny [9–10]. Such evidence has been used in support of hypotheses for Euclidean geometry as one of many domains of core knowledge that predate linguistic and symbolic thought [11–13], [16–17].

Central to these hypotheses is the proposal for an isolated modular cognitive system dedicated to processing geometric information [3], [6], [11–13], [18–19]. Tests of such hypotheses with normally functioning adults who display well-developed linguistic systems have presented numerous challenges, and one of the greatest challenges relates to the difficulty in eliminating the combination of geometric and non-geometric information (e.g., color) through the use of language [20], c.f., [21–22]. In short, an adult's ability to encode spatial relations linguistically (e.g., "the dog is in front of the tree") precludes researchers' ability to investigate the use of pure geometric information isolated from linguistic processing. Although researchers have attempted to disrupt the encoding of spatial relations in a linguistic fashion through the implementation of distractor tasks [20–22], evidence for such a modular cognitive system isolated from linguistic processing remains elusive in normal functioning adults. As a result, any evidence of such an isolation of geometric processing from linguistic processing in adult participants with well-developed linguistic systems would assist in illuminating potential evolutionary and developmental origins of spatial and linguistic processes.

In the present experiment, we modified an interference paradigm to probe the isolation of geometric and linguistic processing [23–24]. By modifying a match-to-sample task that has previously provided evidence for semantic interference [24], we were able to present bi-dimensional samples composed of a two-dimensional geometric shape and a shape word (see Figure 1) and manipulate whether the shape and shape word were congruent (e.g., "circle" in a circle) or incongruent (e.g., "circle" in a square). After a delay, we probed each sample dimension independently during target presentation via shape targets (i.e., two shapes) or word targets (i.e., two words; Figure 1A/C and Figure 1B/D, respectively) and manipulated whether the incorrect response option was related or unrelated to the irrelevant sample dimension.

We believe this approach is uniquely suited to probe the isolation of geometric and linguistic processing because it allows for determination of the extent to which a two-dimensional

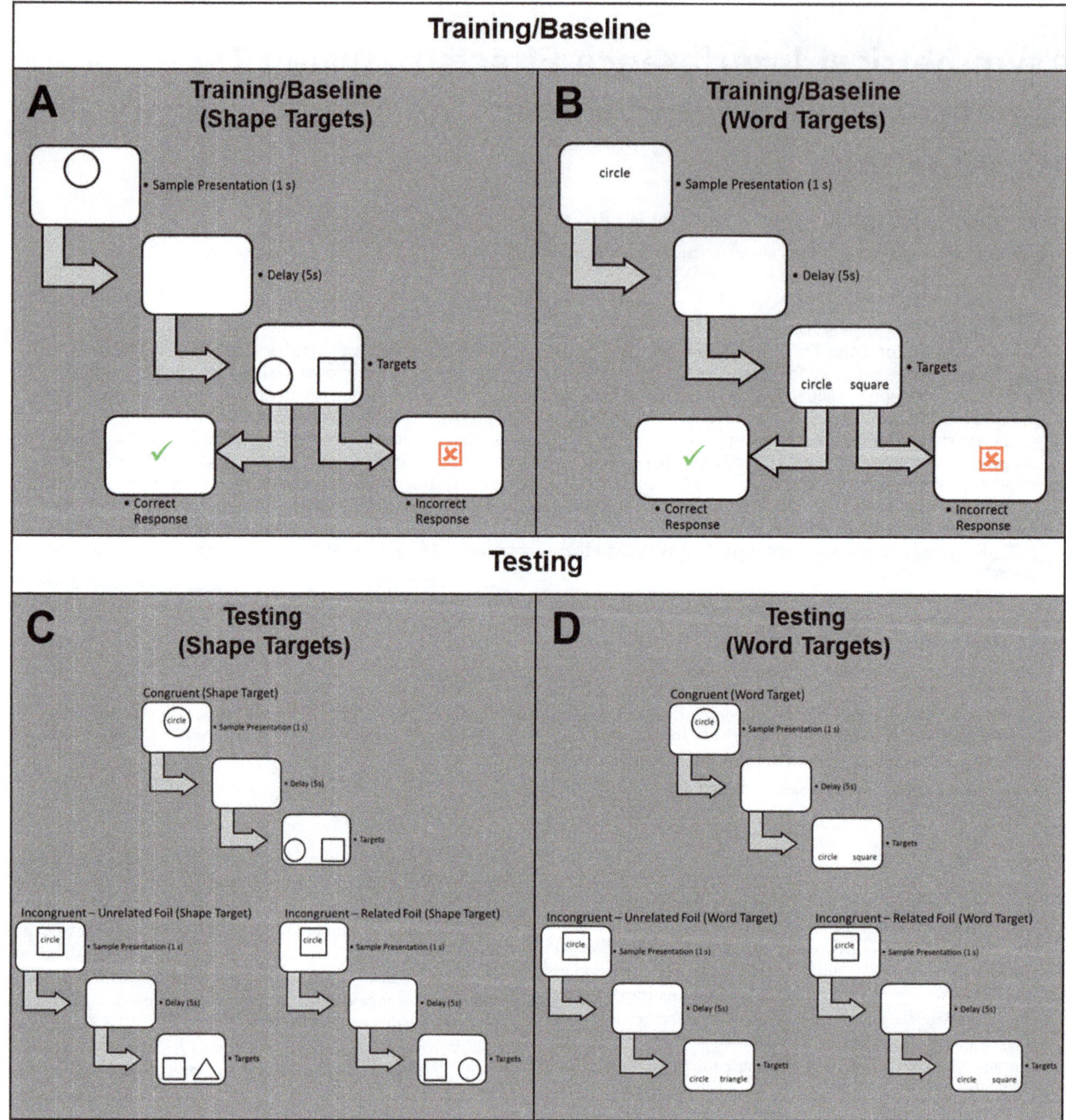

Figure 1. Sample trial types and trial structures for the Delayed Match-to-Sample (DMTS) task. One sample Baseline/Training trial is illustrated for Shape Targets (**A**) and Word Targets (**B**), and one sample Congruent, Incongruent – Unrelated Foil, and Incongruent – Related Foil trial is illustrated for Shape Targets (**C**) and Word Targets (**D**). For illustrative purposes, all correct matches are shown as the left target even though correct target and foil target locations were balanced (see text for details).

geometric shape interferes with identification of a shape word as well as determination of the extent to which a shape word interferes with identification of a two-dimensional geometric shape. In the presence of a bi-dimensional stimulus composed of a two-dimensional geometric shape and a shape word that are incongruent (e.g., "circle" in a square), the sample word dimension could activate a spatial representation of the shape, and this spatial representation could interfere with identifying the sample shape dimension in the presence of two shape targets. Reciprocally, the sample shape dimension could activate a linguistic representation of the shape, and this linguistic representation could interfere with identifying the sample word dimension in the presence of two word targets.

From a strict empiricist perspective, interference effects should be symmetrical for shape and word targets such that RTs on trials in which a shape and shape word are presented in isolation (i.e., Baseline trials, see Figure 1A/B) or are congruent (i.e., Congruent trials, see Figure 1C/D) should not differ, but both of these trial types should be faster than trials in which the shape and shape word are incongruent (i.e., Incongruent – Unrelated Foil and Incongruent – Related Foil trials, see Figure 1C/D). Importantly, these trial type RT effects should hold for *both* shape and word targets, and interference by the irrelevant sample dimension should be reflected in accuracy measures of *both* shape and word targets – especially in the presence of an incorrect response option that is related to the irrelevant sample dimension (due to a greater probability of error in the presence of two potential matches on the basis of linguistic or spatial representations of the sample dimensions).

In contrast, under the assumption of a modular cognitive system dedicated to processing geometric information isolated from linguistic processing [11–13], [16–18], interference effects should be asymmetrical. A shape should not activate a linguistic representation (i.e., a circle should not activate the word "circle"). As a result, an incongruent shape should not interfere with identifying the relevant sample word dimension in the presence of two word targets. To the extent that geometric processing is isolated from linguistic processing, the predictions outlined above regarding the trial type RT and accuracy effects should hold for shape targets but not word targets.

Methods

Participants

Twenty-four undergraduate students at Georgia Southern University (12 males; 12 females) served as participants. Participants had normal or corrected-to-normal vision and received extra class credit or participated as part of a course requirement.

Ethics statement

The research was conducted following the relevant ethical guidelines for human research. We obtained written informed consent from all participants, and all procedures were approved by Georgia Southern University's Institutional Review Board.

Apparatus

We constructed and implemented a delayed match-to-sample task (see Figure 1) on a personal computer with a 22-inch flat-screen liquid crystal display (LCD) monitor (1,680 × 1,050 pixels). Responses occurred via the "c" (left target) and "m" (right target) keys on a standard keyboard. Experimental events were controlled and recorded using E-Prime (Psychology Software Tools, Inc., www.pstnet.com).

Stimuli

There were two stimulus types: Shapes (Figure 1 A/C) and Words (Figure 1 B/D). Shape stimuli were circles, squares, and triangles each presented in a 5 pixel width black outline measuring 312 pixels in diameter (circle), 312 pixels in height and width (square), and 440 pixels in base width and 312 pixels in height (triangle) subtending 7.3° visual angle horizontally and vertically (circle and square) and 10.3° horizontally and 7.3° vertically (triangle). Word stimuli were "circle", "square", and "triangle" presented in bold 40 point Courier New font and were 187 ("circle" and "square") and 250 ("triangle") pixels in width, subtending 4.4° ("circle" and "square"), and 5.9° ("triangle") visual angle horizontally, and 34 ("circle" and "square") or 44 ("triangle") pixels in height, subtending 0.8° or 1.0° visual angle vertically. Words were presented in black font color. All stimuli were presented on a white background. Samples were presented in the horizontal center of the screen 25% down from its top edge. Targets were presented on opposite sides of the screen, 50% of screen width apart, and 25% up from its bottom edge.

Procedure

We provided participants with instructions that they would complete a memory test in which one of several shapes and words would appear on the screen, would disappear, and then either a pair of shapes or words would appear. Instructions also informed them that their task would be to select the shape that matched the sample shape (if shape pairs) or select the word that matched the sample word (if word pairs).

The experimental protocol consisted of 120 total trials for each participant composed of 24 Training Trials and 96 Testing Trials. All trials presented samples for 1 s, followed by a 5 s blank screen retention interval delay, followed by target stimuli for 1.5 s. A response to the correct target (i.e., match) resulted in the presentation of a green check mark; a response to the incorrect target (i.e., foil) resulted in the presentation of a red "X", and failure to respond during the 1.5 s target presentation produced a "No Response" statement. Feedback was presented for 1 s, and served as the inter-trial interval (ITI).

Training. To familiarize participants with the task, we provided them with 24 training trials composed of two 12-trial blocks. One block included 12 unique shape training trials in which participants matched a sample shape to its corresponding shape target (Figure 1A), and the other block included 12 unique word training trials in which participants matched a sample word to its corresponding word target (Figure 1B). We balanced for gender and counterbalanced the training blocks order of presentation.

Testing. Testing consisted of 96 trials composed of 12 eight-trial blocks. Each trial block was composed of two trials of each of four trial types (see Figure 1): Baseline (Training), Congruent (sample shape with corresponding shape word), Incongruent – Unrelated Foil (sample shape with non-corresponding shape word and a foil unrelated to the irrelevant sample dimension), and Incongruent – Related Foil (sample shape with non-corresponding shape word and a foil related to the irrelevant sample dimension). Baseline trials were identical to Training trials. For all trial types, when shape targets were presented (e.g., circle and square), the corresponding sample shape was the correct response. When words targets were presented (e.g., "circle" and "square"), the corresponding sample word was the correct response.

We presented one trial with shape targets and one trial with word targets for each trial type within each block in randomized sequences. The left/right location of the correct target (i.e., match) and foil were counterbalanced, which resulted in each unique combination of each trial type being presented once, without replacement, for a total of 96 trials during Testing (24 Baseline trials, 24 Congruent trials, 24 Incongruent – Unrelated Foil trials, and 24 Incongruent – Related Foil trials). Feedback was identical to Training.

Results

We analyzed Testing data via RTs and proportions correct.

Response time

We analyzed correct trials (error rates opposite of proportion correct shown Figure 2B). Figure 2A shows the mean RTs (in ms) plotted by Target Type for each Trial Type. A two-way repeated measures analysis of variance (ANOVA) on RT with Target Type (shape, word) and Trial Type (baseline, congruent, incongruent – unrelated foil, incongruent – related foil) as factors revealed a main effect of Trial Type $F(3, 69) = 11.47$, $p<.001$, $\eta_p^2 = 0.33$, but a non-significant effect of Target Type, $F(1, 23) = 0.77$, $p = .39$. These results were qualified by a significant Target Type x Trial Type interaction, $F(3, 69) = 13.47$, $p<.001$, $\eta_p^2 = 0.37$. To illuminate the source of the interaction, we conducted two separate one-way repeated measures ANOVAs for each Target Type with Trial Type (baseline, congruent, incongruent – unrelated foil, incongruent – related foil) as a factor. For the Shape Targets, there was a main effect of Trial Type, $F(3, 69) = 15.30$, $p<.001$, $\eta_p^2 = 0.40$. Post hoc tests revealed that Baseline and Congruent trials were not significantly different from each other ($p = .07$), but both of these trial types were significantly faster than Incongruent – Unrelated Foil and Incongruent – Related Foil trials (*p*s <.01). Incongruent – Unrelated Foil and Incongruent – Related Foil trials were not significantly different from each other ($p = .09$). For Word Targets, the main effect of Trial Type was not significant, $F(3, 69) = 0.95$, $p = .42$.

Proportion correct

We eliminated trials in which participants failed to respond (30/2304; 1.3%). Figure 2B shows the mean proportion correct plotted by Target Type for each Trial Type. A two-way repeated measures ANOVA on proportion correct with Target Type (shape, word) and Trial Type (baseline, congruent, incongruent – unrelated foil, incongruent – related foil) as factors revealed a main effect of Trial Type, $F(3, 69) = 10.8$, $p<.001$, $\eta_p^2 = 0.32$, but a non-significant effect of Target Type, $F(1, 23) = 2.36$, $p = .14$. These results were qualified by a significant Target Type x Trial Type interaction, $F(3, 69) = 8.79$, $p<.001$, $\eta_p^2 = 0.28$. To illuminate the source of the interaction, we conducted two separate one-way repeated measures ANOVAs for each Target Type with Trial Type (baseline, congruent, incongruent – unrelated foil, incongruent – related foil) as a factor. For the Shape Targets, there was a main effect of Trial Type, $F(3, 69) = 12.92$, $p<.001$, $\eta_p^2 = 0.36$. Post hoc tests revealed that Baseline, Congruent, and Incongruent – Unrelated Foil trials were not significantly different from each other (*p*s >.64), but all three of these trial types were significantly more accurate than Incongruent – Related Foil trials (*p*s <.01). For Word Targets, the main effect of Trial Type was not significant, $F(3, 69) = 2.38$, $p = .08$. All mean proportions correct were significantly greater than chance (0.5), one-sample *t*-tests, $ts(23) > 7.8$, *p*s <.001.

It should be noted that an analysis identical to that reported above including errors of omission yielded qualitatively identical statistical results. It should also be noted that we conducted a follow-up experiment with 24 new participants. This follow-up experiment substituted nonsense words for shape words and confirmed that the trial type differences for shape targets did not result from facilitation of redundant sample cues. Congruent trials did not differ from Incongruent trials in measures of RT or accuracy. These results provide converging evidence that an incongruent shape word interferes with the identification of a shape but an incongruent shape does not interfere with the identification of a shape word - providing further support for an interpretation that shape words activate spatial representations of shapes but shapes do not activate linguistic representations of shape words.

Discussion

Although RTs did not differ across trial types for Word Targets, RTs for both incongruent trial types were significantly slower than Baseline and Congruent trials for Shape Targets. Accuracy analyses indicated a decrement in performance only for Shape Targets on trials in which the foil was related to the irrelevant sample dimension (i.e., Incongruent – Related Foil trials) suggesting that interference by the sample word dimension resulted in two potential relevant matches on the basis of spatial representations during target presentations. These results appear to be opposite of a speed-accuracy trade-off and corroborate an interpretation of shape words activating spatial representations of shapes, but shapes not activating linguistic representations of shape words.

Collectively, we provide evidence for an asymmetry in shape and shape word interference such that, in the presence of a bi-dimensional stimulus composed of a shape and shape word, an incongruent shape word interferes with identifying a shape, but an incongruent shape does not interfere with identifying a shape word. Although our inclusion of adult participants with well-developed linguistic systems prevents us from drawing definitive conclusions about the evolutionary and developmental origins of

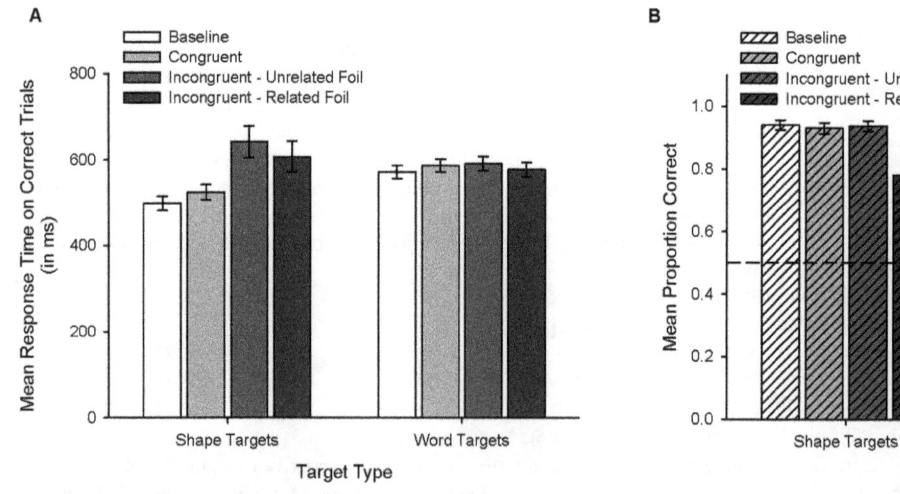

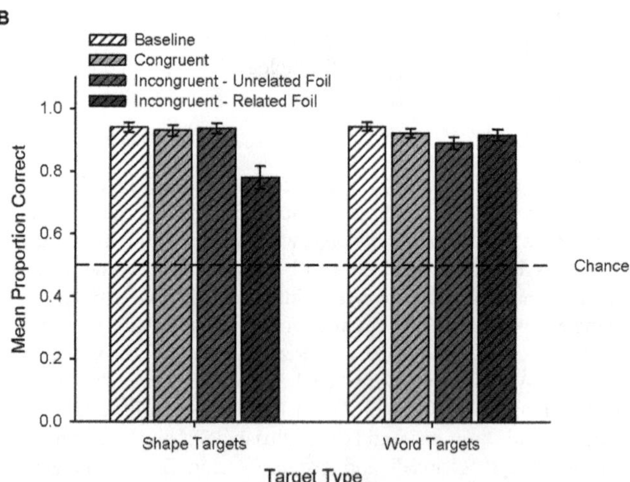

Figure 2. Performance During Testing. (**A**) Mean response time on correct trials during Testing (in milliseconds) plotted by Target Type for each Trial Type. (**B**). Mean proportion correct during Testing plotted by Target Type for each Trial Type. Dashed line represents chance performance (0.5). Error bars represent standard errors of the means.

spatial and linguistic processes, these findings are consistent with recent evidence that shape recognition processes emerge before, interact with, and enable the development of linguistic shape categories [25–26]. We believe our results are also consistent with nativists approaches that would suppose asymmetrical effects for spatial and linguistic processes, and, by extension, our results appear to provide evidence consistent with hypotheses concerning core knowledge for geometric properties of space that predates linguistic and symbolic thought [11–13], [16–18]. In contrast, our results appear largely inconsistent with strict empiricist approaches that would suppose symmetrical effects for the emergence of spatial and linguistic processes. Future research should be able to utilize the current paradigm coupled with neural imaging techniques to isolate associated brain regions and further substantiate our behavior results while providing converging evidence for a cognitive system dedicated to processing geometric information isolated from linguistic processing.

Author Contributions

Conceived and designed the experiments: BRS JEE TWB. Performed the experiments: JEE. Analyzed the data: BRS JEE TWB. Wrote the paper: BRS TWB JEE.

References

1. Hatfield G (1990) The natural and the normative: Theories of spatial perception from Kant to Helmholtz. Cambridge, MA: MIT Press.
2. Brown AA, Spetch ML, Hurd PL (2007) Growing in circles: Rearing environment alters spatial navigation in fish. Psychol Sci 18: 569–573.
3. Cheng K (1986) A purely geometric module in the rat's spatial representation. Cognition 23: 149–178.
4. Chiandetti C, Vallortigara G (2008) Is there an innate geometric module? Effects of experience with angular geometric cues on spatial re-orientation based on the shape of the environment. Anim Cogn 11: 139–146.
5. Chiandetti C, Vallortigara G (2010) Experience and geometry: Controlled rearing studies with chicks. Anim Cogn 13: 463–470.
6. Hermer L, Spelke E (1994) A geometric process for spatial representation in young children. Nature 370: 57–59.
7. Vallortigara G, Pagni P, Sovrano VA (2004) Separate geometric and non-geometric modules for spatial reorientation: Evidence from a lopsided animal brain. J Cognitive Neurosci 16: 390–400.
8. Dakin SC, Watt RJ (2009) Biological "bar codes" in human faces. J Vis 9: 1–10.
9. Cheng K, Huttenlocher J, Newcombe NS (2013) 25 years of research on the use of geometry in spatial reorientation: A current theoretical perspective. Psychon Bull Rev 20: 1033–1054.
10. Cheng K, Newcombe NS (2005) Is there a geometric module for spatial orientation? Squaring theory and evidence. Psychon Bull Rev 12: 1–23.
11. Lee SA, Spelke ES (2010) Two systems of spatial representation underlying navigation. Exp Brain Res 206: 179–188.
12. Lee SA, Spelke ES (2010) A modular geometric mechanism for reorientation in children. Cogn Psychol 61: 152–176.
13. Spelke ES, Lee SA (2012) Core systems of geometry in animal minds. Philos Trans R Soc Lond B Biol Sci 367: 2784–2793.
14. Doeller CF, Burgess N (2008) Distinct error-correcting and incidental learning location relative to landmarks and boundaries. Proc Natl Acad Sci USA 105: 5909–5914.
15. Doeller CF, King JA, Burgess N (2008) Parallel striatal and hippocampal systems for landmarks and boundaries in spatial memory. Proc Natl Acad Sci USA 105: 5915–5920.
16. Dehaene S, Izard V, Pica P, Spelke E (2006) Core knowledge of geometry in an Amazonian indigene group. Science 311: 381–384.
17. Spelke E, Lee SA, Izard V (2010) Beyond core knowledge: Natural geometry. Cogn Sci 34: 863–884.
18. Lee SA, Shusterman S, Spelke ES (2006) Reorientation and landmark-guided search by young children: Evidence for two systems. Psychol Sci 17: 577–582.
19. Gallistel CR (1990) The organization of learning. Cambridge, MA: MIT Press.
20. Hermer-Vazquez L, Spelke E, Katsnelson A (1999) Source of flexibility in human cognition: Dual task studies of space and language. Cogn Psychol 39: 3–36.
21. Ratliff KR, Newcombe NS (2008) Is language necessary for human spatial reorientation? Reconsidering evidence from dual task paradigms. Cogn Psychol 56: 142–163.
22. Ratliff KR, Newcombe NS (2008) Reorienting when cues conflict: Evidence for an adaptive combination view. Psychol Sci 19: 1301–1307.
23. Compton RP, Flowers JH (1977) Pictorial interference with geometric shapes and achromatic shades. Percept Psychophys 22: 303–309.
24. Sturz BR, Green ML, Locker L, Boyer TW (2013) Stroop interference in a delayed match-to-sample task: Evidence for semantic competition. Front Psychol 4: 842. DOI: 10.3389/fpsyg.2013.00842
25. Smith LB (2003) Learning to recognize objects. Psychol Sci 14: 244–250.
26. Smith LB (2009) From fragments to geometric shape: Changes in visual object recognition between 18 and 24 months. Curr Dir Psychol Sci 18: 290–294.

10

Before the Endless Forms: Embodied Model of Transition from Single Cells to Aggregates to Ecosystem Engineering

Ricard V. Solé[1,2,3,4]*, **Sergi Valverde**[1,3,4]

1 ICREA-Complex Systems Lab, Universitat Pompeu Fabra, Barcelona, Spain, **2** Santa Fe Institute, Santa Fe, New Mexico, United States of America, **3** Institut de Biologia Evolutiva, UPF-CSIC, Barcelona, Spain, **4** European Centre for Living Technology, C Foscari University of Venice, Venice, Italy

Abstract

The emergence of complex multicellular systems and their associated developmental programs is one of the major problems of evolutionary biology. The advantages of cooperation over individuality seem well known but it is not clear yet how such increase of complexity emerged from unicellular life forms. Current multicellular systems display a complex cell-cell communication machinery, often tied to large-scale controls of body size or tissue homeostasis. Some unicellular life forms are simpler and involve groups of cells cooperating in a tissue-like fashion, as it occurs with biofilms. However, before true gene regulatory interactions were widespread and allowed for controlled changes in cell phenotypes, simple cellular colonies displaying adhesion and interacting with their environments were in place. In this context, models often ignore the physical embedding of evolving cells, thus leaving aside a key component. The potential for evolving pre-developmental patterns is a relevant issue: how far a colony of evolving cells can go? Here we study these pre-conditions for morphogenesis by using CHIMERA, a physically embodied computational model of evolving virtual organisms in a pre-Mendelian world. Starting from a population of identical, independent cells moving in a fluid, the system undergoes a series of changes, from spatial segregation, increased adhesion and the development of generalism. Eventually, a major transition occurs where a change in the flow of nutrients is triggered by a sub-population. This ecosystem engineering phenomenon leads to a subsequent separation of the ecological network into two well defined compartments. The relevance of these results for evodevo and its potential ecological triggers is discussed.

Editor: Johannes Jaeger, Centre for Genomic Regulation (CRG), Universitat Pompeu Fabra, Spain

Funding: This work has been supported by grants of the James McDonnell Foundation (RVS, SV), Fundación Marcelino Botín (RVS, SV), FIS2004-05422 (RVS) and by the Santa Fe Institute. The funders had no role in study design, data collection and analysis, decision to publish, or preparation of the manuscript.

Competing Interests: The authors have declared that no competing interests exist.

* E-mail: ricard.sole@upf.edu

Introduction

A key problem in evolutionary biology is the emergence of complex life forms under the cooperation of several interacting cells [1,2]. Multicellularity emerged through evolution several times (at least 25) and has been a prerequisite for the generation of complex types of development [3–6]. This major transition brought division of labour and opened the door for the emergence of development and body plans [7–9]. But for many reasons, and in spite of its obvious importance, the evolution of multicellularity is not yet well understood. The fossil traces of the transition are still incomplete, although rapidly improving. However, dedicated efforts to unravel the phylogeny of multicellular living forms, the analysis of special model organisms and the cues provided by the presence of potential genetic toolkits predating the emergence of complex metazoans are defining the potential minimal requirements for the transition towards complex multicellular life forms.

This transition is particularly relevant for the critical changes that took place around 560 Myr ago, associated to the so called Cambrian event [1,10] but its roots predate a much earlier time window, as indicated by the analysis of ancestral genomes. Moreover, the picture gets more complicated as we consider additional components related to the physical environment and the constraints and opportunities posed by ecological interactions. Actually, the multiple facets of the debate on the origins of multicellular organisms have to do with the role played by the different potential shapers of the event. These multiple factors are not independent, and are likely to have interacted in complex ways.

In general terms, the multicellular state is characterized by the existence of cell-cell interactions of some sort that provide a source for collective adaptation to energy limitations, physical fluctuations and eventually division of labor. In multicellular organisms, lower-level entities (cells) have relinquished their ability to reproduce as independent units and instead replicate exclusively as part of the larger whole. But long before a developmental body plan was even defined, in what has been dubbed the "pre-Mendelian world" [11], several layers of complexity were required. This as a particularly relevant problem deeply tied with the problem of hierarchies in evolution [12–13].

Before developmental programs allowed true multicellular organisms to emerge, single cells developed into monomorphic aggregates and later on into differentiated aggregates [13]. Moreover, cell adhesion mechanisms required for the emergence of multicellularity have a much early origin [14]. In this context, long before complex metazoans appeared, some key components

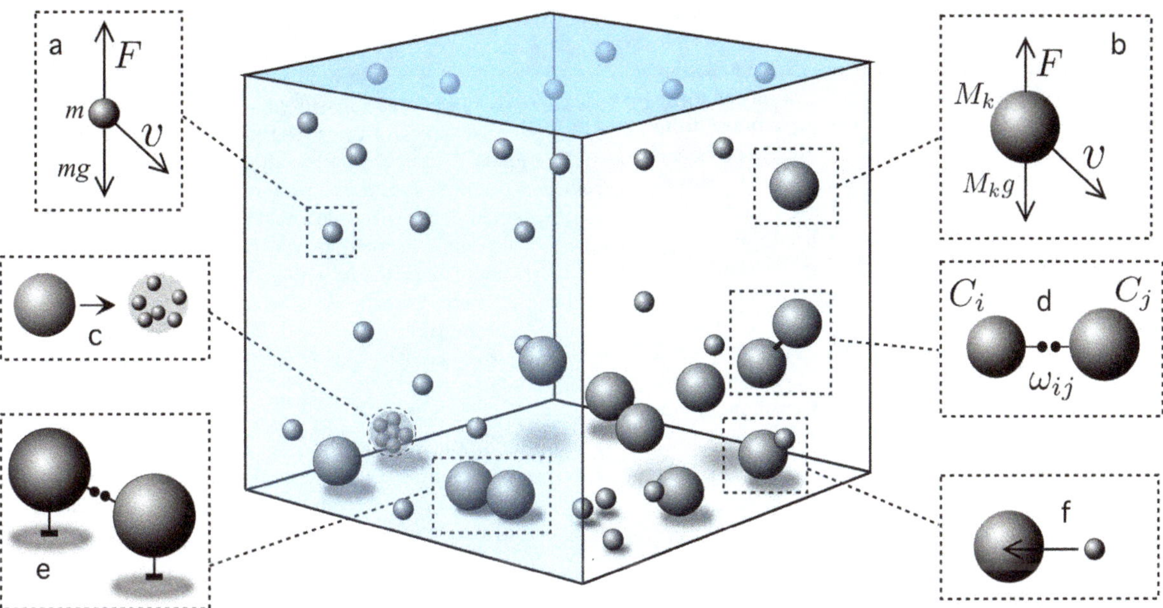

Figure 1. Basic scheme of the components of the CHIMERA model. The system is confined within a rigid cube with a floor where particles fall and to which cells can attach. Five additional square boundaries are also present which cannot be crossed. The upper boundary allows a flow of nutrient particles (here small spheres) at a constant rate. Particles fall under gravity (a,b) and experience local turbulence as a random velocity field. As they reach the floor, they can keep moving under the same flow and also disappear as they degrade (c) into detritus particles. Cells can evolve adhesion among them (d) as well as with the substrate (e). Finally, cells and particles (f) ineract through collisions. If the cell is able to exploit that particular type of energy, the particle involved disappears and is transformed into cell's biomass (see text).

of the toolkit were already in place. How did these components affected the transition to multicellularity is an open question, and theoretical models can help to address it.

Most mathematical and computational models dealing with early evolution of development assume that either genetic networks or even body plans are already in place or instead deal with pattern-forming colonies and their potential to form structures under given spatial and nutrient constraints [15–17]. However, less attention has been given to the physics associated to these processes, particularly in relation with early scenarios lacking fine-tuned genetic regulation of development. By physics we refer to two different levels. One includes diffusion, excitability, oscillations or even cellular interaction forces, which can be captured by cell sorting models based on energy minimization functionals [11,18,19,20]. In this context, it is possible to evolve morphologies and observe the interplay between cell differentiation, growth and communication [21–23] with a properly defined optimization algorithm. The other level deals with the embedding of cells and organisms within a physical medium. Here, forces are closer to standard physics, i.e., the way cells might displace in the three-dimensional environment, how cells interact with the substrate and how cells find nutrients in a fluctuating medium. Previous work on physically-embedded artificial systems was pioneered by a number of researchers, who evolved artificial "organisms" [24,25]. Related work has considered the interplay between genetic networks and morphogenesis [26–30].

In this paper we would like to address, under a well defined framework, some questions related to the pre-multicellular world where cell aggregates (but not true organisms or body plans) could develop, providing some key preconditions for multicellularity to emerge. In particular, we would like to understand how the physical context, cell-cell adhesion properties, ecological and epigenetic factors concur to favor the emergence of cell aggregates. Moreover, we would like to understand how the environment influences (and can be influenced by) the evolution of cell diversity and cell-cell interactions. These are, we believe, important pieces in the extended evolutionary synthesis provided by evo-devo [31].

We also try to see how this explicit three-dimensional embodiment can play a role in favoring the emergence of innovations. We explore such a pre-body plan scenario, and its potential for generating complexity, by allowing a physically-embedded model of a cellular community to freely evolve. The interaction between our simple evolved aggregates and the environment is shown to trigger the emergence of *ecosystem engineering* (EEN) an important component in macroevolutionary patterns [32]. EEN can be defined as a modification of the abiotic environment by a species that affects resource availability for other species [33–37]. Because it involves a persistent ecological

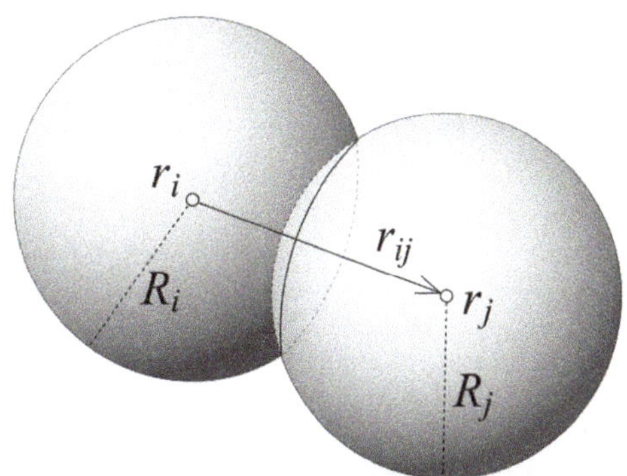

Figure 2. Simplified geometry of cell-cell collisions. Here, $\mathbf{r}_i$ and $\mathbf{r}_j$ are the cell positions, $\mathbf{r}_{ij}$ is the offset vector, R_i and R_j are the cell radius, and $R_i + R_j - |\mathbf{r}_{ij}|$ is the interpenetration depth.

modification, its presence implies the existence of ecological inheritance.

As we will shown below, a simple model of physically interacting cells with adhesion properties starting from a set of independent, genetically identical cells exploiting a single energy resource (from a given repertoire) evolves in time towards a spatially segregated community involving a trophic chain. The ecological network includes both a population of generalists feeding on all available food sources along with a population of specialized detritivores. The transition from the original monomorphic population to the spatially organized aggregate with ecological structure takes place through the emergence of an innovation grounded in evolving adhesion between cells and walls as well as cell-cell adhesion. In spite of its simplicity, it fairly well illustrates the potential of this type of model to explore the emergence of major transitions in pre-Mendelian scenarios.

Environment, Cells and Physical Simulation

In our model, evolution takes place within a spatially confined environment (the spatial domain is a cube with floor and walls) where physical forces play a role as external constraints. Organisms are spatially embedded structures and their embodiment is relevant as it provides the proper link with the external world and the biotic scenario where other organisms inhabit.

Physical models of cell interactions have been developed for a broad range of problems involving multicellular assemblies [38–41] and our study follows some of the standard methods of computational physics [42]. Figure 1 displays a basic scheme of the system considered here along with the different components of the physical interactions that will be taken into account.

1.1 Cells and particles

Our starting point is a population of single-cell organisms, where each cell in the initial population is identical. Cells and particles are simulated with rigid bodies moving within a fluid-like environment. A cell (particle) has spherical geometry with radius R_i, mass M_i, spatial position r_i and velocity $\mathbf{v}_i = dr_i/dt$. The motion of a cell is described by the standard second law:

$$M_i \frac{d\mathbf{v}_i}{dt} = \mathbf{F}_i \tag{1}$$

Numerical integration gives cell velocity at time $t + \Delta t$:

$$\mathbf{v}_i(t+\Delta t) = \mathbf{v}_i(t) + \frac{\mathbf{F}_i}{M_i}\Delta t \tag{2}$$

where Δt is the size of the integration step, and the total force acting on M_i will be:

$$\mathbf{F}_i = \mathbf{F}_e + \mathbf{F}_i^c + \mathbf{F}_i^n + \mathbf{F}_i^l - k_d\mathbf{v}_i + M_i\mathbf{g} \tag{3}$$

applied to any cell is the sum of environmental forces $\mathbf{F}_e$, the gravitational field $\mathbf{g}$, the collision force $\mathbf{F}_i^c$, the cell-wall adhesion $\mathbf{F}_i^n$ and the cell-cell adhesion $\mathbf{F}_i^l$ term.

1.2 Environmental forces

For simplicity, we asume that the external environment exerts the same force to every cell or particle, i. e.:

$$\mathbf{F}_e = k_e\mathbf{n} \tag{4}$$

where n is a random vector with $|\mathbf{n}| = 1$ and $k_e > 0$ is a constant parameter that indicates the strength of the external field. This choice implies that external fluctuations in the fluid medium are homogeneous, thus affecting all parts of the system uniformly. This approximation thus neglects potential effects played by small-scale eddies, which might actually play a role in adaptational changes.

Movement of particles in a fluid is subject to dissipation, as defined by a viscous drag $(-k_d\mathbf{v}_i)$ where k_d is the drag coefficient associated to the surrounding fluid. The effect of drag is to resist motion, making the particle gradually come to rest in the absence of other influences.

1.3 Cell-cell collisions

We apply the discrete element method [43] to the computation of collision forces between the $i-th$ cell and all its interpenetrating cells, that is, cells located at a distance below a given threshold (see figure 2):

$$\mathbf{F}_i^c = \sum_{i \neq j} \nu \left[\mathbf{v}_{ij} - \frac{\mathbf{r}_{ij}}{\|\mathbf{r}_{ij}\|}\left(\mathbf{v}_{ij}\frac{\mathbf{r}_{ij}}{\|\mathbf{r}_{ij}\|}\right)\right] + \xi\mathbf{v}_{ij} - \nabla\Phi_{ij}(\mathbf{r}_{ij}; R_i, R_j) \tag{5}$$

where ν is the shear coefficient, ξ is the damping coefficient, k is the spring coefficient, and $\mathbf{r}_{ij} = \mathbf{r}_j - \mathbf{r}_i$ is the offset vector between the $\mathbf{r}_i$ and $\mathbf{r}_j$ cell positions, and $\mathbf{v}_{ij} = \mathbf{v}_j - \mathbf{v}_i$ is the difference between cell velocities. The last term in the right-hand side gives the force resulting from a potential function associated to a soft-core interaction, namely.

$$\Phi_{ij}(\mathbf{r}_{ij}; R_i, R_j) = \frac{1}{2}k\left(R_i + R_j - \|\mathbf{r}_{ij}\|\right)^2 \tag{6}$$

Computation of spatial interaction forces requires $N(N-1)/2$ possible collision checks for N bodies in the worst-case scenario. In order to reduce this computational load we will use a spatial partitioning scheme. Here, a 3-D uniform grid subdivides the entire container volume in equally-sized square voxels [42]. All the bodies (cells and particles) are sorted by a mapping function $\mathbf{h(r)}$ that computes their voxel index:

$$\mathbf{h(r)} = \frac{1}{d}(\mathbf{r} - \mathbf{o}) \tag{7}$$

where o is the grid origin and d is the voxel size. For each body, we only check their collisions with different particles and cells belonging to a neighboring voxel, i.e., having indexes $\mathbf{h(r)}$ or $\mathbf{h(r)} \pm 1$. By adjusting the size of the voxel partition, we can minimize the total number of collision tests and enable the real-time simulation of large numbers of physically interacting bodies.

1.4 Cell-substrate adhesion

Attachment of cells to surfaces may provide a favorable micro-environment (e.g., biofilms). Cell-wall distance $D(\mathbf{r}_i)$ is:

$$D(\mathbf{r}_i) = \frac{ax_i + by_i + cz_i + d}{\sqrt{a^2 + b^2 + c^2}} \tag{8}$$

where $ax + by + cz + d = 0$ is the plane equation for the closest wall to the cell located at $\mathbf{r}_i = (x_i, y_i, z_i)$. When a cell with adhesion strength to the substrate $J_i^f > 0$ is closer than the adhesion range $\delta_w > 0$, i.e., $D(\mathbf{x}_i) < \delta_w$, we attach a spring connecting the cell $\mathbf{x}_i$ with its projection on the wall $\mathbf{x}_i^w$ (see figure 3). Now, the wall spring exerts the following attraction force:

$$\mathbf{F}_i^n = -k_s\left(\|\mathbf{r}_i - \mathbf{r}_i^w\| - d_s\right)\frac{\mathbf{r}_i - \mathbf{r}_i^w}{\|\mathbf{r}_i - \mathbf{r}_i^w\|} \tag{9}$$

where d_s is the spring equilibrium distance, k_s is the spring constant and $\mathbf{F}_n^i = 0$ when the cell is not attached to any spring. Existing cell-wall springs can be removed with certain probability $1 - q(i)$ or when the spring length is above the maximal length, i.e., $\|\mathbf{r}_i - \mathbf{r}_i^w\| > d_s^m$. As we will see, cells can evolve cell-wall adhesion J_i^f in order to maximize the intake of nutrient particles.

1.5 Cell-cell adhesion

Cells can form aggregates by attaching to other cells. Each cell has an intrinsic probability J_i^c to create a new adhesion link. Given two close cells located at $\mathbf{r}_i$ and $\mathbf{r}_j$, we will set an adhesion string connecting them with probability $(J_i^c + J_j^c)/2$. The adhesion force to any cell is the sum of forces contributed by all the active cell-cell adhesion springs:

$$\mathbf{F}_i^l = -\sum_j k_l\left(\|\mathbf{r}_i - \mathbf{r}_j\| - d_l\right)\frac{\mathbf{r}_i - \mathbf{r}_j}{\|\mathbf{r}_i - \mathbf{r}_j\|} \tag{10}$$

where d_l is the spring equilibrium distance, and k_l is the spring constant. Adhesion springs break spontaneously with rate $\delta \approx 0.001$ or when the spring is very large, i.e., $\|\mathbf{r}_i - \mathbf{r}_j\| > d_l$.

On the other hand, interpenetration collisions between cells and walls are not resolved with explicit forces (see figure 3A). When a body interpenetrates a wall, it bounces along its negative velocity direction, i.e., $\mathbf{v}_i(t + \Delta t) = \mu \mathbf{v}_i(t)$ by a constant factor $\mu < 0$. In addition, we relocate the body outside the wall (see below). This approximation ensures that particles will never move beyond container limits and does not change the main results presented here. The final cell (or particle) position is obtained from:

$$\mathbf{r}_i(t + \Delta t) = \mathbf{r}_i(t) - \delta\mathbf{n} + \mathbf{v}_i(t + \Delta t)\Delta t \tag{11}$$

where $\delta > 0$ is the cell-wall interpenetration distance and n is the unit normal vector of the wall at the collision point with $|\mathbf{n}| = 1$ (see figure 3A).

Finally, the computation of particle force $\mathbf{f}_i$ is more simple than in the case of cells. Food particles can collide with boundary walls, other particles and with cells (see below). But particles cannot adhere to the walls, or to any other body. The total force exerted on a particle is

$$\mathbf{f_i} = \mathbf{F}_e + \mathbf{f}_i^p + \mathbf{f}_i^c - k_d\mathbf{v}_i + \mathbf{g} \tag{12}$$

where $\mathbf{F}_e$ is the environmental force (as defined for cells, see above), $\mathbf{v}_i$ is the particle velocity and $\mathbf{g}$ is the gravity field. Particle collision force $\mathbf{f}_i^p$ is similar to the equation used to compute cell collision responses (5) but using particle mass (M) and radius (r). Similarly, the term $\mathbf{f}_i^c$ accounting for the collision response between any pair of interpenetrating particles mirrors equation (5).

We have calibrated the parameters associated to the physics as described above in order to avoid numerical instabilities. We have used small integration steps, i.e., $\Delta t = 0.03$ and fixed several physical parameters, including the cell and particle masses and the spring constants, to suitable values. An exhaustive exploration of the physical parameter space will be investigated elsewhere.

Evolutionary Rules

Once our embodied, physical description of how interactions take place within our environment, we need to further extend our model by including evolutionary rules. Mechanical interactions, for example, can be understood in terms of predefined mechanisms (constrains derived from the nature of physical laws) together with sets of parameters that tune their relevance in terms of how different forces influence dynamics. In this paper we explore the impact of such parameter changes once the basic laws are already in place.

The introduction of physical constraints as part of the framework defining our cells and their interactions allows us to integrate different factors emerging at the boundaries between physics and biology. Moreover, the explicit nature of selection pressures associated with gathering nutrients from a spatially explicit, fluctuating environment allows to explicitly introduce ecological factors into play. In summary, the model takes into account physical embodiment, ecological and evolutionary constraints and allows interaction parameters among cells to change. Although no developmental programs or gene regulatory modules can emerge at this level, we will see how the degrees of freedom included in our approximation allow cell aggregates to emerge, along with innovation and the creation of niches.

At any given time, there is a set of cells $\{C_i\}$ and a set of energy particles of identical mass m belonging to different energy types, here indicated as E_k, with $k = 1,...,n_p$ (here we use $n_p = 4$ sources). Each cell C_i can in principle metabolize a subset of different sources with efficiency $\eta_{ik} \in [0,1]$. At the beginning, we start from a homogenous population of N_0 cells, each having mass $M_i = M_0$, all feeding on the same resource, namely E_1 i.e., $\eta_{i1} > 0$ and $\eta_{ij} = 0$ for all $j > 1$. At every simulation step, we introduce N_p new particles in the system (here we use $N_p = 80$, uniformly distributed over each class) starting at the top of our world. This number provides a quantitative value to the intake energy flow Φ of particles entering the system.

Eventually, cells collide with nutrient particles and consume them according to the cell-specific efficiency η_{ij} (a given nutrient particle is consumed only when this is non-zero, otherwise the particle is deflected and does not enter into the cell body). A particle of class E_k will be successfully grazed contributing to a linear mass increase:

$$M_i(t + \Delta t) = M_i(t) + \eta_{ik} m \tag{13}$$

Once consumed, the particle is removed from the system. Nutrient particles have a characteristic lifetime and degrade at rate μ_p. Cells consume a constant amount of energy $e_s > 0$ per time step to sustain themselves. A cell dies when mass M_i falls below a critical threshold $M_i < M_c$. Cell death leads to disintegration and the release of $(M_i(t)/m) \approx (M_c/m)$ residual particles to the medium. These detritus particles will be consumed by detritivores when their associated efficiency is nonzero. Detritus particles have also a characteristic lifetime and degrade with a slower degradation rate, here fixed to $\mu_p/2$.

Cells divide once their total mass is at least twice their initial mass, $M_i(t) \geq 2M_0$. After reproduction, the mother cell C_i reduces its mass by M_0, which is transferred to the daughter cell, say C_k. In other words, we have

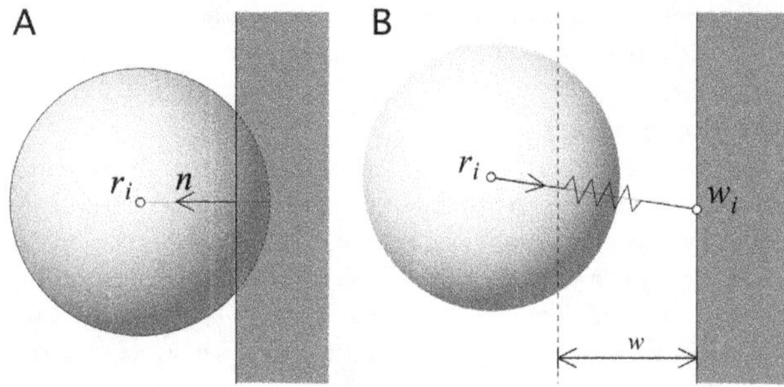

Figure 3. Possible interactions between cells and boundary walls. (A) Bouncing collision takes place when the cell interpenetrates the nearest wall any distance $\delta > 0$ along the unit normal vector n of the wall. (B) Cell-wall adhesion occurs when cell-wall distance is below some threshold w and according to adhesion probability (which depends on the cell genome, see text). In this case, a spring attaches the cell to a fixed point w_i on the wall (see text).

$$M_i \rightarrow M_i - M_0$$

$$M_k = M_0$$

However, cell division is limited by the available surrounding space. Here, we allow the mother cell to reproduce only when its number of neighboring cells (i.e., cells within a radius of size $2R_i$) is below a given density threshold N_c. The new offspring is placed at a random location close to the mother cell, and specific rules ensure that this new cell is within the boundaries of the simulated medium.

The offspring cell C_k inherits metabolic efficiencies from the mother cell C_i. Metabolic efficiencies can be slightly changed, i. e.,

$$\eta_{jk} = \eta_{ik} + v$$

where v is non-zero with probability of mutation $\mu \approx 0.02$ and zero otherwise. Here, the random v perturbation follows a Gaussian distribution with zero mean and standard deviation $\sigma_m = 0.3$. In addition, the offspring cell also inherits both cell-floor adhesion and cell-cell adhesion coefficients, respectively. Again, mutations are allowed to occur, and we have now:

$$J_k^f = J_i^f + v \tag{14}$$

$$J_k^c = J_i^c + v \tag{15}$$

The values of these parameters are always normalized between zero and one. If a given mutation is accepted and the parameters are either lower than zero or higher than one, they are fixed to zero or one, respectively. Similarly, efficiencies are normalized so that the condition $\sum_j \eta_{ij} = 1$ is always satisfied.

Results

As a result of the previous set of rules and initial condition, we have a simple ecological food web involving a set of resources and a single specialist grazer, which takes E_1 particles and grows at the expense of metabolizing them. Further degradation of particles leads to waste that is removed from the virtual tank and moreover cell death also generates an additional resource (the detritus compartment, E_d). Ecologically, this is our starting point, which will evolve as organisms change their feeding preferences and as a consequence of interactions between organisms and their physical environment.

Given the potential for evolving physical parameters as well as nutrient intake-related parameters, our cells will be able to evolve within a range of possible adaptations. For example, given the number of potential food sources to be consumed, we will observe a spread from the original parameter set defined by

$$\eta_{i1} = 1 \cap \eta_{i,j>1} = 0 \tag{16}$$

to a continuum space of efficiencies

$$\eta_i \equiv \{(\eta_{i1}, \ldots, \eta_{in}) \mid \eta_{ij} \in [0,1]; \ \sum_j \eta_{ij} = 1 \ \} \tag{17}$$

As we will see below, these spread leads to a predictable outcome in a first phase of the evolution process, while it leads to an unexpected innovation later on.

1.1 Transition to generalism epoch

The first trend observed in all our simulations is a tendency from the starting specialization (all cells exploiting one source with maximal efficiency) to generalism: mutations allow to exploit other resources with less efficiency but overall this is a better strategy given the finite amount of energy particles. In our model, $n_p = 4$ different types of energy particles are used. Additionally, since detritus particles resulting from cell death can also be consumed, a total of $n_p + 1$ types of particles are available.

Nutrients enter the system from the upper layer at a constant rate p and degrade into waste at a rate δ_k. Under the absence of grazers, the time evolution of the number N_k of energy particles of class k will follow a linear model $dN_k/dt = p - \delta_p N_k$ (for $k = 1, \ldots, n_p$) and thus each component will have an average steady value of $N_k^* = p/\delta_p$. Then, the overall number of particles at the beginning (before grazing starts) will be $N^* = \Phi p/\delta$ with $\Phi = n_p p$ is the total flow of incoming particles. Right at the beginning, our grazers exploit only one source particle (say E_1). If no mutations were involved, we would observe a trend towards an equilibrium population of both cells and class-one particles over time.

The above describes the basic dynamics that occurs at the very early stages in our model, when only one source is being exploited. However, since we are interested in the long-term evolution, we would like to study the particle-grazer population dynamics. One of the first trends easily observed in our model is the tendency towards a broader range of exploitable resources, that is, grazers become generalists. Such trend towards a more opportunistic behavior occurs when the payoff given by the exploitation of a variety of resources outweighs the loss of efficiency as more resources are grazed.

The degree of generalism of a given individual is measured by means cell entropy $H(C_i)$ defined as as the normalized diversity of efficiencies:

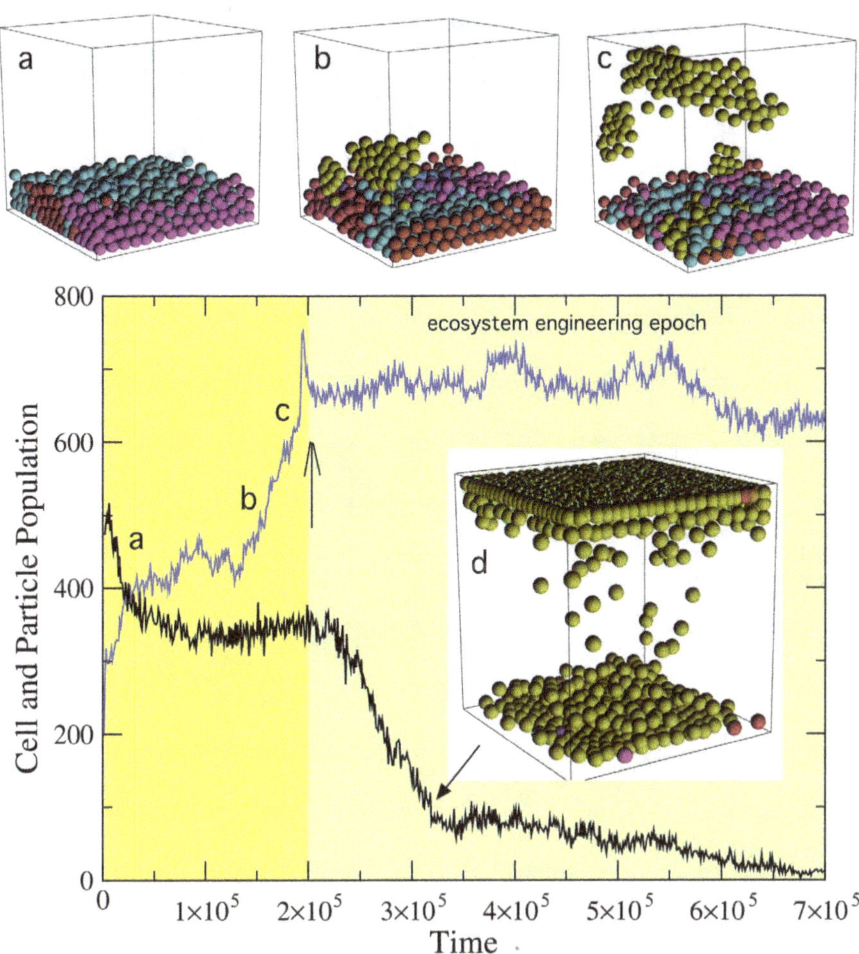

Figure 4. The time evolution of cells (blue) and particles (black) can be described with a sequence of processes. Before the transition to the inverted population state (inset) cells grow and divide as they also evolve their characteristic parameters. At the beginning, the cell population grows by adapting to the variety of energy sources and we can see a thick layer of cells (a) plotted at $t=17000$. After a while, an increase in the surface displayed by the population increases its efficiency to gather particles. Such increase is reached by evolving cell-substrate and cell-cell adhesion. The state shown in (b) is found at $t=170000$ where we can appreciate a cluster of cells which appear to be climbing the surface of the cube. In (c) the cluster close to reaching the upper floor (here $t=200000$) at this point the cell population does not experience further significant increase, but the number of particles decays to very low values. Here bright cells have high floor adhesion (where yellow indicates maximum adhesion) and darker colors correspond to cells with low adhesion or free-moving. Notice that adhesion evolves first in cells close to the wall boundaries. Here: $\Phi=25$ particles per timestep, $\Delta t=0.03, \delta_w=0.05, d_s=0.01, v=80, \xi=80, k=400, M_i=20, R_i=0.08, m_i=3.35, r_i=0.025, k_e=2, k_d=0.001$ (see text).

$$H(C_i;\eta_i)=-\frac{1}{\log(n_p+1)}\sum_{1\leq j\leq n_p+1}\eta_{ij}\log[\eta_{ij}] \qquad (18)$$

where η_{ij} is the relative efficiency of the i-th cell when feeding on the $j-th$ nutrient resource and n_p is the number of different resources. Notice that, in the initial configuration, all cells have entropy $H(C_i)=0$. As defined, the degree of generalism will be zero when only one source is used (specialized diet) and will reach a maximum value for individuals grazing on multiple sources. Similarly, since the average efficiency for each cell is given by

$$\langle\eta_{ij}\rangle=\frac{1}{n_p(i)}\sum_j\eta_{ij} \qquad (19)$$

where $n_p(i)$ indicates the number of nutrient sources exploited by C_i, given the normalization of efficiencies we have simply $\langle\eta_{ij}\rangle=1/n_p(i)$.

The simulation shows a steady increase of cell-wall attachment and cell-cell adhesion followed by a steady increase in number of cells. This is a consequence of the fact that, in order to get access to falling particles, a larger surface is needed. Cells who attach to the surface can climb up the walls. Having a moderate cell-cell adhesion also helps in avoiding them to fall down. In this way, we have formed aggregates that are moving up as cells divide. The impact of this, along with the simultaneous tendency towards a generalist grazing behavior, can be seen in figure 4. The cell population (blue curve) grows rapidly as the aggregates start to emerge and expand, eventually covering the top layer.

During the process, groups of cells, often forming layers parallel to the floor, become increasingly larger. Three snapshots involving transient steps (a-c) are also displayed in figure 4, where we have used color-coded spheres to indicate the strength of the cell-wall

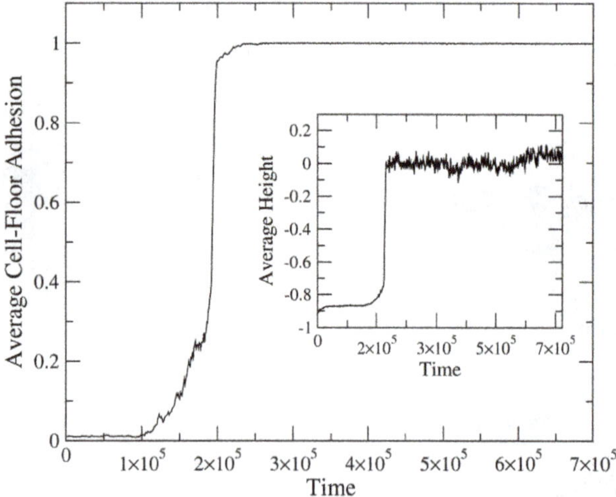

Figure 5. Sharp transition in the time evolution of average cell-floor adhesion. This transition is associated to the increasing height (inset) of the cell populations as further cells climb the walls towards the upper layer. Here the top floor is taken as the zero coordinate, whereas the bottom floor is taken as the $z = -1$ (normalized) minimum (see text).

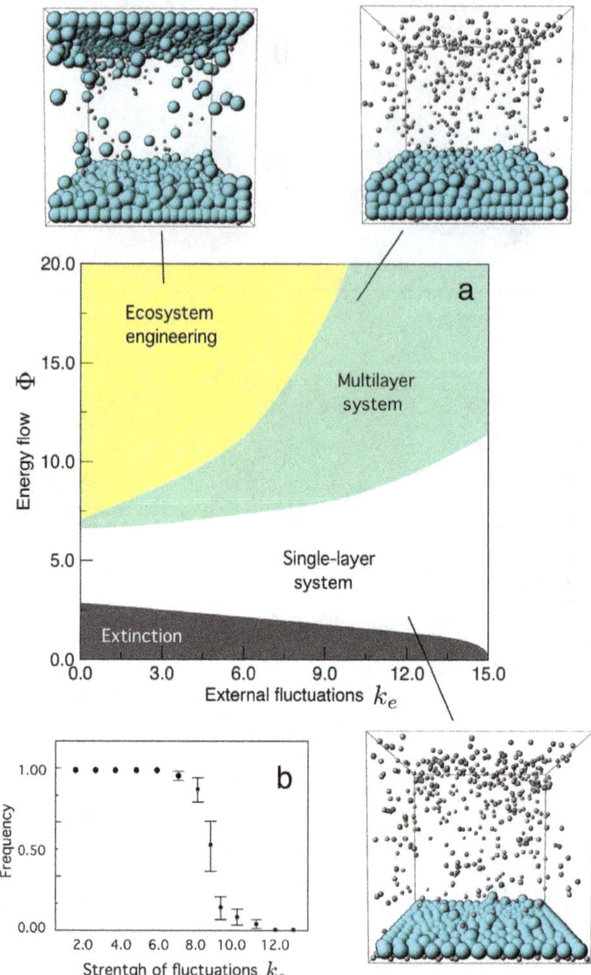

Figure 6. Interplay between energy intake and environmental fluctuations on the emergence of ecosystem engineering in CHIMERA's model. Here we have run CHIMERA using the same parameters of the previous figure but with different combinations of (k_e,Φ). Four phases are found (a) including extinction (lower part), cellular monolayers as well as multilayered systems. Examples of each scenario are indicated with 3D snapshots where the adhesion strength between cells and the boundaries is color coded. A quantitative analysis of the transition from the EEN to the multilayer phase is indicated in (b), where $\Phi = 20$ has been used along with different levels of fluctuation. Ten replicas of each parameter combination were used and 2×10^5 steps used to determine the final state. The standard deviation is also shown as error bars.

adhesion. Lighter and spheres indicate higher and lower levels of adhesion, respectively.

1.2 Transition to ecosystem engineering epoch

As cells occupy the upper layer, the flow of nutrients further declines (filled arrow, figure 4) to reach very low levels. The kinetics of this process and how it is connected with evolving adhesion rates is summarized in figure 5, where we plot the average cell-floor adhesion (averaging over all the cell population) and the corresponding average height of cells within the cube (inset). Here we have used the top floor as the ground height ($z = 0$) and the bottom one as the minimum. For convenience, we have normalized the later to $z = -1$. We can appreciate in these plots how rapidly the selection for higher adhesion occurs. As a consequence of this redistribution of cells through the upper part of the system, a whole redistribution of flows takes place, effectively triggering the emergence of a new ecological organization.

The most relevant result of our study is a key innovation made by our evolving cell populations as they "discover" that gathering particles is easier if they attach to the top floor of the system. As discussed above, cells approach the nutrient source by evolving cell-floor adhesion. This allows the population to expand its effective area for gathering nutrients, but it also favors more frequent interactions with particles, which are trapped between cells and the walls. This makes intake slightly higher than far from the boundary increases (and thus a higher fitness) because the sustained intake of nutrients. Eventually, cells with high floor adhesion ($J_i^f \approx 1$) colonize the source of nutrients in a brief burst of super-exponential growth. This is illustrated by figure 4d, where we can see a snapshot of our system soon after cells have "discovered" the roof. In terms of the population dynamics, it is also observed that the number of free nutrient particles declines as they are more efficiently found and removed from the system.

The emergence of EEN is common but requires both a high enough intake of energy (the flow of particles, Φ) and a moderate level of environmental fluctuations. The analysis of these two parameters reveals several phases in the potential types of structures that the system generates. This is illustrated in figure 6 where we present a summary of our exploration of the parameter space defined by (k_e,Φ). The choice of these two parameters is based on considering two relevant aspects associated to adaptation in evolving systems. The amplitude of the external fluctuations (as given by k_e, see equation 4) provides a measure of the random movements experienced by both cells and energy particles. In order to approximately determine the boundaries of these phases, the parameter space was partitioned into 25×25 pairs and evolution experiments were done for each pair.

The phase diagram reveals four phases. As it should be expected, at low flow levels the initial population cannot sustain itself. However, it is interesting to notice that the domain of extinction gets smaller for higher levels of perturbation. This is understandable when we realize what that means. Since we use a closed system, both cells and particles are influenced by external noise and, at high levels, there is a coherent movement of both types of elements, which end to aggregate in the corners of the lower level. This introduces a natural increase of encounter rates and thus allows cell survival more feasible. In a way, the high environmental noise leads to a predictable outcome due to the presence of boundary conditions.

The next phase (indicated in white color) involves a more or less stable population of cells that are confined to the floor of the system. Adhesion is high for low k_e and decreases to small (but non-zero) levels at higher levels. Here the population presents the largest area compatible with the incoming flow. On top of this phase, several layers can be stabilized. Well-defined smooth multilayers are observed on the left part of the diagram whereas they become a bit rugged on the right side, due to the presence of noise. A moderate level of cell-cell interactions allows to stabilize the system. It is within this phase that the initially multilayered system can develop a more complex organization and EEN.

An example of the type of transition that occurs between phases is illustrated in figure 6b. Here we have fixed the incoming flow to $\Phi=20$ and determined the presence or absence of a top population of cells after 2×10^5 steps, using different k_e values. Averaging over ten replicas for each k_e, we obtain a phase transition close to the reported boundary. Around $k_e^*\approx8$ we observe a phase change separating two clear domains associated to the presence $k_e<k_e^*$ or absence $k_e>k_e^*$ of a top layer. The error bars provide the standard deviation and are highest at the transition point, consistently with so called second-order transitions [44] where fluctuations strongly increase at criticality.

1.3 Emergence of detritivores

A major consequence is derived from this innovation. As cells in the top layer start to dominate the whole flow of nutrients, something new happens. After the colonization is completed, cells die and their waste material falls to the bottom. The resulting detritus stays for a while and is eventually removed by degradation. But some cells have already developed a mild efficiency to exploit these particles. After falling from the roof, these cells will find themselves inhabiting a niche that is rich in a given type of energy source. After a while, they develop a higher efficiency and eventually become specialists. This is illustrated in figure 7a–b, where we can see that the efficiency of detritivores grows fast after the top floor population has been established.

In the inset of figure 7a we display an example of the stationary state This transition illustrates the nonlinear impact of ecosystem engineering [31–33]. By changing the flow of nutrients, the top floor population causes a physical change in the environment due to their spatial distribution. As it occurs with freshwater phytoplankton organisms, which intercept light by placing themselves on the top of the water column or with higher plants, producing litter form dead leaves, our ecosystem provides a clever example of this scenario. By modulating the flow of nutrients, which eventually shifts from nutrient-rich to detritus-dominated flows, an effective asymmetry is generated.

The process of niche creation and the emergence of ecosystem engineering promote a new feedback towards the way detritivores behave. If cell-cell adhesion is allowed to evolve, it can be shown that the detritivore compartment evolves significant adhesion among individual cells, which can be described as a new form of

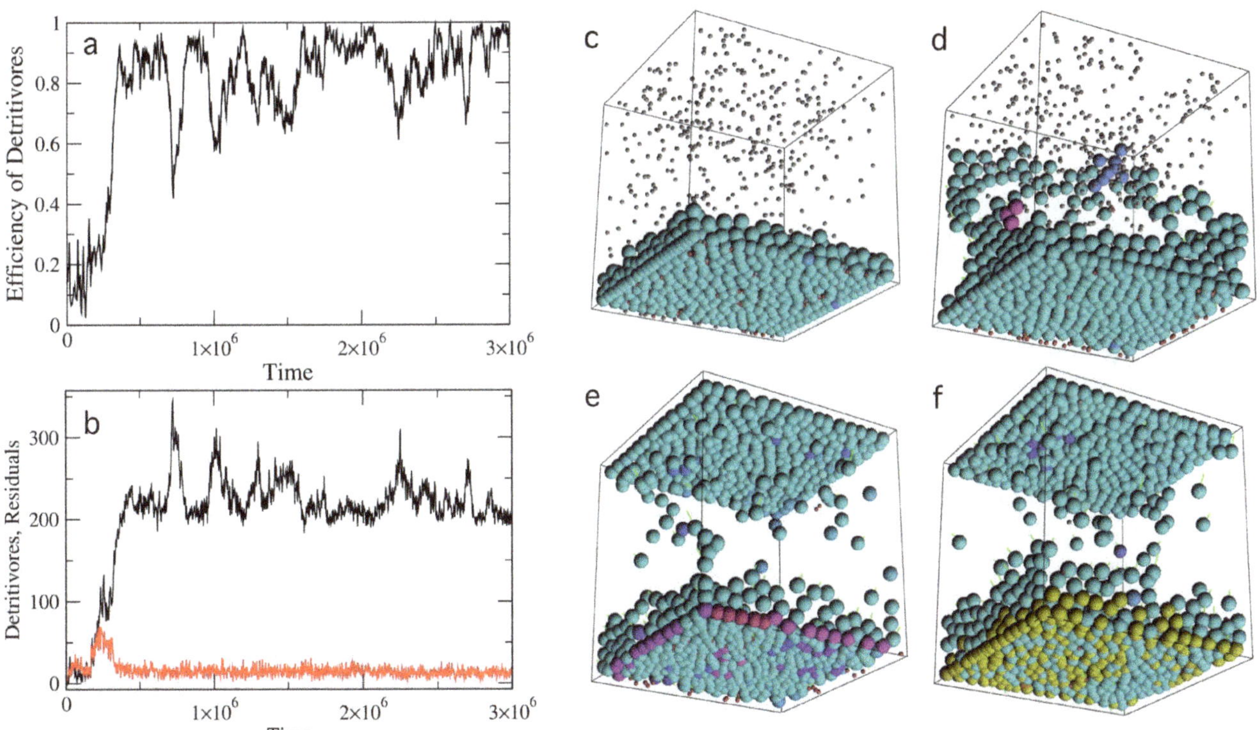

Figure 7. Dynamics of detritus grazing. In (a) the increase in efficiency rate of detritus-grazing cells is shown. In (b) we display the population dynamics of both detritivores (number of cells exhibiting some level of detritus feeding behavior) and detritus particles. In (c–f) four snapshots of the evolved system are shown. Detritivores evolve initially at the corners between wall boundaries thanks to the increased concentration of residuals. Afterwards, detritivores spread along the bottom plane and consume residuals produced by the disintegration of generalist cells coming from above. Four snapshots are shown in (c-f) at different times in the evolution of detritivores. Hot colors indicate the relative detritivore efficiency. Here: $\Phi=25$ particles per timestep, $\Delta t=0.03$, $\delta_w=0.05$, $d_s=0.01$, $v=80$, $\xi=80$, $k=400$, $M_i=20$, $R_i=0.08$, $m_i=3.35$, $r_i=0.025$, $k_e=2$, $k_d=0.001$ (see text).

cell aggregate. It is interesting to notice that our aggregates do not cooperate in terms of sharing resources, but instead they emerge from selection constraints imposed by the requirement of an expanded area, which can only be obtained by attaching to the vertical surfaces. An additional advantage is obtained by living in higher locations: energy particles are falling near there and in fact, in the presence of fluctuations, it is likely that particles hitting the walls are captured by cells adhered to them. But the process has a discontinuity: by covering the roof, an accelerated transformation takes place, modifying the whole organization of the ecological assembly and allowing the emergence of an additional trophic compartment.

In figure 8 we summarize the ecological transitions experienced by our system as we cross through the different regimes. Here we have indicated the flows of matter from the external energy sources to waste. In fig. 8a, the initial state of our system is shown, with several sources of particles but only one being exploited by the single specialist. Such scenario is slowly replaced by a heterogeneous one (fig. 8b) marked by an increasing tendency to generalism: mutations affect efficiency rates and the potential for exploiting several resources. As evolution proceeds and cells develop adhesion (here indicated as small protrusions) along with a complete generalism (c). Here we indicate with $\{E_i\}$ that cells use equally all types of nutrients with the same coupling. Eventually (d) the new niche of detritus-rich particles triggers the evolution of a specialized population of detritivores . This result is predictable, provided that the abundance of incoming particles is large enough.

Discussion

In this paper we have presented a spatially-explicit model of evolution that incorporates ecology, physical embodiment and a very simple description of individuals. Our motivation was to provide an in silico platform to explore early stages of multicellular evolution under a pre-Mendelian scenario. Such pre-Mendelian world would be characterized by a higher morphological plasticity, far from the gene-regulated one of the Cambrian world [11,18]. In

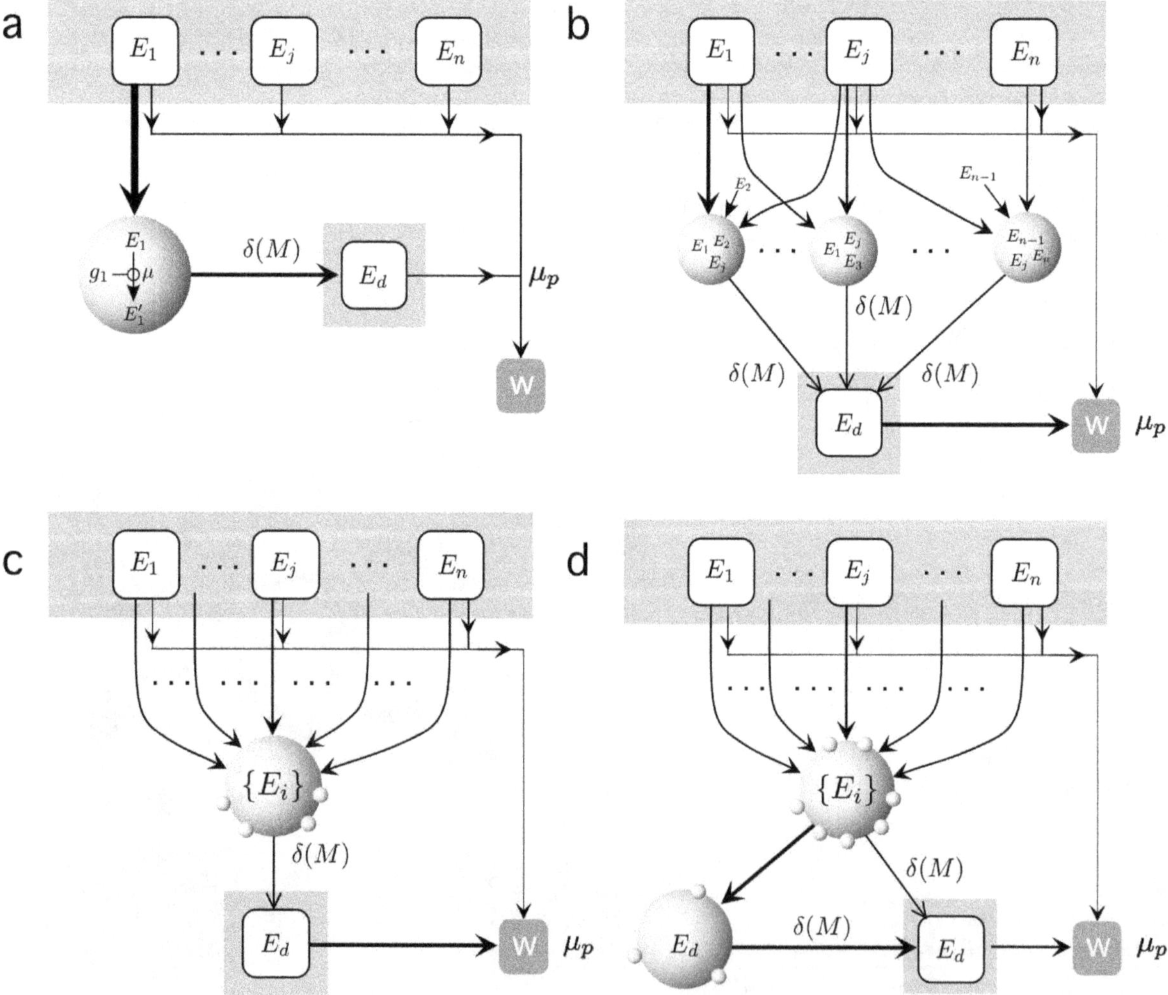

Figure 8. Chimera's food webs. (a) Flows in the initial food web, limited to a population of single-cell elements eating only one energy source (E_1) as defined by the CHIMERA model. The 1st nutrient is processed by the cells and transformed into cellular components high efficiency. At the end of the life cycle, the cell disintegrates into a number of residual particles. All molecules have similar degradation rates and get removed (right, lower box) from the system. After a while (b) an intermediate, diverse ecosystem is observed, with a large variety of cell exploiting different resources. This process proceeds until most cells become generalists, able to exploit all sources (c) whereas adhesion (indicated with smaller spheres) increases. Finally, once the population attached to the top floor has developed, another specialized population emerges at the bottom, fully composed by detritivores.

order to explore the potential repertoire of multicellular patterns that can emerge under these conditions, we have used a specific setting where a vertical flow of nutrients is introduced at the upper boundary of our system.

If the initial steps towards life took place in water, an appropriate model considering the role of physical interactions must take into account (a) events taking place within a fluid and (b) those associated to physical exchanges between individuals and individuals with the boundaries of the system (such as the sea bottom or a pond floor). By considering these basic forces, we introduce a minimal set of (possible) rules of interaction that can evolve through time. Such rules allow our artificial cells to explore their environment, interact and make decisions. But it also provides a very basic framework to explore the potential for finding patterns of pre-developmental pathways based on interactions among single-celled entities.

Despite the simplified nature of our simulation approach, which prevents (in particular) the emergence of complex regulatory networks and introduce limitations to the spatial organization of cell aggregates, several nontrivial results emerge. The evolutionary transition from the floor to the top layer leads to ecosystem engineering [27] and, as far as we know, this is the first example of such event happening in an artificial life system. By changing the actual flow of nutrients, they cause a physical change in the environment due to their physical distribution. Moreover, by doing so they also allow the emergence of a specialized, spatially segregated compartment of detritivores. Following the classification scheme from [35], our artificial creatures are autogenic engineers: they change their environment mainly via their own physical structures. Here the main structure created is a cell barrier associated to cell-floor interactions. The tempo of the transition exhibits a marked acceleration as the cell aggregates approach the top of the world. We can actually see that the cell population moves through its world by increasing in size until they detect the asymmetry associated to the top-down and start exploiting it. As it occurs with any other system out of equilibrium, our system is sustained by a gradient linked to a given flow of energy. In terms of the innovation process, the punctuated nature of the change is due to the transient time required in order to physically reach the energy-rich boundary and exploit it.

Future work should consider the explicit introduction of minimal genetic networks, the potential for cooperation in terms of nutrient sharing, an active role of our organisms in modifying their environmental conditions (thus adding an additional layer to the potential for ecosystem engineering) and variability of spatial conditions. These in silico experiments could be compared in some cases with evolutionary sequences of growth and selection observed in experiments involving transitions from single-cell to multicellular aggregates [45–49]. In this context, it can help designing and interpreting experimental approaches based on the introduction of selection favoring given traits. All these extensions of the CHIMERA model will allow us to approach relevant questions concerning the tempo and mode of the transition to multicellularity and how the different players (ecology, genetic interactions, physical embodiment and path dependence) influence the final outcome.

Finally, CHIMERA can be used to explore the effect of external events involving large extinctions and their aftermath. Recovery from mass extinction provides a unique insight into how communities are rebuild and the role played by different biotic and abiotic components in the reconstruction of a paleocommunity [50,51]. This is very well illustrated by the end-Permian extinction event [52,53] which devastated most species, to the point of near annihilation of complex life forms. The existing fossil record of the process and the patterns of change is very rich and well established [54] an illustrates the power of looking at macroevolutionary dynamics using multiple perspectives, from geological data to multispecies interactions. Some theoretical and computer models [55–57] have shown that a systems approach to these events can help to determine potential causal scenarios of recovery. But these models lack the physical embodiment that is characteristic of our model both at the individual and ecosystem levels. Since we also can trace the patterns of change in sediment particles and have a well-established segregation between trophic levels, we have a unique way of evaluating the relative contributions of ecological processes of competition or cooperation as well as the underlying evolutionary changes taking place at the individual level. These in silico experiments can help understanding the role played by altruism and cooperation and how they emerge and the nature of the resulting cooperative units [58,59].

Acknowledgments

This work started on a flight between Barcelona and New York and developed over the last years between Santa Fe, Venice, New York and the Complex Systems Lab. We thank the members of the Lab for so many useful discussions. SV thanks David Baraff for useful advices on constrained rigid-body simulation. We also thank the hospitality of the European Centre for Living Technology (ECLT) and the Santa Fe institute.

Author Contributions

Conceived and designed the experiments: RVS SV. Performed the experiments: RVS SV. Analyzed the data: RVS SV. Wrote the paper: RVS.

References

1. Carroll SB (2005) Endless forms most beautiful: The New Science of Evo Devo and the Making of the Animal Kingdom. W. W. Norton and Co.
2. Grossberg RK, Strahmann RR (2007) The Evolution of Multicellularity: A Minor Major Transition? Annual Reviews of Ecology and Evolution and Systematics 38: 621654.
3. Knoll AH (2011) The Multiple Origins of Complex Multicellularity Annual Reviews in Earth and Planetary Science 39: 217–239.
4. Lipps JH, Signor PH (1992) Origin and early evolution of the Metazoa. Plenum Press, New York.
5. Rokas A (2008) The molecular origins of multicellular transitions. Current Opinion Genetics and Development 18: 472–478.
6. King N (2004) The unicellular ancestry of animal development. Developmental Cell. 7: 313325.
7. Maynard Smith J, Szathmary E (1995) The major transitions in evolution. Oxford U. Press.
8. Niklas K (1997) The Evolutionary Biology of Plants. University of Chicago Press. Chicago.
9. Bonner JT (2001) First signals: the evolution of multicellular development. Princeton U. Press.
10. Fedonkin MA (2007) The Rise of Animals: Evolution and Diversification of the Kingdom Animalia. JHU Press.
11. Newman SA, Forgacs, G Muller GB (2006) Before programs: the physical origination of multicellular forms. International Journal of Developmental Biology 50: 289–299.
12. Eldredge N (1985) Unfinished Synthesis: Biological Hierarchies and Modern Evolutionary Thought. Oxford University Press.
13. McShea DW (2001) The hierarchical structure of organisms: a scale and documentation of a trend in the maximum. Paleobiology 27: 405–423.
14. Sebe-Pedros A, Roger AJ, Lang FB, King N, Ruiz-Trillo I (2010) Ancient origin of the integrinmediated adhesion and signaling machinery. Proceedings of the National Academy of Sciences USA. 107: 10142–10147.
15. Solé RV, Fernandez P, Kauffman SA (2003) Adaptive walks in a gene network model of morphogenesis: insights into the Cambrian explosion. International Journal of Developmental Biology 47: 685–693.

16. Fujimoto K, Ishihara S, Kaneko K (2008) Network evolution of body plans. PLoS ONE 3: e2772.
17. Tusscher KH, Hogeweg P (2011) Evolution of networks for body plan patterning; interplay of modularity, robustness and evolvability. PLoS Computational Biology 7: e1002208.
18. Newman SA, Baht R (2008) Dynamical patterning modules: physico-genetic determinants of morphological development and evolution. Physical Biology 5: 015008.
19. Graner F, Glazier JA (1992) Simulation of biological cell sorting using a two-dimensional extended Potts model. Physical Review Letters 69: 2013–2016.
20. Glazier JA, Graner F (1993) Simulation of the differential adhesion driven rearrangement of biological cells. Physical Review E 47: 21282154.
21. Hogeweg P (2000) Evolving Mechanisms of Morphogenesis: on the Interplay between Differential Adhesion and Cell Differentiation. Journal of Theoretical Biology 203: 317333.
22. Hogeweg P (2000) Shapes in the shadow: Evolutionary dynamics of morphogenesis. Artificial Life 6: 85–101.
23. Savill NJ, Hogeweg P (1997) Modeling morphogenesis: from single cells to crawling slugs. Journal of Theoretical Biology 184: 229–235.
24. Sims K (1994) Evolving 3D morphology and behavior by competition. Artificial Life 1: 353–372.
25. Pfeifer R, Lungarella M, Lida F (2007) Self-Organization, Embodiment, and Biologically Inspired Robotics. Science 318: 1088–1093.
26. Eggenberger P (1997) Evolving morphologies of simulated 3d organisms based on differential gene expression. In. Fourth Europ. Conf. Artificial Life, P. Husbands and I. Harvey, editors. 205–213. MIT Press.
27. Doursat R (2008) Organically grown architectures: creating decentralized, autonomous systems by embryomorphic engineering. In: Organic Computing, R. P. Würtz, ed: 167–200. Springer-Verlag.
28. Cummings F (2006) On the origin of pattern and form in early metazoans. International Journal of Developmental Biology 50: 193–208.
29. Kaandorp JA, Blom JG, Verhoef J, Filatov M, Postma M, et al. (2008) Modelling genetic regulation of growth and form in a branching sponge. Proceedings of the Royal Society B 275: 2569–2575.
30. Coen E, Rolland-Lagan AG, Matthews M, Bangham JA, Prusinkiewicz P (2004) The genetics of geometry.Proceedings of the National Academy of Sciences USA 101: 4728–4735.
31. Müller GB (2007) Evo-devo: extending the evolutionary synthesis. Nature Reviews Genetics 8: 943–949.
32. Forgacs G, Newman SA (2005) Biological physics of the developing embryo. Cambridge Univ. Press, Cambridge.
33. Erwin DH (2008) Macroevolution of ecosystem engineering, niche construction and diversity. Trends Ecol Evol. 23: 304–310.
34. Jones CG, Lawton JM, Shachak M (1994) Organisms as ecosystem engineers. OIKOS 69: 373–370.
35. Jones CG, Lawton JM, Shachak M (1997) Positive and negative effects of organisms as physical ecosystem engineers. Ecology 78: 1946–1957.
36. Erwin DH,Tweedt S (2012) Ecological drivers of the Ediacaran-Cambrian diversification of Metazoa. Evolutionary Ecology 26: 417–433.
37. Hastings A (2006) Ecosystem engineering in space and time. Ecology Letters 10: 153–164.
38. Palsson E (2001) A Three-dimensional model of cell movement in multicellular systems. Future Generation Computer Systems17: 835–852.
39. Palsson E, Othmer HG (2000) A model for individual and collective cell movement in Distyostlium discoideum. Proceedings of the National Academy of Sciences USA 97: 10448–10453.
40. Newman TJ (2005) Modeling Multi-cellular Systems using Sub-cellular Elements. Mathematical biosciences and engineering 2: 611–622.
41. Sandersius SA, Newman TJ (2008) Modelling cell rheology with the subcellular element model. Physical Biology 5: 015002.
42. Bell N, Yu Y, Muchs PJ (2005) Particle-Based Simulation of Granular Materials. In: SCA '05 Proceedings of the 2005 ACM SIGGRAPH Eurographics symposium on Computer animation. 77–86.
43. Ericson C (2005) Real-Time Collision Detection. Morgan Kaufmann.
44. Solé RV (2011) Phase Transitions. Princeton U. Press. Princeton.
45. Greig D, Louis EJ, Borts RH, Travisano M (2002) Hybrid speciation in experimental populations of yeast. Science 298: 1773–1775.
46. Rainey PB, Travisano M (1998) Adaptive radiation in a heterogeneous environment. Nature 394: 69–72.
47. Travisano M, Mongold JA, Bennett AF, Lenski RE (1995) Experimental tests of the roles of adaptation, chance, and history in evolution. Science 267: 87–90.
48. Lenski RE, Travisano M (1994) Dynamics of adaptation and diversification: a 10,000-generation experiment with bacterial populations. Proceedings of the National Academy of Sciences USA 91: 6808–6814.
49. Ratcliff WC, Denison RF, Borrello M, Travisano M (2012) Experimental evolution of multicellularity. Proceedings of the National Academy of Sciences USA 109: 1595–1600.
50. Erwin DH (1998) The end and the beginning: recoveries from mass extinctions. Trends in Ecology and Evolution 13: 344349.
51. Erwin DH (2001) Lessons from the past: biotic recoveries from mass extinction. Proceedings of the National Academy of Sciences USA 98: 53995403.
52. Erwin DH (1994) The PermoTriassic extinction. Nature 367: 231236.
53. Benton MJ, Twitchett RJ (2003) How to kill (almost) all life: the end-Permian extinction event. Trends in Ecology and Evolution 18: 358365.
54. Chen ZQ, Benton MJ (2012) The timing and pattern of biotic recovery following the end-Permian mass extinction. Nature Geoscience 5: 375383.
55. Solé RV, Montoya JM, Erwin DH (2002) Recovery after mass extinction: evolutionary assembly in large-scale biosphere dynamics. Philosophical Transactions of the Royal Society B 357: 697707.
56. Roopnarine PD (2006) Extinction cascades and catastrophe in ancient food webs. Paleobiology 32: 1–19.
57. Yedid G, Stredwick J, Ofria CA, Agapow PM (2012) A comparison of the e_ects of random and selective mass extinctions on erosion of evolutionary history in communities of digital organisms. PLoS ONE 7: e37233.
58. Queller DC (1997) Cooperation since life began. Quarterly Review of Biology 72: 184–188.
59. Queller DC (2000) Relatedness and the fraternal major transitions. Philosophical Transactions of the Royal Society B 355: 1647–1655.

11

Multiple Determinants of Whole and Regional Brain Volume among Terrestrial Carnivorans

Eli M. Swanson[1,2]*, Kay E. Holekamp[1,2], Barbara L. Lundrigan[1,2,3], Bradley M. Arsznov[5], Sharleen T. Sakai[4,5]

1 Department of Zoology, Michigan State University, East Lansing, Michigan, United States of America, **2** BEACON Center for the Study of Evolution in Action, Michigan State University, East Lansing, Michigan, United States of America, **3** Michigan State University Museum, Michigan State University, East Lansing, Michigan, United States of America, **4** Neuroscience Program, Michigan State University, East Lansing, Michigan, United States of America, **5** Department of Psychology, Michigan State University, East Lansing, Michigan, United States of America

Abstract

Mammalian brain volumes vary considerably, even after controlling for body size. Although several hypotheses have been proposed to explain this variation, most research in mammals on the evolution of encephalization has focused on primates, leaving the generality of these explanations uncertain. Furthermore, much research still addresses only one hypothesis at a time, despite the demonstrated importance of considering multiple factors simultaneously. We used phylogenetic comparative methods to investigate simultaneously the importance of several factors previously hypothesized to be important in neural evolution among mammalian carnivores, including social complexity, forelimb use, home range size, diet, life history, phylogeny, and recent evolutionary changes in body size. We also tested hypotheses suggesting roles for these variables in determining the relative volume of four brain regions measured using computed tomography. Our data suggest that, in contrast to brain size in primates, carnivoran brain size may lag behind body size over evolutionary time. Moreover, carnivore species that primarily consume vertebrates have the largest brains. Although we found no support for a role of social complexity in overall encephalization, relative cerebrum volume correlated positively with sociality. Finally, our results support negative relationships among different brain regions after accounting for overall endocranial volume, suggesting that increased size of one brain regions is often accompanied by reduced size in other regions rather than overall brain expansion.

Editor: Andrew Iwaniuk, University of Lethbridge, Canada

Funding: This research was funded by National Science Foundation grants IOB0618022 and IOS1121479 (Dr. Holekamp, Dr. Lundrigan) and an Intramural Research Grants Program (IRGP) grant from Michigan State University (Dr. Sakai, Dr. Holekamp, Dr. Lundrigan). Dr. Swanson was supported on an National Science Foundation predoctoral fellowship. The funders had no role in study design, data collection and analysis, decision to publish, or preparation of the manuscript.

Competing Interests: The authors have declared that no competing interests exist.

* E-mail: eliswanson@gmail.com

Introduction

The considerable brain size variation evident among mammals is thought to result primarily from variation in body size [1–3] and secondarily from variation in encephalization, which involves changes in brain size independent of body size [1]. Although body size often explains as much as 95% of the variance in absolute brain size, brain sizes at a given body size can nevertheless range over an order of magnitude [4], and a number of different factors have been proposed to explain this variation. Some of the most prominent factors proposed to explain variation in encephalization include social complexity [5,6], life history [7], recent evolutionary changes in body size [8], and complexity in the non-social environment as indicated by such variables as home range size, manual dexterity required during food processing, and factors related to diet such as complexity of foraging behavior [9,10].

The 'social brain hypothesis', which argues that degree of encephalization increases with the complexity of the intraspecific social environment [6,11–13], is one of the most popular hypotheses proposed to explain variation in encephalization. This hypothesis is strongly supported by data gathered from primates (see [14]). However, its generality among non-primate mammals is poorly understood, as it has only been tested in a few taxa [15,16], and different studies have yielded conflicting results even when such results were based on the same data (e.g. compare [17,18]). While the social brain hypothesis is probably the most widely studied explanation for encephalization, life history traits have also been suggested to influence encephalization. Gestation length, for example, has been linked to degree of encephalization because, relative to the rate of body size growth, prenatal brain growth is far more rapid than postnatal brain growth [1,19,20]. Prolonged lactation has also been suggested to result in increased brain size, as the nutritional benefits of extended access to milk may often be required to help offset the high metabolic cost of neural tissue [21–23]. Finally, longevity has been proposed to increase degree of encephalization for adaptive reasons; specifically, species with larger brains may be able to respond better to environmental changes requiring resource shifts during an extended lifespan [24,25]. In addition to social complexity and life history, characteristics of the physical environment might influence brain size independent of body size. For example, home range size has been suggested to relate to brain size because larger home ranges require species to utilize complex information about food location and distribution that would not be necessary for species more constantly in contact with their food sources [9]. Use of the forelimb in food processing has been suggested to relate to brain size [26], reflecting the link between manual dexterity and motor or somatosensory cortex [27–29]. Finally, diet may relate to degree of encephalization either as an energetic constraint due to the metabolic 'expense' of brain tissue [30,31], or because some diets require more complex foraging or processing techniques [9,32]. Despite the fact that these hypotheses relating to sociality, the non-social environment, and life history are generally viewed in a competitive framework, it is highly likely that more than one of these factors operate in a given species to shape brain volume [33,34]. Although some research has considered multiple factors (e.g. [35–37]), it is still common to examine only one of these potential sources of variation, despite the demonstrated importance of considering multiple hypotheses simultaneously [38].

Most of the hypotheses purporting to explain encephalization generalize specific functions to the entire brain, yet, different brain functions are often associated with neural activity in different areas of the brain [39]. Thus, many hypotheses proposed to explain overall encephalization should perhaps preferentially be applied to specific brain regions (e.g. [40–42]). This is rarely done, likely due in part to the difficulty of identifying and separating brain regions in a large set of taxa. Moreover, there is considerable controversy regarding the extent to which different brain regions can evolve independently. Expansion of particular brain regions may be the result of concerted change due to developmental linkages among brain regions [43]. Conversely, there is evidence that brain regions evolve independently, known as 'mosaic evolution' [44]. Both processes undoubtedly play roles in brain evolution ([45], pgs. 157–159), but their relative importance critically affects our ability to recognize adaptive variation in specific brain regions; we may be able to identify such adaptive variation in brain regions only if mosaic evolution is common. An additional consideration is simply that the brain must fit within the skull ([45], pg. 131). This simple requirement means that if antagonistic selection or developmental factors constrain skull size evolution, then increases in one brain region must be accompanied by concomitant decreases in other regions rather than overall increases in encephalization (e.g. [40,42]).

In addition to these adaptive explanations for variation in encephalization, other hypotheses have been proposed that are not directly adaptive. For example, some variation in the degree of encephalization has commonly been hypothesized to arise from an evolutionary 'lag', where body size evolves first, later followed by brain evolution [20,46–48]. The primary evidence for this hypothesis has been that the slope of a regression of brain volume on body size is much greater among distantly- than closely-related species [1,20,49]. If the relationship between brain and body size arose solely due to selection on body size, the two slopes should be the same [47], but they are not. Brain size has been observed to change more slowly than body size among closely related species pairs in which one species exhibits rapid recent body size change, such as those containing 'phyletic dwarfs' (e.g. [50]). However, there is currently no empirical evidence directly supporting the hypothesis that such lags persist over long evolutionary time periods or that this operates as a general mechanism in brain:body size evolution [51]. In fact, the only test of the 'lag' hypothesis across a large taxonomic group, the primates, failed to support this hypothesis [8].

Here we use computed tomography (CT) techniques to create virtual brain endocasts from the skulls of 36 terrestrial species in the order Carnivora to assess the relative importance of social, ecological, and life history traits on both overall encephalization and the relative volumes of specific brain regions. Carnivores offer an excellent model for these tests because they exhibit great variation in brain and body size, and their social and physical environments both span broad ranges of complexity. Nevertheless, in part because research on brain evolution still focuses mainly on primates (e.g. [9,20,42]) and birds (e.g. [35,37,52]) we still lack a complete understanding of the environmental correlates of encephalization in Carnivora (e.g. [17,18,53]). Previous work on brain evolution has demonstrated the importance of simultaneously assessing predictors in a framework integrating multiple hypotheses [5,38]. We therefore implement phylogenetically-corrected generalized least squares (PGLS) models to account for shared evolutionary history, while simultaneously assessing the importance of the social and physical environment in encephalization. Finally, we evaluate the effects of recent evolutionary changes in overall body size on relative brain volume, as well as effects of recent evolutionary changes in overall endocranial volume on the relative volumes of specific brain regions.

Methods

Data Collection

Details of phylogeny. We used the phylogeny for the order Carnivora presented by Bininda-Emonds et al. [54] due to its broad taxonomic coverage, but supplemented it with updated family-level molecular phylogenies for Felidae [55] and Hyaenidae [56]. Branch lengths are poorly known for Carnivora, so we used Pagel's arbitrary branch lengths [57], which is probably the most common approach (e.g. [18]). We estimated Blomberg's k [58], a measure of the degree of phylogenetic autocorrelation among our raw brain measurements using the 'picante' package [59] in R v.2.12.1 [60]. Branch length differences generally have little effect on regression analyses [61]. Pagel's branch lengths were calculated in Mesquite [62] using the PDAP:PDTree module [63]. All other transformations and analyses were performed in R.

Specimens and measurements. Skull specimens from multiple adult members of each of 36 carnivore species were obtained from the collections of the Michigan State University Museum, Field Museum of Natural History, Natural History Museum of Los Angeles County, National Museum of Natural History, and University of Michigan Museum of Zoology (see Table S1). We used mean values from each species, averaging male and female values when they differed (Table S2).

All skulls were scanned using a General Electric Lightspeed 4 slice CT or General Electric Discovery ST 16 slice CT scanner in the Department of Radiology at Michigan State University. Each skull was aligned in the scanner rostrocaudally to replicate the natural anatomical position of the head. Parameters for each scan were as follows: 0.625 mm slice thickness, 30 cm field of view, 5.62 mm/rotation table speed, and 0.562:1 pitch. CT images were saved in DICOM (Digital Imaging and Communications in Medicine) Centricity Version 2.2 format. Virtual endocasts were created using the software package MIMICS 11.02 (Materialise, Inc., Ann Arbor, MI, USA). The skull was separated from air space by setting a grayscale pixel value as the threshold for filling in the endocranial air space. This space was filled in each coronal slice starting rostrally where the cribiform plate forms the floor of the intracranial cavity and continuing caudally through the foramen magnum. The resulting coronal sections were combined to create a three-dimensional reconstruction (virtual endocast) of the intracranial cavity using the MIMICS 3D object operation. Smoothing algorithms were applied to enhance the image and eliminate uneven surfaces. Detailed external brain morphology, including gyral and sulcal patterns, could be seen clearly in all virtual endocasts utilized here [64].

Skull basal length was defined as the distance from the anterior border of the median incisive alveolus to the mid-ventral border of the foramen magnum. Skull height was measured as the greatest height of the cranium perpendicular to the plane of the basioccipital and basisphenoid bones ([65]: measurement 18), excluding the sagittal crest, when one was present [66]. Finally, zygomatic arch breadth was measured as the greatest width of the skull. A single observer (B.M.A) collected all linear skull measurements from the CT images. Species means for skull measures are given in Table S2.We used species-wide averages obtained for body mass from the literature because we did not have individual mass values for many of the specimens in our data set, and wanted to avoid introducing two separate sources of error (see Table S3).

Virtual Endocasts (VE). Assessments of total endocranial and regional volumes were obtained using the MIMICS 3D volume measurement operation. Total endocranial volume was defined as the volume extending from the rostral tip of the olfactory bulbs caudally to the foramen magnum. CT files were coded by animal number only, and analysis and demarcation of brain regions were conducted blind with regard to the identity of individuals or species. All volumetric measurements and analyses were made by a single observer (B.M.A). Although relevant comparisons among carnivorans have not been published, previous work suggests that differences between endocranial and brain volumes are either very small [67] or nonexistent [68]. Moreover, our CT measures of endocranial volume were comparable to actual brain volume measurements from Rohrs' data [69,70] given in Dunbar and Bever [71] (paired t-test: $t = 0.826$, number of species pairs $= 12$, $p = 0.425$). We also compared the log transformed endocranial volumes based on the present CT method with previously reported log transformed brain volumes primarily estimated using the bead method [17] and found no difference ($t = 0.320$, number of species pairs $n = 29$, $p = 0.751$). Because the species compositions of these two comparisons were different, log transformation was used to improve normality for the second comparison, but not the first. Log transformation did not qualitatively affect the result in either case.

For each endocast, sulcal patterns and/or bony landmarks were used to delineate 3 principal regions: anterior cerebrum (Ac) as a measure of frontal cortex, posterior cerebrum (Pc) as a measure of the remaining cerebrum, and the combination of the cerebellum and the brainstem (Cb+Bs) as an approximate rhombencephalic measure. For the volumetric analysis, anterior cerebrum (Ac) was defined as the region rostral to the junction of the cruciate sulcus and midline, and caudal to the olfactory bulbs. For species lacking cortical maps, we relied on identification of the anterior cortical areas in other carnivores, and applied the same criteria. In primates, frontal cortex includes cortex rostral to the central sulcus, but the likely homologue to the central sulcus in carnivorans is too subtle to provide a suitable landmark to use in endocasts [27]. The boundary between motor and somatosensory cortex is the post-cruciate sulcus, which is not reliably present in all carnivores [27]. Instead, the cruciate sulcus was used as a landmark for demarcating brain regions, as this is a prominent feature that demonstrates less intra- and interspecific variation than the post-cruciate sulcus [72]. The cruciate sulcus is coincident with the rostral-most portion of motor cortex in the cat (*Felis catus*: [73]), dog (*Canis lupus*: [74]), and raccoon (*Procyon lotor*: [75]). In each of these species, the dorsal bank of the cruciate sulcus is coincident with cytoarchitectonic area 4. This has not been confirmed for other carnivoran families, but seems likely given the phylogenetic spread of the families for which it has been confirmed (see Fig. 1 for families included in our analysis). Anterior cerebral volume was calculated from the endocranial slices and was thus comprised of frontal cortex and subcortical structures, including a small portion of the rostral-most head of the caudate nucleus, ventral pallidum, olfactory tubercle and prepiriform cortex. Posterior cerebrum (Pc) included the endocranial volume posterior to the cruciate sulcus, but anterior to the tentorium cerebelli. Lastly, the cerebellum and brainstem (Cb+Bs) are housed within the intracranial cavity in the posterior cranial fossa; this region was defined as the portion of the intracranial cavity that begins at the most anterior border of the tentorium cerebelli and extends posteriorly to the foramen magnum of the occipital bone.

Transformations and brain variables. Total endocranial volume included all of the measures of interest, as well as the olfactory bulbs, which were not used in other analyses. Overall endocranial volume relative to body size was calculated in two different ways, each using a different correction factor. First, we calculated overall endocranial volume relative to species mean body mass. Second, we calculated overall endocranial volume relative to the size of the particular skull from which each virtual endocast was generated; skull size was calculated as the first principal component axis (PC1) from a phylogenetically-corrected principal component analysis (PCA) on three skull measurements: basal length, skull height, and zygomatic arch breadth [76]. Here we use the acronym OEV to represent overall endocranial volume. Thus mass-corrected OEV is noted as MCOEV and size-corrected OEV as SCOEV. MCOEV and SCOEV were calculated as residuals from phylogenetically-corrected regressions of endocranial volume on body mass and skull size, respectively, by regressing endocranial volume on each measure using PGLS (PGLS: [77–80]).

In addition to OEV, we also estimated the volume of each of the measured brain regions relative to OEV, including cerebrum anterior to the cruciate sulcus (Ac), cerebrum posterior to the cruciate sulcus (Pc), total cerebrum (Ac+Pc), and hindbrain, which includes both cerebellum and brainstem (Cb+Bs). The relative volume of each brain region was calculated by taking residuals from a regression of the brain region on 'brain rest' (overall endocranial volume minus the volume of the region of interest; e.g. [40,71]); the resulting variables are henceforth referred to as the "relative" region (e.g. relative Ac, Pc, Cb+Bs, total cerebrum). All morphological data were log-transformed prior to analysis.

Ancestral reconstruction and SCI calculation. We performed maximum likelihood (ML) ancestral reconstruction [81] on multivariate skull size, body mass and endocranial volume using the 'geiger' package in R [82]. Indices of recent evolutionary size change (SCIs) were calculated by subtracting a reconstructed ancestral species value (ASV) from an extant species value (ESV). For this calculation, we used the value reconstructed for the node representing the most recent hypothetical ancestor shared by any other extant species in the phylogeny (see [83]). This calculation represents the magnitude of change rather than the rate, bypassing the problem associated with using arbitrary branch lengths when calculating rate of change.

Social, ecological and life history variables. We collected social, life history and ecological data from a variety of sources (detailed in Table S3). We used these variables to generate one composite variable representing social complexity and another representing life history traits. Ecological variables were not combined because we had no a priori expectation that those

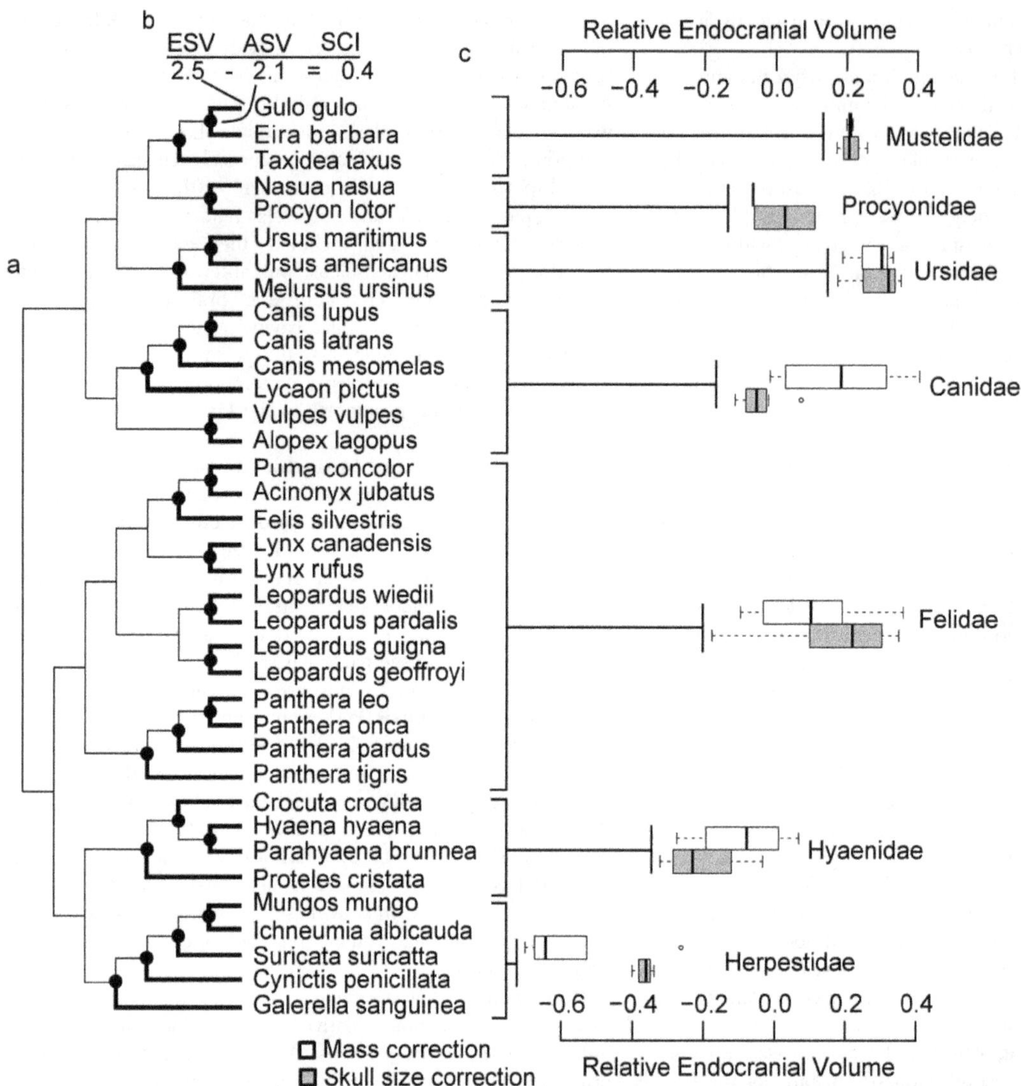

Figure 1. Carnivore phylogeny, demonstration of size-change indices, and relative endocranial volumes by family. a) Carnivore phylogeny with Pagel's arbitrary branch lengths. Filled circles represent the hypothetical ancestors or nodes at which the ancestral traits were estimated. Heavy lines link each extant species to the ancestral node that was subtracted from the value for the extant species to obtain size change indices (SCIs). b) Demonstration of how SCIs were calculated. Most recent estimated ancestral size values (ASV) were subtracted from the associated value for extant species size (ESV), and the difference is equal to the SCI. c) Box-and-whisker plot displaying degree of variation in relative brain size within each family. Relative MCOEV is indicated by a white box and relative SCOEV by a grey box. Boxes indicate interquartile range, and whiskers spread to the furthest points outside the interquartile range, but within 1.5 times the interquartile range from the median.

variables were conceptually related. For the variable representing social complexity, group size, feeding/hunting group size and degree of social cohesion were included in a phylogenetically-corrected PCA [76], and each species' score from the first principal component axis (PC1) was used as a composite measure of social complexity. A composite measure of sociality is useful because it has been suggested by a number of sources that complexity of the social system is more important than the number of individuals, especially given the high prevalence of diffuse grouping patterns [37,84,85]. A composite measure provides information not only on the size of groups, but their complexity, encompassing the degree of social cohesion and possible interaction during feeding/hunting. Similarly, age at weaning, gestation length, and longevity were included as life-history variables in a phylogenetically corrected PCA to calculate an overall life history axis for the sample (PC1). Life history variables have widely been documented as varying strongly along a multivariate axis, often termed the 'fast-slow' life history axis, which explains a great deal of life history variation even after correcting for body size [86–92]. Life history variables were corrected here for skull size, used as a proxy for overall body size, before performing the PCA. Ecological variables included diet (primarily carnivorous, insectivorous or omnivorous) and log home range size corrected for skull size. Finally, we scored species on an index describing degree of forepaw use in food processing (see Table S3). All data were collected in accordance with National Institutes of Health guidelines, and have been approved under Michigan State University's Animal Care and Use Protocol #AUF 07/-08-099-00.

Data Analysis

Influence of phylogeny. We estimated Blomberg's k for each relative brain measure using Pagel's branch lengths. In addition,

Moran's I, a general measure of autocorrelation, was estimated at three different taxonomic levels (suborder, family, genus) for each relative brain measure [93–95]. Moran's I was calculated on cophenetic distances among traits, both for the entire phylogeny and at lower taxonomic levels, using the R package 'ape' [96].

PGLS analysis. Phylogenetic comparative methods are potentially powerful tools for studying adaptation that are now commonly used to avoid the problem of phylogenetic non-independence [97–99]. Although there are alternative methods for addressing this problem, PGLS techniques allow simultaneous consideration and estimation of the degree of phylogenetic non-independence using Pagel's lambda (λ). λ describes a continuous variable in which zero represents a trait that displays no phylogenetic signal and one describes a trait that has evolved under brownian motion. We used PGLS to fit a series of six models with the ecological, social and life history variables of interest as predictor variables, as well as diet (see Table S3 for variables included). We fit each of the six models three different ways: fixing λ to zero, allowing it to assume its maximum likelihood estimate, and fixing it to one. Model selection was performed using sample-size corrected Akaike's Information Criterion (AICc). To avoid overfitting, a potential consequence of stepwise model selection procedures [100], we estimated only full models, comparing models that vary only in the degree of phylogenetic autocorrelation among residuals. Thus every model we estimated contained all possible predictor variables.

Results

Effect of Phylogeny

Phylogenetic autocorrelation based on Blomberg's k was strong and statistically significant for both measures of relative overall endocranial volume, and moderate and significant for relative Ac volume and relative Cb+Bs volume (Table 1). The strength of phylogenetic signal was moderate for relative Pc and relative total cerebrum volume, but did not differ significantly from zero. Estimates of phylogenetic signal using Moran's I generally supported the results obtained using Blomberg's k (Fig. 2). Specifically, traits that exhibited strong, statistically significant phylogenetic signal using Blomberg's k also did so using Moran's I, but this pattern was clear only at the taxonomic level of family, suggesting that phylogenetic autocorrelation at or below the level of family drives the observed overall phylogenetic autocorrelation. Autocorrelation at the level of genus was non-significant, even when of large magnitude, although this may simply reflect the small sample sizes within most genera in our data set (Fig. 1). Finally, autocorrelation at the level of suborder was statistically significant for all brain measures, with Moran's I close to 0 for all except SCOEV, for which it was negative (Fig. 2). Negative autocorrelation is exhibited by traits for which closely related species differ more than do distantly related species, and can result from character displacement [101].

Composite Social Complexity and Life History Variables

The first principal component (PC) axis resulting from the PCA on the three social variables used as our proxy for social complexity, explained 78.5% of the variance. All three univariate social variables exhibit strong positive loadings with PC1 (social group size: 0.932, feeding group size: 0.764, social cohesion: 0.950), indicating a large contribution from each. The composite life history variable (the scores from the first PC axis) resulting from the PCA on gestation length, weaning age, and maximum lifespan explained 44.2% of the variance in those data. Gestation length, weaning age, and maximum lifespan all increase with PC1 (loadings were 0.828, 0.499, and 0.627 respectively). As has been documented in many studies on life history covariation (e.g. [86,91,92,102,103]), the first multivariate axis of covariation among life history variables in our data appears to correspond to a 'fast-slow' life history axis, where species develop more slowly, reproduce more slowly, but live longer as PC1 increases. This 'fast-slow' axis is observed both with raw life history variables and also after correcting for mass [86,89,91,92,102,103].

Table 1. Phylogenetic autocorrelation among brain measures.

	K	z	p
MCOEV	1.188	−4.036	<0.001
SCOEV	1.169	−4.212	<0.001
Relative Ac	0.774	−3.293	<0.001
Relative Pc	0.4	−0.862	0.207
Relative Cb+Bs	0.491	−1.821	0.027
Relative Cerebrum	0.364	−0.415	0.367

Degree of phylogenetic autocorrelation in relative brain volume measures using Blomberg's k. K is the degree of phylogenetic signal, Z is the position in the Z distribution estimated from a tip rearrangement test using 100,000 iterations, and p is the p-value estimated from the tip rearrangement test.

Overall Encephalization

For the PGLS regressions and ANOVA, we present in each case the results from the best model, chosen from models with lambda fixed at zero or one, or allowing lambda to take its MLE (see Table S4). After accounting for phylogeny, both measures of relative endocranial volume were influenced by diet (Fig. 3; Corrected for skull size: $F_{2,28} = 4.57$, $p = 0.019$, Corrected for mass: $F_{2,28} = 12.01$, $p<0.001$). Flesh-eating species had the largest relative endocranial volumes, omnivores were intermediate, and insectivores had the smallest relative brain volumes, although the difference between insectivores and omnivores was statistically significant only for MCOEV (Table 2). Both MCOEV and SCOEV were negatively related to recent changes in the respective body size measure (SCIs), suggesting an evolutionary lag, during which body size has evolved, but brain size has yet to catch up and return to the basal brain:body allometry (Table 3).

Regional Brain Volumes

Sociality was positively related to relative total cerebrum volume, and negatively related to Cb+Bs volume, but not significantly related to other brain measures (Table 3). Both forelimb use and home range size predict relative Ac volume negatively, and exhibit nearly significant positive trends with Pc (Table 3). The composite life history variable was not significantly related to any response variable. Although the effect of diet on relative Cb+Bs volume was observable only as a non-significant trend ($F_{2,28} = 2.549$, $p = 0.097$), insectivores have significantly larger relative Cb+Bs than omnivores (Table 2). Diet did not have an influence on the relative volume of the other brain regions (Cerebrum: $F_{2,28} = 1.77$, $p = 0.189$; Ac: $F_{2,28} = 0.19$, $p = 0.828$; Pc: $F_{2,28} = 0.889$, $p = 0.422$; for contrasts see Table 2) Finally, relative Ac volume was positively associated with recent evolutionary changes in brain size (indicated by SCI values), suggesting that increased overall encephalization has been accompanied by a disproportionate increase in size of the frontal brain in mammalian carnivores (Table 3).

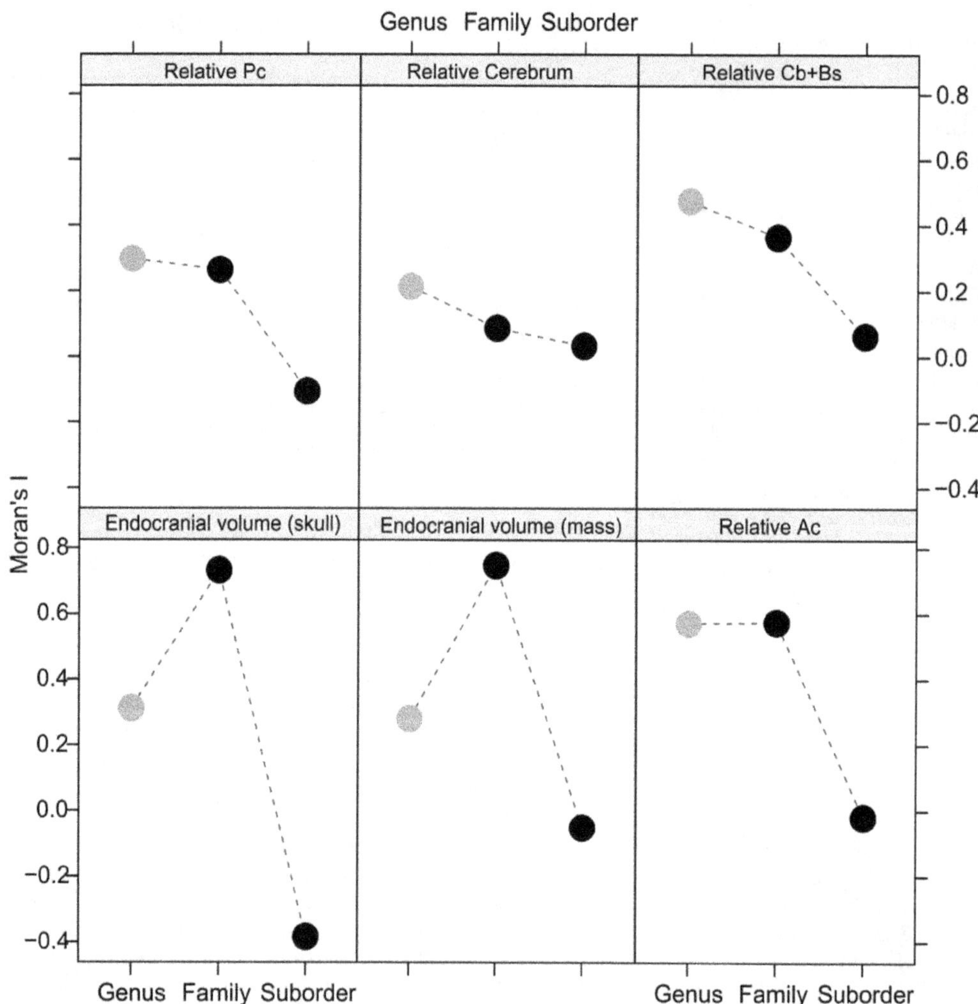

Figure 2. Phylogenetic autocorrelation as measured using Moran's I, which ranges between −1 and 1. Black circles indicate statistically significant autocorrelation at $\alpha \leq 0.05$, and grey circles indicate measures that are not significant at $\alpha \leq 0.05$.

Discussion

Although sociality plays an important role in primate brain evolution [42], our data failed to support the social brain hypothesis as an explanation for overall encephalization in Carnivora. Previous studies have demonstrated a relationship between relative brain volume and sociality in this order (e.g. [18,71]), but those findings are controversial. In particular, Perez-Barberia et al. [18] suggested that carnivore species with large brains for their body size are more commonly social than other species. However, Finarelli and Flynn (2009) showed that this trend in the Perez-Barberia et al. data disappeared when the family Canidae was removed from the analysis, and noted further that both Ursidae and Mustelidae are largely asocial, yet relatively large-brained. Although our results did not support the social brain hypothesis as it pertains to overall brain volume, we did identify a positive relationship between relative cerebrum volume and sociality (Table 3). This relationship suggests that simply excluding the brain stem, cerebellum and olfactory bulbs, regions of the brain that are likely to be less critical for social cognition, allows us to identify a pattern not observed using relative endocranial volume. Interestingly, the opposing relationship demonstrated between relative cerebellum volume and sociality most likely suggests that either reduced cerebellum and brain stem volume accompanies increased sociality, or there are additional factors opposing increases in overall brain volume, thus necessitating a decrease in cerebellum and brain stem volume with increases in cerebrum volume. The factors most likely to act to constrain evolutionary increases in brain size are selection to maintain skull size or shape, or antagonistic selection on overall brain size due to the energetic costs of neural tissue [30].

Although we failed to demonstrate a relationship between sociality and overall encephalization in Carnivora, we did identify a relationship between overall encephalization and diet (Table 2). Gittleman [53] had suggested this relationship, but was unable to support it statistically. In our data set, species that are primarily carnivorous have larger relative endocranial volumes than omnivorous or insectivorous species, and omnivorous species have larger relative endocranial volumes than insectivorous species, though the latter difference is statistically significant only when endocranial volume is corrected for mass, not skull size (Table 2). Two main hypotheses have been put forward to explain the relationship between diet and degree of encephalization. The first is that some diets are more energetically efficient than others, allowing the evolution of metabolically expensive brain tissue

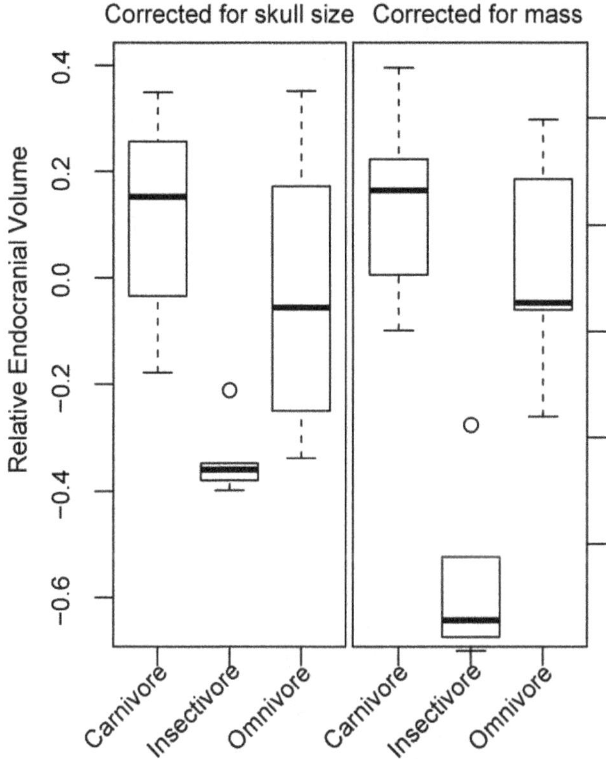

Figure 3. Box plot showing relationship between diet and relative endocranial volume. Boxes indicate interquartile ranges, and whiskers spread to the furthest points outside the interquartile range, but within 1.5 times the interquartile range from the median.

Table 2. Effects of diet on relative endocranial measures.

Brain region	Comparison	*ß*	*SE*	*T*	*p*
MCOEV	**I vs. C**	−0.586	0.13	−4.506	**<0.001**
	O vs. C	−0.251	0.076	−3.306	**0.003**
	I vs. O	−0.335	0.127	−2.638	**0.014**
SCOEV	**I vs. C**	−0.308	0.129	−2.392	**0.024**
	O vs. C	−0.193	0.075	−2.567	**0.016**
	I vs. O	−0.115	0.124	−0.928	0.361
Cerebrum	**I vs. C**	−0.134	0.073	−1.822	0.079
	O vs. C	−0.051	0.043	−1.183	0.247
	I vs. O	−0.083	0.068	−1.222	0.232
Ac	**I vs. C**	0.216	0.444	0.487	0.63
	O vs. C	−0.043	0.264	−0.164	0.871
	I vs. O	0.26	0.421	0.616	0.543
Pc	**I vs. C**	−0.286	0.23	−1.243	0.224
	O vs. C	−0.018	0.17	−0.108	0.915
	I vs. O	−0.267	0.219	−1.224	0.231
Cb + Bs	**I vs. C**	0.15	0.085	1.762	0.089
	O vs. C	−0.046	0.069	−0.675	0.505
	I vs. O	0.196	0.088	2.232	**0.034**

Effects of diet from ANOVAs, and contrasts from multiple regressions. For the contrasts, C represents carnivore, O represents omnivore and I represents insectivore. In each case, the first species is the one being contrasted. For example, I vs. C means that the estimate of effect under β is the change in the response variable due to insectivory with carnivores as the intercept. Note that diet effects were estimated as part of the multiple regression results in Table 3, and are in a separate table only for presentation purposes.

[30,31]. The second is that some diets require more complex cognitive processing to acquire and process food items [9,32].

In addition to its relationship with overall encephalization, diet also appears to be related to the relative volume of one of our measured brain regions, the cerebellum and brain stem, which is significantly larger in insectivores than in carnivores (Table 2). Other ecological traits are also associated with the volume of specific brain regions. The relative volume of the cerebrum anterior to the cruciate sulcus (Ac) is negatively related to home range size and degree of forelimb use in food processing, but positively related to the magnitude of recent size change in overall brain volume (Table 3). Our Ac measure consists primarily of frontal cortex. In primates, frontal cortex has been implicated in social cognition and executive function [104]. Interestingly, recent work has revealed a relationship between Ac and social group size within the family Hyaenidae [64], a finding not supported by our larger data set representing the entire order Carnivora. This suggests that some of the patterns observed here might be evident among species in broader taxonomic groups, whereas other patterns may become apparent only when considering species at lower taxonomic levels, such as families.

Estimates of recent evolutionary changes in body size in our data set, indicated by SCI values, were negatively associated with relative brain volume (Table 3). In other words, species in lineages characterized by a recent increase in body size have relatively small brains for their body size, while those in lineages characterized by a recent decrease in body size have relatively large brains for their body size. Thus, our results provide the first direct empirical support for the 'lag' hypothesis over longer evolutionary time periods, which suggests that brain and body size co-evolve, but that body size changes first, followed later by changes in brain size that return the relationship to its basal allometry [1,8,20,47]. Interestingly, an earlier analysis using slightly different methods found no relationship between SCIs and brain volume in primates [8]. Although it may be that the lag is simply weaker or absent among primates, it is also possible that the pattern becomes evident only when other important variables, such as diet, are accounted for statistically. The eventual changes in brain size required to return to the basal brain:body size allometry observed among distantly related species may occur either due to selection directly on brain size, or to changes in the genetic and developmental mechanisms underlying body size evolution at greater taxonomic distances [105].

A similar relationship, though positive, was found between relative Ac volume and recent changes in overall brain volume, indicating that increases in brain size result in disproportionate increases in frontal brain. This pattern suggests that developmental mechanisms similar to those suggested by Finlay and Darlington [43] might explain some variation in relative Ac volume. However, the fact that we were able to reveal that specific brain regions are significantly related to ecological and social traits requires that brain regions also evolve independently to some extent. In addition, some environmental factors in our data set are positively related to one brain region but negatively related to another (Table 3). For example, sociality positively predicts relative total cerebrum volume, but negatively predicts Cb+Bs, which represents the remaining endocranial volume minus the olfactory bulbs. Similarly, Ac volume is negatively related to forelimb dexterity and home range size, while Pc volume is positively related to forelimb dexterity and home range size.

Table 3. PGLS regression outputs for all variables other than diet.

Trait	Predictor	*ß*	*SE*	*T*	*p*
MCOEV	**Intercept**	0.110	0.129	0.853	0.401
	Body SCI	−0.102	0.034	−3.045	**0.005**
	Sociality	−0.025	0.027	−0.919	0.366
	Home Range	−0.005	0.015	−0.350	0.729
	Forelimb Use	0.022	0.053	0.405	0.689
	Life History	0.039	0.037	1.045	0.305
SCOEV	**Intercept**	0.003	0.126	0.026	0.979
	Brain SCI	−0.170	0.074	−2.295	**0.029**
	Sociality	−0.046	0.026	−1.756	0.090
	Home Range	0.000	0.015	−0.023	0.982
	Forelimb Use	0.050	0.052	0.958	0.346
	Life History	0.029	0.036	0.790	0.436
Cerebrum	**Intercept**	0.067	0.045	1.470	0.153
	Brain SCI	−0.032	0.063	−0.501	0.620
	Sociality	0.081	0.022	3.711	**0.001**
	Home Range	0.017	0.015	1.143	0.263
	Forelimb Use	−0.014	0.024	−0.555	0.583
	Life History	−0.004	0.020	−0.224	0.825
Ac	**Intercept**	0.989	0.427	2.314	0.028
	Brain SCI	0.573	0.233	2.464	**0.020**
	Sociality	−0.036	0.088	−0.404	0.689
	Home Range	−0.117	0.051	−2.305	**0.029**
	Forelimb Use	−0.554	0.178	−3.120	**0.004**
	Life History	0.034	0.122	0.278	0.783
Pc	**Intercept**	−0.285	0.209	−1.363	0.184
	Brain SCI	−0.163	0.156	−1.047	0.304
	Sociality	0.061	0.063	0.961	0.345
	Home Range	0.067	0.035	1.891	0.069
	Forelimb Use	0.183	0.102	1.785	0.085
	Life History	0.043	0.080	0.540	0.594
Cb + Bs	**Intercept**	−0.092	0.076	−1.214	0.235
	Brain SCI	0.084	0.080	1.050	0.303
	Sociality	−0.106	0.031	−3.487	**0.002**
	Home Range	−0.031	0.018	−1.690	0.102
	Forelimb Use	0.048	0.041	1.171	0.252
	Life History	0.043	0.030	1.419	0.167

Multiple regression output for the best model for each of the different response variables, not including effects of diet, because diet is categorical. Body SCI is the size change index for body mass or skull size. Brain SCI is the size change index for brain volume. Sociality is PC1 from the PCA of the variables describing social complexity. Home range is log home range size corrected for body size in the same way as brain volume. Forelimb use is our measure of forelimb dexterity. Finally, life history is PC1 from a PCA of the three life history variables we included in our analysis.

Although these latter relationships were not statistically significant, the trade-off effect between the Ac and Pc may be due to an increased demand for processing forelimb tactile information in the somatosensory cortex [27] and enhanced spatial memory processing in the hippocampus [106] respectively, within Pc. Since the endocast method does not permit analysis of subcortical features, we were unable to separate the relative contributions of these areas within Pc. Such tradeoffs among regions of the brain have been seen in previous studies [42], and suggest that increases in some brain regions are accompanied by concomitant decreases in other brain regions. These patterns are suggestive of negative microevolutionary tradeoffs among brain regions due to either space or energetic limitations, and indicate that variation in overall encephalization might reflect not only selection to increase the volume of the brain, but also antagonistic selection acting to oppose further encephalization.

Interestingly, the high degree of variation apparent among families within the order Carnivora suggests that there are some factors operating at the level of the family that prevent or slow evolutionary change, at least over the time scale considered here. Specifically, species within some families share large brains or bodies, whereas species in others share small brains or bodies (Fig. 1), and significant phylogenetic autocorrelation appears to be highest at the family level for most traits (Fig. 2). In many cases, intrafamily variation in morphological and ecological traits is fairly low, resulting in clear diagnostic characters for some families. Additionally, it is clear from our data set that much of the variation in several different measures is explained simply by the family to which a species belongs. This strong family-level variation echoes our earlier suggestion that analyses within families may in some cases uncover clearer patterns than analyses targeting broader taxonomic ranges.

Selection of a body size correction factor for use in studies of brain size is clearly a more complicated issue than is commonly assumed. Although correcting for skull size or mass yielded similar results in our regression analyses, some differences were apparent. For instance, canids have much larger values for MCOEV than SCOEV, whereas the converse is true for herpestids. Although there are arguments in favor of using both mass and skull size as correction factors, the proper correction factor depends on why brain volume covaries with body size, which is not known. Until we understand why brain volume scales to body size, the best scaling factor cannot be known unequivocally, if there is in fact a 'best' scaling factor. It is therefore worth noting that most studies do not replicate their analyses using more than one measure of body size. Although such replication may complicate analyses of the environmental predictors of encephalization, more explicit consideration of the size measure used as a correction factor is clearly warranted. Indeed, whether the volume of a brain region is corrected for whole-brain volume or the volume of another specific region of the brain can lead to different conclusions, with very little reason to presume that one correction factor is superior to another [107].

In conclusion, despite some lack of consensus in the study of the evolutionary forces acting on brain volume within the order Carnivora, a number of points are clear from our analysis. Most importantly, several different variables influence encephalization or regional brain volume in our data set, including sociality, ecology, phylogeny and recent evolutionary changes in body size and overall brain size, measured by SCIs. However, the composite life history trait we included in our analysis was unrelated to encephalization or the volume of any brain region (Table 3). Previous analyses have found relationships within Carnivora between encephalization and neonate mass, but not weaning age or gestation length [7], and our results support these conclusions. It is clearly important to consider multiple factors simultaneously, including not only those that are most likely adaptive, but also those that may have no adaptive value. In addition, it is interesting that analyses even of large, fairly crude subdivisions of overall brain volume (e.g. 'total cerebrum') can reveal relationships not found in analyses of whole brains. This may be explained by a combination of mosaic processes, developmental linkages and external factors such as antagonistic selection on skull size and shape constraining brain evolution. Thus, although total brain size may not solely reflect selection pressures on specific brain regions, the effects of such selection are still seen in large subdivisions of the brain. Our results provide at least circumstantial evidence that processes leading to concerted change throughout the brain, and those influencing only specific brain regions, both play a role in brain evolution among mammalian carnivores.

Supporting Information

Table S1 Details for specimens used in analysis. Field Museum (FMNH); Los Angeles County Museum of Natural History (LACM); Michigan State University Museum (MSUM); University of Michigan Museum of Zoology (UMMZ). All skulls were scanned using a General Electric Lightspeed 4 slice CT or General Electric Discovery ST 16 slice scanner in the Department of Radiology at Michigan State University. Scanner type is indicated in the final column.

Table S2 Cranial and endocranial measures used in analysis including: endocranial volume in mm^3 (brain volume), combined cerebellum and brainstem volume in mm^3 (Cb + Bs), cerebrum anterior to the cruciate sulcus in mm^3 (Ac), cerebrum posterior to the cruciate sulcus in mm^3 (Pc), skull basal length in mm (BL), zygomatic arch breadth in mm (ZB), skull height in mm (SH). Details of measurements are given in the methods section of the main document. BL, ZB and SH were included in a principal components analysis to create the skull size variable used for analyses. Cerebrum volume is equal to Ac + Pc.

Table S3 Data used in analysis including: group size (GS), social cohesion (Cohesion), FGS (Feeding group size), mass (in kg), gestation length (Gest. Len.; in days), weaning age (WA; in months), maximum recorded longevity (Longevity; in years), home range (in sq. km), diet, and degree of forelimb processing of food (Forelimb). Group size and home range size are arithmetic species means excluding any values that are more than 5 standard deviations from the mean value to avoid including extremely influential outliers. Social cohesion is scored as either a 1 (solitary except during the breeding season), 2 (primarily pair-living), 3 (fission-fusion sociality) or 4 (obligately social). Mass was taken as the mean value given, pooling males and females. Gestation length is given as the mean value, excluding periods of delayed implantation (embryonic diapause). Diet was coded as primarily insectivorous, carnivorous or omnivorous. Finally, degree of forelimb use during food processing was coded based on descriptions of hunting or food processing, or where unavailable, on the type of food consumed. Forelimb processing was scored from 1 to 4, with 1 being no use of forelimbs in food processing, 2 representing use of forepaws with no grasping or independent use of digits, 3 representing grasping behavior and fairly complex using during processing, and 4 representing highly dextrous use of forepaws during food processing including grasping behavior (only raccoons were placed in this category among our sample of species as per [36]). Superscripts indicate the source of the data: [1]Wilson and Mittermeier [108]; [2]Sunquist and Sunquist [109]; [3]Watts [110]; [4]Nowak et al. [111]; [5]Holekamp and Dloniak [112]; [6]Mech [113]; [7]Baker [114]; [8]Calculated from family-specific regression on mass. [9]Gestation length for sea otters was taken from the AnAge database [115], [10]Mills and Hofer [116]. A subscript next to a column header indicates that all values in the column specified are taken from the source noted unless otherwise specified in an individual cell. A subscript in the 'Genus and species' column indicates that all values in the row specified are taken from the source noted unless otherwise specified in an individual cell or next to a column header.

Table S4 Model selection for best PGLS models, comparing models fitting a fixed brownian motion model (equivalent to using independent contrasts), a fixed 'no effect of phylogeny' model, and a model allowing lambda to take its MLE.

Acknowledgments

The authors are most grateful to Dr. Kevin Berger and the Department of Radiology, Michigan State University, and the University of Wisconsin Comparative Mammalian Brain Collections (supported by the National Science Foundation). We thank two anonymous reviewers for very helpful comments on a previous version of this manuscript as well as Shana Bussa, Kelly Wagner and Jaime Tanner for their help in the CT analysis. We also thank the staff of the following museums for making their specimens available for this study: Michigan State University Museum, Field Museum of Natural History, Natural History Museum of Los Angeles County, National Museum of Natural History, and University of Michigan Museum of Zoology.

Author Contributions

Conceived and designed the experiments: KEH STS BLL BMA. Performed the experiments: BMA BLL EMS KEH. Analyzed the data: EMS. Contributed reagents/materials/analysis tools: STS BLL. Wrote the paper: EMS KEH STS.

References

1. Jerison HJ (1973) Evolution of the Brain and Intelligence. New York and London: Academic Press.
2. Changizi MA (2009) Brain scaling laws. In: Squire LR, ed. Encyclopedia of Neuroscience: Academic Press.
3. Jerison HJ (1985) Animal intelligence as encephalization. Philosophical Transactions of the Royal Society of London Series B, Biological Sciences 308: 21–35.
4. Finlay BL, Darlington RB, Nicastro N (2001) Developmental structure in brain evolution. Behavioral and Brain Sciences 24: 263–308.
5. Dunbar RIM, Shultz S (2007) Evolution in the social brain. Science 317: 1344–1347.
6. Humphrey NK (1976) The social function of intellect. In: Bateson PPG, Hinde RA, eds. Growing points in ethology. Cambridge: Cambridge University Press. pp 303–317.
7. Finarelli JA (2010) Does encephalization correlate with life history or metabolic rate in Carnivora? Biology Letters 6: 350–353.
8. Deaner RO, Nunn CL (1999) How quickly do brains catch up with bodies? A comparative method for detecting evolutionary lag. Proceedings of the Royal Society of London Series B: Biological Sciences 266: 687–694.
9. Clutton-Brock T, Harvey PH (1980) Primates, brains and ecology. Journal of Zoology 190: 309–323.
10. Parker ST, Gibson KR (1977) Object manipulation, tool use, and sensorimotor intelligence as feeding adaptations in cebus monkeys and great apes. Journal of Human Evolution 6: 623–641.
11. Byrne RW, Whiten A (1988) Machiavellian intelligence: Social expertise and the evolution of intellect in monkeys, apes, and humans. Oxford: Clarendon Press.
12. Whiten A, Byrne RW, eds (1997) Machiavellian intellegence II. Cambridge: Cambridge University Press.
13. Dunbar RIM (2003) The social brain: mind, language, and society in evolutionary perspective. Annual Review of Anthropology 32: 163–181.

14. Byrne RW (1994) The evolution of intelligence. In: Slater PJB, Halliday TR, eds. Behaviour and Evolution. Cambridge: Cambridge University Press. pp 223–265.
15. Harcourt AH, de Waal FBM (1992) Coalitions and alliances in humans and other animals. In: Harcourt AH, de Waal FBM, eds. Coalitions and alliances in humans and other animals. Oxford, United Kingdom: Oxford Science.
16. Holekamp KE, Sakai ST, Lundrigan BL (2007) The spotted hyena (*Crocuta crocuta*) as a model system for study of the evolution of intelligence. Journal of Mammalogy 88: 545–554.
17. Finarelli JA, Flynn JJ (2009) Brain-size evolution and sociality in Carnivora. Proceedings of the National Academy of Sciences of the United States of America 106: 9345–9349.
18. Perez-Barberia FJ, Shultz S, Dunbar RIM (2007) Evidence for coevolution of sociality and relative brain size in three orders of mammals. Evolution 61: 2811–2821.
19. Sacher GA, Staffeldt EF (1974) Relation of gestation time to brain weight for placental mammals - implications for theory of vertebrate growth. American Naturalist 108: 593–615.
20. Gould SJ (1975) Allometry in primates, with emphasis on scaling and the evolution of the brain. Contributions to Primatology 5: 244–292.
21. Weisbecker V, Goswami A (2010) Brain size, life history, and metabolism at the marsupial/placental dichotomy. Proceedings of the National Academy of Sciences of the United States of America 107: 16216–16221.
22. Isler K, van Schaik CP (2009) The Expensive Brain: A framework for explaining evolutionary changes in brain size. Journal of Human Evolution 57: 392–400.
23. Isler K, van Schaik CP (2009) Why are there so few smart mammals (but so many smart birds)? Biology Letters 5: 125–129.
24. Allman J, McLaughlin T, Hakeem A (1993) Brain-weight and life-span in primate species. Proceedings of the National Academy of Sciences of the United States of America 90: 118–122.
25. van Schaik CP, Deaner RO (2003) Life history and cognitive evolution in primates. In: de Waal FBM, Tyack PL, eds. Animal Social Complexity: Intelligence, Culture, and Individualized Societies. Cambridge, MA: Harvard University Press. pp 5–25.
26. Iwaniuk AN, Pellis SM, Whishaw IQ (1999) Brain size is not correlated with forelimb dexterity in fissiped carnivores (Carnivora): A comparative test of the principle of proper mass. Brain Behavior and Evolution 54: 167–180.
27. Welker WI, Campos GB (1963) Physiological significance of sulci in somatic sensory cerebral cortex in mammals of the family Procyonidae. The Journal of Comparative Neurology 120: 19–36.
28. Merzenich MM, Nelson RJ, Stryker MP, Cynader MS, Schoppmann A, et al. (1984) Somatosensory cortical map changes following digit amputation in adult monkeys. Journal of Comparative Neurology 224: 591–605.
29. Donoghue JP, Leibovic S, Sanes JN (1992) Organization of the forelimb area in squirrel-monkey motor cortex - representation of digit, wrist, and elbow muscles. Experimental Brain Research 89: 1–19.
30. Aiello LC, Wheeler P (1995) The expensive-tissue hypothesis: The brain and the digestive system in human and primate evolution. Current Anthropology 36: 199–221.
31. Fish JL, Lockwood CA (2003) Dietary constraints on encephalization in primates. American Journal of Physical Anthropology 120: 171–181.
32. Eisenberg JF, Wilson DE (1978) Relative brain size and feeding strategies in the Chiroptera. Evolution 32: 740–751.
33. Lefebvre L, Sol D (2008) Brains, Lifestyles and Cognition: Are There General Trends? Brain Behavior and Evolution 72: 135–144.
34. Holekamp KE (2007) Questioning the social intelligence hypothesis. Trends in Cognitive Sciences 11: 65–69.
35. Sol D, Garcia N, Iwaniuk A, Davis K, Meade A, et al. (2010) Evolutionary Divergence in Brain Size between Migratory and Resident Birds. PLoS ONE 5: e9613.
36. Iwaniuk AN, Pellis SM, Whishaw IQ (2000) The relative importance of body size, phylogeny, locomotion, and diet in the evolution of forelimb dexterity in fissiped carnivores (Carnivora). Canadian Journal of Zoology-Revue Canadienne De Zoologie 78: 1110–1125.
37. Shultz S, Dunbar RIM (2010) Social bonds in birds are associated with brain size and contingent on the correlated evolution of life-history and increased parental investment. Biological Journal of the Linnean Society 100: 111–123.
38. Dechmann DKN, Safi K (2009) Comparative studies of brain evolution: a critical insight from the Chiroptera. Biological Reviews 84: 161–172.
39. Young RM (1990) Mind, brain and adaptation in the nineteenth century. Oxford: Clarendon Press.
40. Corfield JR, Wild JM, Hauber ME, Parsons S, Kubke MF (2008) Evolution of brain size in the Paleognath lineage, with an emphasis on New Zealand ratites. Brain, Behavior and Evolution 71: 87–99.
41. Dunbar RIM (1992) Neocortex size as a constraint on group size in primates. Journal of Human Evolution 20: 469–493.
42. Dunbar RIM, Shultz S (2007) Understanding primate brain evolution. Philosophical Transactions of the Royal Society B-Biological Sciences 362: 649–658.
43. Finlay BL, Darlington RB (1995) Linked regularities in the development and evolution of mammalian brains. Science 268: 1578–1584.
44. Barton RA, Harvey PH (2000) Mosaic evolution of brain structure in mammals. Nature 405: 1055–1058.
45. Striedter GF (2005) Principles of brain evolution. Sunderland, MA: Sinauer Associates.
46. Gould SJ (1991) Of kiwi eggs and the liberty bell. Bully for Brontosaurus. New York, NY: W.W. Norton & Company, Inc 109–123.
47. Lande R (1979) Quantitative genetic analysis of multivariate evolution, applied to brain:body size allometry. Evolution 33: 402–416.
48. Deacon TW (1990) Rethinking mammalian brain evolution. American Zoologist 30: 629–705.
49. Martin RD, Harvey PH (1985) Brain size allometry: ontogeny and phylogeny. In: Jungers WL, ed. Size and scaling in primate biology. New York: Plenum Press. pp 147–173.
50. Weston EM, Lister AM (2009) Insular dwarfism in hippos and a model for brain size reduction in Homo floresiensis. Nature 459: 85–U88.
51. Barton RA (1998) Visual specialisation and brain evolution in primates. Proceedings of the Royal Society of London Series B: Biological Sciences 265: 1933–1937.
52. Iwaniuk AN, Lefebvre L, Wylie DR (2009) The Comparative Approach and Brain-Behaviour Relationships: A Tool for Understanding Tool Use. Canadian Journal of Experimental Psychology-Revue Canadienne De Psychologie Experimentale 63: 150–159.
53. Gittleman JL (1986) Carnivore brain size, behavioral ecology, and phylogeny. Journal of Mammalogy 67: 23–36.
54. Bininda-Emonds ORP, Gittleman JL, Purvis A (1999) Building large trees by combining phylogenetic information: a complete phylogeny of the extant Carnivora (Mammalia). Biological Reviews 74: 143–175.
55. Johnson WE, Eizirik E, Pecon-Slattery J, Murphy WJ, Antunes A, et al. (2006) The Late Miocene radiation of modern Felidae: A genetic assessment. Science 311: 73–77.
56. Koepfli K-P, Jenks SM, Eizirik E, Zahirpour T, Van Valkenburgh B, et al. (2006) Molecular systematics of the Hyaenidae: Relationships of a relictual lineage resolved by a molecular supermatrix. Molecular Phylogenetics and Evolution 38: 603–620.
57. Pagel MD (1992) A method for the analysis of comparative data. Journal of Theoretical Biology 156: 431–442.
58. Blomberg SP, Garland T, Ives AR (2003) Testing for phylogenetic signal in comparative data: Behavioral traits are more labile. Evolution 57: 717–745.
59. Kembel SW, Cowan PD, Helmus MR, Cornwell WK, Morlon DD, et al. (2010) Picante: R tools for integrating phylogenies and ecology. Bioinformatics 26: 1463–1464.
60. R Development Core Team (2010) R: A language and environment for statistical computing. Vienna, Austria: R Foundation for Statistical Computing.
61. Díaz-Uriarte R, Garland T Jr. (1998) Effects of branch length errors on the performance of phylogenetically independent contrasts. Systematic Biology 4: 654–672.
62. Maddison WP, Maddison DR (2004) Mesquite: a modular system for evolutionary analysis. 1.05 ed.
63. Midford PR, Garland T Jr., Maddison WP (2005) PDAP package of Mesquite.
64. Sakai ST, Arsznov BM, Lundrigan BL, Holekamp KE (2011) Brain size and social complexity: A computed tomography study in Hyaenidae. Brain, Behavior and Evolution 77: 91–104.
65. Young RW (1959) The influence of cranial contents on postnatal growth of the skull in the rat. American Journal of Anatomy 105: 383–415.
66. Finarelli JA (2006) Estimation of endocranial volume through the use of external skull measures in the Carnivora (Mammalia). Journal of Mammalogy 87: 1027–1036.
67. Ricard AS, Desbarats P, Laurentjoye M, Montaudon M, Caix P, et al. (2010) On two equations about brain volume, cranial capacity and age. Surgical and Radiologic Anatomy 32: 989–995.
68. Iwaniuk AN, Nelson JE (2002) Can endocranial volume be used as an estimate of brain size in birds? Canadian Journal of Zoology-Revue Canadienne De Zoologie 80: 16–23.
69. Rohrs M (1986) Cephalization in Canidae. Zeitschrift Fur Zoologische Systematik Und Evolutionsforschung 24: 300–307.
70. Rohrs M, Ebinger P, Weidemann W (1989) Cephalization in Viverridae, Hyaenidae, Procyonidae and Ursidae. Zeitschrift Fur Zoologische Systematik Und Evolutionsforschung 27: 169–180.
71. Dunbar RIM, Bever J (1998) Neocortex size predicts group size in carnivores and some insectivores. Ethology 104: 695–708.
72. Radinsky L (1969) Outlines of canid and felid brain evolution. Ann NY Acad Sci 167: 277–288.
73. Hassler R, Muhs-Clement K (1964) Architektonischer aufbau des sensomotorischen und parietalen cortex der katze. J Hirnforsch 6: 377–420.
74. Gorska T (1974) Functional organization of cortical motor areas in adult dogs and puppies. Acta Neurobiol Exp 34: 171–203.
75. Sakai ST (1982) The thalamic connectivity of the primary motor cortex (MI) in the raccoon. Journal of Comparative Neurology 204: 238–252.
76. Revell LJ (2009) Size-correction and principal components for interspecific comparative studies. Evolution 63: 3258–3268.
77. Martins EP, Hansen TF (1997) Phylogenies and the comparative method: a general approach to incorporating phylogenetic information into the analysis of interspecific data. The American Naturalist 149: 646–667.
78. Grafen A (1989) The Phylogenetic Regression. Philosophical Transactions of the Royal Society of London Series B, Biological Sciences 326: 119–157.

79. Harvey PH, Pagel MD (1991) The comparative method in evolutionary biology. New York: Oxford University Press.
80. Garland T Jr., Ives AR (2000) Using the past to predict the present: confidence intervals for regression equations in phylogenetic comparative methods. The American Naturalist 155: 346–364.
81. Pagel M (1999) The maximum likelihood approach to reconstructing ancestral character states of discrete characters on phylogenies. Systematic Biology 48: 612–622.
82. Harmon L, Weir J, Brock C, Glor R, Challenger W, et al. (2009) geiger: Analysis of evolutionary diversification. 1.3–1 ed. pp. R package version.
83. Webster AJ, Gittleman JL, Purvis A (2004) The life history legacy of evolutionary body size change in carnivores. Journal of Evolutionary Biology 17: 396–407.
84. Shultz S, Dunbar RIM (2007) The evolution of the social brain: anthropoid primates contrast with other vertebrates. Proceedings of the Royal Society B-Biological Sciences 274: 2429–2436.
85. Emery NJ, Seed AM, von Bayern AMP, Clayton NS (2007) Cognitive adaptations of social bonding in birds. Philosophical Transactions of the Royal Society B-Biological Sciences 362: 489–505.
86. Oli MK (2004) The fast-slow continuum and mammalian life-history patterns: an empirical evaluation. Basic and Applied Ecology 5: 449–463.
87. Bielby J, Mace GM, Bininda-Emonds ORP, Cardillo M, Gittleman JL, et al. (2007) The fast-slow continuum in mammalian life history: An empirical reevaluation. American Naturalist 169: 748–757.
88. Gaillard J-M, Pontier D, Allaine D, Lebreton J-D, Trouvilliez J, et al. (1989) An analysis of demographic tactics in birds and mammals. Oikos 56: 59–76.
89. Harvey PH, Zammuto RM (1985) Patterns of mortality and age at first reproduction in natural populations of mammals. Nature 315: 319–320.
90. Stearns SC (1983) The influence of size and phylogeny on patterns of coviariation among life-history traits in the mammals. Oikos 41: 173–187.
91. Harvey PH, Read AF, Promislow DEL (1989) Life-history variation in placental mammals: unifying data with theory. Oxford Surveys in Evolutionary Biology 6: 13–31.
92. Promislow DEL, Harvey PH (1990) Living fast and dying young: a comparative analysis of life-history variation among mammals. Journal of Zoology 220: 417–437.
93. Moran PAP (1950) Notes on continuous stochastic phenomena. Biometrika 37: 17–23.
94. Paradis E (2006) Analysis of Phylogenetics and Evolution with R. New York: Springer.
95. Gittleman JL, Kot M (1990) Adaptation - statistics and a null model for estimating phylogenetic effects. Systematic Zoology 39: 227–241.
96. Paradis E, Claude J, Strimmer K (2004) APE: analyses of phylogenetics and evolution in R language. Bioinformatics 20: 289–290.
97. Felsenstein J (1985) Phylogenies and the Comparative Method. The American Naturalist 125: 1.
98. Clutton-brock TH, Harvey PH (1977) Primate ecology and social organization. Journal of Zoology 183: 1–39.
99. Gittleman JL, Luh HK (1992) On comparing comparative methods. Annual Review of Ecology and Systematics 23: 383–404.
100. Burnham KP, Anderson DR (2002) Model Selection and Multimodel Inference.New York: Springer-Verlag.
101. Losos JB (1999) Uncertainty in the reconstruction of ancestral character states and limitations on the use of phylogenetic comparative methods. Animal Behaviour 58: 1319–1324.
102. Dobson FS, Oli MK (2007) Fast and slow life histories of mammals. Ecoscience 14: 292–299.
103. Read AF, Harvey PH (1989) Life-history differences among the eutherian radiations. Journal of Zoology 219: 329–353.
104. Adolphs R (2001) The neurobiology of social cognition. Current Opinion in Neurobiology 11: 231–239.
105. Riska B, Atchley WR (1985) Genetics of growth predict patterns of brain-size evolution. Science 229: 668–671.
106. Healy SD, Krebs JR (1992) Food storing and the hippocampus in Corvids: amount and volume are correlated. Proceedings of the Royal Society of London Series B: Biological Sciences 248: 241–245.
107. Deaner RO, Nunn CL, van Schaik CP (2000) Comparative tests of primate cognition: different scaling methods produce different results. Brain, Behavior and Evolution 55: 44–52.
108. Wilson DE, Mittermeier R, eds. Handbook of the Mammals of the World - Volume 1: Lynx Edicions.
109. Sunquist M, Sunquist F (2002) Wild Cats of the World. Chicago: University of Chicago Press.
110. Watts H, Tanner JB, Lundrigan BL, Holekamp KE (2009) Post-weaning maternal effects and the evolution of female dominance in the spotted hyena. Proceedings of the Royal Society B 276: 2291–2298.
111. Nowak RM, Macdonald DW, Kays RW (2005) Walker's Carnivores of the World. Baltimore, Maryland: The Johns Hopkins University Press.
112. Holekamp KE, Dloniak SM (2010) Intra-specific variation in the behavioral ecology of a tropical carnivore, the spotted hyena. In: Macedo R, ed. Behavioral Ecology of Tropical Animals: From Ants to Zebras. pp 189–229.
113. Mech LD (1970) The Wolf: The Ecology and Behavior of an Endangered Species. New York: Doubleday Publishing Co.
114. Baker RH (1983) Michigan mammals. East Lansing, MI: Michigan State University Press.
115. de Magalhaes JP, Costa J (2009) A database of vertebrate longevity records and their relation to other life-history traits. Journal of Evolutionary Biology 22: 1770–1774.
116. Mills MGL, Hofer H, eds (1998) Hyenas: status survey and conservation action plan. IUCN, Switzerland.

12

Predation Life History Responses to Increased Temperature Variability

Miguel Barbosa[1,2]*, Joao Pestana[1], Amadeu M. V. M. Soares[1,3]

1 Center for Environmental and Marine Studies (CESAM), Departamento de Biologia, Universidade de Aveiro, Campus de Santiago, Aveiro, Portugal, 2 Centre for Biological Diversity and Scottish Oceans Institute, University of St Andrews, St Andrews, Fife, United Kingdom, 3 Programa de Pós-Graduação em Produção Vegetal, Universidade Federal de Tocantins, Campus de Gurupi, Gurupi, Brasil

Abstract

The evolution of life history traits is regulated by energy expenditure, which is, in turn, governed by temperature. The forecasted increase in temperature variability is expected to impose greater stress to organisms, in turn influencing the balance of energy expenditure and consequently life history responses. Here we examine how increased temperature variability affects life history responses to predation. Individuals reared under constant temperatures responded to different levels of predation risk as appropriate: namely, by producing greater number of neonates of smaller sizes and reducing the time to first brood. In contrast, we detected no response to predation regime when temperature was more variable. In addition, population growth rate was slowest among individuals reared under variable temperatures. Increased temperature variability also affected the development of inducible defenses. The combined effects of failing to respond to predation risk, slower growth rate and the miss-match development of morphological defenses supports suggestions that increased variability in temperature poses a greater risk for species adaptation than that posed by a mean shift in temperature.

Editor: Pauline Ross, University of Western Sydney, Australia

Funding: This study was funded by a Postdoctoral fellowship to MB (SFRH/BPD/82259/2011) Fundacao para a Ciencia e Tecnologia (www.fct.pt/index.phtml.pt) and with a "Bolsista CAPES/BRASIL", (Project A058/2013) to AMVMS. The funders had no role in study design, data collection and analysis, decision to publish, or preparation of the manuscript.

Competing Interests: The authors have declared that no competing interests exist.

* Email: mb334@st-andrews.ac.uk

Introduction

Temperature directly affects metabolic rate and consequently energy expenditure. As a result, changes in temperature are often accompanied by both physiological and behavioural responses. One example of temperature change stress-induced disruption is the loss of the ability to recognize and respond to predation threat [1]. Recent global change data reveal that the environment is changing at an unprecedented pace, with temperature predicted to become increasingly stochastic [2–4]. While organisms are typically able to cope with a natural rate of temperature change, increased variability in temperature is likely to impose additional physiological stress, potentially affecting the way organisms respond to environmental conditions. Here we address this issue by examining the effects of increased variation in temperature on life history responses to predation risk.

The effects of predation risk on prey life history traits have been well established [5–7]. Under high predation risk, selection favours the production of more and smaller sized offspring, and fewer, bigger offspring when predation risk is low. Predation risk also induces responses in terms of the onset of reproduction. Generally, under high predation risk individuals mature and start reproducing sooner [8,9]. Nevertheless, despite these life history expectations, the optimal reproductive response to predation risk is expected to be dynamic and primarily determined by energetic constraints [10].

Both predation risk and thermoregulation have associated energetic costs. The costs incurred by increased predation risk [11] can, together with thermoregulation requirements, interact and lead to a situation of greater stress, which can ultimately affect optimal life-history strategies. Further, individuals have a finite amount of energy to invest between growth, reproduction and maintenance [12]. Increased variability in mean temperatures, as predicted by global change models, are likely to influence how much energy is allocated to growth and thermoregulation. Specifically, under such circumstances of variability, organisms may be required to allocate more energy towards maintenance, because of thermoregulation, at the expense of growth or reproduction [13]. There is, therefore, the potential for a conflict between the energetic costs of thermoregulation and the energetic costs of life history responses to predation level.

Here we test the hypothesis that optimal life history responses to predation are impaired by increased variation in temperature using the waterflea *Daphnia magna*. Numerous studies have demonstrated that predation risk is an important driver of *Daphnia spp* life history [14]. Namely, *Daphnia spp* start reproducing sooner and produce more neonates when exposed to chemical cues released by fish [15]. It has also been shown that the presence of fish kairomones induces changes in the pattern of energy allocation, causing more energy to be directed towards reproduction at the expense of growth [16]. Besides predation risk, temperature also affects how *Daphnia spp* allocate resources to reproduction [12].

There is an extensive literature on the synergetic effects of temperature and predation in shaping life history traits [17–19]. But while life history responses to predation under constant temperature regimes are well known, it is less clear how *Daphnia spp* respond to predation risk under increased temperature variability. With recent global change models forecasting an increase in the variability around mean temperature [2], it is important to understand how this increased variability can affect life history responses.

The goal of this study is, therefore, to examine the effects of increased temperature variability on life history responses to different levels of predation risk. Predation may induce a response in some life history traits but not in others [14]. Hence, response to predation was examined using multiple traits. Specifically, brood size, neonate length at birth, time between broods and time to first brood were compared between individuals reared under constant and variable temperatures while exposed to high or low levels predation risk. Predation risk can also promote the evolution of inducible defenses, such as spines [20]. Therefore to complement the analysis of life history traits, the effect of predation risk on the development of defense traits was examined under both constant and variable temperatures. Finally, because of the link between temperature and predation on population dynamics [21–22], the synergetic effect of increased temperature variability and predation risk on population growth rate was investigated.

Methods

The effect of increased variation in temperature in life history responses to predation was examined following the Organization for Economic and Co-operation and Development (OECD) guidelines for assessing the influence of stressors on *Daphnia magna* reproductive responses [23].

All F0 individuals used in this study were 3rd brood neonates generated from *Daphnia magna* clone F [24] raised at the constant temperature of 20°C in a 16: 8 hour light: dark photoperiod in ASTM (American Society for Testing Materials) and fed with green algae *Pseudokirchneriella subcapitata*, at a concentration of 3.0×10^{5} cells mL^{-1}. We decided to use this clone based on their responsiveness to fish kairomones detected in previous studies [25].

Immediately after birth, individuals were randomly allocated to either a constant (20°C; n = 30) or variable (15 to 25°C; n = 30) temperature treatment. There were no significant differences in length at birth between treatments (Table S1). Temperatures in the variable treatment varied randomly between 15 and 25°C. However, to avoid unrealistic temperatures, the temperature varied according to 2 sub sections - 00:00 to 08:00/18:00 to 24:00 (dawn-morning/late afternoon) and from 08:00 to 18:00 (morning and afternoon). In the dawn-morning/late afternoon section, the temperature varied between 15°C and 20°C. In the morning and afternoon section it varied between 20°C and 25°C. The daily temperature in the variable treatment varied by ~10°C. Numerous studies have reported similar [26], as well as even greater [27,28] daily variations of water temperatures. The mean temperature in the variable treatment was 19.8°C. Because the mean temperatures in the variable and in the constant treatment were similar, any effect observed in the variable treatment can be unambiguously attributed to differences in stochasticity rather than to spending different amount of time at different mean temperatures. Temperatures in the treatments were obtained using temperature controlled chambers. The temperature was checked daily.

Within each temperature treatment, 10 individuals were randomly allocated to one of three predation treatments: 1) high concentration of predator cues 2) low concentration of predator cues, or 3) substrate with no predator cues (control). We used the tropical zebrafish (*Danio rerio*), as a model vertebrate predator. Zebrafish came from laboratory cultures and we used 3-month-old individuals of similar size.

To prepare the kairomone solution, we held 20 zebrafish in 20 L aerated ASTM water for 24 hours. During this period fish were allowed to consume 400 *D. magna* of various sizes. After 24 hours, the water containing fish kairomones was filtered (0.45 mm Whatman acetate cellulose filter) and frozen at −20°C. We thawed these kairomone stock solutions 1–2 h before each medium renewal and diluted kairomones, in ASTM hard water for the three predation treatments; 1) 0.2 fish/L for high predation risk, 2) 0.05 fish/L for low predation risk and 3) ASTM for control. We renewed medium and food every other day. All individuals remained in their original conditions until the fifth generation was produced. Vials where F0's were allocated were checked daily for neonates (F1), after birth each single neonate was photographed and its total length and spine length measured using ImageJ.

The effects of increased variation of temperature on life history traits and inducible defenses in response to different levels of predation risk were analyzed using a Generalized Linear Model (GLM). Each response variable (i.e. brood size, neonate length at birth, time between broods, time to first brood and relative spine length) was analyzed separately. For all response variables the full model included two fixed factors (temperature and predation). Maternal standard length was included in the model as a covariate.

Brood size and neonate length at birth showed little departure from normality, hence they were modelled using a Gaussian error distribution. Time between broods, on the other hand, was modelled with a Poisson error distribution. Time to first brood was modelled using a negative binomial error distribution (i.e. number of days until a success [29]). Relative difference in spine length (i.e. the proportion of total length) is bounded, which makes it particularly difficult to be modeled. Because of this, the effect of increased temperature variability and predation risk in the relative difference in spine was analysed using a GLM with a Gamma error distribution and a Log link function, avoiding the problems associated to percentages being are either smaller (<20%) or bigger (>80%) [30].

For models with a known dispersion parameter (φ) (e.g. Poisson and binomial fits), Chi-squared tests are most appropriate, whereas for those with dispersion estimated by moments (e.g. Gaussian and quasi fits) the F test is best suited [31]. Also, because the value of deviance given by the Chi-squared test is analogous to the sum of squares, the proportion of total explained variance attributed to each factor of interest can be validly compared.

For each model we tested whether all factors were needed in the minimal adequate model using Akaike's Information Criterion [32]. Specifically, we calculated ΔAIC, the difference between the AIC of each model and that of the estimated best model (the model with the lowest AIC). We also calculated Akaike weights, which are estimates of the probability that each model is the best in the model set. Post-hoc multiple pairwise comparisons between groups were conducted using the multcomp package [33]. In order to estimate the strength of the association between each factor and the response variable we partitioned the explained variance of each factor within each model [34].

Finally, we investigate the interaction between increased variability in temperature and predation risk on population growth rate using the Malthusian Growth Model (eq.1).

Table 1. Summary of the results for Generalized Linear Model for the effect of temperature and predation on brood size, neonate length at birth, time between broods, time to first brood and relative spine length.

Response variable: brood size		**df**	**Deviance**	**Test**	***p*-value**
Minimal adequate model ΔAIC – <0.001	Temperature	1	75.682	0.921	0.336
	Predation	2	58.490	8.596	<0.001
	Temperature * Predation	2	53.792	2.348	0.095
Response variable: Neonate length at birth					
Minimal adequate model ΔAIC – <0.001	Temperature	1	1.567	587.2	<0.001
	Predation	2	0.058	11.01	<0.001
	Temperature * Predation	2	0.012	2.384	0.092
Response variable: Time between broods					
Maximal model ΔAIC – 14.87	Temperature	1	4.03	16.1	<0.001
Minimal adequate model ΔAIC – <0.001					
Response variable: Time to first reproduction					
Maximal model ΔAIC – 12.3	Temperature	1	40.38	157.6	<0.001
	Predation	2	2.81	5.475	0.007
Minimal adequate model ΔAIC – <0.001					
Response variable: Relative spine length					
	Temperature	1	0.080	35.64	<0.001
Minimal adequate model ΔAIC – <0.001	Predation	2	0.029	6.566	0.011
	Temperature * Predation	2	0.019	4.267	0.025

Only best minimal adequate models are presented. The model with the lowest ΔAIC was selected as being the minimal adequate model.

$$p_n = (1+r)^n p_0 \quad \text{(eq.1)}$$

The integer p_n represents the number of neonates produced until time n, and r represents the average growth rate. P_0 represents the starting population size (i.e. 10 individuals per treatment combination). The solution given by eq.1 is an exponential function with base (1+ r) raised to the power of n. Because the number of individuals is discrete, we assume that observations follow a Poisson distribution with the mean described as a Malthusian growth curve. Regression coefficients for the Malthusian growth rate can, therefore, be estimated using a GLM with a Poisson error distribution with a log link.

All p-values were subjected to Bonferroni correction. All analyses were performed using R [35].

Results

Our results indicate that predation risk has a significant effect on brood size (Table 1, Figure 1A). Surprisingly, we failed to detect an effect of variable temperature on brood size (Table 1, Table 2, Figure 1A). In contrast, individuals reared under constant temperature, responded to increased predation risk by producing more neonates (Table 2, Figure 1A). Analysis of total proportion of variation explained by the model revealed that temperature accounted for 21.2% and predation to 13.2% of the total model variation (Table S2).

In terms of neonate length at birth, model selection included the interaction between temperature and predation as factors in the minimal adequate model (Table 1). Temperature significantly affected neonate length at birth (Table 1, Figure 1B), with smaller neonate sizes at variable temperatures (Table 2, Figure 1B). Under a constant temperature, individuals produced smaller neonates in response to increased predation risk (Table 1, Figure 1B). Under variable temperature, however, there were no significant differences in length at birth between predation treatments (Table 2, Figure 1B). Analysis of total proportion of variation explained by the model revealed that temperature accounted for 21.2% of total variation whereas predation to only 7.2% (Table S2).

Temperature, but not predation risk, significantly affected time between broods (Table 1, Figure 1C). Model selection indicated that the variation in time between broods was best explained by temperature alone (Table 1). Individuals reared at a constant temperature took less time between broods than those reared under a variable temperature regime (Table 2, Figure 1C).

There were significant differences between temperature treatments for time to first brood (Table 1, Figure 1D). Variable temperatures led to a longer time to first brood than a constant temperature (Table 2, Figure 1D). On the other hand, individuals were only able to respond to different levels of predation when at a constant temperature and not when under a variable temperature regime (Table 2, Figure 1D). Under constant temperature, time to first brood was significantly shorter when predation risk was high. Results also show that 97.4% of the total variance of the model can be attributed to temperature and only 0.8% to predation (Table S2).

Neonates produced by F0 reared at a constant temperature responded to high levels of predation risk by developing longer spines relative to their body size (Table 1, 2, Figure 1E). In contrast, neonates born to F0 reared under a variable temperature regime developed longer spines in all predation treatments (Table 2, Figure 1E). Analysis of partitioning of explained

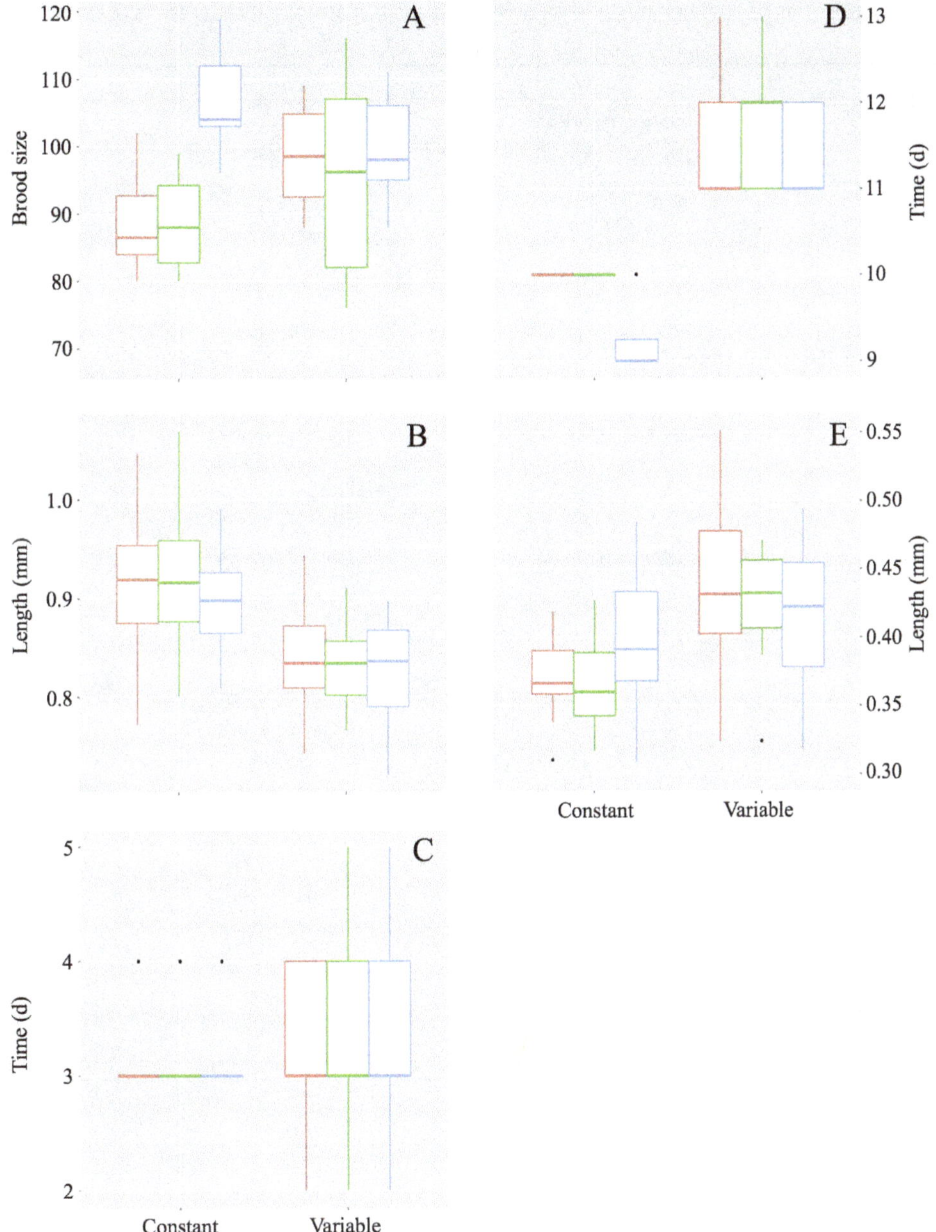

Figure 1. Effect of constant and variable temperatures on (A) brood size, (B) neonate length at birth, (C) time between broods, (D) time to first brood and (E) relative spine length, under no predation (Red), low predation (green) and high predation (blue) cues. Whiskers indicate 95% confidence intervals.

variance showed that temperature accounts for 8.47% of total variation in relative spine length and predation accounts for 0.57% (Table S2).

Population growth rate trajectories as function of time were different between temperature treatments (Table 3, Figure 2). Individuals reared at a constant temperature displayed faster growth rates in comparison to those individuals reared at a variable temperature regime (Table 3, Figure 2). Growth rate was faster under high predation risk than it was at both low predation risk and under control conditions (Table 3, Figure 2).

Discussion

In this study we test the hypothesis that stress caused by increased variability in temperature leads to erroneous life history responses to predation risk. Our results support this prediction and reveal that increased temperature variability restricts life history responses to predation risk. While under a constant temperature individuals responded to high predation risk by producing more neonates at smaller sizes, and by shortening the time to produce a first brood, when temperature varied, these responses were not detected. Interestingly, predation had no effect on time between broods. This may indicate a cost-benefit ratio that exceeded the

Table 2. Summary of the results of post-hoc multiple pairwise comparisons after significant results obtained from the Generalized Linear Model.

	Multiple comparisons		
Response variable	**Treatment**	**Adjusted *p*-value**	
	Constant	Control vs. Low	0.999
		Control vs. High	0.005
		Low vs. High	0.007
Brood size			
	Variable	Control vs. Low	0.999
		Control vs. High	0.880
		Low vs. High	0.761
	Constant	Control vs. Low	0.319
		Control vs. High	0.041
		Low vs. High	<0.001
Neonate length at birth			
	Variable	Control vs. Low	0.924
		Control vs. High	0.149
		Low vs. High	0.822
Time between broods	Constant vs. Variable		<0.001
	Constant	Control vs. Low	0.991
		Control vs. High	0.03
		Low vs. High	0.041
Time to first reproduction			
	Variable	Control vs. Low	0.997
		Control vs. High	0.971
		Low vs. High	0.718
	Constant	Control vs. Low	0.999
		Control vs. High	0.047
		Low vs. High	0.041
Relative spine length			
	Variable	Control vs. Low	0.999
		Control vs. High	0.781
		Low vs. High	0.999

potential fitness advantages of such adjustment. Importantly, our results also show that variable temperatures have a negative effect on population growth rate and on the development of morphological predator-inducible defenses. Taken together, our results strongly support the suggestion that increased variability in temperature has far greater consequences than shifts in mean temperature, namely by leading to an inability to respond appropriately to predation risk.

Despite the intensive debate on whether global temperature is becoming more variable or not [2–4,36,37], it is acknowledged that variable temperature poses a greater threat than shifts in mean temperature [13,38]. Our results provide strong evidence that *Daphnia magna* can detect predation risk and respond accordingly when temperatures are constant, but not when temperature varies. In most species the physiological responses to changes in temperature are often preceded by behavioural adjustments in ways that optimize fitness [39]. This process of adjustment, however, implies changes in how energy is allocated between the needs of basal rate and development of ecological traits [40]. Further, variability in temperature increases metabolic sensitivity [13]. It is conceivable that the intense conditions experienced in the variable treatment generated conflicting energetic requirements between metabolic rate and the development of predation response traits. Such a possibility is supported by suggestions that adjustments in energetic demands in response to changes in temperature may cause mismatches in traits [41,42]. Our result confirms that increased temperature variability has a greater effect on life history responses to predation risk than mean temperatures, and lends support to hypotheses that organisms are likely to face greater difficulties in adapting to increased variability in temperature.

It has been reported that some species respond to predation risk by reducing time between broods [43]. In our case, however, time between broods was the only trait for which predation risk had no effect. The evolution of life history traits is mediated by a cost-benefit of investing in a specific trait in a given environment [6]. It is possible that the physiological cost of adjusting time between broods exceeds the potential fitness advantages of such an adjustment. Changes in basal metabolic rate in response to different temperatures have been linked to changes in time

Table 3. Parameter estimates for fitting regression lines for the effect of temperature and predation on growth rate using Malthusian growth rate.

Temperature	Predation	Malthusian estimate	Std. Error	95%CI	*p*-value
Constant	Control	0.09471	0.00111	0.09250–0.09688	<0.001
	Low	0.09496	0.00122	0.09254–0.09735	<0.001
	High	0.10181	0.00110	0.09963–0.10396	<0.001
Variable	Control	0.08966	0.00095	0.08778–0.09152	<0.001
	Low	0.08552	0.00095	0.08363–0.08738	<0.001
	High	0.09472	0.00101	0.09271–0.09669	<0.001

between broods [44]. Time between broods was longer in variable temperatures than under constant temperatures. Further, model selection showed that temperature was the only significant factor that best explained the variation in time between broods. This result, in combination with the link between metabolic rate and time between broods, provides a strong explanation for why we failed to detect an effect of predation risk on time between broods.

In many species the presence of predators favours the growth of 'inducible defenses' [20]. These are chemical and/or morphological traits that evolved as adaptations to deter predators. In the specific example of *Daphnia spp*, it has been demonstrated that chemical cues from predators induce the development of spines as morphological defenses [45]. Our results are in agreement with this study, in that they demonstrate that exposure to high, but not low, predation cues prompts the development of relatively longer spines. As with any trait, the development of inducible defenses comes at a cost [46]. The development of longer spines in response to predation risk was recorded for constant temperatures but was not observed when temperature varied. Surprisingly, regardless of the predation risk treatment, neonates developed longer spines when temperature variability was greater. This result is not consistent with the expectation that increased variation in temperature causes an energetic disequilibrium affecting the development of some traits [13]. Producing many larger offspring when the presence of predators is uncertain is likely to generate fitness costs. However, the fitness costs of producing offspring without inducible defenses when predation risk is uncertain may be fatal, especially if predators end up being present. It is possible that the extreme stress caused by variable temperature prevents individuals from assessing predation level, and given the cost-benefits of inducible defenses, it is always better to produce defenses in the absence of predators than failing to produce them under high predation.

Classical population dynamics models predict that under variable conditions and when food provisioning is constant, selection favours an increase in Malthusian growth rate, also referred to as an exponential growth [47]. Our results do not support such a prediction, as population growth rate was slowest when temperature was variable. It has also been suggested that high levels of predation risk are expected to suppress breeding [48]. Again, we failed to detect an effect of predation on breeding suppression. Population growth rate was faster under high predation cues than under low predation cues or in the control. This result, although unexpected, is in line with previous studies on

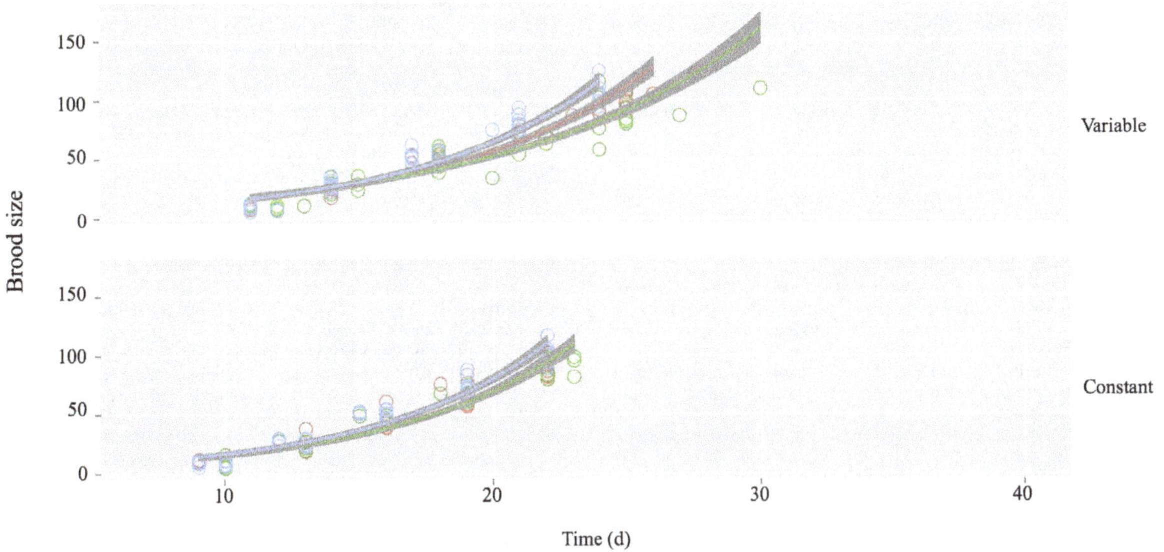

Figure 2. Exponential rate of population growth until the fifth brood in *Daphnia magna* reared at constant and variable temperatures and exposed to no predation (red), low predation (green) or high predation (blue) cues. Population growth rates estimated using Malthusian Growth Rate. Shaded areas indicate 95% confidence intervals.

population growth rate (eg. [25]). We used zebrafish to produce kairomones, and zebrafish prey upon all size classes of *Daphnia spp*. Effects of predation risk in suppressing breeding have been reported for when predators prey upon a particular size class (smaller sized) [48,49]. Given this, the absence of size-specific predation risk may explain why population growth rate was faster under high predation risk. It is also possible that an increase in population growth rate is an intrinsic consequence of an earlier time to first reproduction and greater number of neonates produced.

Our work adds to the body of knowledge regarding the significant role of predators in shaping all levels of the organizational system of a prey population [9]. Demographic selection caused by predator-induced prey mortality, may pass along the entire network causing a cascade effect within the trophic web [50]. The consequences of failing to respond to predation go, therefore, beyond the direct effects on the target species and may result in shifts at the community, and even the ecosystem, level. Our results provide conclusive evidence that increased variability in temperature disrupts the ability to respond to predation risk. Moreover, there are suggestions that species may be more vulnerable to variability in temperature than upward shifts in mean temperature [13,38]. Our results empirically support such concerns and highlight the importance of considering interactions among multiple stressors on life history responses.

Acknowledgments

We are grateful to Maria Dornelas, Hideyasu Shimadzu and Amy Deacon for providing helpful comments on early drafts and for statistical advice on model selection. We are also thankful to the Biodiversity and Behaviour Group for the rewarding discussions and comments on the manuscript.

Author Contributions

Conceived and designed the experiments: MB JLT AMVMS. Performed the experiments: MB JLT. Analyzed the data: MB. Contributed reagents/materials/analysis tools: JLT AMVMS. Wrote the paper: MB. Responsible for funding: JLT AMVMS.

References

1. Broomhall SD (2004) Egg temperature modifies predator avoidance and the effects of the insecticide endosulfan on tadpoles of an Australian frog. J. Anim. Ecol. 41: 105–113.
2. Mora C, Frazier AG, Longman RJ, Dacks RS, Walton MM, et al. (2013) The projected timing of climate departure from recent variability. Nature 502: 183–187.
3. Morice CP, Kennedy JJJ, Rayner NA, Jones PD (2012) Quantifying uncertainties in global and regional temperature change using an ensemble of observational estimates: The HadCRUT4 dataset. Geophys. Res-Atmos. 117: 1–22.
4. Hansen J, Sato M, Ruedy R (2012) Perception of climate change. Proc Natl Acad Sci U S A 109: E2415–E2423.
5. Reznick D, Endler J (1982) The impact of predation on life history evolution in Trinidadian guppies (Poecilia reticulata). Evolution 36: 160–177.
6. Stearns SC (1992) The evolution of life histories. Oxford: Oxford University Press.
7. Roff AD (1992) The evolution of life histories: theory and analysis. New York: Chapman & Hall. 548 p.
8. Law R (1979) Optimal life histories Under age-specific predation. Am Nat 114: 399–417.
9. Crowl TA, Covich AP (1990) Predator-induced life-history shifts in a freshwater snail. Science 247: 949–951.
10. Ludwig D, Rowe L (1990) Life-history strategies for energy gain and predator avoidance under time constraints. Am Nat 135: 686–707.
11. Milinski M, Heller R (1978) Influence of a predator on the optimal foraging behaviour of sticklebacks (*Gasterosteus aculeatus* L.). Nature 275: 642–644.
12. Kooijman SALM (2000) Dynamic energy and mass budgets in biological systems: Cambridge University Press. 490 p.
13. Hawkins AJS (1995) Effects of temperature change on ectotherm metabolism and evolution: Metabolic and physiological interrelations underlying the superiority of multi-locus heterozygotes in heterogeneous environments. J Therm Biol 20: 23–33.
14. Boersma M, Spaak P, De Meester L (1998) Predator-mediated plasticity in morphology, life history, and behavior of Daphnia: The uncoupling of responses. Am Nat 152: 237–248.
15. Stibor H (1992) Predator induced life-history shifts in a freshwater cladoceran. Oecologia 92: 162–165.
16. Macháček J (1991) Indirect effect of planktivorous fish on the growth and reproduction of *Daphnia galeata*. Hydrobiologia 225: 193–197.
17. Sakwińska O (1998) Plasticity of *Daphnia magna* life history traits in response to temperature and information about a predator. Freshwater Biol 39: 681–687.
18. Weetman D, Atkinson D (2002) Antipredator reaction norms for life history traits in *Daphnia pulex*: dependence on temperature and food. Oikos 98: 299–307.
19. Doksaeter A, Vijverberg J (2001) The effects of food and temperature regimes on life-history responses to fish kairomones in Daphnia hyalina x galeata. Hydrobiologia 442: 207–214.
20. Tollrian R, Harvell CD, editors (1999)The ecology and evolution of inducible defenses: Princeton University Press. 395 p.
21. Werner EE, Gilliam JF, Hall DJ, Mittelbach GG (1983) An experimental test of the effects of predation risk on habitat use in fish. Ecology 64: 1540–1548.
22. Clark RA, Fox CJ, Viner D, Livermore M (2003) North Sea cod and climate change – modelling the effects of temperature on population dynamics. Glob Change Biol 9: 1669–1680.
23. OECD (2008) *Daphnia magna* reproduction test. OECD.
24. Baird DJ, Barber I, Bradley M, Soares AMVM, Calow P (1991) A comparative study of genotype sensitivity to acute toxic stress using clones of *Daphnia magna* straus. Ecotox Environ Safe 21: 257–265.
25. Pestana JLT, Baird DJ, Soares AMVM (2013) Predator threat assessment in Daphnia magna: the role of kairomones versus conspecific alarm cues. Mar Freshwater Res 64: 679–686.
26. Jacobs AFG, Heusinkveld BG, Kraai A, Paaijmans KP (2008) Diurnal temperature fluctuations in an artificial small shallow water body. Int J Biometeorol 52: 271–280.
27. Dallas H (2008) Water temperature and riverine ecosystems: An overview of knowledge and approaches for assessing biotic responses, with special reference to South Africa. Water SA 34 393–404.
28. Ward JV (1985) Thermal characteristics of running waters. In: Davies BR, Walmsley RD, editors. Perspectives in Southern Hemisphere Limnology: Springer Netherlands. pp.31–46.
29. Pitman J (1993) Probability. New York: Springer Verlag. 576 p.
30. Crawley MJ, editor (2007) The R book: John Wiley & Sons, Ltd. 942 p.
31. McCullagh P, Nelder JA, editors (1989) Generalized linear models. 2nd ed. New York: Chapman & Hall. 532 p.
32. Burnham KP, Anderson DR (2002) Model selection and multi-model inference: A practical information- Theoretic approach. New York: Springer-Verlag.
33. Hothorn T, Bretz F, Westfall P (2008) Simultaneous inference in general parametric models. Biom J 50: 346–363.
34. Hays WL (1994) Statistics: Holt, Rinehart and Winston.
35. Team RDC (2013) R: A language and environment for statistical computing. R. R Foundation for Statistical Computing. R 2.15.3 ed. Vienna, Austria: R Foundation for Statistical Computing.
36. Huntingford C, Jones PD, Livina VN, Lenton TM, Cox PM (2013) No increase in global temperature variability despite changing regional patterns. Nature 500: 327–330.
37. Rhines A, Huybers P (2013) Frequent summer temperature extremes reflect changes in the mean, not the variance. Proc Natl Acad Sci U S A 110: E546.
38. Rahmstorf S, Coumou D (2011) Increase of extreme events in a warming world. Proc Natl Acad Sci U S A 108: 17905–17909.
39. Angilletta MJ (2009) Thermal Adaptation: A Theoretical and Empirical Synthesis: Oxford, University Press.
40. Dell AI, Pawar S, Savage VM (2011) Systematic variation in the temperature dependence of physiological and ecological traits. Proc Natl Acad Sci U S A 108: 10591–10596.

41. Gillooly JF, Brown JH, West GB, Savage VM, Charnov EL (2001) Effects of size and temperature on metabolic rate. Science 293: 2248–2251.
42. Gillooly JF, Charnov EL, West GB, Savage VM, Brown JH (2002) Effects of size and temperature on developmental time. Nature 417: 70–73.
43. Evans JP, Gasparini C, Pilastro A (2007) Female guppies shorten brood retention in response to predator cues. Behav Ecol Sociobiol 61: 719–727.
44. Harvey PH, Pagel MD, Rees JA (1991) Mammalian metabolism and life histories. Am Nat 137: 556–566.
45. Stabell OB, Ogbebo F, Primicerio R (2003) Inducible defences in daphnia depend on latent alarm signals from conspecific prey activated in predators. Chem Senses 28: 141–153.
46. Harvell CD (1990) The ecology and evolution of inducible defenses. Q Rev Biol 65: 323–340.
47. MacArthur RH (1962) Some generalized theorems of natural selection. Proc Natl Acad Sci U S A 48: 1893–1897.
48. Ruxton GD, Lima SL (1997) Predator-induced breeding suppression and its consequences for predator-prey population dynamics. Proc R Soc Lond B 264: 409–415.
49. Urban MC (2007) The Growth-predation risk trade-off under a growing gape-limited predation threat. Ecology 88: 2587–2597.
50. Schmitz OJ, Beckerman AP, O'Brien KM (1997) Behavioural mediated trophic cascades: effects of predation risk on food web interactions. Ecology 78: 1388–1399.

13

When David Beats Goliath: The Advantage of Large Size in Interspecific Aggressive Contests Declines over Evolutionary Time

Paul R. Martin[1]*, Cameron K. Ghalambor[2]

1 Department of Biology, Queen's University, Kingston, Ontario, Canada, **2** Department of Biology and Graduate Degree Program in Ecology, Colorado State University, Fort Collins, Colorado, United States of America

Abstract

Body size has long been recognized to play a key role in shaping species interactions. For example, while small species thrive in a diversity of environments, they typically lose aggressive contests for resources with larger species. However, numerous examples exist of smaller species dominating larger species during aggressive interactions, suggesting that the evolution of traits can allow species to overcome the competitive disadvantage of small size. If these traits accumulate as lineages diverge, then the advantage of large size in interspecific aggressive interactions should decline with increased evolutionary distance. We tested this hypothesis using data on the outcomes of 23,362 aggressive interactions among 246 bird species pairs involving vultures at carcasses, hummingbirds at nectar sources, and antbirds and woodcreepers at army ant swarms. We found the advantage of large size declined as species became more evolutionarily divergent, and smaller species were more likely to dominate aggressive contests when interacting with more distantly-related species. These results appear to be caused by both the evolution of traits in smaller species that enhanced their abilities in aggressive contests, and the evolution of traits in larger species that were adaptive for other functions, but compromised their abilities to compete aggressively. Specific traits that may provide advantages to small species in aggressive interactions included well-developed leg musculature and talons, enhanced flight acceleration and maneuverability, novel fighting behaviors, and traits associated with aggression, such as testosterone and muscle development. Traits that may have hindered larger species in aggressive interactions included the evolution of morphologies for tree trunk foraging that compromised performance in aggressive contests away from trunks, and the evolution of migration. Overall, our results suggest that fundamental trade-offs, such as those associated with body size, are more likely to break down over evolutionary time, changing the rules that govern species interactions and structure ecological communities.

Editor: James A. R. Marshall, University of Sheffield, United Kingdom

Funding: The authors' work was funded by the Natural Sciences and Engineering Research Council of Canada (Discovery Grant RGPIN/355519-2008; www.nserc-crsng.gc.ca), a Baillie Family Chair Endowment, and a Good Family Fellowship. The funders had no role in study design, data collection and analysis, decision to publish, or preparation of the manuscript.

Competing Interests: The authors have declared that no competing interests exist.

* Email: pm45@queensu.ca

Introduction

Phylogenetic perspectives have changed the way we view ecological communities by incorporating evolutionary history into explanations of patterns of coexistence and resource use [1,2]. Because closely-related species are more likely to share traits and be ecologically similar due to recent, shared ancestry, phylogenetic relationships may influence the degree to which species can coexist [1,2]. One particular way that evolutionary relatedness can influence community structure is by altering the trade-offs that constrain species interactions and distributions [3,4]. For example, small-sized animals benefit from reduced energetic and water requirements for survival and reproduction, reduced developmental times, reduced costs of locomotion, greater ability to quickly shed or absorb heat, greater maneuverability, and faster response time [5,6]. However, small-sized animals typically lose aggressive contests for resources because larger animals generate greater force for a given acceleration, require a greater opposing force to overcome their inertia or change their momentum, and have greater muscle mass and strength, stronger defensive coverings, and larger and stronger traits used as weapons (e.g., teeth, bills, claws) [5,7,8]. Yet, the dominance of larger body size is not universal. Smaller species can overcome the advantages of large size in aggressive contests with the evolution of novel traits or trait values (e.g., weapons, enhanced maneuverability, or social behavior) that offset the disadvantage of small size (e.g., [9–11]). Such novel traits should accumulate over evolutionary time [3,12], leading to the hypothesis that the advantage of large size in aggressive interactions should decline with evolutionary distance among the species.

Here, we test this hypothesis using data from 23,362 interactions among 246 species pairs, representing three phylogenetically and ecologically distinct groups of birds that have been studied extensively with respect to aggressive contests for shared food resources. These three groups are: 1) New World vultures (Accipitriformes: Cathartidae) and Old World vultures (Accipitriformes: Accipitridae) interacting at carcasses, 2) hummingbirds (Apodiformes: Trochilidae) interacting at nectar sources, and 3)

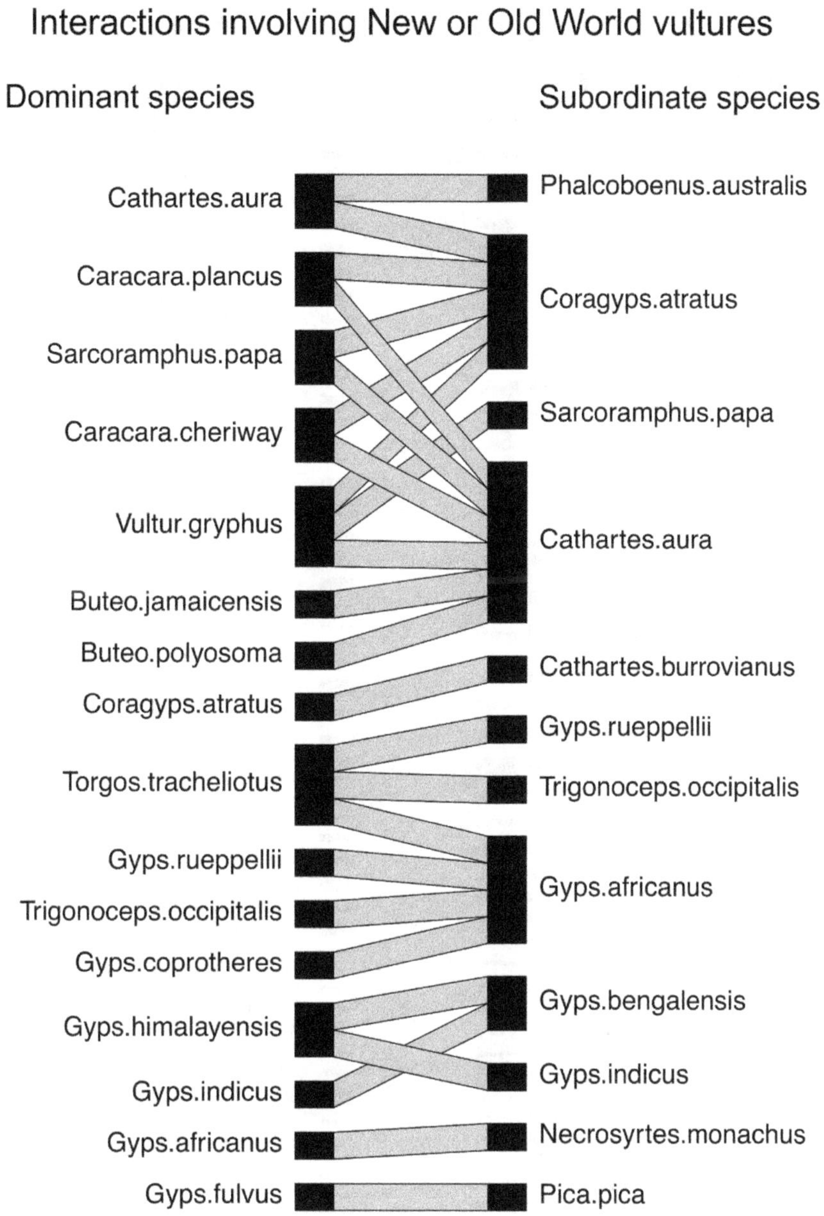

Figure 1. An interaction web for aggressive interactions involving New World vultures (Cathartidae) or Old World vultures (Accipitridae) at a carrion food source. Lines connect species pairs for which we include data on aggressive interactions in this study. Dominant species (left column) were defined as species winning the majority of aggressive interactions with the subordinate species (right column).

antbirds (Passeriformes: Thamnophilidae) and woodcreepers (Passeriformes: Dendrocolaptidae) competing for invertebrates and small vertebrates fleeing from army ant swarms (Hymenoptera: Formicidae) [13].

Materials and Methods

Aggressive Interactions

We compiled published quantitative data on aggressive interactions between our focal species (vultures, hummingbirds, and woodcreepers and antbirds) and any other species of bird, but restricted our analysis to pairs of species with at least 6 interactions where each interaction was won by one of the two species (Figures 1–3; Tables S1, S2). We compiled data on aggressive interactions that were associated with a shared resource (following [14]), including (a) chases, where one species actively pursued the other species, (b) supplants and displacements, where one species actively flew at, lunged, pecked, or otherwise aggressively engaged another species, causing the other species to retreat, and (c) physical attacks, where one species fought with another species (e.g., pecking, grabbing, hitting with wings, pinning to ground), resulting in the losing species retreating from the altercation [14]. We included only observations where one species was a clear winner over the other, as described in the original reference. We excluded observations that could be viewed as defense of eggs, young, or nests because these interactions did not involve a shared resource, and because the fitness costs of losing offspring are higher for the parent species. We excluded interactions that

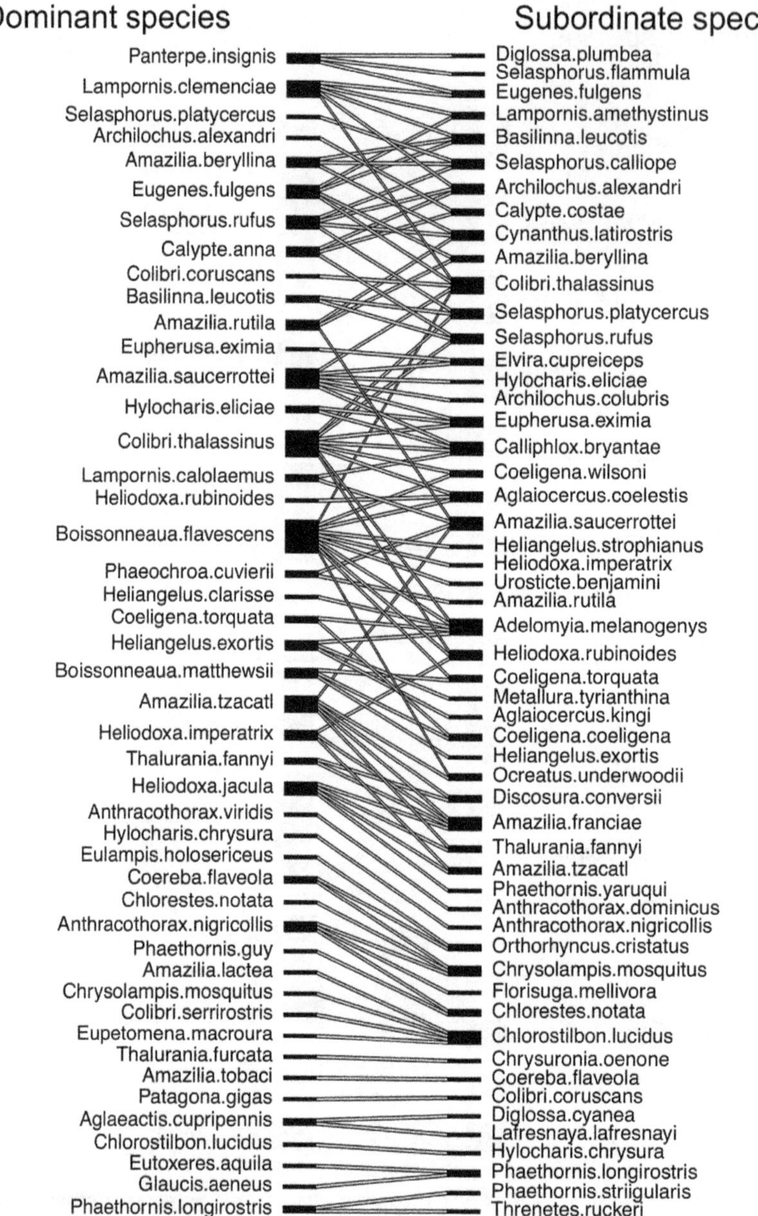

Figure 2. An interaction web for aggressive interactions involving hummingbirds (Trochilidae) at a nectar food source. Lines connect species pairs for which we include data on aggressive interactions in this study. Dominant species (left column) were defined as species winning the majority of aggressive interactions with the subordinate species (right column). One species pair (*Hylocharis chrysura* — *Thalurania furcata*) was omitted from this figure because each species won the same number of interactions with the other.

involved more than one individual of each of two species. When possible, we excluded interactions involving young birds; however, details of age were often unavailable. We supplemented our interaction dataset with our own unpublished natural history observations of hummingbirds interacting at feeders in South America.

Overall, we compiled data on 23,362 aggressive interactions among 246 species pairs (Tables S1, S2). In total, 145 different species were represented in our dataset, including 99 species as dominant (i.e., winning the majority of aggressive interactions with another species) and 99 species as subordinate (i.e., losing the majority of aggressive interactions with another species) (Figures 1–3). Twenty-seven pairs of interacting species were in the same genus, 157 pairs were in different genera but in the same family, and 62 pairs represented different taxonomic families. On average, each of the 145 species was represented in 3.4 pairwise interactions (range 1–18).

Body Mass

Whenever possible, we obtained average adult mass for each species in our study from the same studies that provided the interaction data. Otherwise, we obtained adult mass data from

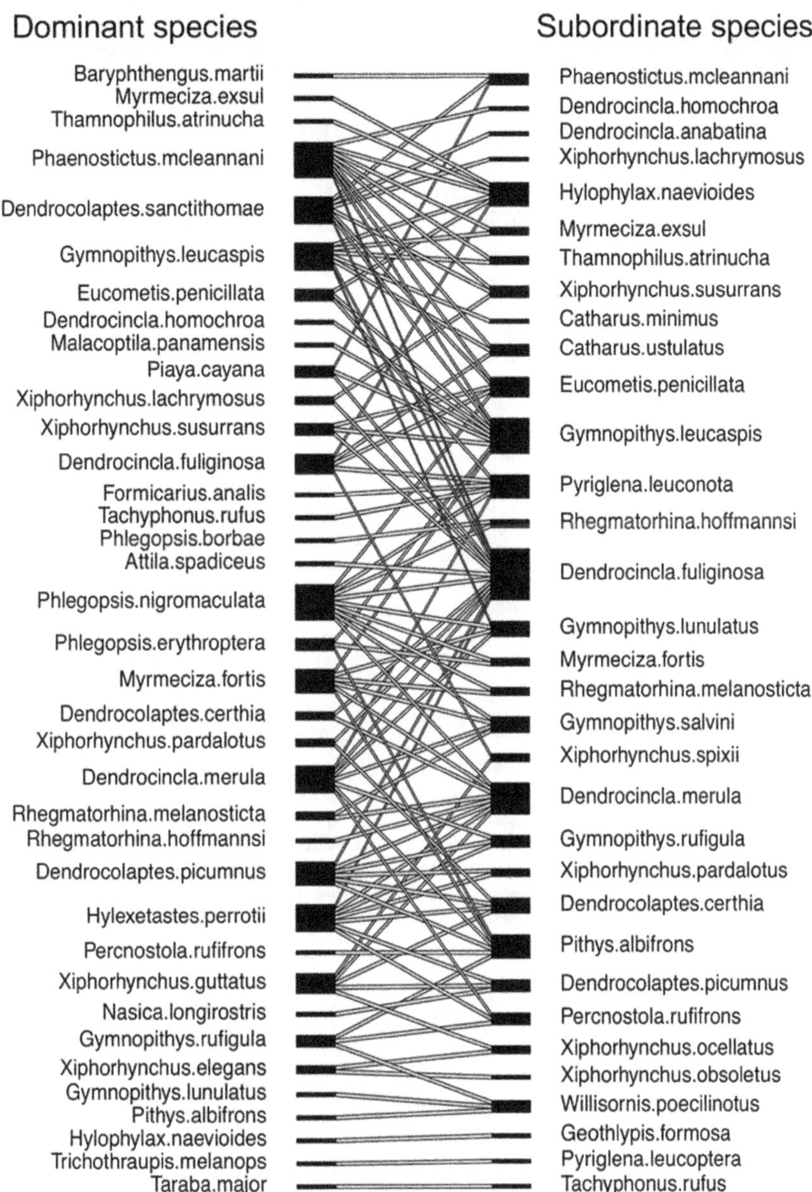

Figure 3. An interaction web for aggressive interactions involving antbirds (Thamnophilidae) or woodcreepers (Dendrocolaptidae) at army ant swarms, where they feed on prey flushed by the ants. Lines connect species pairs for which we include data on aggressive interactions in this study. Dominant species (left column) were defined as species winning the majority of aggressive interactions with the subordinate species (right column). One species pair (*Pyriglena leucoptera — Tachyphonus coronatus*) was omitted from this figure because each species won the same number of interactions with the other.

areas as geographically close as possible to the populations where the interaction data originated. We calculated the mean mass of each species as the mean of males (mean male mass) and females (mean female mass), if males and females had different masses. We provide the sources for mass data for each species pair in Table S2.

Genetic Distance

We compiled mitochondrial genetic sequence data from Genbank (accession numbers are provided in Text S1). Once aligned, we measured genetic distance between the sequences of focal species pairs using MEGA version 5.0 [15]. We calculated between-group mean Tamura-Nei genetic distances because this measure corrects for multiple substitutions at one site, incorporates differences in substitution rates between nucleotides, and does not assume equal nucleotide frequencies [16]. Species interactions lacking relevant genetic sequence data were excluded from our analysis.

We preferentially used the mitochondrial gene cytochrome *b* (cyt*b*) to estimate genetic distance between pairs of interacting species because this gene appears to evolve in a clock-like fashion in birds [17]. However, too few hummingbird species had cyt*b* sequences available, so we examined the mitochondrial gene

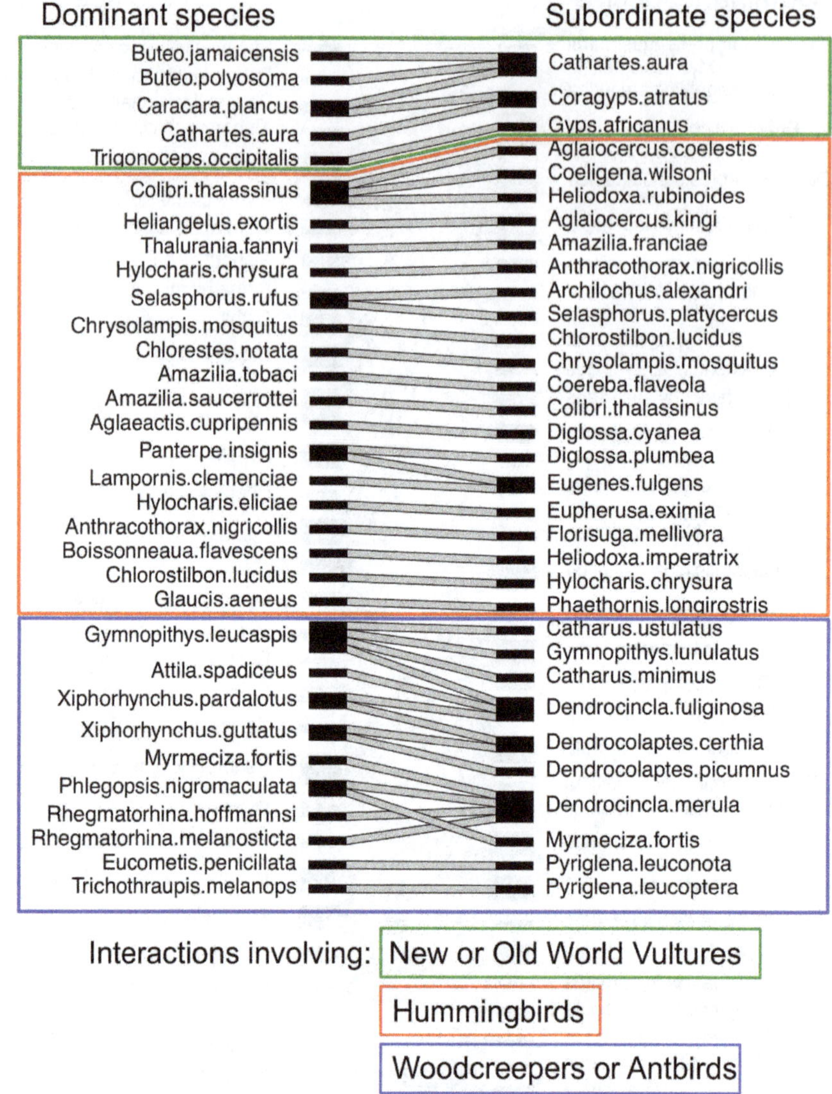

Figure 4. An interaction web for aggressive interactions where the smaller (lighter) species was dominant to the heavier species. Each interacting species pair where the smaller species was dominant is connected by a line ($n = 43$ species pairs). Dominant species (left column) were defined as species winning the majority of aggressive interactions with the subordinate species (right column). Species pairs that include a vulture are indicated by a green box (top), hummingbird by a red box (middle), and antbird or woodcreeper by a blue box (bottom).

NADH dehydrogenase subunit 2 (ND2) for hummingbirds. We aligned sequences of each gene with the same gene from the chicken (*Gallus gallus*) [18] using Clustal X [19], visually inspected the sequences using MacClade version 4.08 [20], and removed sequences that did not align with the chicken sequence. We included transitions and transversions, all codon positions, assumed uniform rates among sites and homogeneous patterns among lineages, and used pairwise deletion to address gaps or missing data [15]. We calculated genetic distances for each pair of interacting species when species were in the same taxonomic family. For species representing different families, we calculated the genetic distances incorporating all sequences of species in those families that were included in our study to improve the accuracy of our longer distance estimates.

Statistical Tests

We conducted all of our statistical tests in R [21]. We tested the hypothesis that the importance of large size in determining the outcome of aggressive contests declined with genetic distance between interacting species using a linear mixed-effects model with sqrt{ln[(wins by dominant species+1)/(wins by subordinate species+1)]} as the response variable, {(mass of dominant species−mass of subordinate species)/[(mass of dominant species+mass of subordinate species)/2]} and genetic distance as predictors within a saturated model, and group (vultures, hummingbirds, or antbirds/woodcreepers) as a random factor, with a Gaussian distribution using the R package nlme [22]. The dominant species was defined as the species that won the majority of aggressive contests. We also tested if smaller bird species were more likely to be dominant when interacting with a more distantly-related

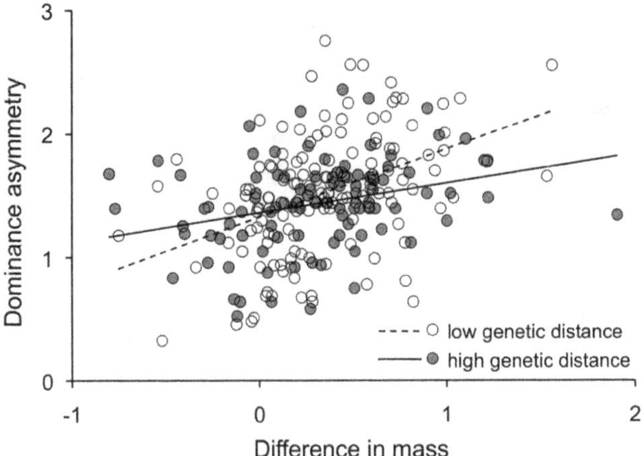

Figure 5. Relatively heavier bird species were more likely to win aggressive contests for resources (positive slope), but the advantage of large size in aggressive contests declined with genetic distance (shallower slope for high genetic distance). Dominance asymmetry (y-axis) = sqrt {ln ((wins by dominant species+1)/(wins by subordinate species+1))}. Difference in mass (x-axis) = (mass of dominant species−mass of subordinate species)/(average mass of dominant and subordinate species). Genetic distance groups are split by the midpoint value for the dataset (low = 0.006–0.179; high = 0.180–0.352).

species using a generalized linear mixed model with the response variable equal to 0 if the larger species was dominant, and 1 if the smaller species was dominant, genetic distance and number of interactions between species as predictors in a saturated model, and group (vultures, hummingbirds, or antbirds/woodcreepers) as a random factor, with a binomial distribution using the R package lme4 [23]. In our generalized linear model, we standardized both genetic distance and the number of interactions between species prior to analysis by subtracting the mean and dividing by 2 standard deviations, using the *rescale* command in the R package *arm* [24]. Two species comparisons were excluded from analyses because each species won the same number of aggressive encounters, and thus we could not designate a dominant and subordinate species.

We used a mixed models approach with group (vultures, hummingbirds, woodcreepers/antbirds) as a random factor because interactions between species within vultures, hummingbirds, and woodcreepers/antbirds often lacked independence (e.g., one species interacted with more than one other species; Figures 1–3). In contrast, we had no overlap of species across our three groups. We also ran our analyses including either the dominant species or the subordinate species as a random factor, nested within group, to ensure that our results were not influenced by one or a few species that were small and dominant or large and subordinate. We did not run our analysis with both dominant and subordinate species as random factors because each interaction among species pairs was unique.

We first ran the saturated model with different random slopes and intercepts and chose the best model as the model with the lowest Akaike Information Criterion (AIC) value [25]. We then checked the fit of the best saturated model following [25]. If saturated models did not adequately fit the data (e.g., residuals were significantly different from normal for linear mixed-effects models), we either modeled heterogeneity and assessed new model performance using AIC values and improved model fit [25], or transformed dependent or independent variables to improve model fit. We then ran models that incorporated all possible combinations of predictor variables using the R package MuMIn [26], and compared the fit among models using AIC values adjusted for small sample sizes (AICc). We identified the best-fit model as the model with the lowest AICc score. We ran our generalized linear mixed models using both the R packages lme4 [23] and MASS [27] to ensure that our results were consistent across packages [25].

We ran two additional generalized linear models to test our hypothesis, with the same predictor and response variables as in the saturated linear mixed-effects and generalized linear mixed models, but with group (vultures, hummingbirds, antbirds/woodcreepers) entered as a predictor variable in a saturated model (without random effects). This model allowed us to test for variation in the effects of genetic distance and body mass on the outcome of aggressive interactions between our 3 focal groups. The model, however, assumes that interactions among species are independent — an assumption that was violated by some species interacting with multiple species within groups. We ran models incorporating all possible combinations of predictor variables and checked the fit of our models as in our previous models.

Number of Interactions among Species Pairs

The number of aggressive interactions over shared resources are expected to decline as species diverge over evolutionary time because more distantly-related species, on average, are ecologically more distinct [8,28]. Small species could win more interactions

Table 1. Comparison of model fit for different linear mixed-effects models that include all possible combinations of predictor variables.

Model ranking	Intercept[1]	Difference in mass	Genetic distance	Interaction between difference in mass and genetic distance	df	logLik	AICc	delta AICc	weight
1	1.17	0.84	0.91	−2.46	6	−120.5	253.3	0.00	0.882
2	1.35	0.37	—	—	4	−124.9	258.0	4.72	0.083
3	1.31	0.37	0.24	—	5	−124.8	259.8	6.49	0.034
4	1.47	—[2]	—	—	3	−140.5	287.1	33.85	0.000
5	1.45	—	0.14	—	4	−140.5	289.1	35.83	0.000

Models tested the prediction that the advantage of large size in aggressive contests for resources declines with genetic distance among interacting bird species ($n = 244$ species pairs).

[1] numbers for predictor variables are effect sizes.

[2] long dash (—) indicates that the predictor variable was absent from the model.

Table 2. Results from the best-fit (lowest AICc) linear mixed-effects model testing the prediction that the advantage of large size in aggressive contests for resources declines with genetic distance among interacting bird species ($n = 244$ species pairs).

Fixed effects[1]	Estimate	SE	t	df	P
Intercept	1.17	0.10	11.91	238	<0.0001
Difference in mass[2]	0.84	0.17	4.91	238	<0.0001
Genetic distance	0.91	0.48	1.89	238	0.0598
Difference in mass[2]×Genetic distance	−2.46	0.84	−2.94	238	0.0036

[1] dependent = sqrt{ln[(wins by dominant species+1)/(wins by subordinate species+1)]}; taxonomic group included as a random effect.
[2] (mass of dominant species − mass of subordinate species)/(average mass of dominant and subordinate species).

with larger species when interactions are rare if they can bluff or if species are more inclined to back down when the other species is poorly known. We did not include the number of interactions among species pairs in our main linear mixed-effects model because the number of interactions was already incorporated into the dependent variable. We did, however, include the total number of interactions among species pairs in our generalized linear mixed model.

The total number of interactions in our generalized linear mixed model may be biased by the inclusion of different studies with different effort and sample sizes. Thus, we ran the analysis again including each study as a random factor nested within group (vultures, hummingbirds, woodcreepers/antbirds). By including study as a random factor, we could address if the number of interactions among species pairs within studies influenced our main results. We included only studies that had at least 2 species pairs, and included only species pairs with at least 6 interactions within a study. We again included the number of interactions and genetic distance as predictors in a saturated model, with the response variable equal to 0 if the larger species was dominant, and 1 if the smaller species was dominant, and a binomial distribution using the R package lme4 [23]. For all analyses, the number of interactions was transformed to improve model fit, as: n = ((log(log(number of interactions)))^0.1).

Results

The larger species won the majority of aggressive interactions with the smaller species (i.e., was behaviorally dominant) in 201 of 246 species pairs (81.7%). Smaller species were behaviorally dominant in 43 species pairs (17.5%) involving 32 different small, dominant species, and 33 different large, subordinate species (Figure 4). Wins and losses were equal in only two species pairs. In some cases where smaller species were behaviorally dominant, differences in weight were small and perhaps insignificant. In 32 species pairs, however, the larger species was>5% heavier than the smaller, dominant species, while 25 species pairs involved larger species that were>10% heavier than the smaller, dominant species.

The advantage of large size in aggressive contests was greatest among closely-related species and declined with increasing genetic distance (Figure 5; Tables 1,2; see Figure S1 for a 3-dimensional plot that includes difference in mass, genetic distance and dominance asymmetry). Similarly, smaller species were more likely to be dominant in aggressive interactions with distantly-related species (Figure 6; Tables 3,4). For example, among interacting species in the same genus, the smaller species was dominant in only 2 of 27 species pairs (7.4%), whereas among interacting species in different taxonomic families, the smaller species was dominant in 18 of 62 species pairs (29.0%). The decline in the importance of large size in aggressive interactions with increased genetic distance was evident in each of the three independent groups and did not differ significantly between them (Tables 5,6).

The decline in the advantage of large size with increased genetic distance was not caused by variation in the number of interactions among species pairs (overall number of interactions: Tables 3,4; number of interactions within a study: Tables S3,S4). The number of interactions among species (overall, or within a study) also did not predict when small bird species would be dominant in aggressive interactions (overall number of interactions: Tables 3,4; number of interactions within a study: Tables S3,S4).

Including species (either the dominant species, or the subordinate species) as a random factor, nested within group, yielded similar results for both linear mixed-effects models (Table S5) and generalized linear mixed models (Table S6), with the interaction between genetic distance and body mass significant in all analyses. We also found similar results in linear mixed-effects models with interactions between *Gyps fulvus* and *Pica pica* (an outlier point) excluded from the dataset (interaction between genetic distance and body mass in our best-fit model, estimate = -2.21 ± 0.86SE, $t = -2.58$ $df = 237$, $p = 0.011$; Figure S2).

Discussion

Small animals flourish in many ecological settings, but they often suffer the recurrent cost of losing aggressive contests for resources with larger animals [5–8]. At the same time, numerous examples exist of smaller species dominating larger species (e.g., [9–11]), but the conditions leading to small species dominating larger species are poorly understood. Here we tested the

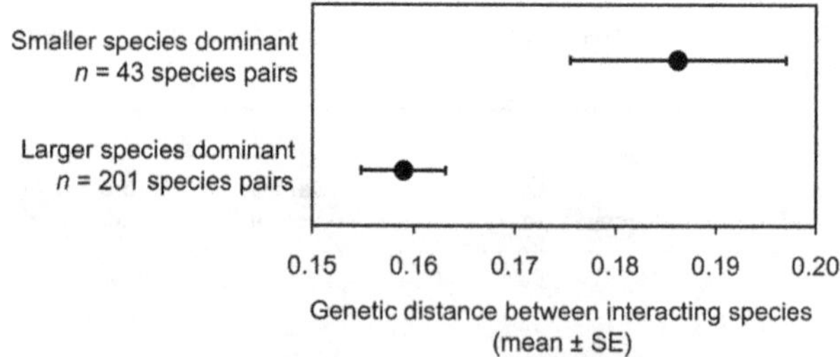

Figure 6. Smaller (lighter) bird species were more likely to be dominant (i.e., win the majority of aggressive contests for resources) when interacting with a more distantly-related species (greater genetic distance between species). Data are from aggressive interactions involving vultures at carcasses, hummingbirds at nectar sources, and woodcreepers and antbirds at army ant swarms.

Table 3. Comparison of model fit for different generalized linear mixed models (binomial distribution) that include all possible combinations of predictor variables.

Model ranking	Intercept[1]	Genetic Distance	Number of interactions	Interaction between genetic distance and number of interactions	df	logLik	AICc	delta AICc	weight
1	−1.60	0.89	—	—	3	−110.1	226.4	0.00	0.587
2	−1.61	0.91	0.07	—	4	−110.1	228.4	2.02	0.213
3	−1.58	0.96	0.00	0.76	5	−109.6	229.4	3.03	0.129
4	−1.54	—[2]	—	—	2	−113.6	231.3	4.88	0.051
5	−1.54	—	−0.10	—	3	−113.6	233.2	6.84	0.019

Models tested the prediction that smaller bird species were more likely to win aggressive contests for resources when interacting with a more distantly-related species ($n = 244$ species pairs).
[1] numbers for predictor variables are effect sizes; genetic distance and number of interactions were standardized prior to analysis by subtracting the mean and dividing by 2 standard deviations.
[2] long dash (—) indicates that the predictor variable was absent from the model.

hypothesis that the disadvantage of being small in aggressive interactions could be overcome over evolutionary time through the accumulation of novel traits that can counteract the advantages of being large. We found support for this hypothesis: larger species were dominant over smaller species during aggressive interactions for shared resources, but the advantage of a larger body size declined with increased evolutionary distance in vultures, hummingbirds, and antbirds/woodcreepers (Figure 5; Tables 1,2). Similarly, small species were more likely to win the majority of aggressive contests when they interacted with more distantly-related species (Figure 6; Tables 3,4). These results suggest that body size and phylogenetic distance jointly shape the outcome of aggressive interactions in birds, which in turn may influence the structure of bird communities.

While our results are consistent with the predictions of our hypothesis, our results are also consistent with three alternative hypotheses. Distantly-related species are, on average, less likely to share preferred resources [8,28], which could explain the patterns observed here, if (i) large species were less willing to invest in aggressive encounters over less preferred resources with distantly-related smaller species, (ii) if the costs of losing an aggressive encounter are lower over less preferred resources, or (iii) if the context of aggressive interactions (e.g., age, experience, condition, hunger level, time of arrival or colonization) shift as resource preferences diverge, allowing some distantly-related species to win interactions that they would normally lose (e.g., [29–31]). If ecological similarity declines with genetic distance [8,28], then these alternatives predict that behaviorally dominant species should lose a larger proportion of interactions with distantly-related subordinates, independent of differences in size. Our results did not support this prediction: while genetic distance was an important predictor of the outcome of aggressive interactions outside of its interaction with body size (Tables 1,2,5,6), dominant species won a greater proportion of their interactions with more distantly-related subordinate species, opposite to the predicted pattern. Detailed data on resource use also suggests that larger species did not lose aggressive contests over less preferred food (Text S2, Table S7), contrary to the predictions of (i) and (ii). While our data do not rule out the contribution of these alternative hypotheses to the decline in the importance of size in aggressive encounters among more distantly-related species, they suggest that these alternative hypotheses cannot explain the pattern by themselves.

What traits offset the importance of size in aggressive interactions?

No study to date has quantified a general set of traits which predictably offset the advantages of large size in aggressive interactions. However, upon reviewing the focal studies used to generate the data in this study (Text S3), in addition to other studies, we find a diverse set of traits that play a key role in offsetting the importance of size in aggressive interactions among birds. These traits include the evolution of well-developed leg musculature and talons (raptors [32]), adaptations that enhance flight acceleration (hummingbirds [33,34]) and maneuverability (hummingbirds [33–37], woodcreepers [38–40]), novel fighting behaviors (woodcreepers [38–40], grouse [41]), and traits directly associated with aggression, such as testosterone and muscle development (hummingbirds [29,42,43], woodcreepers [44], blackbirds [45,46]), The evolution of social behavior, where smaller individuals perform coordinated attacks on individuals of larger species, may shift the outcomes of aggressive interactions among species (e.g., mammalian carnivores [8]), although we excluded interactions involving multiple individuals of each species

Table 4. Results from the best-fit (lowest AICc) generalized linear mixed model (binomial distribution) testing the prediction that smaller bird species were more likely to win aggressive contests for resources when interacting with a more distantly-related species ($n = 244$ species pairs).

Fixed effects[1]	Estimate	SE	z	P
Intercept	−1.60	0.18	−9.08	<0.0001
Genetic distance[2]	0.89	0.35	2.59	0.0098

[1]dependent = 0 if larger species was dominant, 1 if smaller species was dominant; taxonomic group included as a random effect.
[2]standardized prior to analysis by subtracting the mean and dividing by 2 standard deviations.

in our study. In addition, intraspecific clustering, leading to high densities of subordinate species, can overwhelm individuals of dominant species, leading to dominants reducing their territory size or abandoning resources altogether (vultures [47,48], hummingbirds [29,49], blackbirds [45,46]). In these cases, social coordination among individuals is unnecessary — simply a high density of subordinate relative to dominant individuals may increase the costs of aggressive defense of a resource for dominants. In all of these cases, the evolution of novel traits (or trait values) in smaller species were thought to allow them to overcome the costs of small size in aggressive contests with larger species.

Focal studies of species interactions also revealed cases where adaptations for other functions compromised the ability of large species to compete aggressively with smaller species — a mechanism that we did not predict. Traits that may have hindered larger species in aggressive interactions included the evolution of specialized morphologies for tree trunk foraging in woodcreepers [50] that in turn compromised their performance in aggressive contests away from trunks [44,51], and the evolution of migratory behavior, that compromised performance in aggressive contests with resident species. In the case of woodcreepers, the specialized adaptations for tree climbing [50] enhanced their performance in aggressive contests with some woodcreeper species on tree trunks [38–40], but compromised their performance in aggressive interactions with antbirds, which typically occur on small saplings and branches [44,50–52]. In the case of migration, a fundamental trade-off appears to constrain the ability of species to excel at both migration and performance in aggressive interactions simultaneously (Text S3), resulting in smaller resident species dominating larger migrant species across many different environments [14,53–55]

Overall, observations of smaller species dominating larger species suggest that diverse adaptations in both the smaller and larger species may offset the disadvantage of small size in aggressive contests among species. While the evolution of novel traits in small species could lead to counter-adaptations in large species, and thus an evolutionary arms race, subordinate species (large or small) may reduce costs of aggressive interactions from dominant species in a myriad of ways, such as by using alternative resources (e.g., [47]), shifting resource use in space and time (e.g., [13,56]), or even by mimicking dangerous species (e.g., mimicry of bumblebees by some subordinate hummingbirds; [33,42]). Trade-offs and other evolutionary constraints are also likely to limit evolutionary arms races — most traits involved in interspecific aggression are used for other functions as well, and changes to these traits could influence organismal performance in many different ways. Most of our focal species also interacted with many species simultaneously (e.g., [43,56]), creating diverse sources of selection that can influence patterns of co-evolution among species [57].

Overcoming trade-offs over evolutionary time

Overall, our results suggest that the phylogenetic relationships among species can influence the rules that govern their interactions. Closely-related species typically share more traits in common [28], intensifying the importance of fundamental trade-offs that constrain their interactions and relative distributions. Thus, any advantages to being small must be balanced against the fitness cost of coexisting with larger, closely-related species [7,58]. Over evolutionary time, however, novel traits may significantly alter the costs associated with trade-offs [3], creating new ecological opportunities and different patterns of community organization. Thus, a phylogenetic perspective of community ecology is important, not just for understanding causal processes or rules that structure communities [1,2], but also for understanding when and how these rules can be broken over evolutionary time. These dynamic interactions provide an example of why phylogenetic perspectives are invaluable for our understanding of the structure and function of ecological communities [1,2].

Supporting Information

Figure S1 A 3-dimensional plot illustrating the relationships between dominance asymmetry, difference in mass, and genetic distance among the focal species pairs in our study. Dominance asymmetry = sqrt {ln ((wins by dominant species+1)/(wins by subordinate species+1))}. Difference in mass = (mass of dominant species−mass of subordinate species)/(average mass of dominant and subordinate species). Genetic distance is the Tamura-Nei genetic distance between interacting species for mtDNA.

Figure S2 Relatively heavier bird species were more likely to win aggressive contests for resources (positive slope), but the advantage of large size in aggressive contests declined with genetic distance (shallower slope for high genetic distance). Dominance asymmetry (y-axis) = sqrt {ln ((wins by dominant species+1)/(wins by subordinate species+1))}. Difference in mass (x-axis) = (mass of dominant species−mass of subordinate species)/(average mass of dominant and subordinate species). Genetic distance groups are split by the midpoint value for the dataset (low = 0.006–0.179; high = 0.180–0.352). Figure S2 is identical to Figure 5, except that an outlier point (interaction between *Gyps fulvus* and *Pica pica*) has been removed.

Table 5. Comparison of model fit for different generalized linear models, with taxonomic group entered as a predictor that include all possible combinations of predictor variables.

Model ranking	Intercept[1]	Taxonomic group	Difference in mass	Genetic distance	Interaction between group and difference in mass	Interaction between group and genetic distance	Interaction between difference in mass and genetic distance	Interaction between group, genetic distance and mass	df	logLik	AICc	delta AICc	weight
1	1.08	+[2]	0.85	0.85	—	—	−2.53	—	7	−117.2	249.0	0.00	0.387
2	0.79	+	0.93	3.26	—	+	−2.92	—	9	−115.3	249.3	0.35	0.325
3	1.11	+	0.78	0.93	+	—	−2.88	—	9	−116.4	251.6	2.65	0.103
4	0.82	+	0.87	3.35	+	+	−3.32	—	11	−114.5	252.1	3.11	0.082
5	1.27	+	0.37	—	—	—	—	—	5	−121.9	254.0	5.07	0.031
6	1.17	—[3]	0.85	0.93	—	—	−2.44	—	5	−122.1	254.5	5.59	0.024
7	0.65	+	1.19	4.51	+	+	−5.35	+	13	−113.6	254.8	5.87	0.021
8	1.25	+	0.37	0.19	—	—	—	—	6	−121.8	255.9	6.99	0.012
9	1.28	+	0.32	—	+	—	—	—	7	−121.4	257.3	8.34	0.006
10	1.10	+	0.37	1.50	—	+	—	—	8	−120.8	258.3	9.31	0.004
11	1.35	—	0.39	—	—	—	—	—	3	−126.4	259.0	10.03	0.003
12	1.26	+	0.33	0.17	+	—	—	—	8	−121.3	259.3	10.32	0.002
13	1.31	—	0.39	0.23	—	—	—	—	4	−126.3	260.7	11.79	0.001
14	1.12	+	0.34	1.46	+	+	—	—	10	−120.4	261.7	12.76	0.001
15	1.41	+	—	—	—	—	—	—	4	−137.2	282.5	33.59	0.000
16	1.40	+	—	0.12	—	—	—	—	5	−137.2	284.6	35.62	0.000
17	1.29	+	—	1.09	—	+	—	—	7	−136.6	287.8	38.81	0.000
18	1.48	—	—	—	—	—	—	—	2	−143.5	291.1	42.19	0.000
19	1.47	—	—	0.09	—	—	—	—	3	−143.5	293.2	44.20	0.000

Models tested the prediction that the advantage of large size in aggressive contests for resources declines with genetic distance among interacting bird species ($n = 244$ species pairs).

[1] numbers for predictor variables are effect sizes.

[2] plus sign (+) indicates that the predictor variable (factor) was included in the model.

[3] long dash (—) indicates that the predictor variable was absent from the model.

Table 6. Results from the best-fit (lowest AICc) generalized linear model, with taxonomic group included as a predictor variable, testing the prediction that the advantage of large size in aggressive contests for resources declines with genetic distance among interacting bird species ($n = 244$ species pairs).

Fixed effects[1]	Estimate	SE	*t*	*df*	*P*
Intercept[2]	1.08	0.11	9.70	243,238	<0.0001
Taxonomic group (hummingbirds)	0.05	0.09	0.58	243,238	0.56
Taxonomic group (antbirds/woodcreepers)	0.20	0.09	2.21	243,238	0.028
Difference in mass[3]	0.85	0.17	4.93	243,238	<0.0001
Genetic distance	0.85	0.49	1.76	243,238	0.08
Difference in mass[3] ×Genetic distance	−2.53	0.84	−3.01	243,238	0.0029

[1] dependent = sqrt{ln[(wins by dominant species+1)/(wins by subordinate species+1)]}.
[2] intercept value for taxonomic group = vultures.
[3] (mass of dominant species−mass of subordinate species)/(average mass of dominant and subordinate species).

Table S1 Description of variables for our dataset (Table S2).

Table S2 Dataset used in our study.

Table S3 Comparison of model fit for different generalized linear mixed models (binomial distribution) that include all possible combinations of predictor variables, controlling for variation in sample sizes of interactions across studies. Models tested the prediction that smaller bird species were more likely to win aggressive contests for resources when interacting with a more distantly-related species.

Table S4 Results from the best-fit (lowest AICc) generalized linear mixed model (binomial distribution), controlling for variation in sample sizes of interactions across studies. The model tests the prediction that smaller bird species were more likely to win aggressive contests for resources when interacting with a more distantly-related species.

Table S5 Results from the best-fit (lowest AICc) linear mixed-effects model, with dominant or subordinate species included as a random effect. The model tested the prediction that the advantage of large size in aggressive contests for resources declines with genetic distance among interacting bird species (n = 244 species pairs).

Table S6 Results from the best-fit (lowest AICc) generalized linear mixed model (binomial distribution), with dominant or subordinate species included as a random effect. The model tested the prediction that smaller bird species were more likely to win aggressive contests for resources when interacting with a more distantly-related species (n = 244 species pairs).

Table S7 Relative diet specialization among species pairs where the smaller species was dominant (i.e., won the majority of aggressive interactions). Focal diets were the food source over which aggressive interactions occurred: carrion for interactions involving vultures, nectar for interactions involving hummingbirds, and prey flushed by army ant swarms for interactions involving woodcreepers and antbirds.

Text S1 Genbank accession numbers for genetic sequences used in our study.

Text S2 Patterns of resource use and preference in cases where small species dominated large species in aggressive contests.

Text S3 A review of traits that have been proposed to explain why small species dominate large species of birds, focusing on specific cases in all three of our taxonomic groups.

Acknowledgments

We thank F. Bonier and an anonymous reviewer for helpful comments on earlier drafts of this manuscript.

Author Contributions

Conceived and designed the experiments: PRM CKG. Performed the experiments: PRM. Analyzed the data: PRM. Contributed reagents/materials/analysis tools: PRM CKG. Wrote the paper: PRM CKG.

References

1. Web CO, Ackerly DD, McPeek MA, Donoghue MJ (2002) Phylogenies and community ecology. Annual Review of Ecology and Systematics 33: 475–505.
2. Cavender-Bares J, Kozak KH, Fine PVA, Kembel SW (2009) The merging of community ecology and phylogenetic biology. Ecology Letters 12: 693–715.
3. Lenski RE (1988) Experimental studies of pleiotropy and epistasis in *Escherichia coli*. II. Compensation for maladaptive effects associated with resistance to virus T4. Evolution 42: 433–440.
4. Martin PR (2014) Trade-offs and biological diversity: integrative answers to ecological questions. In Integrative organismal biology (eds. Martin LB, Ghalambor CK, Woods HA), pp. 291–308. John Wiley and Sons, Inc., New York, New York, USA.
5. Peters RH (1983) The ecological implications of body size. Cambridge University Press, Cambridge, UK.
6. Bonner JT (2006) Why size matters. Princeton University Press, Princeton, New Jersey, USA.
7. Morse DH (1974) Niche breadth as a function of social dominance. American Naturalist 108: 818–830.
8. Donadio E, Buskirk SW (2006) Diet, morphology, and interspecific killing in Carnivora. American Naturalist 167: 524–536.

9. Burger J (1978) Competition between Cattle Egrets and native North American herons, egrets, and ibises. Condor 80: 15–23.
10. Gill FB, Mack AL, Ray RT (1982) Competition between hermit hummingbirds Phaethorninae and insects for nectar in a Costa Rican rain forest. Ibis 124: 44–49.
11. Hino T (2005) Resident males of small species dominate immigrants of large species in heterospecific, winter bird flocks. Ornithological Science 4: 89–94.
12. Harvey PH, Pagel MD (1991) The comparative method in evolutionary biology. Oxford University Press, Oxford, UK.
13. Willis EO, Oniki Y (1978) Birds and army ants. Annual Review of Ecology and Systematics 9: 243–263.
14. Freshwater C, Ghalambor CK, Martin PR (2014) Repeated patterns of trait divergence between closely related dominant and subordinate bird species. Ecology 95: 2334–2345.
15. Tamura K, Peterson D, Peterson N, Stecher G, Nei M, et al. (2011) MEGA5: Molecular Evolutionary Genetics Analysis using maximum likelihood, evolutionary distance, and maximum parsimony methods. Molecular Biology and Evolution 28: 2731–2739.
16. Tamura K, Nei M (1993) Estimation of the number of nucleotide substitutions in the control region of mitochondrial DNA in humans and chimpanzees. Molecular Biology and Evolution 10: 512–526.
17. Weir JT, Schluter D (2008) Calibrating the avian molecular clock. Molecular Ecology 17: 2321–2328.
18. Desjardins P, Morais R (1990) Sequence and gene organization of the chicken mitochondrial genome. Journal of Molecular Biology 212: 599–634.
19. Larkin MA, Blackshields G, Brown NP, Chenna R, McGettigan PA, et al. (2007) Clustal W and Clustal X version 2.0. Bioinformatics 23: 2947–2948.
20. Maddison DR, Maddison WP (2005) MacClade, version 4.08. Sinauer Associates, Sunderland, Massachusetts, USA.
21. R Core Team 2014 R: A language and environment for statistical computing. R Foundation for Statistical Computing, Vienna, Austria; http://www.R-project.org/. Accessed 2014.
22. Pinheiro J, Bates D, DebRoy S, Sarkar D, R Development Core Team (2012) nlme: linear and nonlinear mixed effects models. R package version 3.1-103.
23. Bates D, Maechler M, Bolker B (2011) lme4: Linear mixed-effects models using S4 classes. R package version 0.999375-42.
24. Gelman A, Su Y-S, Yajima M, Hill J, Pittau MG, et al. (2014) arm: Data analysis using regression and multilevel/hierarchical models. R package version 1.7-03.
25. Zuur AF, Ieno EN, Walker NJ, Saveliev AA, Smith GM (2009) Mixed effects models and extensions in ecology with R. Springer, New York, New York, USA.
26. Bartoń K (2013) MuMIn: Multi-model inference. R package version 1.9.5.
27. Ripley B, Hornik K, Gebhardt A, Firth D (2012) Support functions and datasets to support Venables and Ripley's 'Modern applied statistics with S' (4th edition, 2002). R package version 7.3.18.
28. Violle C, Nemergut DR, Pu Z, Jiang L (2011) Phylogenetic limiting similarity and competitive exclusion. Ecology Letters 14: 782–787.
29. Stiles FG (1973) Food supply and the annual cycle of the Anna Hummingbird. University of California Publications in Zoology 97: 1–109.
30. Cole BJ (1983) Assembly of mangrove ant communities: patterns of geographical distribution. Journal of Animal Ecology 52: 339–347.
31. Tanner CJ, Adler FR (2009) To fight or not to fight: context-dependent interspecific aggression in competing ants. Animal Behaviour 77: 297–305.
32. del Hoyo J, Elliott A, Sargatal J, eds. (1994) Handbook of the birds of the world. Volume 2. Lynx Edicions, Barcelona, Spain.
33. Feinsinger P, Chaplin SB (1975) On the relationship between wing disc loading and foraging strategy in hummingbirds. American Naturalist 109: 217–224.
34. Altshuler DL (2006) Flight performance and competitive displacement of hummingbirds across elevational gradients. American Naturalist 167: 216–229.
35. Feinsinger P, Colwell RK, Terborgh J, Chaplin SB (1979) Elevation and the morphology, flight energetics, and foraging ecology of tropical hummingbirds. American Naturalist 113: 481–497.
36. Kodric-Brown A, Brown JH (1978) Influence of economics, interspecific competition, and sexual dimorphism on territoriality of migrant Rufous Hummingbirds. Ecology 59: 285–296.
37. del Hoyo J, Elliott A, Sargatal J, eds. (1999) Handbook of the birds of the world. Volume 5. Lynx Edicions, Barcelona, Spain.
38. Willis EO (1982) The behavior of Black-banded Woodcreepers (*Dendocolaptes picumnus*). Condor 84: 272–285.
39. Willis EO (1983) Cis-Andean *Xiphorhynchus* and relatives (Aves, Dendrocolaptidae) as army ant followers. Revista Brasileira de Biologia 43: 133–142.
40. Willis EO (1992) Comportamento e ecologia do arapaçu-barrado *Dendrocolaptes certhia* (Aves, Dendrocolaptidae). Boletim do Museu Paraense Emílio Goeldi, Série Zoologia 8: 151–216.
41. Sharp WM (1957) Social and range dominance in gallinaceous birds — pheasants and prairie grouse. Journal of Wildlife Management 21: 242–244.
42. Feinsinger P, Colwell RK (1978) Community organization among neotropical nectar-feeding birds. American Zoologist 18: 779–795.
43. Feinsinger P, Swarm LA, Wolfe JA (1985) Nectar-feeding birds on Trinidad and Tobago: comparison of diverse and depauperate guilds. Ecological Monographs 55: 1–28.
44. Willis EO (1972) The behavior of Plain-brown Woodcreepers, *Dendrocincla fuliginosa*. Wilson Bulletin 84: 377–420.
45. Orians GH (1961) The ecology of blackbird (*Agelaius*) social systems. Ecological Monographs 31: 285–312.
46. Orians GH, Collier G (1963) Competition and blackbird social systems. Evolution 17: 449–459.
47. König C (1983) Interspecific and intraspecific competition for food among Old World vultures. In Vulture biology and management (Wilbur SR, Jackson JA, eds.), pp. 153–171. University of California Press, Berkeley, California, USA.
48. Wallace MP, Temple SA (1987) Competitive interactions within and between species in a guild of avian scavengers. Auk 104: 290–295.
49. Lyon DL (1976) A montane hummingbird territorial system in Oaxaca, Mexico. Wilson Bulletin 88: 280–299.
50. del Hoyo J, Elliott A, Christie DA, eds. (2003) Handbook of the birds of the world. Volume 8. Lynx Edicions, Barcelona, Spain.
51. Willis EO (1979) Behavior and ecology of two forms of White-chinned Woodcreepers (*Dendrocincla merula*, Dendrocolaptidae) in Amazonia. Papéis Avulsos de Zoologia 33: 27–66.
52. Willis EO (1967) The behavior of Bicolored Antbirds. University of California Publications in Zoology 79: 1–132.
53. Willis EO (1966) The role of migrant birds at swarms of army ants. Living Bird 5: 187–231.
54. DesGranges J-L, Grant PR (1980) Migrant hummingbirds' accommodation into tropical communities. In Migrant birds in the Neotropics: ecology, behavior, distribution, and conservation (eds. Keast A, Morton ES), pp. 395–409. Smithsonian Institution, Washington, DC, USA.
55. Leisler B (1992) Habitat selection and coexistence of migrants and Afrotropical residents. Ibis 134: S77–S82.
56. Feinsinger P (1976) Organization of a tropical guild of nectarivorous birds. Ecological Monographs 46: 257–291.
57. Thompson JN (2005) The geographic mosaic of coevolution. University of Chicago Press, Chicago, Illinois, USA.
58. Brown JH, Maurer BA (1986) Body size, ecological dominance and Cope's rule. Nature 324: 248–250.

14

The Human Semicircular Canals Orientation Is More Similar to the Bonobos than to the Chimpanzees

Marwan El Khoury[1], José Braga[1], Jean Dumoncel[1], Javotte Nancy[2], Remi Esclassan[1], Frederic Vaysse[1]*

1 University of Toulouse, Molecular Anthropology and Image Synthesis Laboratory (Centre National de la Recherche Scientifique), Toulouse, France, **2** University of Bordeaux, Faculty of dentistry, Bordeaux, France

Abstract

For some traits, the human genome is more closely related to either the bonobo or the chimpanzee genome than they are to each other. Therefore, it becomes crucial to understand whether and how morphostructural differences between humans, chimpanzees and bonobos reflect the well known phylogeny. Here we comparatively investigated intra and extra labyrinthine semicircular canals orientation using 260 computed tomography scans of extant humans (*Homo sapiens*), bonobos (*Pan paniscus*) and chimpanzees (*Pan troglodytes*). Humans and bonobos proved more similarities between themselves than with chimpanzees. This finding did not fit with the well established chimpanzee – bonobo monophyly. One hypothesis was convergent evolution in which bonobos and humans produce independently similar phenotypes possibly in response to similar selective pressures that may be associated with postural adaptations. Another possibility was convergence following a "random walk" (Brownian motion) evolutionary model. A more parsimonious explanation was that the bonobo-human labyrinthine shared morphology more closely retained the ancestral condition with chimpanzees being subsequently derived. Finally, these results might be a consequence of genetic diversity and incomplete lineage sorting. The remarkable symmetry of the Semicircular Canals was the second major finding of this article with possible applications in taphonomy. It has the potential to investigate altered fossils, inferring the probability of post-mortem deformation which can lead to difficulties in understanding taxonomic variation, phylogenetic relationships, and functional morphology.

Editor: Luca Bondioli, Museo Nazionale Preistorico Etnografico 'L. Pigorini', Italy

Funding: The research was funded by the University of Toulouse (http://www.univ-tlse3.fr/) and the Molecular Anthropology and Image Synthesis laboratory (http://www.anthropobiologie.fr/). The funders had no role in study design, data collection and analysis, decision to publish, or preparation of the manuscript.

Competing Interests: The authors have declared that no competing interests exist.

* E-mail: frederic.vaysse@univ-toulouse.fr

Introduction

Phenotypic traits have been used for decades for the purpose of reconstructing the evolutionary history of humans (*Homo sapiens*) and their closest relatives [1], chimpanzee and bonobo species (*Pan troglodytes* and *Pan paniscus*, respectively). More recently, phenotypic data have been supplemented by growing evidence from the genome wide sequencing analysis [2–4].

The comparison between human and chimpanzee genomes revealed genetic differences accumulated since these species diverged from their common ancestor [5]. The hominin-*Pan* split date was lately recalibrated to at least 7–8 million years [6]. It is often argued that chimpanzee subspecies and bonobos carry no or marginal genetic differences, when compared to the corresponding differences seen in humans from different continents [7–9]. Besides, common chimpanzees show the greatest population stratification when compared to all other great ape lineages, while humans and western chimpanzees show a remarkable dearth of genetic diversity when compared to other great apes. It was also found that the rate of gene loss in the human branch is not different as compared to other internal branches in the great ape phylogeny [10]. Recently, more extensive comparisons revealed that bonobo and chimpanzee genomes were not necessarily more closely related to each other than to the human one [2].

The phenotype of bonobos has received less attention than that of chimpanzees, despite several studies investigating dental development [11] and morphology [12,13], craniometry [14,15], intralabyrinthine angles [16–19], cranial development and vascularization [20–22], endocranial ontogeny and morphology [23]. For some traits, bonobos appear to be less diverse than chimpanzees in both their phenotype [12,14,20], and DNA [9,24–27]. Interestingly, the variation of some phenotypic traits has been shown to correlate more closely with genetic data than others [5].

For instance, the inner ear morphology has proven to be useful to assess diversity among extant and fossil primates [18,28]. However, a few studies have focused on variability within the genus *Pan* and have compared such variations with the extant human figures [16,18,19,29]. Studies from the early 70's [16–18] included large samples focused on angles taken from 2D radiographs. They concluded that the bonobo and the chimpanzee labyrinths are more similar to each other than to the human one.

Here, we investigated differences in the orientation of semicircular canals (SCC) starting from the null hypothesis (H0) that not only *Homo* and *Pan* differ in SCC orientation, but also that the two *Pan* species are more similar as compared to humans. Based on a sample of 260 medical X-ray computed tomography (CT) scans, we applied a mathematical model verified with microcomputed tomography (μCT) scans [30] and measurement error quantification. We subsequently investigated how these features discriminate the three species and we discussed our results in the context of their evolutionary relationships. Besides, we explored the existence of eventual sexual dimorphism related to human SCCs.

Material and Methods

Ethics statement

The human CT scans were provided from medical CT. The data reported here involved no experimentation on human subjects but only reprocessing of existing anonymized scan data. The use of these data for the present purpose was in respect of bioethical laws in France. Written consent was given by the patients for their information to be stored in the hospital database and used for research purpose. The "comité de protection des personnes – Bordeaux" (French IRB) approved the use of these data for the present purpose.

The *Pan* sample was composed of a set of inner ear reconstructed from dry skulls. We obtained permission from the Royal Museum of Central Africa of Tervuren (Belgium) and the Museum of Comparative Zoology at Harvard University (Cambridge, MA, USA) to access the collections. The collections were elaborated in the early twentieth century from mostly wild-shot animals donated to the museums. These collections were widely used in several studies including Lieberman et al. [31] and Durrleman et al. [23]. The samples were donated to the museums, and the parties involved in the hunting of the animals held the proper permits.

Samples

Our sample was composed of 137 anonymized human clinical records (*H. sapiens*: 70 females and 67 males), 61 *P. paniscus* and 62 *P. troglodytes* (8 *P. troglodytes verus* and 54 *P. troglodytes schweinfurthi*) (see Table 1 and Table S1). The three species were represented by subadult and adult individuals.

The human sample was composed of patients from the Pasteur Hospital (Toulouse, France) and the Faculty of Dentistry at the University of Toulouse (France), scanned between 2007 and 2010. These individuals had been referred for cranial trauma, inflammation of maxillary sinuses or neonatal distress but were found to be free of reportable abnormalities having any direct or indirect impact on inner ear morphology. The pixel size ranged from 0.3 to 0.49 mm and the slice thickness from 0.3 to 0.8 mm (for detailed information see Table S1). The human CT scans were provided from medical CT.

Table 1. Description of the sample.

Class abbreviation	Human	Bonobo		Chimpanzee	Apes	Total
infant	NJ1	1	3	7	10	11
infant stage 2	J1	0	11	5	16	16
young juvenile	J2	35	14	11	25	60
old juvenile	J3	50	10	10	20	70
sub-adult	A1	16	8	15	23	39
adult	A2	35	15	14	29	64
Total		137	61	62	123	260

Number of individuals according to Age classes [30] and species.

The *Pan* sample was composed of wild animals. The pixel size ranged from 0.27 to 0.49 mm and the slice thickness from 0.5 to 1 mm (for detailed information see Table S1).

The Maturational Status (MS) was assessed using dental stages [32] and reported in Table S1.

Data collection

Data were saved initially as Digital Imaging and Communications in Medicine (DICOM) format files, and then as Tagged Image File Format (TIFF) files.

Thirty landmarks (Table 2) were placed on the CT images using Amira software to best represent each SCC, as well as the midsagittal (MSP) and horizontal (HP) planes of the skull, used as references (Table 3). Anterior, posterior and lateral SCCs were referred to as ASCC, PSCC and LSCC, respectively. Each SCC was represented by three landmarks located at the center of its lumen [33]. The vestibule was also represented by one single landmark (see Method S1 for detailed information). Each SCC plane coordinate was then calculated from its landmarks as well as from the vestibular one. Then we calculated the 3D angles between planes (Figure 1) using the dot product of the two plane normal vectors [30]. For the main results, significance level was set at $p = 0.01$.

Body size is important to consider when comparing the size of morphological structures between species. However it was not considered in our study as it has no real impact on angular measurements [19]. The radii of the SCC were not reported as they did not meet our validation criteria.

Statistics

Statistical tests were performed with R software. Before angular comparisons, the normality was tested using either Shapiro-Wilk's test ($N<50$) or D'Agostino's test ($N>50$). The normal probability plot and frequency histogram were also established to visually check the normality of the distribution. The equality of variances was tested with Levene's tests.

The angular measurements were compared using a one-way ANOVA and Tukey's HSD post-hoc test providing a correction due to multiple comparisons. The Kruskall-Wallis test for multiple comparisons was used when normality criteria were not completed. A between-group principal component analysis (bgPCA) [34] of angular measurements was also applied on angles presented in Table 4. The bgPCA was previously successfully used to discriminate labyrinthine shape differences between predefine groups of chimpanzees or fossil hominins [35,36]. Measurements were tested also for sexual dimorphism, fluctuating asymmetry, anti-symmetry and directional asymmetry [37,38] using a two-way anova and student's t-tests.

Measurement error and validation

In order to validate the reproducibility of our results, we ran a test to assess the extent of intra-observer error. We used 60 randomly selected subjects. All measurements were taken twice by one observer (MEK), on separate days and without any knowledge on sex, age and species attributions [39]. Mean error was then analyzed using the angular mean error and a two-way ANOVA (sides*individuals) to compare Measurement Error relative to fluctuating asymmetry [40]. We did not find significant differences between the two sets of measurements over the 60 selected subjects ($p>0.80$) resulting in a mean error of $0.48° \pm 0.30$. ME was found to be significantly lower than fluctuating asymmetry ($p<0.02$).

Additionally, we ran other tests to validate the accuracy of angular calculation by using μCT as a comparison. To this end, Four *P. troglodytes* were scanned using both clinical Cone Beam

Computed Tomography (K9500, Trophy, KODAK) and µCT (Xtreme CT by SCANCO - Switzerland). Non parametric Kruskal-Wallis tests were used to detect differences between conventional-CT and µCT groups. Since µCT images covered only the petrous bones, only inter-SCC angles were computed. No significant differences were detected between the two imaging procedures ($p>0.63$). Similar evidence resulted from an independent additional test run on three *Papio anubis*, and one *Gorilla gorilla* using the same procedure as the four *P. troglodytes*.

Results

Medical CT images are contentious to measure very small dimensions [41] due to their spatial resolution [42]. The intra-observer error was found negligible in the present study and no statistical differences were found between the CT and µCT imaging procedures providing an indication of the method reliability [30].

Intra-specific comparisons

The possible influence of sex was not assessed in *Pan*, as sex attribution was not available for all specimens. In humans, we found sexual dimorphism in ASCC/PSCC and LSCC/PSCC angles. Both of them were more open in females than in males (respectively +2.7°, $p<0.001$ and +2.9°, $p<0.001$). Previously, sexual dimorphism was already pointed out as reflecting allometry only [19].

No differences were found between MS confirming that SCC orientations undergo no further important changes after birth [19,43].

Only chimpanzees proved evidence for directional asymmetry in the MSP/LSCC angle (right: 82.6±6.4°, left: 78.6±6.7°, $p<0.002$). In all other instances, we observed a remarkable symmetry (asymmetry ranging from 0.09% to 3.05%).

Inter-specific comparisons

As shown in Table 4, significant differences were found between chimpanzees and the two other species. Humans and Bonobos shared more similarities. The bgPCA (Figure 2) highlight this result: the chimpanzee group was quite separated from both humans and bonobos, with a weak overlap seen on the graph, while a tighter overlap between humans and bonobos was observed.

Between humans and chimpanzees, MSP/ASCC and MSP/PSCC angles were significantly different ($p<0.001$), indicating in humans ASCC and PSCC respectively 4.2° and 3.2° closer to the MSP. Thus, the angle between ASCC and PSCC was more open in humans (+5.5°, $p<0.001$). LSCC was 2.9° more horizontal in chimpanzees (HP/LSCC was 2.9° more open in humans) ($p<0.001$), 2.8° closer to ASCC in humans ($p<0.001$), and 8.1° closer to PSCC in chimpanzees ($p<0.001$). The angle between ASCC and the contra lateral PSCC was 7.6° more open in humans than in chimpanzees ($p<0.001$). Between chimpanzees and Bonobos, LSCC was more horizontally oriented in chimpanzees ($p<0.001$). The ASCC/PSCC angle was 3.8° more open in bonobos ($p<0.001$). PSCC was more closely related to MSP in bonobos (4° closer, $p<0.001$). The angle between ASCC and the contra lateral PSCC was 5.1° flatter in bonobos ($p<0.001$). In chimpanzees, as compared to bonobos, LSCC was more horizontally oriented ($p<0.001$). Only the MSP/ASCC angle

Table 2. Definition of the landmarks used in the present study.

N°	Name	Definition	Bookstein landmarks type
1	Frontal crest (Fc)	Summit of the Fc	II
2	Crista galli (Cr)	Summit of Cr	I
3	Internal occipital crest (iOc)	Medial most eminent point on the iOc	II
4	Vomer (Vm)	Point on the posterior border of the Vm	II
5	Nasopalatine foramen (NPf)	Central point of the NPf	I
6	Foramen Magnum (fMo)	Midpoint on the anterior border of the fMo	II
7,8	Infraorbital foramina (IOf)	Midpoint of the IOf	I
9,10	Supraorbital foramen (SOf)	Cranial part of the notch of the SOf	II
11,12	Vestibule (Vb)	Center of the lumen of the Vb	I
13,16	ASCC (middle)	Superior-most point at the center of the ASCC lumen	II
14,17	ASCC (anterior)	Anterior point at the center of the ASCC lumen before its ampulla	II
15,18	ASCC (posterior)	Posterior point at the center of the ASCC lumen before the common crus	II
19,22	LSCC (anterior)	Anterior point at the center of the LSCC lumen before its ampulla	II
20,23	LSCC (middle)	Lateral-most point at the center of the LSCC lumen	II
21,24	LSCC (posterior)	Posterior point at the center of the LSCC lumen before joining the vestibule	II
25,28	PSCC (inferior)	Inferior point at the center of the PSCC lumen before its ampulla	II
26,29	PSCC Right (middle)	Posterior-lateral-most point at the center of the PSCC lumen	II
27,30	PSCC Right (superior)	Superior point at the center of the PSCC lumen before the common crus	II

Anterior, posterior and lateral SCCs (semicircular canals) were noted respectively ASCC, PSCC and LSCC. Landmarks of **Type I** were well defined locally; their homology from individual to another was strongly supported. **Type II** landmarks was corresponding to points which position was first defined locally using specific structures but it was also depending on less specific factors such as the maximum or minimum of a curve. When using type II landmarks the individual to individual homology was only supported geometrically, to calculate plane coordinates for example.

Table 3. Definition of the reference planes used in the study.

planes	landmarks (Table 2)
Mid-Sagittal Plane	MSP: 1, 2, 3, 4, 5, 6
Horizontal Plane	HP: 7, 8, 11, 12
Semi Circular Canal planes	ASCC right: 11, 13, 14, 15
	ASCC left: 12, 16, 17, 18
	LSCC right: 11, 19, 20, 21
	LSCC left: 12, 22, 23, 24
	PSCC right: 11, 25, 26, 27
	PSCC left: 12, 28, 29, 30

Anterior, posterior and lateral SCC (semicircular canals) were noted respectively ASCC, PSCC and LSCC.

exhibited a significant difference ($p<0.001$) between humans and bonobos.

Discussion and Conclusions

Differences between humans and *Pan* species were previously noticed in semicircular canal size and shape [19,35,44], but not specifically between chimpanzees and bonobos, usually found to be similar (see Table S2). In this regard, we found two significantly distinct groups: humans and bonobos on one side and chimpanzees on the other side. This new finding could have three explanations. First, earlier studies excluded the vestibule from calculations of the SCC planes. Second, the present study used 3D angles instead of projected angles. Finally, previous intra-SCC angle measurements in *Pan* were based on very limited sample sizes. We used the vestibule as it contains the utricule and the saccule that are parts of the vestibular system. The fluids filling the vestibule and SCC play a major role in their physiology by the detection of head rotational movement which contributes to balance and maintaining visual fixation during head movements. Because the utricule and the saccule are parts of this system to detect motion and orientation, their use is functionally relevant in SCC study. The precision and reliability of the vestibule landmark is presented in additional data (Method S1).

A previous bgPCA analysis of overall labyrinth shape revealed significant discrimination among two subspecies of *P. troglodytes* [35] demonstrating subtle, yet significant, differences between *P. troglogytes troglodytes and P. t. verus* and a clear separation between humans and the two chimpanzees groups. However Gunz et al. [35] focused on the total morphological pattern of the labyrinth (through a Geometric Morphometric approach) whereas our study concentrated only on a limited number of angular measurements of the labyrinth. Despite this, our results demonstrate differences between the two species of *Pan* and are in line with Gunz et al. [35] study. We did not find a clear separation between the two subspecies of chimpanzees represented in our sample (*P. t. schweinfurthi* and *P. t. verus*), even though genetical and morphological data suggest that among subspecies of common chimpanzee, *P.t. verus* is the most distinct [24]. As regard the labyrinth, Gunz et al. [35] observed only subtle differences in canal radii. This may explain why our angular values failed to discriminate *P. t. schweinfurthi* and *P. t. verus*.

We found more similarities between humans and bonobos than between chimpanzees and bonobos. The differences recorded

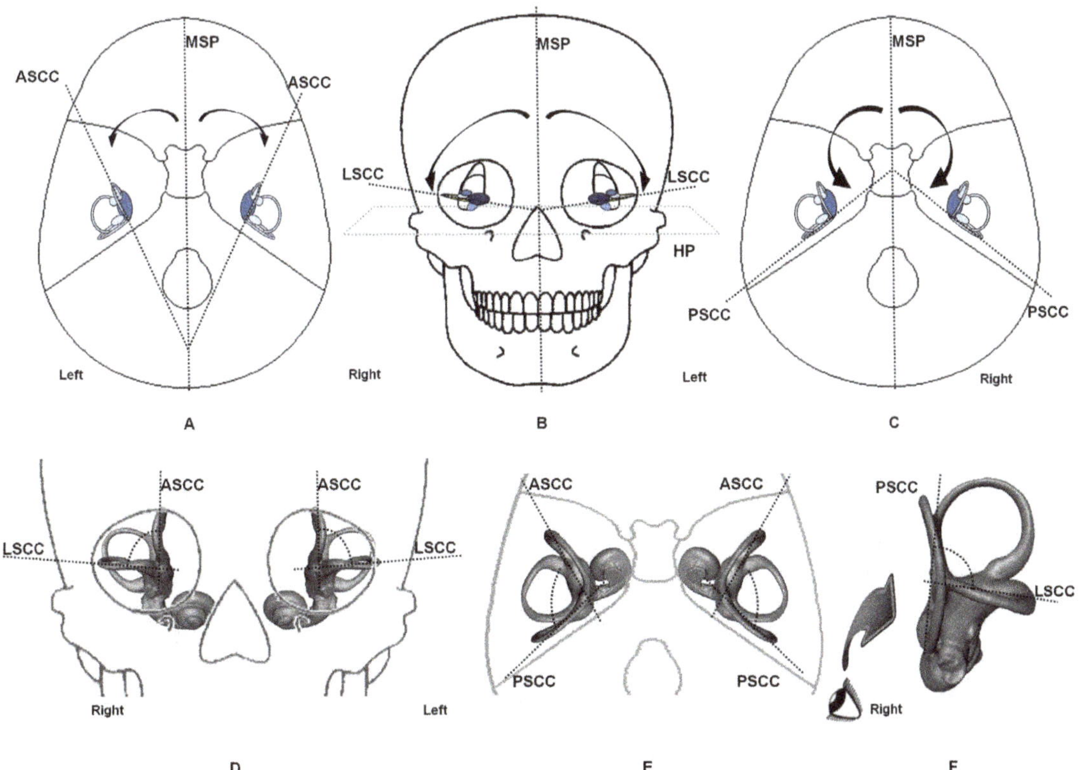

Figure 1. SCC angle representations. (a) MSP/ASCC angles. (b) Orientation of LSCC with MSP and HP (c) MSP/PSCC angles (d) ASCC/LSCC angles (e) ASCC/PSCC angles (f) LSCC/PSCC angles.

Table 4. Statistical comparisons of the SCC orientation between humans, chimpanzees and bonobos.

	Humans	← p →	Chimpanzees	← p →	Bonobos	← p →	Humans
Canal pair	N = 137	H vs C	N = 62	C vs B	N = 61	B vs H	N = 137
ASCC/LSCC*	**74.2±4.4°**	<0.001	**77.0±7.4°**	NS	**74.8±7.1°**	NS	**74.2±4.4°**
ASCC/PSCC*	**111.2±6.4°**	<0.001	**105.7±7.6°**	<0.001	**109.5±6.1°**	NS	**111.2±6.4°**
LSCC/PSCC**	**88.2±6.2°**	<0.001	**80.1±10.8°**	<0.001	**86.9±10.5°**	NS	**88.2±6.2°**
LSCC right/left**	**19.7±8.9°**	NS	**22.0±9.5°**	NS	**21.3±15.2°**	NS	**19.7±8.9°**
ASCC right/PSCC left**	**10.6±5.4°**	<0.001	**19.2±9.5°**	<0.001	**12.8±7.8°**	NS	**10.6±5.4°**
ASCC left/PSCC right**	**11.4±6.4°**	<0.001	**18.0±10.6°**	<0.001	**14.3±9.2°**	NS	**11.4±6.4°**
MSP/ASCC*	**34.9±5.1°**	<0.001	**39.1±5.9°**	NS	**38.1±4.6°**	<0.001	**34.9±5.1°**
MSP/LSCC*	**80.9±5.5°**	NS	**80.6±6.6°**	NS	**82.9±8.6°**	NS	**80.9±5.5°**
MSP/PSCC*	**145.3±6.2°**	<0.001	**142.1±8.0°**	<0.001	**146.1±7.4°**	NS	**145.3±6.2°**
HP/LSCC*	**25.9±6.7°**	<0.001	**23.0±7.2°**	<0.001	**26.1±7.5°**	NS	**25.9±6.7°**

Angles measurements and angles comparisons between species showing subtle differences between humans and bonobos and marked differences with chimpanzees (NS = not significant, H = humans, B = Bonobos, C = Chimpanzees). Anterior, posterior and lateral SCC were referred to as ASCC, PSCC and LSCC. (*) parametric tests were used since angular measurements showed normal distribution. (**) non parametric tests were used since angular measurements did not show normal distribution; however parametric tests results were consistent with those of non parametric tests. All angles in degrees.

between the present study and that of Spoor and Zonneveld (1998) [44] could result from differences in resolution, as the smallest available slice thickness used by the latter (1.5 mm) was higher than our highest available slice thickness (1 mm) (Table S1). Also, we used a sample of *Pan* wild specimens of known geographical origin much larger than in Spoor and Zonneveld (1998) (61 vs 6 *P. paniscus* and 62 vs 7 *P. troglodytes*).

The extent of differences and similarities between *P. paniscus*, *P. troglodytes*, and *H. sapiens* were illustrated in Figure 3 and 4 by the μCT-based reconstruction of 2 humans, 2 chimpanzees and 2 bonobos labyrinths. A simple visual inspection (Figure 3 and 4) showed important intra-specific differences. We remarked morphological differences between *Pan* and humans. The thickness was greater in humans, especially for ASCC and PSCC whereas the 3D curvature did not clearly separate human and *Pan* species. However, these differences did not suffice to identify clear-cut inter-SCC angular inter-specific differences

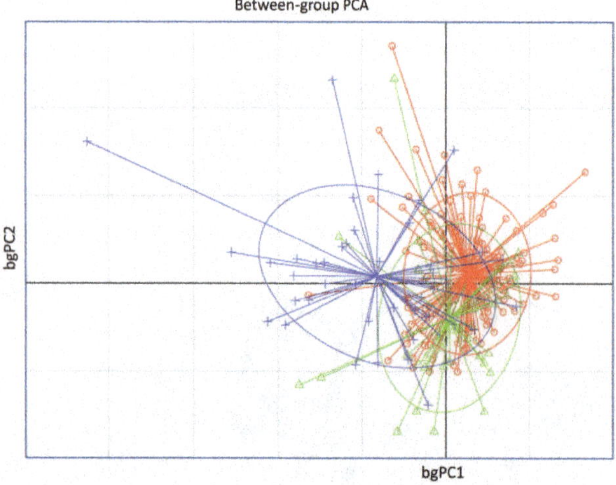

Figure 2. Between group principal component analysis (bgPCA) of the angular measurements. Humans (●) and bonobos (▲) were very closed to each other and distant from chimpanzees (+). The ellipses graphically represent the scatter plots encompassing approximately 67% of the subjects. The bgPCA showed a strait overlap of humans with bonobos and a weaker overlap of the latter two species with the chimpanzees.

In the context of our observed significant intra-SCC angular differences between chimpanzees and bonobos, it is important to add that basicranial differences between the two species of *Pan* were demonstrated by Cousin et al. [29] using 83 and 179 specimens of *P. troglodytes* and *P. paniscus* respectively. In their comparative study, the skulls were oriented using the lateral semicircular canal (the so-called «vestibular orientation method»). They found inter-specific differences in basicranial geometry. For some measurements and at each stage of its growth, *H. sapiens* appeared closer in shape to one of the two *Pan* species than they were to each other. Bonobos and humans were closer in the angle between LSCC and a line rejoining Nasion-Vestibule (Na-Op/V). Chimpanzees and humans were closer in the angle between LSCC and Nasion-sella turcica. Even though we could not directly compare our results with those taken from Cousin et al. [29], more studies are needed to investigate in detail, with appropriate comparative samples, how and which labyrinthine features show different orientations when compared to basicranial structures in humans, chimpanzees and bonobos.

We observed that chimpanzees have more horizontally-oriented LSCC. However, to our knowledge, there is currently no satisfactory and consensual explanation based on comparative physiological studies between the *Pan* species and modern humans. There is evidence that the use of bipedality is different between the two species [45]. Using feet contact time and hind limb joint angles, Pontzer et al. [46] suggested that bipedal locomotion might have been relatively costly for bonobos as compared to chimpanzees. This cost may reflect a difference in gait mechanics whereas none of the two *Pan* species is more bipedal than the other [45]. There is still no detailed explanation based on the relationship between locomotor patterns and SCC orientation.

Directional asymmetry was also found in chimpanzees with a more open MSP/LSCC angle on the right side. While gait asymmetry is well documented in chimpanzees with predominance of the right to make ground contact first [47], data are

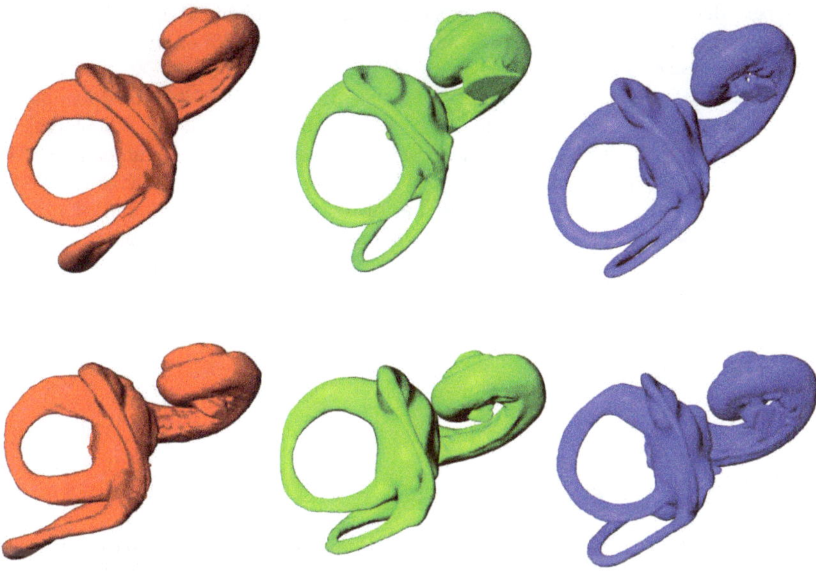

Figure 3. MicroCT-based rendering of the left bony labyrinth superior aspect in: *Homo sapiens* (red, on the left), *Pan paniscus* (green, in the middle) and *Pan troglodytes* (blue, on the right).

missing on bonobos as well as for bipedality. Our unexpected results for more horizontally oriented LSCC in chimpanzees need to be confirmed by further studies.

The characterization of functional relationships between canal morphology and locomotion is limited by the lack of data, especially in vivo. Only the MSP/ASCC angle clearly discriminates humans from the two *Pan* species. This may be due to the fact that bonobos and chimps show anatomical features that favor versatility [48]. The more open MSP/ASCC angle in *Pan* may contribute to their greater locomotor repertoire, particularly to head rotation through a transverse axis. A second possible explanation may lie in gait modalities of the three species. Bipedal and quadrupedal walking among chimpanzees and bonobos is similar but drastically different from that in humans.

The labyrinth morphology is open to external developmental influences only prior to ossification of the otic capsule, in humans at the end of the second trimester of gestation [28]. No differences were found between MS confirming that SCC orientations undergo no further important changes after birth [19,43].

Our main result about similarity between bonobos and humans, as compared to chimpanzees, does not fit with our null hypothesis. One explanation is convergent evolution in which bonobos and

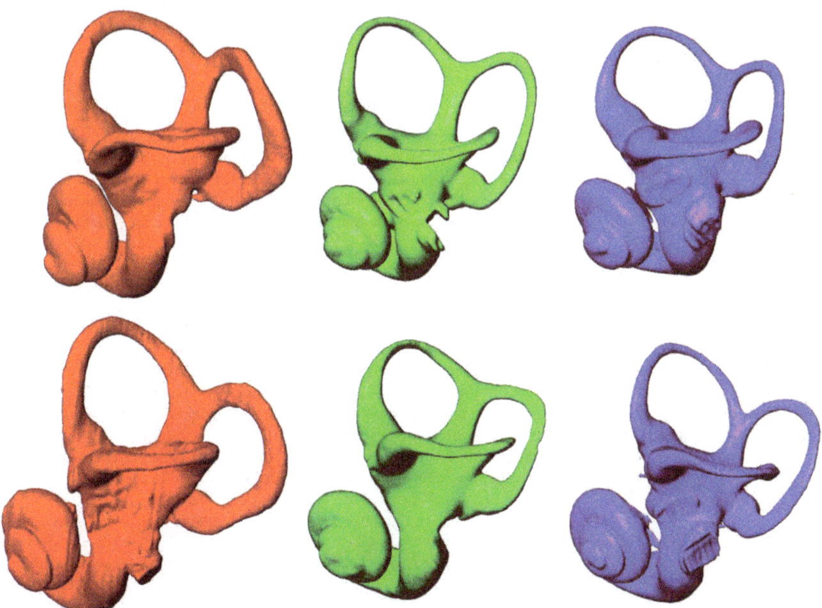

Figure 4. MicroCT-based rendering of the left bony labyrinth lateral aspect in: *Homo sapiens* (red, on the left), *Pan paniscus* (green, in the middle) and *Pan troglodytes* (blue, on the right).

humans produce independently similar phenotypes due to selection. Such an homoplasic pattern has already been argued for genetic data [2,49]. Another possibility is convergence following a "random walk" (Brownian motion) evolutionary model (e.g. [50,51]). Moreover, as demonstrated by complete genome analysis of gorillas, chimpanzees and bonobos [2,4], incomplete lineage sorting (ILS) [52] may influence phenotypic similarities that humans share with one of these three species but not the others.

Two species might share more similarities in a single trait whereas the phylogenetic tree reveals a different overall pattern of speciation. This might be a consequence of a complex intertwining of genetic diversity from ancestral population, selection and "Brownian walk". Human and chimpanzee genomes reveals genetic differences accumulated since the divergence of these two species from their common ancestor [5], dating at least to 7–8 million years [6]. The genetically-based estimated time between *P. troglodytes* and *P. paniscus* split is 2.1–1.5 million years [53]. In order to identify the respective role of each process, it would be interesting to investigate the labyrinthine morphology in fossil specimens attributed to *Ardipithecus* (ARA-VP-6/500) [54], and *Sahelanthropus* (TM 266-01-060-1) [55].

A final and more parsimonious explanation is that the bonobo-human labyrinthine shared morphology, more closely represents the *Pan-Homo* ancestral condition with chimpanzees being subsequently derived. Interestingly, the MSP/LSCC angle does not show the same trend as the other angles, showing no significant differences between the three species. This may reflect a primitive shared condition.

The petrous bone is often well preserved in fossil specimens [44,56–59]. However, deformation occurring from compaction and other diagenetic processes [60,61] makes it sometimes difficult to infer phylogenetic relationships [62] as well as missing data on extant specimens. The remarkable symmetry of the SCC is another important result of this study and may have implications in taphonomy. Further observations on fossil hominids are needed to investigate the potential usefulness of asymmetry to evaluate post-mortem deformation.

Most studies which have attempted to find sexual dimorphism at the inner ear level provided inconsistent results. However our observed sexual dimorphism in ASCC/PSCC and LSCC/PSCC angles are in agreement with results obtained in modern humans either for the bony labyrinth as a whole [63] or on the cochlea only [64,65].

In conclusion, we have used the largest known sample of individuals to measure the three dimensional orientation of semicircular canals in humans, bonobos and chimpanzees. We have demonstrated in this sample that bonobos SCC orientation is closer to humans than to chimpanzees. This finding may have crucial implications in hominid evolution that still need to be addressed. Additionally, the low intra-individual asymmetrical signal of the inner ear in our sample could open a most interesting track for the study of paleoanthropological records.

Acknowledgments

We thank W. Van Neer and E. Gilissen as the former and current curator of the Musée Royal de l'Afrique Centrale at Tervuren (Belgium) and J.L. Kahn of the Musée d'Anatomie Normale et Pathologique of the University of Strasbourg (France). We are also indebted to ME Rutzmoser of the Department of Mammalogy at the Museum of Comparative Zoology, Harvard University (Cambridge, MA, USA). We thank Jean-Noel Vergnes for his kind support in statistics. The helpful comments of the two referees and PLOS ONE academic editor improved remarkably the manuscript.

Author Contributions

Conceived and designed the experiments: MEK JB FV. Performed the experiments: MEK JN JD RE FV. Analyzed the data: MEK JB FV JN. Contributed reagents/materials/analysis tools: JD RE. Wrote the paper: MEK JB RE FV. Designed and performed the statistics: FV JD.

References

1. Wood B, Harrison T (2011) The evolutionary context of the first hominins. Nature, 470: 347–352.
2. Prüfer K, Munch K, Hellmann I, Akagi K, Miller JR, et al. (2012) The bonobo genome compared with the chimpanzee and human genomes. Nature 486(7404): 527–531.
3. Locke DP, Hillier LW, Warren WC, Worley KC, Nazareth LV (2011) Comparative and demographic analysis of orang-utan genomes. Nature 469(7331): 529–533.
4. Scally A, Dutheil JY, Hillier LW, Jordan GE, Goodhead I, et al. (2012) Insights into hominid evolution from the gorilla genome sequence. Nature 483(7388): 169–175.
5. Mikkelsen TS, Hillier LW, Eichler EE, Zody MC, Jaffe DB, et al. (The Chimpanzee Sequencing and Analysis Consortium) (2005) Initial sequence of the chimpanzee genome and comparison with the human genome. Nature 437: 69–87.
6. Langergraber KE, Prüfer K, Rowney C, Boesch C, Crockford C, et al. (2012) Generation times in wild chimpanzees and gorillas suggest earlier divergence times in great ape and human evolution. Proc Natl Acad Sci 109(39): 15716–15721.
7. Enard W, Paabo S (2004) Comparative primate genomics. Annu Rev Genomics Hum Genet 5: 351–378.
8. Prado-Martinez J, Sudmant PH, Kidd JM, Li H, Kelley JL, et al. (2013) Great ape genetic diversity and population history. Nature 499: 471–475.
9. Jensen-Seaman MI, Deinard AS, Kidd KK (2001) Modern African ape populations as genetic and demographic models of the last common ancestor of humans, chimpanzees, and gorillas. J Hered 92(6): 475–480.
10. Kaessmann H, Wiebe V, Pääbo S (1999) Extensive nuclear DNA sequence diversity among chimpanzees. Science 286(5442): 1159–1162.
11. Kinzey WG (1984) The dentition of the pygmy chimpanzee, *Pan paniscus*. In: The Pygmy Chimpanzee: Evolutionary Biology and Behavior (Ed. Susman RL), pp. 65–88. Plenum Press, New York.
12. Uchida A (1996) Craniodental variation among great apes. Peabody Mus Bull 4. Cambridge: Peabody Museum of Archaeology and Ethnology, Harvard University. 198p
13. Skinner MM, Gunz P, Wood BA, Boesch C, Hublin JJ (2009) Discrimination of extant *Pan* species and subspecies using the enamel–dentine junction morphology of lower molars. Am J Phys Anthropol 140: 234–243
14. Shea BT, Inouye SE (1993) Knuckle-walking ancestors. Science 259: 293–294.
15. Robinson C (2012) Geometric morphometric analysis of mandibular shape diversity in *Pan*. J Hum Evol 63: 191–204.
16. Dardenne J (1970) Etude comparative des principaux paramètres sagittaux de la face et du crane, chez l'homme et les chimpanzés, par la méthode vestibulaire d'orientation. Thèse de doctorat. Université de Lille.
17. Deblock R (1973) Craniométrie comparée de *Pan paniscus* et *Pan troglodytes*. Thèse de doctorat. Université de Lille.
18. Fenart R, Dufresnoy P, Deblock R (1973) Use of boundary lines in the search for neurocranial sexual differences in man and chimpanzees. Arch Anat Pathol 21(4): 341–348.
19. Spoor F (1993) The comparative morphology and phylogeny of the human bony labyrinth. PhD thesis. Monograph Utrecht University.
20. Braga J (1995) Skeletal variation and measure of divergence among chimpanzees - contribution of the study of discrete traits. C R Acad Sci Paris II 320(10): 1025–1030.

21. Braga J (1998) Chimpanzee variation facilitates the interpretation of the incisive suture closure in South African Plio-Pleistocene hominids. Am J Phys Anthropol 105(2): 121–135.
22. Braga J, Boesch C (1997) Further data about venous channels in South African Plio-Pleistocene hominids. J Hum Evol 33(4): 423–447.
23. Durrleman S, Pennec X, Trouvé A (2011) Comparison of the endocranial ontogenies between chimpanzees and bonobos via temporal regression and spatiotemporal registration. J Hum Evol 62(1): 74–88.
24. Morin PA, Moore JJ, Chakraborty R, Jin L, Goodall J, et al. (1994) Kin selection, social structure, gene flow, and the evolution of chimpanzees. Science 265: 1193–1201.
25. Gonder MK, Locatelli S, Ghobrial L, Mitchell MW, Kujawski JT, et al. (2011) Evidence from Cameroon reveals differences in the genetic structure and histories of chimpanzee populations. Proc Natl Acad Sci 108(12): 4766–4771.
26. Deinard AS, Kidd K (2000) Identifying conservation units within captive chimpanzee populations. Am J Phys Anthropol 111(1): 25–44.
27. Fischer A, Prüfer K, Good JM, Halbwax M, Wiebe V, et al. (2011) Bonobos Fall within the Genomic Variation of Chimpanzees. PLoS One 6(6): e21605.
28. Lebrun R, Ponce de León MS, Tafforeau P, Zollikofer CPE (2010) Deep evolutionary roots of strepsirrhine primate labyrinthine morphology. J Anat 216: 368–380.
29. Cousin RP, Fenart R, Deblock R (1981) Variation ontogéniques des angles basicraniens et faciaux: Etudes comparatives chez *Homo* et chez *Pan*. Bull Mem Soc Anthropol Paris 8(2): 189–212.
30. Bradshaw AP, Curthoys IS, Todd MJ, Magnussen JS, Taubman DS, et al. (2010) A mathematical model of human semicircular canal geometry: a basis for interpreting vestibular physiology. J Assoc Res Otolaryngol 11: 145–159.
31. Lieberman DE, Carlo J, Ponce de León M, Zollikofer CP (2007) A geometric morphometric analysis of heterochrony in the cranium of chimpanzees and bonobos. J Hum Evol 52(6): 647–662.
32. Shea BT (1989) Heterochrony in Human Evolution: The Case for Neoteny Reconsidered. Yearb Phys Anthropol 32: 69–101.
33. Aoki S, Takei Y, Suzuki K, Masukawa A, Arai Y (2012) Planer orientation of the bilateral semicircular canals in dizzy patients. Auris Nasus Larynx 39(5): 451–454.
34. Mitteroecker P, Bookstein F (2011) Linear Discrimination, Ordination, and the Visualization of Selection Gradients in Modern Morphometrics. Evol Biol 38(1): 100–114.
35. Gunz P, Ramsier M, Kuhrig M, Hublin JJ, Spoor F (2012) The mammalian bony labyrinth reconsidered, introducing a comprehensive geometric morphometric approach. J Anat 220: 529–543.
36. Braga J, Thackeray JF, Dumoncel J, Descouens D, Bruxelles L, et al. (2013) A new partial temporal bone of a juvenile hominin from the site of Kromdraai B (South Africa). J Hum Evol 65: 447–456.
37. Palmer AR (1994) Fluctuating asymmetry analyses: A primer. In: Developmental Instability: Its Origins and Evolutionary Implications. (ed. MarKow TA), pp. 335–364. Kluwer, Dordrecht, The Netherlands.
38. Palmer AR, Strobeck C (2003) Fluctuating asymmetry analyses revisited. In: Developmental Instability (DI): Causes and Consequences, (ed. M . Polak). pp. 279–319. Oxford University Press, Oxford.
39. Crespi B, Vanderkist B (1997) Fluctuating asymmetry in vestigial and functional traits of a haplodiploid insect. Heredity 79: 624–630.
40. Palmer AR, Strobeck C (1986) Fluctuating asymmetry: Measurement, analysis, patterns. Annu Rev Ecol Evol Syst 17: 391–421.
41. Lee JY, Shin KJ, Kim JN, Yoo JY, Song WC, et al. (2013) A morphometric study of the semicircular canals using micro-ct images in three-dimensional reconstruction. Anat Rec 296(5): 834–839.
42. Spoor F, Zonneveld F (1995) Morphometry of the primate bony labyrinth: a new method based on high-resolution computed tomography. J Anat 186: 271–286.
43. Richard C, Laroche N, Malaval L, Dumollard JM, Ch Martin, et al. (2010) New insight into the bony labyrinth: a microcomputed tomography study. Auris Nasus Larynx 37(2): 155–161.
44. Spoor F, Zonneveld F (1998) Comparative review of the human bony labyrinth. Am J Phys Anthropol 107(27): 211–251.
45. Videan EN, McGrew WC (2001) Are bonobos (*Pan paniscus*) really more bipedal than chimpanzees (*Pan troglodytes*)? Am J Primatol 54: 233–239.
46. Pontzer H, Raichlen DA, Rodman PS (2014) Bipedal and quadrupedal locomotion in chimpanzees. J Hum Evol 66: 64–82.
47. Arcadi AC, Wallauer W (2011) Individual-level lateralization in the asymmetrical gaits of wild chimpanzees (*Pan troglodytes*): implications for hand preference and skeletal asymmetry? Behaviour 148: 1419–1441.
48. D'Août K, Vereecke E, Schoonaert K, De Clercq D, Van Elsacker L, et al. (2004) Locomotion in bonobos (*Pan paniscus*): differences and similarities between bipedal and quadrupedal terrestrial walking, and a comparison with other locomotor modes. J Anat 204(5): 353–61.
49. Patterson N, Richter DJ, Gnerre S, Lander ES, Reich D. (2006) Genetic evidence for complex speciation of humans and chimpanzees. Nature 441(7097): 1103–1108.
50. Hunt G (2007) The relative importance of directional change, random walks, and stasis in the evolution of fossil lineages. Proc Natl Acad Sci 104(47): 18404–18408.
51. Gómez-Robles A, Bermúdez de Castroc JM, Arsuagad JL, Carbonell E, P. Polly D (2013) No known hominin species matches the expected dental morphology of the last common ancestor of Neanderthals and modern humans. Proc Natl Acad Sci In press. doi:10.1073/pnas.1302653110.
52. Hobolth A, Dutheil JY, Hawks J, Schierup MH, Mailund T (2011) Incomplete lineage sorting patterns among human, chimpanzee, and orangutan suggest recent orangutan speciation and widespread selection. Genome Res 21(3): 349–356.
53. Stone AC, Battistuzzi FU, Kubatko LS, Perry GH Jr, Trudeau E, et al. (2010) More reliable estimates of divergence times in *Pan* using complete mtDNA sequences and accounting for population structure. Philos Trans R Soc Lond B Biol Sci 365(1556): 3277–3288.
54. Suwa G, Asfaw B, Kono RT, Kubo D, Lovejoy CO, et al. (2009) The Ardipithecus ramidus skull and its implications for hominid origins. Science 326(5949): 68e1–7. doi:10.1126/science.1175825.
55. Brunet M, Guy F, Pilbeam D, Mackaye HT, Likius A (2002) A new hominid from the Upper Miocene of Chad, Central Africa. Nature 418(6899):801.
56. Spoor F, Wood BA, Zonneveld F (1994) Implications of early hominid labyrinthine morphology for evolution of human bipedal locomotion. Nature 369: 645–648.
57. Spoor F, Garland T, Krovitz G, Ryan TM, Silcox MT, et al. (2007) The primate semicircular canal system and locomotion. Proc Natl Acad Sci 104(26): 10808–10812.
58. Walker A, Ryan TM, silcox MT, Simons EL, Spoor F, et al. (2008) The semicircular canal system and locomotion: The case of extinct lemuroids and lorisoids. Evol Anthropol 17: 135–145.
59. Silcox MT, Bloch JI, Boyer DM, Godinot M, Ryan TM, et al. (2009) Semicircular canal system in early primates. J Hum Evol 56: 315–327.
60. Shipman P (1981) Life history of a fossil: an introduction to taphonomy and paleoecology. Harvard University Press. Cambridge, Massachusetts and London
61. Lyman RL (1994) Vertebrate taphonomy. Cambridge Manuals in Archaeology. Cambridge University Press. Cambridge.
62. Boyd A, Motani R (2008) Three-Dimensional Re-Evaluation of the Deformation Removal Technique Based on "Jigsaw Puzzling". palaeontologia electronica, 11: 7A–7p. Available: http://palaeo-electronica.org/2008_2/131/index.html.
63. Osipov B, Harvati K, Nathena D, Konstantinos Spanakis, Karantanas A, et al. (2013) Sexual dimorphism of the bony labyrinth: a new age-independent method. Am J Phys Anthropol 151(2): 290–301.
64. Sato H, Sando I, Takahashi H (1991) Sexual dimorphism and development of the human cochlea. Computer 3-D measurement. Acta Otolaryngol 111, 1037–1040.
65. Marcus S, Whitlow CT, Koonce J, Zapadka ME, Chen MY, et al. (2013) Computed tomography supports histopathologic evidence of vestibulocochlear sexual dimorphism. Int J Pediatr Otorhinolaryngol 77(7): 1118–1122.

15

Bayesian Analysis Using a Simple Likelihood Model Outperforms Parsimony for Estimation of Phylogeny from Discrete Morphological Data

April M. Wright*, David M. Hillis

Department of Integrative Biology, University of Texas at Austin, Austin, Texas, United States of America

Abstract

Despite the introduction of likelihood-based methods for estimating phylogenetic trees from phenotypic data, parsimony remains the most widely-used optimality criterion for building trees from discrete morphological data. However, it has been known for decades that there are regions of solution space in which parsimony is a poor estimator of tree topology. Numerous software implementations of likelihood-based models for the estimation of phylogeny from discrete morphological data exist, especially for the Mk model of discrete character evolution. Here we explore the efficacy of Bayesian estimation of phylogeny, using the Mk model, under conditions that are commonly encountered in paleontological studies. Using simulated data, we describe the relative performances of parsimony and the Mk model under a range of realistic conditions that include common scenarios of missing data and rate heterogeneity.

Editor: Art F. Y. Poon, British Columbia Centre for Excellence in HIV/AIDS, Canada

Funding: AMW received financial support for material purchases related to this project from the National Science Foundation's Doctoral Dissertation Improvement Grant (number 201203267-001; http://www.nsf.gov/funding/pgm_summ.jsp?pims_id=5234). The funder had no role in study design, data collection and analysis, decision to publish, or preparation of the manuscript.

Competing Interests: The authors have declared that no competing interests exist.

* Email: wright.aprilm@gmail.com

Introduction

For many decades, parsimony methods have been the most widely used approaches for estimation of phylogeny from discrete phenotypic data, despite the availability of likelihood-based methods for phylogenetic analysis. Maximum likelihood and Bayesian methods are commonly used in data sets combining molecules and morphology [1–5], but are used less frequently in morphology-only data sets [6]. As such, the efficacy of these methods under a range of conditions is not well-explored. In particular, the conditions that are investigated in most paleontological studies (many characters missing across sampled taxa, and rate heterogeneity among different sampled characters) lead some investigators to raise questions about the applicability of model-based approaches under these conditions [6–9].

At the present, the most widely implemented (in both pure likelihood and Bayesian contexts) model for estimating phylogenetic trees from discrete phenotypic data is the Mk model proposed by Lewis [10]. This model is a generalization of the 1969 Jukes-Cantor model of nucleotide sequence evolution [11]. The Mk model assumes a Markov process for character change, allowing for multiple character-state changes along a single branch. The probability of change in this model is symmetrical; in other words, the probability of changing from one state to another is the same as change in the reverse direction. This assumption can be relaxed in Bayesian implementations through the use of a hyperprior allowing variable change probabilities among states [12–14]. As many morphologists collect only variable or parsimony-informative characters (i.e., characters that can be used to discriminate among different tree topologies under the parsimony criterion), the distribution of characters collected does not reflect the distribution of all observable characters. This sampling bias can lead to poor estimation of the rate of character evolution within a data set, as well as inflated estimates of character change along branches of the estimated tree. To counteract this bias, Lewis [10] introduced versions of the Mk model that correct for biases in character collection. These versions were subsequently shown to have the desirable quality of statistical consistency [15].

Sampled characters within data sets typically evolve under different rates, developmental processes, and modes of evolution [7,16,17]. Although heterogeneity in the underlying evolutionary processes can present challenges to the application of evolutionary models [18], a distribution of different evolutionary rates of characters can be helpful for resolving branches at different levels in the tree. Extremely labile characters, for example, are useful for resolving recently diverged lineages, whereas slowly evolving characters may be more useful for resolving deep divergences in the tree. Likelihood-based methods can benefit from this heterogeneity by accounting for different rates of character evolution and the amount of time available for change (based on the estimated branch lengths in the tree; [19]). In contrast, high levels of rate heterogeneity among characters can be more problematic for parsimony methods, especially if all character changes are weighted equally [20].

The ability to estimate branch lengths in numbers of changes per site or character is also useful for estimating divergence times. The Mk model, for example, is implemented in the software packages BEAST [21] and MrBayes [12,13,22] for use in divergence dating. Trees with explicit divergence dates are useful for a variety of comparative methods for answering evolutionary questions at a large scale. Methods for time-scaling parsimony trees and quantifying the uncertainty of these scaling methods exist [23,24,25], although at present, there is no thorough comparison of the performance of maximum likelihood, Bayesian, and parsimony-based approaches for morphological data.

Though there are many positive aspects of the Mk model (statistical consistency, ability to accept superimposed changes, explicit modeling of rate heterogeneity with a gamma distribution), paleontologists have been slow to adopt model-based approaches. Comparisons between the Mk model and parsimony analyses have provided interesting and illuminating results. For example, Xu et al. [26] found a controversial result when they added a new fossil taxon to an existing theropod data set and reanalyzed this expanded data set using parsimony. The reanalysis by Xu et al. supported a grouping of *Archeoptyeryx* with deinychosaurians—a change that has broad implications for the evolution of flight. In contrast, a further reanalysis of this data set with the Mk model by Lee and Worthy [6] yielded trees in which *Archeopteryx* was grouped in a more traditional placement with birds. An analysis of the characters supporting each topology demonstrated that the parsimony tree tended to be supported by characters with low consistency indices [6]. The Mk model has also been applied in co-estimation of phylogeny and divergence dates using fossils as terminal taxa in combined molecular–morphological data sets by several authors [22,27,28].

Here, we investigate the relative performance of parsimony and Bayesian analyses using the Mk model, under a variety of conditions applicable to paleontological investigations. We based simulations on empirically estimated trees so that we could sample realistic branch lengths and tree topologies. We then designed the simulations to investigate a range of factors associated with accuracy of phylogenetic estimation, including missing data, rate heterogeneity, and overall character change rate.

Methods

Simulations

To investigate the efficacy of the Mk model for phylogenetic estimation, we simulated data sets in the R package GEIGER [29]. We simulated characters under the discrete model of evolution—a modification of the Juke–Cantor model [11] for binary characters. Under this continuous-time Markov process, characters are simulated under a user-specified rate of change per character. For the single-rate data sets, one rate was drawn from a gamma distribution, and all characters were simulated according to this rate. For data sets with rate heterogeneity, each character had a rate of change drawn independently from the same gamma distribution. This approximates a condition under which each character has an independent evolutionary rate, which can be binned into discrete rates during phylogenetic analyses.

We simulated data sets of two sizes. The first data set size was 350 characters. This number of characters is representative for data sets of phenotypic data, as many published data sets are this size or smaller. We also simulated comparatively larger data sets of 1000 characters to investigate the effects of character sample sizes. The empirical tree along which data were simulated was based on the tree presented by Pyron [27] and was chosen for its complexity. This tree (Figure 1) contains many short branches, which is representative of many analyses that include fossil specimens.

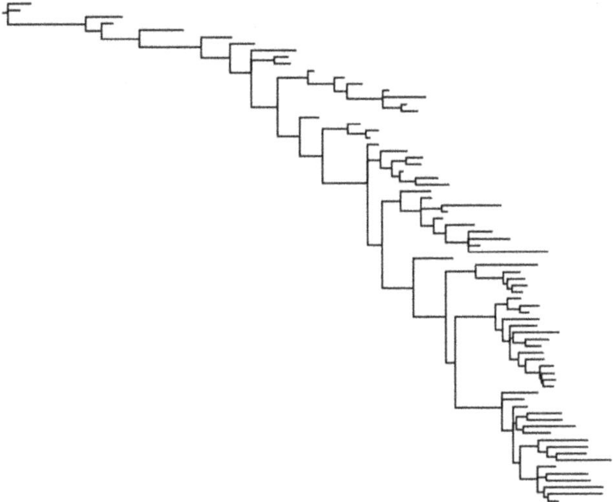

Figure 1. Tree used for simulations. This tree was obtained from a combined molecular–phenotypic data set analyzed by Pyron [27].

Ascertainment bias in morphological characters

Phenotypic data are often filtered by an observer-defined scheme. Characters that do not vary or vary in a parsimony uninformative way (such as autapomorphies) are usually excluded from analysis. In contrast to molecular sequence data, this means that there are rarely invariant sites in paleontological data sets. This bias can result in inflation of the estimated rate of evolutionary change in the data set, increasing the estimated branch lengths on the tree [10]. Under likelihood-based methods, branch lengths are estimated alongside tree structure, and unrealistically-inflated branch lengths can lead to topological error. MrBayes incorporates three versions of the Mk model. The uncorrected model (Mk) does not account for any form of sampling bias. Two corrected models account for the bias of collecting only variable characters (Mkv) and the bias of collecting only parsimony-informative characters (Mk-pars). To examine the effects of character acquisition bias, we filtered data sets according to different data acquisition schemes. The unfiltered data sets contained invariant characters, variable characters that were not parsimony-informative (e.g., autapomorphies), and variable characters that were parsimony-informative. Intermediate data sets excluded invariant sites, but retained variable sites that were not parsimony-informative. The least inclusive data sets contained only parsimony-informative characters.

Each character filtration scheme was parameterized appropriately in MrBayes. We did not explore the effects of model misspecification or incorrectly accounting for acquisition bias in this study. Data files can be found in the online supporting material, along with scripts for assembling MrBayes and PAUP blocks.

Missing Data

To assess the effects of missing data on phylogenetic estimation, we used several schemes for character deletion. We sorted the characters by rate of change, and divided them into three categories: fast-, intermediate-, and slow-evolving sites. Within each class of sites, we created data sets in which we removed between 10% and 100% of sites to investigate the effects of

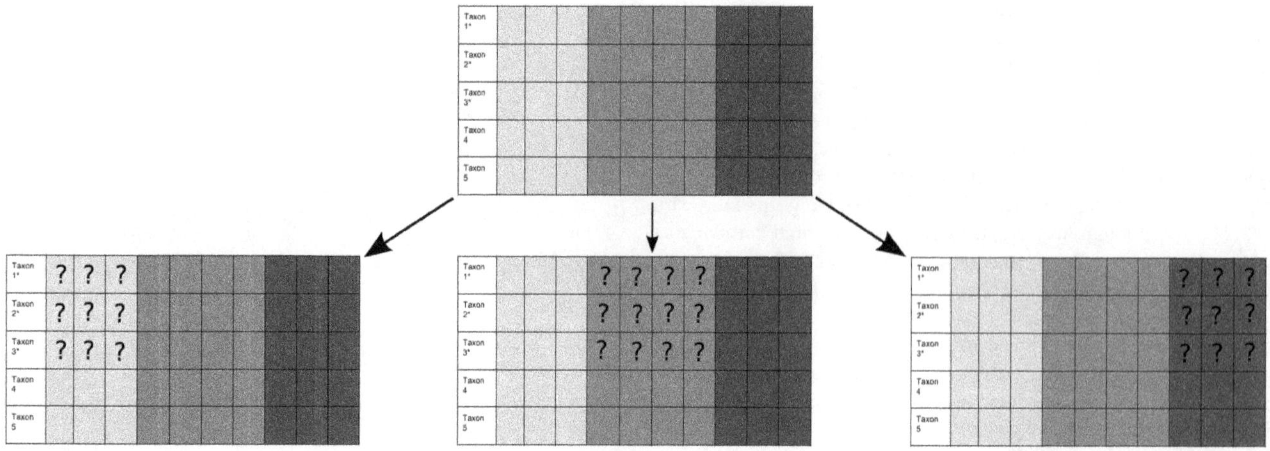

Figure 2. A schematic representing different missing data distributions. Columns represent characters. In the taxon-names column, an asterisk represents fossil taxa. Characters with the slowest rate of change are represented in light grey; intermediate-rate characters are represented in medium grey; characters with highest rate of change are represented in dark grey. In the top matrix, all characters are present for all taxa. The bottom matrices illustrate the missing data conditions that we simulated in this paper.

underrepresentation of certain classes of characters. Missing data were concentrated in fossil taxa, as seen in Figure 2.

Estimating Phylogenetic trees

We estimated Bayesian phylogenetic trees in MrBayes 3.2.2 [12,13] on the Lonestar server of the TACC computing facility at the University of Texas–Austin. We used the majority-rule consensus tree returned by MrBayes in all calculations and comparisons.

We used PAUP* for parsimony analyses. In PAUP* [30], we estimated phylogenetic trees using the TBR swapping algorithm with random branch addition and one thousand replicates. Estimation was performed on a ROCKS v4.1 computing cluster.

Analysis of Estimated Trees

There are many ways to categorize how well a tree has been estimated. Given that these data were simulated under a tree, we can compare the estimated phylogenetic trees to the true phylogenetic tree. We used a script written in Python, making use of the Dendropy library [31], to calculate the symmetric distance (the unweighted Robinson–Foulds distance [32]) between the estimated trees and the phylogenetic tree under which the data were generated. For unrooted trees of N taxa, there are N–3 bipartitions of the taxa (excluding bipartitions involving single taxa, which are the same for all trees). The Robinson–Foulds distance considers both the presence of incorrect bipartitions as well as the absence of correct bipartitions, so the maximum symmetric distance between two trees is 2(N–3). Therefore, for a 75-taxon tree, the maximum Robinson–Foulds distance is 144 symmetric distance units. For ease of interpreting graphs, we rescaled these values so that the total error is 100% (which would indicate all bipartitions in the tree are estimated incorrectly).

In a Bayesian analysis, the posterior sample of trees is not comprised of equally optimal solutions. Instead, each tree in the sample typically has a different likelihood score. A majority-rule consensus tree can be used to summarize the variation across the posterior sample, and this consensus tree is often taken as a summary estimate of the phylogeny. Therefore, we used the symmetric distance from the majority-rule consensus tree of the posterior sample to the model tree to evaluate the performance of the Bayesian analyses. In contrast, under the parsimony criterion, equally parsimonious trees are each considered optimal alternative solutions. Therefore, in parsimony analyses, we calculated the symmetric distance from each equally parsimonious solution to the model tree, and then averaged these scores within each data set to obtain an average symmetric distance score. We also used a majority-rule consensus tree to evaluate the parsimony analyses, and found the results were almost identical with the two measures (Fig. S2). All code to replicate results can be found in the online Supplemental Information.

Results

Character Filtration

Sampling bias does not affect Bayesian estimation when appropriate corrections are implemented. Correcting for ascertainment bias in MrBayes [12–13] is described by Lewis [10] based on the unobserved character counting method of Felsenstein [33]. In this approach, a likelihood for the data set is calculated conditional on only variable or parsimony informative characters present in the data. This conditional likelihood is then combined with the likelihood of a hypothetical constant character to arrive at a correction for acquisition bias. As shown in Figure S1, all parameterizations of the Mk model in MrBayes returned the same distributions of error. This demonstrates that corrections for sampling schemes are effective.

Single-Rate Simulations

As seen in Figure 3, at the lowest evolutionary rates, the amount of error in phylogenetic trees estimated compared to the true tree is fairly high, with nearly one in five branches being incorrectly estimated for both Bayesian and parsimony estimation. We would expect this to be true, as in this region of the graph, there are few character changes in the matrix. As evolutionary rate is increased, topological error reaches a minimum in error for both types of estimation. This minimum occurs at about one expected change per character. As more changes per character occur, there is an increase in topological error. This increase in error is seen more sharply in parsimony than Bayesian estimation, as Bayesian methods account for superimposed and parallel changes. Among different corrections of the Mk model for acquisition bias, performance is very similar (Figure S1).

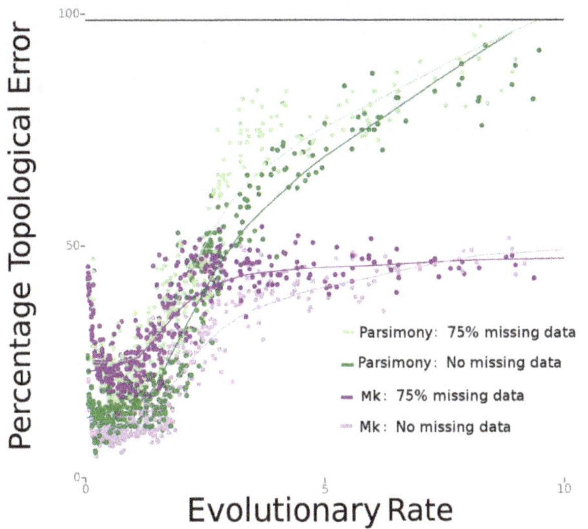

Figure 3. Results from simulations with a single rate of character evolution. Bayesian-Mk outperforms parsimony most strongly when the rate of character evolution (and hence homoplasy) is high.

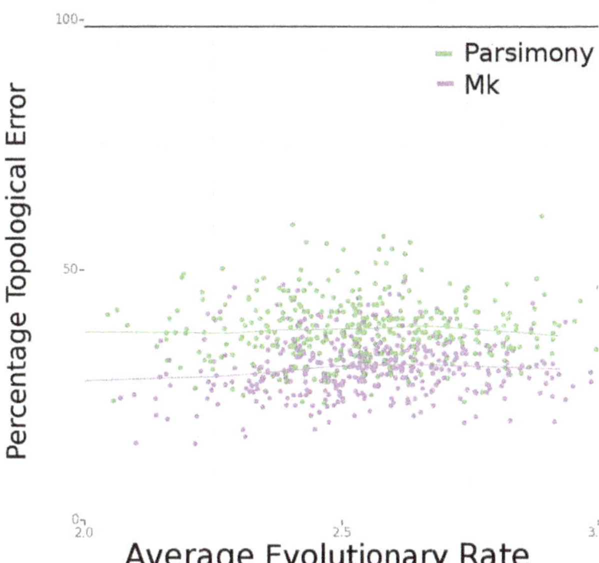

Figure 4. In data sets with character rate heterogeneity and with no missing data, Bayesian-Mk results in lower error compared to parsimony analyses. Note that, unlike Figure 3, the X-axis is the average rate of change across all characters in the data set, as opposed to one single rate applied uniformly to all characters.

As the amount of missing data increases in these data sets, the amount of error also increases. With 75% of data missing, as seen on Figure 3, parsimony and the Bayesian implementation of the Mk model perform very similarly at low rates of character change. However, at high rates of character change, the Bayesian Mk method outperforms parsimony strongly. In these regions of sample space, the characters show a poorer fit to the tree, with many characters exhibiting parallelisms and reversals.

Rate Heterogeneity

In data sets with rate heterogeneity among the characters, the Mk model continues to outperform parsimony, as shown in Figure 4. We also examined the effects of structured missing data in these data sets. Figure 5 compares the effects of removing various classes of characters (of different evolutionary rates) in the Bayesian Mk and parsimony analyses.

Both Bayesian Mk analyses and parsimony show degraded performance when characters of different rate classes are removed from the analysis, although the negative effects of missing data are much greater for parsimony than for the Bayesian analyses (especially for deletion of the slowest-evolving characters). Part of this effect is related to reduction in the overall number of characters available for analysis. Increasing the total number of characters in the analysis improves the performance for both Bayesian and parsimony analyses, although the Bayesian analyses continue to exhibit higher accuracy compared to parsimony in the 1000-character analyses (Figure 6).

Discussion

Our results suggest that Bayesian methods of analysis are likely to exhibit lower error rates compared to parsimony analyses in phylogenetic analyses of morphological and paleontological data sets. Moreover, researchers should carefully consider character-sampling design, as error rates can increase if characters are evolving too rapidly (Figure 3). As seen in Figure 3, before missing data or rate heterogeneity are introduced, phylogenetic estimation is most accurate for characters with relatively slow rates of change, as long as they are evolving fast enough to produce some phylogenetic signal. In these regions of the sample space, parsimony and Bayesian methods perform very similarly.

However, it is unlikely that empirical data sets will have only one rate of evolution across the whole data set. Rather, they are likely to be made up of characters that have been subjected to different selective pressures, different developmental constraints, and different evolutionary processes [16,17]. Rate heterogeneity in morphological data sets is well documented [7]. Therefore, the relationship between topological error and the location of missing data within a data set is of interest to researchers who build trees, as systematically under-representing certain classes of characters may produce different effects. Slowly-evolving characters include some characters that have too little change to be parsimony-informative; the fastest-evolving characters in these data sets include some characters with reversals and parallelism. In likelihood-based analyses, both parsimony-noninformative and parsimony-misinformative characters are still useful, as they provide information about the average rate of evolution in a data set. Rapidly-evolving characters can mislead parsimony analyses, which are unable to account for superimposed changes on a given branch. It would be expected that removing slowly-evolving characters (even those that are not parsimony-informative) would inflate the estimated average evolutionary rate, potentially leading to branch-length overestimation, and removing characters that change many times on the tree would result in underestimation of the average evolutionary rate. Figure 5 supports this conclusion, demonstrating that removing either of these classes of characters does result in higher topological error. Removing any class of characters (but especially the slowest-evolving characters) also results in lower performance of the parsimony analyses (Figure 5), presumably due to loss of information in an already small data set. Concerns about missing data have been cited as a reason to choose parsimony over likelihood-based methods [9]. Our results suggest that incomplete matrices do not necessitate the use of parsimony.

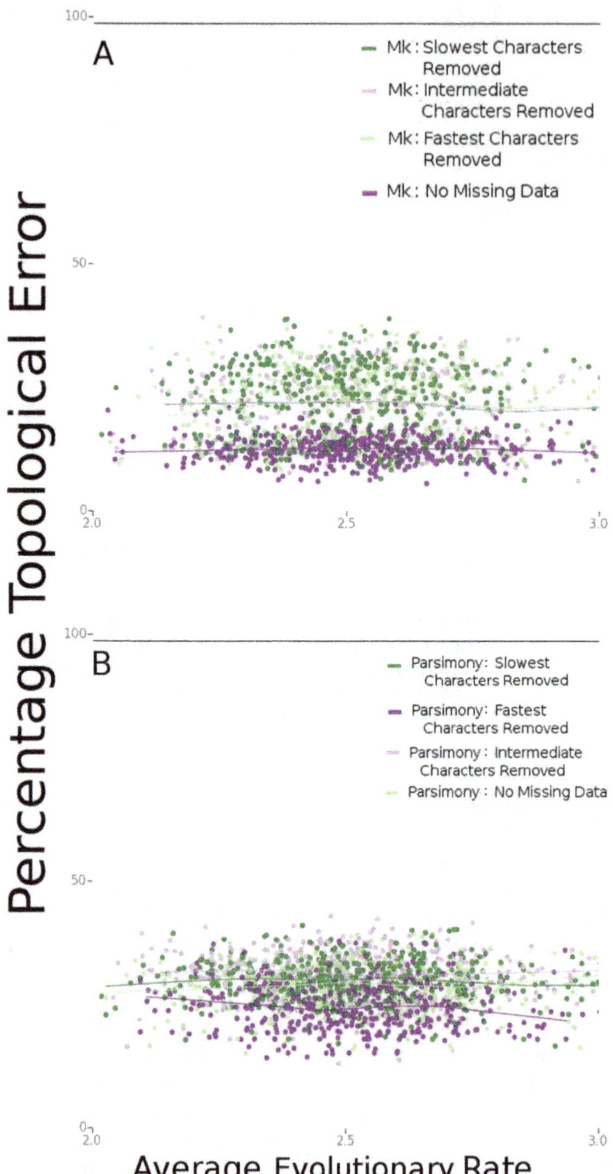

Figure 5. The effects of missing data vary with the rate of character evolution. This figure compares the effect of deleting one-third of the characters from three different rate classes. (A) Comparisons of Bayesian-Mk analyses. (B) Comparisons of parsimony analyses.

Increasing the size of the data set improves estimation for both parsimony and Bayesian methods. However, even in large data sets with no missing data, the Bayesian analyses using a simple likelihood model of character change typically outperform parsimony analyses (Figure 4). Paleontologists may be strongly constrained in how many characters or taxa they can add to a data set, due to a lack of specimens, a lack of observed homologous characters across a clade of interest, or poor specimen quality. Our results suggest that the use of Bayesian methods is even more important when relatively few characters are analyzed, and that even a simple probabilistic model can considerably improve the accuracy of tree estimation.

The benefits of adding fossil taxa to a data set are numerous. Earlier research has argued that fossil taxa can alleviate the issue of

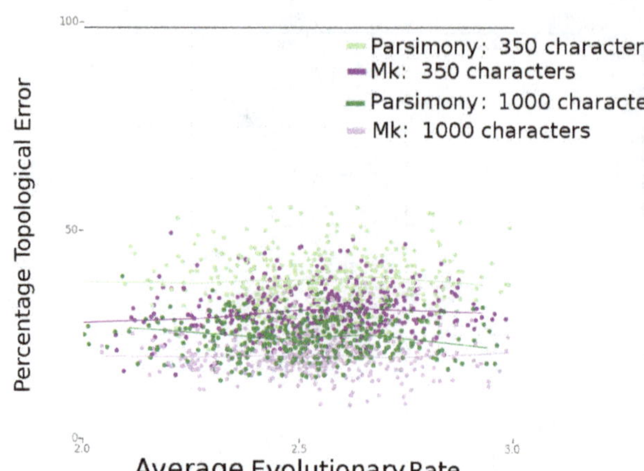

Figure 6. Comparison of 350- and 1000-character data sets.

long-branch attraction (LBA), particularly when additional extant taxa cannot be added to break up long branches [34,35]. Previous simulations have also suggested that, in combined analysis, even highly incomplete fossils can help alleviate the affects of LBA [36]. Empirical studies have confirmed these results, indicating that fossils with up to 75% missing data can help improve resolution in parsimony analysis [37] and result in vastly different topologies compared to molecular-only analyses [38]. Our results indicate that a model-based analysis is an even more effective way to gain performance improvements from such additions of fossil taxa.

In addition to exhibiting lower error rates, model-based methods offer another important advantage over parsimony: the ability to estimate time based on branch lengths of the phylogenetic tree. The Mk model, for example, is implemented in the software packages BEAST [21] and MrBayes [22] for use in divergence dating (although in BEAST, characters that are not variable or parsimony-informative must be explicitly listed by the author; see [38] for a discussion of counting unobserved site patterns). In turn, trees with explicit divergence dates are useful with a variety of comparative methods [39]. Methods for time-scaling parsimony trees exist [24,40,41,42], although at the present, there is no thorough investigation of the performance of model-based versus parsimony-based approaches for estimating time with morphological data.

Our results demonstrate that Bayesian methods are more accurate than parsimony for estimating trees from discrete morphological data under a wide set of realistic conditions. Even when there are large amounts of missing data (as is common in paleontological studies), a simple likelihood model consistently produces less error in tree estimation compared to parsimony. Although there is considerable room for models of morphological character evolution to be improved, even simple model-based methods can result in considerable improvement of phylogenetic analyses of morphological data sets.

Supporting Information

Figure S1 The effect of filtering characters before estimating phylogenies in a Bayesian context. MrBayes has three parameterizations of the Mk model, which account for sampling bias. As seen above, these methods estimate trees with the same degree of accuracy under the conditions we examined.

Figure S2 Parsimony analyses return sets of equally optimal trees. A symmetric difference score to the true (model) tree can be calculated either by creating a consensus tree and using this tree to calculate the symmetric difference, or by calculating the symmetric difference for every tree in the solution set and averaging this score. In our study, these two methods produce very similar results.

Acknowledgments

We thank Craig Dupree and Ming Cheng for technical support on this project. The Texas Advanced Computing Center (TACC) at The University of Texas at Austin provided HPC resources in support of the research results reported within this paper. Luke Harmon and Joseph Brown also provided very useful discussion on this project in its early stages, and Nicholas Matzke and Paul Lewis made numerous helpful suggestions in reviews of this manuscript.

Author Contributions

Conceived and designed the experiments: AMW DMH. Performed the experiments: AMW. Analyzed the data: AMW DMH. Contributed reagents/materials/analysis tools: AMW DMH. Wrote the paper: AMW DMH.

References

1. Nylander JAA, Ronquist R, Huelsenbeck JP, Nieves-Aldrey J (2004) Bayesian phylogenetic analysis of combined data. Syst Biol 53: 47–67. DOI: 10.1080/10635150490264699
2. Fenwick AM, Evans JA, Parkinson CL (2009) Morphological and molecular evidence for phylogeny and classification of South American pitvipers, genera *Bothrops*, *Bothriopsis*, and *Bothrocophias* (Serpentes: Viperidae). Zool J Linn Soc 156: 617–640. DOI: 10.1111/j.1096–3642.2008.00495.x
3. Wiens JJ, Kuczynski CA, Townsend T, Reeder TW, Mulcahy DG, et al. (2010) Combining phylogenomics and fossils in higher-level squamate reptile phylogeny: molecular data change the placement of fossil taxa. Syst Biol 59: 674–688. DOI: 10.1093/sysbio/syq048
4. Asher RJ, Hofreiter M (2006) Tenrec phylogeny and the noninvasive extraction of nuclear DNA. Syst Biol 55: 181–194. DOI: 10.1080/10635150500433649
5. O'Leary MA, Bloch JI, Flynn JJ, Gaudin TJ, Giallombardo A, et al. (2013) The placental mammal ancestor and the post-K-Pg radiation of placentals. Science 339: 662–667. DOI: 10.1126/science.1229237
6. Lee MSY, Worthy TH (2012) Likelihood reinstates *Archaeopteryx* as a primitive bird. Biol Lett 8: 299–303. DOI:10.1098/rsbl.2011.0884
7. Wagner PJ (2011) Modelling rate distributions using character compatibility: implications for morphological evolution among fossil invertebrates. Biol Lett 8: 143–146. DOI: 10.1098/rsbl.2011.0523
8. Goloboff PA, Pol D (2005) Parsimony and Bayesian phylogenetics. In: Albert, V, editor. Parsimony, Phylogeny, and Genomics. New York: Oxford University Press. pp: 148–159.
9. Livesey BC, Zusi RL (2007) Higher-order phylogeny of modern birds (Theropoda, Aves: Neornithes) based on comparative anatomy. Zool J Linn Soc 149: 1–95.
10. Lewis PO (2001) A likelihood approach to estimating phylogeny from discrete morphological character data. Syst Biol 50: 913–925. DOI:10.1080/106351501753462876.
11. Jukes TH, Cantor CR (1969). Evolution of Protein Molecules. In: Munro HN, editor. Mammalian Protein Metabolism. New York: Academic Press. pp: 21–132.
12. Huelsenbeck JP, Ronquist F (2001) MRBAYES: Bayesian inference of phylogeny. Bioinformatics 17:754–755.
13. Ronquist F, Huelsenbeck JP (2003) MRBAYES 3: Bayesian phylogenetic inference under mixed models. Bioinformatics 19:1572–1574.
14. Ronquist F, Huelsenbeck JP, Teslenko M (2011) Draft MrBayes version 3.2 manual: tutorials and model summaries. Distributed with the software from mrbayes.sourceforge.net/mb3.2_manual.pdf
15. Allman EA, Holder MT, Rhodes JA (2010) Estimating trees from filtered data: identifiability of models for morphological phylogenetics. J Theor Biol 263: 108–119.
16. Clarke JA, Middleton KM (2008) Mosaicism, modules, and the evolution of birds: results from a Bayesian approach to the study of morphological evolution using discrete character data. Syst Biol 57: 185–202. DOI:10.1080/10635150802022231
17. de Beer GR (1954) *Archaeopteryx lithographica*. A study based upon the British Museum specimen. London: British Museum (Natural History).
18. Kolaczkowski B, Thornton JW (2004) Performance of maximum parsimony and likelihood phylogenetics when evolution is heterogeneous. Nature 431: 980–984. DOI:10.1038/nature02917
19. Paradis E, Claude J, Strimmer K (2004) APE: analyses of phylogenetics and evolution in R language. Bioinformatics 20: 289–290.
20. Kuhner MK, Felsenstein J (1994) A simulation comparison of phylogeny algorithms under equal and unequal evolutionary rates. Mol Biol Evol 11: 459–468.
21. Bouckaert R, Heled J, Kühnert D, Vaughan TG, Wu C-H, et al. (2014) BEAST2: a software platform for Bayesian evolutionary analysis. PLOS Comp Biol 10: e1003537. DOI:10.1371/journal.pcbi.1003537
22. Ronquist F, Klopfstein S Vilhelmsen L, Schulmeister S, Murray DL, et al. (2012) A total-evidence approach to dating with fossils, applied to the early radiation of the Hymenoptera. Syst Biol 61: 973–999. DOI: 10.1093/sysbio/sys058
23. Pol D, Norell MA (2006) Uncertainty in the age of fossils and the stratigraphic fit to phylogenies. Syst Biol 55: 512–521.
24. Bapst DW (2013) A stochastic rate-calibrated method for time-scaling phylogenies of fossil taxa. Meth Ecol Evol 4: 724–733. DOI:10.1111/2041–210X.12081
25. Bapst DW (2014) Assessing the effect of time–scaling methods on phylogeny–based analyses in the fossil record. Paleobiology 40: 331–351.
26. Xu X, You H, Du K, Han F (2011) An *Archaeopteryx*-like theropod from China and the origin of Avialae. Nature 475: 460–465. DOI:10.1038/nature10288.
27. Pyron RA (2011) Divergence time estimation using fossils as terminal taxa and the origins of Lissamphibia. Syst Biol 60: 466–481. DOI:10.1093/sysbio/syr047
28. Wood HM, Matzke NJ, Gillespie RG, Griswold CE (2013) Treating fossils as terminal taxa in divergence time estimation reveals ancient vicariance patterns in the Palpimanoidea spiders. Syst Biol 62: 264–284. DOI:10.1093/sysbio/sys092
29. Harmon LJ, Weir JT, Brock CD, Glor RE, Challenger W (2008) GEIGER: investigating evolutionary radiations. Bioinformatics 24:129–131.
30. Swofford DL (2002) PAUP*: phylogenetic analysis using parsimony (*and other methods). Version 4.0b10. Sunderland, Massachusetts: Sinauer. DOI:10.1234/12345678
31. Sukumaran J, Holder MT (2010) DendroPy: a Python library for phylogenetic computing. Bioinformatics 26: 1569–1571.
32. Robinson DR, Foulds LR (1981) Comparison of phylogenetic trees. Math Biosci 53: 131–147. DOI:10.1016/0025–5564(81)90043–2
33. Felsenstein J (1992) Phylogenies from restriction sites: a maximum likelihood approach. Evolution 46: 159–173
34. Gauthier J, Kluge AG, Rowe T (1988) Amniote phylogeny and the importance of fossils. Cladistics 4: 105–209. DOI: 10.1111/j.1096–0031.1988.tb00514.x
35. Huelsenbeck JP (1991) When are fossils better than extant taxa in phylogenetic analysis? Syst Biol 40: 458–469. DOI:10.1093/sysbio/40.4.458
36. Wiens JW (2005) Can incomplete taxa rescue phylogenetic analyses from long–branch attraction? Syst Biol 54: 731–742. DOI: 10.1080/10635150500234583
37. Santini F, Tyler JC (2004) The importance of even highly incomplete fossil taxa reconstructing the phylogenetic relationships of the Tetraodontiformes (Acanthomorpha: Pisces). Integr Comp Biol 44: 349–357.
38. Rothwell GW, Nixon KC (2006) How does the inclusion of fossil data change our conclusions about the phylogenetic history of euphyllophytes? Int J Plant Sci 167: 737–749.
39. Slater GJ, Harmon LJ, Alfaro ME (2012) Integrating fossils with molecular phylogenies improves inference of trait evolution. Evolution 66: 3931– 3944. DOI:10.1111/j.1558–5646.2012.01723.x
40. Smith AB (1994) Systematics and the Fossil Record: Documenting Evolutionary Patterns. Oxford: Blackwell Scientific.
41. Sanderson MJ (1997) A nonparametric approach to estimating divergence times in the absence of rate constancy. Molec Biol Evol 14: 1218–1232.
42. Sanderson MJ (2002) Estimating absolute rates of molecular evolution and divergence times: a penalized likelihood approach. Molec Biol Evol 19: 101–109.

16

Primitive Duplicate Hox Clusters in the European Eel's Genome

Christiaan V. Henkel[1]*, Erik Burgerhout[1], Daniëlle L. de Wijze[2], Ron P. Dirks[1], Yuki Minegishi[2], Hans J. Jansen[1], Herman P. Spaink[2], Sylvie Dufour[3], Finn-Arne Weltzien[4], Katsumi Tsukamoto[5], Guido E. E. J. M. van den Thillart[2]

1 ZF-screens B.V., Leiden, The Netherlands, **2** Institute of Biology, Leiden University, Leiden, The Netherlands, **3** UMR BOREA, CNRS 7208, Muséum National d'Histoire Naturelle, Paris, France, **4** Norwegian School of Veterinary Science, Oslo, Norway, **5** Atmosphere and Ocean Research Institute, The University of Tokyo, Kashiwa, Chiba, Tokyo, Japan

Abstract

The enigmatic life cycle and elongated body of the European eel (*Anguilla anguilla* L., 1758) have long motivated scientific enquiry. Recently, eel research has gained in urgency, as the population has dwindled to the point of critical endangerment. We have assembled a draft genome in order to facilitate advances in all provinces of eel biology. Here, we use the genome to investigate the eel's complement of the Hox developmental transcription factors. We show that unlike any other teleost fish, the eel retains fully populated, duplicate Hox clusters, which originated at the teleost-specific genome duplication. Using mRNA-sequencing and *in situ* hybridizations, we demonstrate that all copies are expressed in early embryos. Theories of vertebrate evolution predict that the retention of functional, duplicate Hox genes can give rise to additional developmental complexity, which is not immediately apparent in the adult. However, the key morphological innovation elsewhere in the eel's life history coincides with the evolutionary origin of its Hox repertoire.

Editor: Michael Schubert, Ecole Normale Supérieure de Lyon, France

Funding: This work was supported by the Norwegian School of Veterinary Science and the Research Council of Norway (184851), by Centre National de la Recherche Scientifique and L'Agence Nationale de la Recherche (08-BLAN-0173), and by private resources from ZF-screens B.V., Leiden University and The University of Tokyo. The funders had no role in study design, data collection and analysis, decision to publish, or preparation of the manuscript.

Competing Interests: The authors have read the journal's policy and have the following conflicts: HPS and GEEJMvdT are founders and shareholders of ZF-screens B.V. CVH, EB, RPD and HJJ are employees of ZF-screens B.V.

* E-mail: henkel@zfscreens.com

Introduction

The life history of the European eel (*Anguilla anguilla* L., 1758) involves two distinct ocean-dwelling larval stages, a protracted juvenile phase in European continental freshwater, and finally sexual maturation coincident with migration to spawning grounds in the Atlantic Ocean, presumably the Sargasso Sea (Figure 1) [1]. The complexity and geographical range of this life cycle have long inspired evolutionary and physiological studies, especially on the structure of the eel's single, randomly mating (panmictic) population [2], interspecific hybridization with the American eel (*A. rostrata*, which shares the same oceanic spawning grounds [3]), its hidden migrations [4–6], and the development of fertility [6].

Its catadromous migratory behaviour, long life, serious habitat reduction, pollution, and overfishing may be amongst the causes of the catastrophic collapse of the European eel population observed over the past decades [7]. So far, *Anguilla* species have resisted efforts directed at efficient and sustainable artificial breeding [8]. As knowledge on the eel's genetic makeup is sparse, physiological studies aimed at understanding maturation, reproduction and the sustenance of successive larval stages have not been able to take full advantage of gene expression profiling. In order to alleviate this shortcoming, we have sequenced and assembled its genome.

While the draft genome will be an important tool in reproduction research, it also offers new perspectives for fundamental studies in eel biology, as well as a resource for the comparative interpretation of model fish genomes (e.g. zebrafish and medaka). Here, we investigate its repertoire of Hox genes in a comparative genomics context.

The Hox genes encode transcription factors, which throughout the animal kingdom are involved in the developmental patterning of the body plan. In vertebrates, Hox genes are tightly organized into clusters which exhibit colinearity between gene position and temporal and spatial expression along the primary body axis: genes at the 3′ ends of clusters are expressed earlier in development, and more anterior, than genes at the 5′ ends of clusters [9]. The organization of Hox clusters has been extensively documented for many groups of vertebrates [10].

A. anguilla is a member of the superorder Elopomorpha [11,12], a major teleost group of 856 species [13]. As such, elopomorphs presumably share the inferred whole-genome duplication at the base of the teleost lineage [14,15]. This teleost-specific genome duplication (TSGD) event is most apparent when considering the Hox genes in extant species [10,16,17]. In tetrapods and coelacanths, approximately forty genes are organized in four ancestral vertebrate clusters. In theory, teleosts could have retained eight duplicate clusters. However, whereas tetrapod Hox loci are relatively stable, teleost genomes show dramatic gene loss, such that all species examined in detail retain at most seven of these clusters, each with on average about half their original gene content [9,10]. A PCR-based survey of

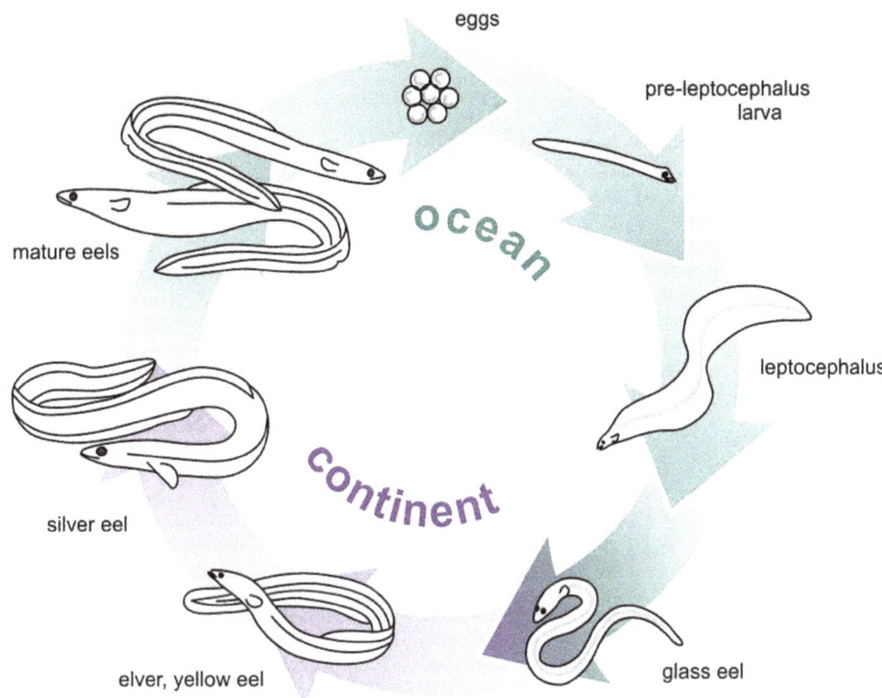

Figure 1. The life cycle of the European eel. After hatching, presumably in the Sargasso Sea, cylindrical larvae develop into leaf-shaped leptocephalus larvae, which after drifting on the Gulf Stream for approximately one year metamorphose into glass eels close to the European coast. The glass eels may stay at the coast or migrate upriver, where they stay as juveniles (elvers and yellow eel) for many years (depending on the region: males 4–6 years, females 8–12 years). Finally, they develop into migrating silver eels; the cause and timing of silvering is not well understood. They mature during or after migration to the spawning grounds.

the Hox clusters of the Japanese eel *A. japonica* found evidence for the conservation of eight clusters and 34 genes [18].

As the Elopomorpha represent an early branch on the teleost tree [12], the eel Hox gene complement may expose constraints on the evolution of morphological complexity in teleost fish, and in vertebrates in general. Furthermore, analysis of the eel's Hox clusters may shed light on the developmental mechanisms and evolutionary history of its life cycle and body plan. In particular, they may provide evidence regarding the evolutionary novelty of the eel's indirect development.

Results

Genome assembly of the European eel

We have sequenced and assembled the genome of a female juvenile *A. anguilla* specimen caught in the brackish Lake Veere, the Netherlands in December 2009. Its haploid genome size was determined to be 1.1 Gbp. Because of the impossibility of breeding *A. anguilla*, no genetic linkage information is available. We therefore employed Illumina Genome Analyzer sequencing technology only in the assembly of a draft genome. Based on a *de novo* genome assembly, we constructed 923 Mbp of scaffolds with a length-weighed median fragment length (N50) of 78 Kbp (Figure S1 and Table S1). An additional 179 Mbp of initial contigs, which are either very small or highly repetitive, were excluded from scaffolding, but included in all further analyses.

Identification of *Hox* transcripts and genes

To identify *A. anguilla* Hox genes, we used a *de novo* assembled transcriptome of a 27 hours post-fertilization (hpf) embryo of the short-finned eel, *A. australis*. This species is closely related to *A. anguilla* [19], yet produces viable embryos more easily [20]. We compared Hox-like sequences from the transcriptome to the genome assembly using Blast [21], which yielded ten genomic scaffolds (Table S2) that were further examined for the presence of additional genes. This resulted in the identification of 73 Hox genes (twice as many as found in *A. japonica* in a previous study using PCR fragments [18]), including three pseudogenes, organized in eight clusters (Figure 2 and Table S3). The flanking regions of these eight clusters contain an additional 24 predicted genes (Figure 2). No further protein-coding genes were found within the Hox clusters.

Conserved microRNAs were discovered using Blast searches with human and zebrafish homologues (Figure 2). miR-10 is present posterior of Hox4 in six clusters (all except Aa and Ab). miR-196 is found between Hox9 and Hox10 in five clusters (all except Bb, Da and Db). This arrangement is consistent with that found in other vertebrates [22,23].

Hox cluster identity

We based preliminary identification of clusters on homology between *A. anguilla* and *Danio rerio* protein sequences and comparisons with all sequences in the NCBI non-redundant protein database (Table S3). Whereas the two *A. anguilla* HoxA clusters can easily be matched to their corresponding HoxAa and HoxAb orthologues in *D. rerio*, each of the two HoxB and HoxC clusters of *A. anguilla* most closely resembles *D. rerio* HoxBa and HoxCa, respectively. Both *A. anguilla* HoxD clusters predictably match *D. rerio* HoxDa only, since the zebrafish HoxDb cluster has lost all protein-coding genes [24].

To more precisely assign the Hox genes to proper cluster orthologues, we generated unrooted maximum likelihood phylogenetic trees for paralogous group 9 (Figures 3 and S2), of which *A.*

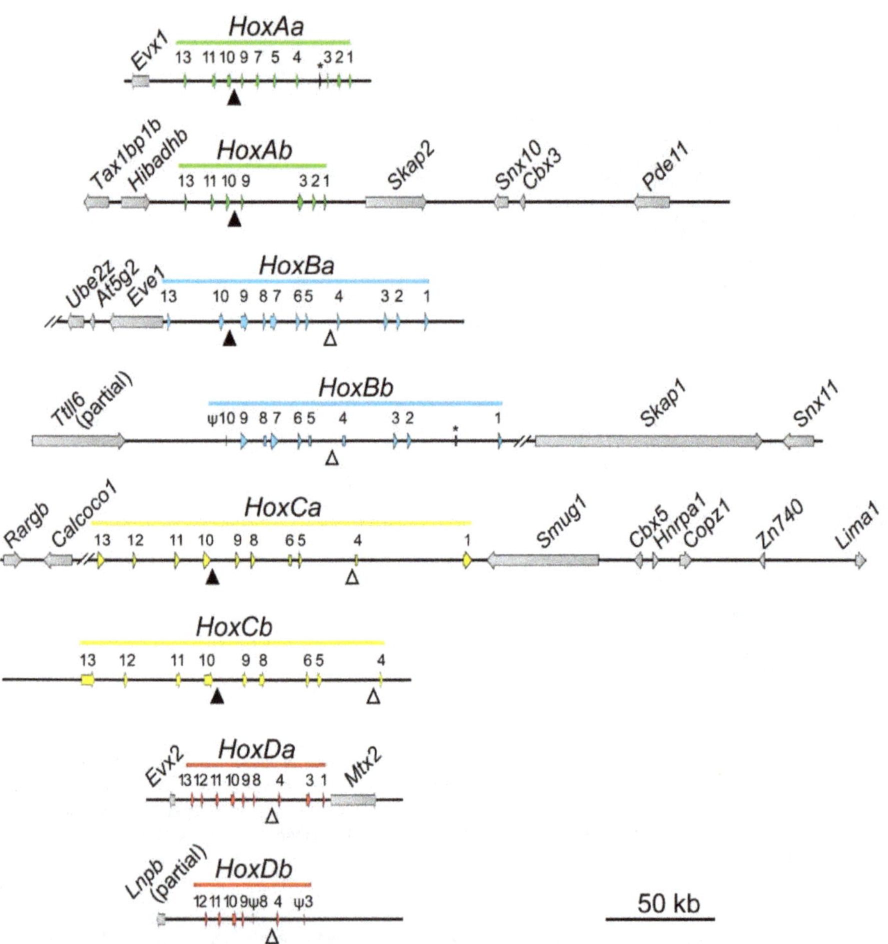

Figure 2. Genomic organization of the Hox gene clusters of the European eel. Scaffolds are indicated by black lines and asterisks represent two gaps between scaffolds. Hox genes are indicated by colored arrows that are numbered according to their paralogous groups. Three pseudogenes are indicated by the symbol ψ. Neighboring genes are indicated by grey arrows. Conserved microRNA genes are indicated by triangles: miR-196 (closed triangles) and miR-10 (open triangles).

anguilla possesses all eight copies. These confirmed the preliminary classification in A, B, C and D paralogous groups, but failed to validate the identity of teleost a and b cluster duplicates (with the exception of HoxAa and HoxAb). Likewise, phylogenetic trees based on multi-gene alignments do not conclusively indicate either a or b cluster membership for HoxB, HoxC and HoxD (Figure 4). In general, there appears to be a lack of sequence divergence between eel Hox gene duplicates, which makes classification based on coding sequence alone inaccurate.

Final orthologous relationships could only be established on the basis of conserved local synteny between Hox clusters and flanking genes (Figure 5). In addition to both HoxA clusters, eel HoxBa and HoxBb appear orthologous with their respective teleost equivalents. This identification is further supported by the absence of miR-196 from both *D. rerio* and *A. anguilla* HoxBb clusters. The affinities of HoxC and HoxD duplicates remain difficult to resolve because of conserved synteny around a and b cluster duplicates, and extensive cluster reduction and deletion in other teleosts (Figure 5c, d).

Hox gene expression

In order to confirm the transcriptional activity of the Hox genes, we determined relative expression levels by aligning transcriptomic reads of the 27 hpf embryo against the Hox protein-coding regions (Figure 6a). Transcriptome reads mapped unambiguously to 71 out of 73 Hox genes, including one pseudogene (*ψHoxD3b*), suggesting that all *A. anguilla* Hox protein-coding genes are functional. The relative expression levels vary over five orders of magnitude with the lowest expression observed for the posterior paralogous groups 12 and 13, and the highest expression for paralogous groups 7–9, but with particularly high expression levels for *HoxB1a*, *HoxB1b*, *HoxB4b* and *HoxD1a*.

Full mRNA-seq read alignment to the entire Hox clusters indicated that transcriptional activity is not restricted to protein coding regions (Figure S3). In fact, intergenic expression sometimes exceeds intragenic levels, supporting the observation that complete Hox clusters function as meta-genes [9,25].

At 27 hpf, expression of posterior Hox genes is very low (Figure 6a). We therefore confirmed transcriptional activity of posterior Hox paralogues by whole mount *in situ* hybridizations (Figure 4b). *HoxB9a* is expressed at 26 and 48 hpf, with an anterior expression boundary coinciding with somite number 5/6. Expression of *HoxD12b* and *HoxC13a* is not yet detectable at 48 hpf, but clearly visible at 96 hpf with anterior expression boundaries located at somite numbers 65/70 and 90/95 for *HoxD12b* and *HoxC13a*, respectively. For these Hox genes, expression in the eel embryo appears to conform to the expected spatio-temporal pattern (colinearity between cluster organization

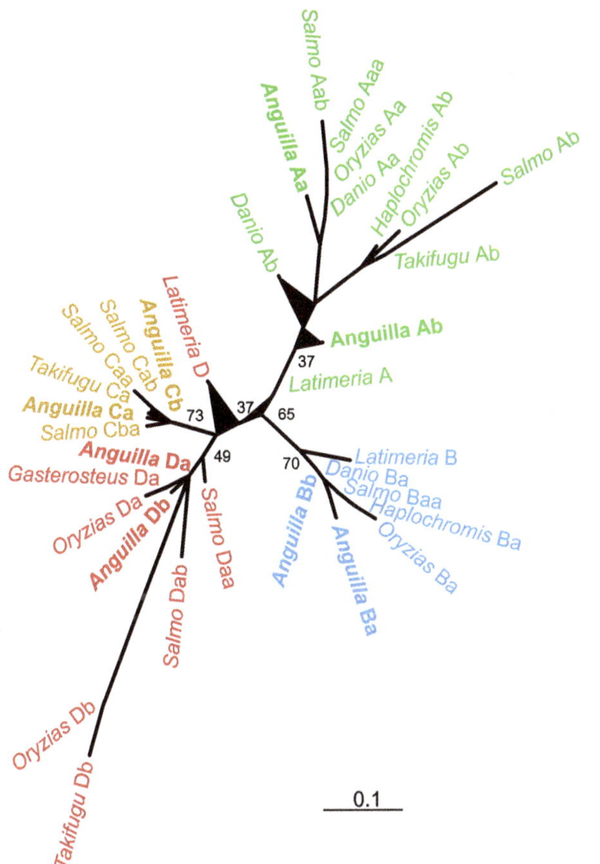

Figure 3. Classification of the European eel Hox clusters. An unrooted phylogenetic tree showing the relationships between *A. anguilla* and fish Hox9 paralogues. Numbers indicate bootstrap support.

and developmental timing and localization), with expression of Hox12 and Hox13 paralogues appearing later in development, and more posterior than Hox9.

The evolution of Hox cluster organization

The early branching of the Elopomorpha from the main teleost trunk allows a new reconstruction of ancestral Hox cluster architectures (Figure 7), which are strongly constrained by the limited organizational divergence between eel HoxB, C and D duplicates.

Since teleost fish are believed to have experienced the TSGD event early in their evolutionary history [14,15], their genomes should in theory possess up to eight cluster duplicates. However, all teleosts examined in detail retain at most seven clusters of protein-coding genes [9]: a HoxC duplicate was lost in the lineage leading to medaka and pufferfish, a HoxD duplicate in the lineage represented by zebrafish. The high number of clusters in salmon is the result of relatively recent further duplications [26].

The main teleost lineages diverged briefly after the TSGD [16]. The reconstruction in Figure 7 demonstrates that nearly all post-duplication gene loss events in the eel's ancestry occurred within this interval, followed by millions of years of stasis. Only the HoxAb cluster appears to have accumulated major changes in pre-branching, post-genome duplication teleosts. Alternative hypotheses, in which a whole-genome duplication is not shared between elopomorphs and advanced teleosts, or in which the genome duplication is followed by successive deletion and duplication of specific clusters in the eel, are less parsimonious and not consistent with local conservation of synteny (Figure 5).

Discussion

Two rounds of Hox cluster duplications in chordates are believed to be responsible for important vertebrate novelties (e.g. brains, heads, jaws) and increases in complexity [27]. A plausible mechanism invokes a temporary relaxation of meta-genic cluster constraints after duplication, paving the way for innovation

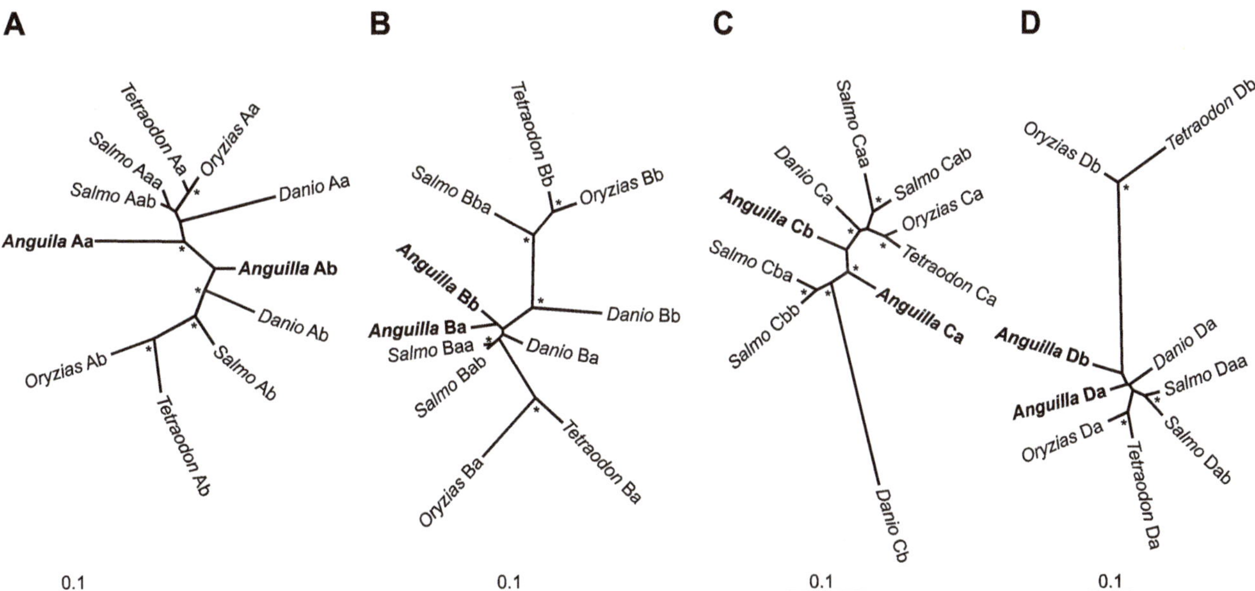

Figure 4. Phylogeny of Hox clusters of the European eel. Unrooted phylogenetic trees based on alignments combining multiple Hox genes per cluster. A) Cluster A relationships, based on HoxA9, HoxA11 and HoxA13 genes. B) Cluster B relationships, based on HoxB1, HoxB5 and HoxB6 genes. C) Cluster C relationships, based on HoxC6, HoxC11, HoxC12 and HoxC13 genes. D) Cluster D relationships, based on HoxD4 and HoxD9 genes. Species included: *A. anguilla*, *Salmo salar* (Atlantic salmon), *Danio rerio* (zebrafish), *Oryzias latipes* (medaka), and *Tetraodon nigroviridis* (green spotted puffer). Asterisks indicate bootstrap support >90%.

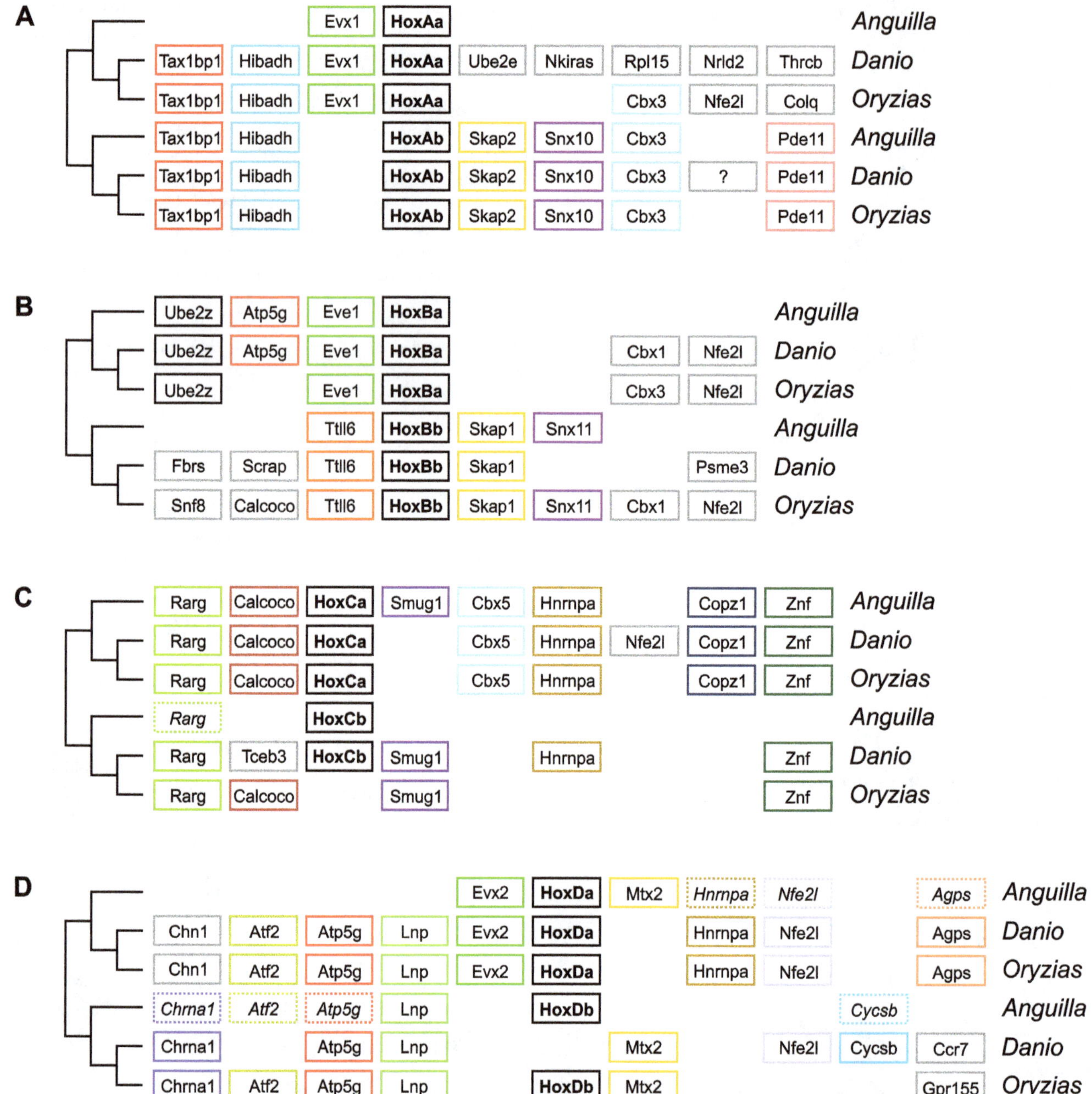

Figure 5. Synteny around Hox clusters. Conservation of flanking genes supports the classification of *A. anguilla* clusters into different orthologous subgroups. The eel clusters and up to seven flanking genes are compared with the genomic organization in zebrafish (*Danio rerio*) and medaka (*Oryzias latipes*). Coloured box outlines indicate preserved synteny between eel and the two other species, dotted outlines denote flanking genes found on extended eel scaffolds (see Methods). Interpretation should take into account residual synteny between a and b paralogous clusters. Limited data is available on HoxCb (lost in *O. latipes*, possibly misassembled in *D. rerio*) and HoxDb (lost in *D. rerio*) clusters.

[28,29]. In contrast, the TSGD-associated third duplication of vertebrate Hox clusters theoretically endowed teleost fish not with additional complexity within individuals, but with increased prospects for morphological diversification between individuals and species [9,10]. In support of this hypothesis, advanced teleosts have extensively pruned their Hox surplus, leading to significant diversity in cluster structure (Figure 7). In all examined representatives (with the exception of salmon [26]), the residual number of Hox genes is not much higher than the non-duplicated count in tetrapods. The resulting teleost Hox cluster architectures have been interpreted as an evolutionary choice for developmental flexibility in a trade-off with robustness [9]. By proving that it is possible for a vertebrate to stably preserve eight densely populated (Figure 2) and functional (Figure 6) Hox clusters, the eel genome presents an exception to these models, and a third alternative in the evolution of vertebrate complexity.

For hundreds of millions of years, *A. anguilla* and its ancestors have maintained the highest ontogenic potential of any vertebrate, indicative of continuous selective pressure. However, as adults, they do not display markedly more complex bodies than other fish or tetrapods. The eel's distinctive life cycle and body plan suggest three (not mutually exclusive) explanations for this cryptic complexity.

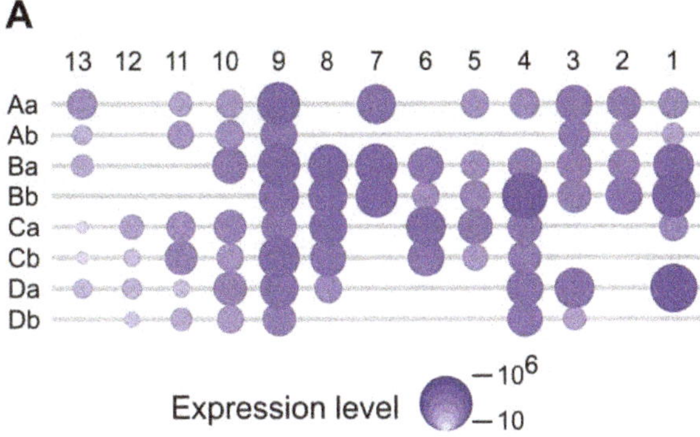

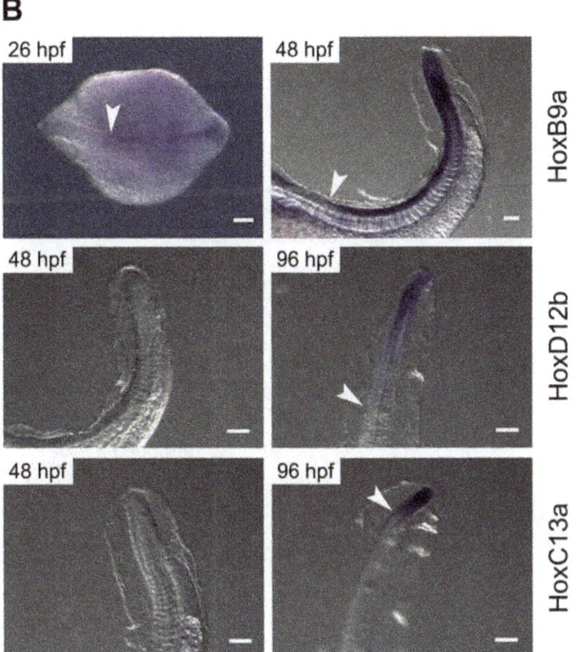

Figure 6. Hox gene expression in *A. australis* embryos. A) mRNA-seq-based gene expression in a 27 hpf embryo. B) Whole mount *in situ* hybridizations showing the expression of *HoxB9a*, *HoxD12b* and *HoxC13a*. *HoxB9a* expression can be detected in 26 hpf (dorsal view) and 48 hpf (lateral tail region view) embryos. *HoxD12b* and *HoxC13a* display expression in the tail region (lateral views) at 96 hpf, but not at 48 hpf. White arrowheads indicate anterior expression boundaries. Scale bars correspond to 100 μm.

Hox genes are involved in the primary patterning of the body axis, which implies a functional role for *A. anguilla*'s Hox surplus in axial elongation. Alterations in Hox genes have been associated with elongated body plans [30,31], however the changes observed are of a regulatory nature, and do not involve extra genes. For example, elongation of the body axis in snakes has been linked to a spatial relaxation in the posterior end of Hox clusters facilitated by the insertion of transposable elements between genes [31]. In addition, even the elongate members of the Elopomorpha (which also includes non-elongated tarpons, bonefish and others) display considerable diversity in the developmental mechanisms resulting in axial lengthening [32]. Hence, the eel's adult body plan cannot explain the preservation of primitive Hox clusters between the TSGD (226–316 million years ago [33]) and the origin of the genus *Anguilla*, estimated at 20–50 million years ago [19]. Similarly, if the European eel may at present experience singular evolutionary forces because of its panmictic population [2], any explanation these offer does not extend beyond the genus *Anguilla* of freshwater eels [34].

Even if for most of their lives eels are eel-shaped, as ocean-dwelling larvae [35] their body plan is radically different (Figure 1). In fact, until the late nineteenth century, these large, long-lived, laterally compressed leptocephali were considered to be autonomous pelagic species [36]. Fully transparent and slowly metabolizing, a leptocephalus provides considerable survival benefits [37,38]. After approximately one year, they undergo a dramatic metamorphosis [39], including extensive tissue remodelling and shortening of the body, resulting in cylindrical juveniles. In the early embryos investigated here (Figure 6), nearly all Hox genes are expressed and presumably functionally involved in determining cell fate. Logically, a high gene and cluster count can be explained by the assumption that the eel's two body plans are simultaneously outlined at this stage.

Leptocephali are the fundamental developmental innovation shared by all Elopomorpha [11–13], and therefore arose either before or soon after the TSGD, or at the base of the lineage (arrows in Figure 7). The last alternative is the most parsimonious (no loss of developmental complexity in advanced teleosts), especially since no member of the Elopomorpha is known to have ever discarded the leptocephalous larval stage [11,13]. Regardless, either of the post-TSGD origins is compatible with an intercalation model of indirect development [40], in which a temporary excess of developmental potential was permanently recruited for the conception of an additional body plan. Although speculative, an explanation invoking the morphological challenges associated with a complex life history is consistent with the stable high Hox gene and cluster count found in the anadromous Atlantic salmon [26].

Further functional studies on eel development will become possible once *A. anguilla*'s life cycle can be completed in captivity. In particular, there exists considerable variation in development (timing, number of somites) between leptocephali of related and interbreeding *Anguilla* species [1,35], which can only be studied when these larvae can be raised under controlled conditions [8,41].

Methods

Eel embryos

Wild female and male silver short-finned silver eels (*A. australis*) from Lake Ellesmere, New Zealand, were held together in a 2,300 L recirculation system with seawater (30 ppt salinity) at 21°C. Sexual maturation was induced as described [20]. Briefly, males received nine weekly injections with 250 IU human chorionic gonadotropin and females were injected once a week with 20 mg salmon pituitary extract. Eggs and milt were stripped and the eggs were dry fertilized. Embryos were reared in glass beakers with UV-sterilized seawater (35 ppt) at 21°C. At 26, 48 and 96 hpf embryos were fixed in 4% paraformaldehyde and stored in 100% methanol.

Total RNA was isolated from 27 hpf embryos using the Qiagen miRNeasy kit according to the manufacturer's instructions (Qiagen GmbH, Hilden, Germany), and analyzed with an Agilent Bioanalyzer 2100 total RNA Nano series II chip (Agilent, Santa Clara). A transcriptome library was prepared from 10 μg total RNA, using the Illumina mRNA-Seq Sample Preparation Kit according to the manufacturer's instructions (Illumina Inc., San Diego, USA).

Genome size determination

Blood samples taken from two eels (*A. anguilla* and *A. australis*) were washed with physiological salt and fixed in cold ethanol. Prior to analysis the cells were collected, resuspended in physiological salt and stained with propidium iodide. After

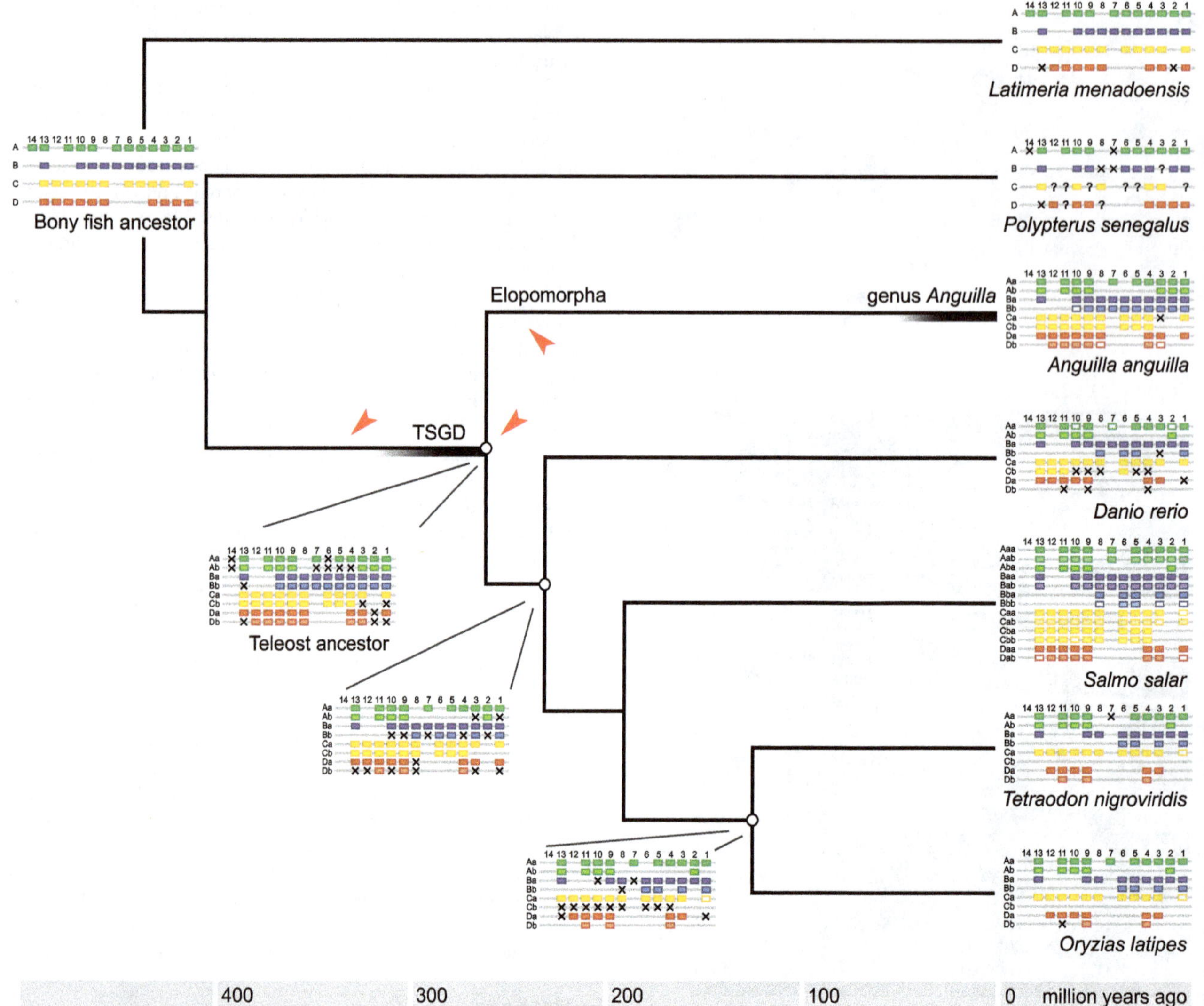

Figure 7. Model for the evolution of teleost Hox gene organization. Schematic Hox clusters [10,26,52] are superimposed on a species phylogeny with estimates of divergence times [53,54] – which vary considerably between studies [33]. Ancestral architectures are inferred on the basis of maximum parsimony, i.e. the number of cluster duplications and gene loss events is minimized. *Salmo salar* (Atlantic salmon) has presumably lost several duplicate clusters [26] (not shown). Deduced gene loss in a lineage is illustrated by a cross, question marks denote uncertainty about cluster gene content in the pre-TSGD actinopterygian *Polypterus senegalus* (bichir). Arrows indicate the possible origins of the leptocephalus body plan.

30 minutes of incubation the cells were analyzed by FACS, using human blood cells as a size reference (3.05 Gbp haploid). The eel genome size was calculated by (human size)/(mean fluorescence human)×(mean fluorescence eel). Both *Anguilla* genomes were determined to be 1.1 Gbp in size (haploid).

Genomic DNA libraries

Genomic DNA was isolated from blood of a female yellow European eel (*A. anguilla*, caught in Lake Veere, The Netherlands) using the Qiagen Blood and Tissue DNeasy kit according to the manufacturer's description. Paired-end libraries were prepared from 5 μg of isolated gDNA using the Paired-End Sequencing Sample Prep kit according to the manufacturer's description. Either a 200 bp band or a 600 bp band was cut from the gel (libraries PE200 and PE600, Table S1). After amplification for 10 cycles the resulting libraries were analyzed with a Bioanalyzer 2100 DNA 1000 series II chip.

Mate pair libraries were prepared from 10 μg of isolated gDNA using the Mate Pair Library Prep Kit v2 (Illumina Inc.). Either a 3,000 bp band or a 10,000 bp band was cut from gel (libraries MP3K and MP10K, Table S1). After the first gel purification the fragment length was analyzed using a Agilent Bioanalyzer 2100 DNA 12000 chip. After circularization, shearing, isolation of biotinylated fragments, and amplification, the 400 to 600 bp fraction of the resulting fragments was isolated from gel. Finally, the libraries were examined with an Agilent Bioanalyzer 2100 DNA 1000 series II chip.

Illumina sequencing

All libraries were sequenced using an Illumina GAIIx instrument according to the manufacturer's description. Genomic paired-end libraries were sequenced with a read length of 2×76 nucleotides (to ~20-fold genome coverage), genomic mate-pair libraries with a read length of 2×51 nucleotides (to ~33-fold genome span), and the mRNA-Seq library with a read length of 2×76 nucleotides (Table S1). Image analysis and base calling was done by the Illumina pipeline.

Genome assembly

Sequencing reads from both paired-end libraries were used in building the initial contigs (Figure S1). Both sets were preprocessed to eliminate low quality and adapter contamination. Whenever possible, PE200 pairs were merged into longer single reads. For initial contig assembly, we employed the De Bruijn graph-based *de novo* assembler implemented in the CLC bio Genomics Workbench version 3.6.5 (CLC bio, Aarhus, Denmark). A run with a k-mer length of 25 nt resulted in an assembly a total length of 969 Mbp and a contig N50 of 1672 bp.

Initial contigs were oriented in larger supercontigs (scaffolds) using SSPACE [42]. In scaffolding the contigs, we decided to exclude low-quality and highly repetitive contigs as much as possible. SSPACE was used in a hierarchical fashion, employing first links obtained from the PE600 library to generate intermediate supercontigs, which were used as input for subsequent runs with links from the MP3K and MP10K libraries, respectively. At each stage, a minimum of three non-redundant links was required to join two contigs. This procedure resulted in a final scaffold set with a total length of 923 Mbp and an N50 of 77.8 Kbp (Table S1). AUGUSTUS (version 2.4) was used to predict genes [43], which were provisionally annotated using Blast2GO (version 2.4.8) [44]. The draft assembly is available at *www.eelgenome.org*.

In order to obtain more information on flanking genes for the analysis of conserved synteny (Figure 5), scaffolds were subjected to a further round of linking by SSPACE using reduced stringency (two instead of three non-redundant links required to join scaffolds). This resulted in extended scaffolds with an N50 of 169 Kbp.

Hox genes

Hox contigs in the short-finned eel embryonic transcriptome (generated using CLC bio's *de novo* assembler) were identified via Blast [21] searches at the NCBI website (*www.ncbi.nlm.nih.gov*). European eel genomic scaffolds were annotated using CLC bio's DNA Workbench. Remaining Hox genes and genes flanking the Hox clusters were identified using Blast, based on AUGUSTUS/Blast2GO predictions. Annotated Hox scaffolds have been submitted to GenBank (accession numbers JF891391–JF891400).

MicroRNAs were identified by Blast using *H. sapiens* and *D. rerio* miR-10 and miR-196 sequences (precursors and mature) retrieved from miRBase release 18 (*www.mirbase.org*, [45]).

Phylogenetic methods

Species and Hox gene accession numbers used are listed in Table S4. Amino acid sequences of Hox genes were aligned using Clustal X [46] and checked manually. After excluding ambiguous alignments, ProtTest 2.4 [47] was used to choose an optimum substitution model, based on the Akaike information criterion. The aligned sequences were subjected to maximum likelihood analysis using RAxML version 7.2.6 [48] with 1000 rapid bootstrap replicates (-f a option).

For the analysis of Hox9 genes (Figure 3), 70 aligned residues were used and analyzed using a JTT+I+Γ model [49]. All other alignments were fitted using a JTT+Γ model. The multi-gene analyses of HoxA, HoxB, HoxC and HoxD (Figure 4) were based on alignments of 427, 493, 935 and 308 amino acid residues, respectively. The phylogenetic trees of sarcopterygian and actinopterygian Hox9 paralogues (Figure S2) were based on 151 (HoxA9), 210 (HoxB9), 248 (HoxC9), and 136 (HoxD9) residues.

Synteny was analyzed using *D. rerio* and *O. latipes* genomic contexts extracted from Ensembl release 65 (*www.ensembl.org*), based on the Zv9 and MEDAKA1 genome assemblies, respectively (Table S5). Pairwise alignments were generated by NCBI tblastx and analyzed using genoPlotR [50].

Whole mount *in situ* hybridization

Chromosomal DNA was isolated from *A. australis* blood using a DNeasy Blood & Tissue Kit (Qiagen). Riboprobe template fragments, including a T7 RNA polymerase promoter, were PCR amplified from chromosomal DNA using the following primer sets: *HoxB9a* forward (5′-TGAAACCGAAGACCCGAC-3′), *HoxB9a* reverse (5′-GAAATTAATACGACTCACTATAGGGCTGAGG-AAGACTCCAA), *HoxD12b* forward (5′-TAATCTTCTCAGTCC-TGGCTATG-3′), *HoxD12b* reverse (5′-GAAATTAATACGAC-TCACTATAGATCCAAGTTTGAAAATTCATATTTGC-3′), *HoxC13a* forward (5′-CACCTTGATGTACGTGTATGAAAA-3′), *HoxC13a* reverse (5′-GAAATTAATACGACTCACTATAGGC-TCCGTGTATTTCTCTGACG-3′). Digoxygenin-labelled riboprobes were made according to standard protocols using T7 RNA polymerase. Whole mount *in situ* hybridization with labelled riboprobes was performed at 70°C, according to a slightly modified version of a standard protocol [51]. Hybridizing riboprobes were made visible using anti-Digoxigenin AP and BM Purple AP substrate (Roche). Stained embryos were bleached using hydrogen peroxide (Sigma-Aldrich) and photographed using a Leica M205 FA stereo microscope.

Supporting Information

Figure S1 Genome assembly pipeline. See Methods section for details.

Figure S2 Unrooted maximum likelihood phylogenetic trees of actinopterygian and sarcopterygian Hox9 genes. See Methods section for details. Sequences used are listed in Table S4.

Figure S3 Meta-genic expression of Hox clusters. mRNA-seq reads of the *A. australis* embryo were aligned to entire Hox-containing scaffolds, demonstrating large amounts of mRNA production from intronic and intergenic regions.

Table S1 Statistics of European eel genome and short-finned embryonic eel transcriptome.

Table S2 Hox transcriptome contigs. All *de novo* assembled Hox contigs of a 27-hour *A. australis* embryo map to ten *A. anguilla* genome scaffolds.

Table S3 *A. anguilla* Hox genes. Complete list of *A. anguilla* Hox genes, predicted protein sizes, matching *A. australis* embryo contigs and best blastp hits.

Table S4 Hox genes used in phylogeny reconstruction. List of the Hox gene sequences used in this study.

Table S5 Hox clusters used in synteny analysis. Genomic locations of *D. rerio* and *O. latipes* Hox clusters. HoxCb is absent from *O. latipes*, and HoxDb from *D. rerio*. However, the genomic loci can still be identified based on the presence of flanking gene duplicates or conserved microRNA (*D. rerio* HoxDb).

Acknowledgments

We thank the following colleagues for generously contributing to our work: Bas Brittijn for production of short-finned eel embryos, Nabila Bardine and Tony Durston for help with *in situ* hybridizations, Tony Durston and Joost Woltering for critical reading of the manuscript, Pieter Slijkerman for project financial management, Marten Boetzer and Walter Pirovano for help with genome assembly.

Author Contributions

Conceived and designed the experiments: CVH EB RPD HPS SD F-AW KT GEEJMvdT. Performed the experiments: EB DLdW HJJ. Analyzed the data: CVH RPD YM. Wrote the paper: CVH RPD.

References

1. Tesch F-W (2003) The eel. Oxford: Blackwell Publishing.
2. Als TD, Hansen MM, Maes GE, Castonguay M, Riemann L, et al. (2011) All roads lead to home: panmixia of the European eel in the Sargasso Sea. Mol Ecol 20: 1333–1346.
3. Avise JC, Nelson WS, Arnold J, Koehn RK, Williams GC, et al. (1990) The evolutionary genetic status of Icelandic eels. Evolution 44: 1254–1262.
4. Schmidt J (1923) Breeding places and migration of the eel. Nature 111: 51–54.
5. Aarestrup K, Okland F, Hansen MM, Righton D, Gargan P, et al. (2009) Oceanic spawning migration of the European eel (*Anguilla anguilla*). Science 325: 1660.
6. Van den Thillart G, Dufour S, Rankin JC, eds. Spawning migration of the European eel. Dordrecht: Springer-Verlag.
7. Freyhof J, Kottelat M (2010) *Anguilla anguilla*. IUCN red list of threatened species, version 2010.4. Available: http://www.iucnredlist.org.
8. Tanaka H, Kagawa H, Ohta H, Unuma T, Nomura K (2003) The first production of glasseel in captivity: fish reproductive physiology facilitates great progress in aquaculture. Fish Physiol Biochem 28: 493–497.
9. Duboule D (2007) The rise and fall of Hox gene clusters. Development 134: 2549–2560.
10. Kuraku S, Meyer A (2009) The evolution and maintenance of Hox gene clusters in vertebrates and the teleost-specific genome duplication. Int J Dev Biol 53: 765–773.
11. Greenwood PH, Rosen DE, Weitsman SH, Myers GS (1966) Phyletic studies of teleostean fishes, with a provisional classification of living forms. Bull Am Mus Nat Hist 131: 339–456.
12. Inoue JG, Miya M, Tsukamoto K, Nishida M (2004) Mitogenomic evidence for the monophyly of elopomorph fishes (Teleostei) and the evolutionary origin of the leptocephalus larva. Mol Phylogenet Evol 32: 274–286.
13. Nelson JS (2006) Fishes of the world, 4th edition. Hoboken: Wiley.
14. Jaillon O, Aury J-M, Brunet F, Petit J-L, Stange-Thomann N, et al. (2004) Genome duplicaton in the teleost fish *Tetraodon nigroviridis* reveals the early vertebrate proto-karyotype. Nature 431: 946–957.
15. Meyer A, Van de Peer Y (2005) From 2R to 3R: evidence for a fish-specific genome duplication (FSGD). BioEssays 27: 937–945.
16. Crow KD, Stadler PF, Lynch VJ, Amemiya C, Wagner GP (2006) The "fish-specific" Hox cluster duplication is coincident with the origin of teleosts. Mol Biol Evol 23: 121–136.
17. Amores A, Force A, Yan Y-L, Joly L, Amemiya C, et al. (1999) Zebrafish *hox* clusters and vertebrate genome evolution. Science 282: 1711–1714.
18. Guo B, Gan X, He S (2010) Hox genes of the Japanese eel *Anguilla japonica* and Hox cluster evolution in teleosts. J Exp Zool (Mol Dev Evol) 314B: 135–147.
19. Minegishi Y, Aoyama J, Inoue JG, Miya M, Nishida M, et al. (2005) Molecular phylogeny and evolution of the freshwater eels genus *Anguilla* based on the whole mitochondrial genome sequences. Mol Phylogenet Evol 24: 134–146.
20. Burgerhout E, Brittijn SA, Kurwie T, Decker P, Dirks RP, et al. (2011) First artificial hybrid of the eel species *Anguilla australis* and *Anguilla Anguilla*. BMC Dev Biol 11: 16.
21. Altschul SF, Gish W, Miller W, Myers EW, Lipman DJ (1990) Basic local alignment search tool. J Mol Biol 215: 403–410.
22. Yekta S, Tabin CJ, Bartel DP (2008) MicroRNAs in the Hox network: an apparent link to posterior prevalence. Nat Rev Genet 9: 789–796.
23. He X, Yan Y-L, Eberhart JK, Herpin A, Wagner TU, et al. (2011) miR-196 regulates axial patterning and pectoral appendage initiation. Dev Biol 357: 463–477.
24. Woltering JM, Durston AJ (2006) The zebrafish hoxDb cluster has been reduced to a single microRNA. Nat Genet 38: 601–602.
25. Mainguy G, Koster J, Woltering J, Jansen H, Durston A (2007) Extensive polycistronism and antisense transcription in the mammalian Hox clusters. PLoS ONE 2: e356.
26. Mungpakdee S, Seo H-S, Angotzi AR, Dong X, et al. (2008) Differential evolution of the 13 Atlantic salmon Hox clusters. Mol Biol Evol 25: 1333–1343.
27. Holland PW, Garcia-Fernàndez J, Williams SA, Sidow A (1994) Gene duplications and the origins of vertebrate development. Development Supplement. pp 123–133.
28. Wagner GP, Amemiya C, Ruddle F (2003) Hox cluster duplications and the opportunity for evolutionary novelties. Proc Natl Acad Sci USA 100: 144603–14606.
29. Lynch VJ, Wagner GP (2008) Resurrecting the role of transcription factor change in developmental evolution. Evolution 62: 2131–2154.
30. Woltering JM, Vonk FJ, Müller H, Bardine N, Tuduce IL, et al. (2009) Axial patterning in snakes and caecilians: evidence for an alternative interpretation of the *Hox* code. Dev Biol 332: 82–89.
31. Di-Poï N, Montoya-Burgos JI, Miller H, Pourquié O, Milinkovitch MC, et al. (2010) Changes in *Hox* genes' structure and function during the evolution of the squamate body plan. Nature 464: 99–103.
32. Mehta RS, Ward AB, Alfaro ME, Wainwright PC (2010) Elongation of the body in eels. Integr Comp Biol 50: 1091–1105.
33. Hurley IA, Lockridge Mueller R, Dunn KA, Schmidt EJ, Friedman M, et al. (2007) A new time-scale for ray-finned fish evolution. Proc R Soc B 274: 489–498.
34. Inoue JG, Miya M, Miller MJ, Sado T, Hanel R, et al. (2010) Deep-ocean origin of the freshwater eels. Biol Lett 6: 363–366.
35. Miller MJ (2009) Ecology of anguilliform leptocephali: remarkable transparent fish larvae of the ocean surface layer. Aqua-BioSci Monogr (ABSM) 2, No 4: 1–94.
36. Cunningham JT (1897) The discovery of the larva of the common eel. Nature 55: 467–468.
37. Pfeiler E (1999) Developmental physiology of elopomorph leptocephali. Comp Biochem Physiol A 123: 113–128.
38. Bishop RE, Torres JJ (1999) Leptocephalus energetics: metabolism and excretion. J Exp Biol 202: 2485–2493.
39. Otake T (2003) Metamorphosis. In: Aida K, Tsukamoto K, Yamauchi K, eds. Eel biology. Tokyo: Springer-Verlag. pp 61–74.
40. Raff RA (2008) Origins of the other metazoan body plans: the evolution of larval forms. Phil Trans R Soc B 363: 1473–1479.
41. Oliveira K, Hable WE (2010) Artificial maturation, fertilization, and early development of the American eel (*Anguilla rostrata*). Can J Zool 88: 1121–1128.
42. Boetzer M, Henkel CV, Jansen HJ, Butler D, Pirovano W (2011) Scaffolding preassembled contigs using SSPACE. Bioinformatics 27: 578–579.
43. Stanke M, Diekhans M, Baertsch R, Haussler D (2008) Using native and syntenically mapped cDNA alignments to improve *de novo* gene finding. Bioinformatics 24: 637–644.
44. Götz S, García-Gómez JM, Terol J, Williams TD, Nagaraj SH, et al. (2008) High-throughput functional annotation and data mining with the Blast2GO suite. Nucleic Acids Res 36: 3420–3435.
45. Kozomara A, Griffiths-Jones S (2011) miRBase: integrating microRNA annotation and deep-sequencing data. Nucl Acids Res 39: D152–D157.
46. Larkin MA, Blackshields G, Brown NP, Chenna R, McGettigan PA, et al. (2007) Clustal W and Clustal X version 2.0. Bioinformatics 23: 2947–2948.
47. Abascal F, Zardoya R, Posada D (2005) ProtTest: selection of best-fit models of protein evolution. Bioinformatics 21: 2104–2105.
48. Stamatakis A, Hoover P, Rougemont J (2008) A rapid bootstrap algorithm for the RAxML Web servers. Syst Biol 57: 758–771.
49. Jones DT, Taylor WR, Thornton JM (1992) The rapid generation of mutation data matrices from protein sequences. Comput Appl Biosci 8: 275–282.
50. Guy L, Kultima JR, Andersson SG (2010) genoPlotR: comparative gene and genome visualization in R. Bioinformatics 26: 2334–2335.
51. Hiroi J, Maruyama K, Kawazu K, Kaneko T, Ohtani-Kaneko R, et al. (2004) Structure and developmental expression of hatching enzyme genes of the Japanese eel *Anguilla japonica*: an aspect of the evolution of fish hatching enzyme gene. Dev Genes Evol 214: 176–184.
52. Raincrow JD, Dewar K, Stocsits C, Prokasha SJ, Amemiya CT, et al. (2011) Hox clusters of the bichir (Actinopterygii, *Polypterus senegalus*) highlight unique patterns in sequence evolution in gnathostome phylogeny. J Exp Zool (Mol Dev Evol) 316B: 451–464.
53. Inoue JG, Miya M, Venkatesh B, Nishida M (2005) The mitochondrial genome of Indonesian coelacanth *Latimera menadoensis* (Sarcopterygii: Coelacanthiformes) and divergence time estimation between the two coelacanths. Gene 349: 227–235.
54. Matschiner M, Hanel R, Salzburger W (2011) On the origin and trigger of the notothenioid adaptive radiation. PLoS ONE 6: e18911.

17

Assessing Trait Covariation and Morphological Integration on Phylogenies Using Evolutionary Covariance Matrices

Dean C. Adams[1]*, Ryan N. Felice[2]

1 Department of Ecology, Evolution, and Organismal Biology, Iowa State University, Ames, Iowa, United States of America, 2 Department of Biological Sciences, Ohio University, Athens, Ohio, United States of America

Abstract

Morphological integration describes the degree to which sets of organismal traits covary with one another. Morphological covariation may be evaluated at various levels of biological organization, but when characterizing such patterns across species at the macroevolutionary level, phylogeny must be taken into account. We outline an analytical procedure based on the evolutionary covariance matrix that allows species-level patterns of morphological integration among structures defined by sets of traits to be evaluated while accounting for the phylogenetic relationships among taxa, providing a flexible and robust complement to related phylogenetic independent contrasts based approaches. Using computer simulations under a Brownian motion model we show that statistical tests based on the approach display appropriate Type I error rates and high statistical power for detecting known levels of integration, and these trends remain consistent for simulations using different numbers of species, and for simulations that differ in the number of trait dimensions. Thus, our procedure provides a useful means of testing hypotheses of morphological integration in a phylogenetic context. We illustrate the utility of this approach by evaluating evolutionary patterns of morphological integration in head shape for a lineage of *Plethodon* salamanders, and find significant integration between cranial shape and mandible shape. Finally, computer code written in R for implementing the procedure is provided.

Editor: Jason M. Kamilar, Midwestern University & Arizona State University, United States of America

Funding: DCA received support from National Science Foundation grants DEB-1257827 and DEB-111884. The funders had no role in study design, data collection and analysis, decision to publish, or preparation of the manuscript.

Competing Interests: The authors have declared that no competing interests exist.

* E-mail: dcadams@iastate.edu

Introduction

A major goal in evolutionary biology is to understand patterns of morphological covariation in biological forms. Advances in geometric morphometrics provide a detailed quantification of two- and three-dimensional shape diversity [1,2], and investigations into the functional, genetic, and developmental causes of morphological variation result in hypotheses regarding how trait covariation evolves. Patterns and magnitudes of morphological trait covariation are referred to as *morphological integration* [3,4], or the extent to which traits, or sets of traits, covary [5–8]. Because organisms must operate as an integrated whole, changes in some traits are accompanied by changes in other traits that are linked to them through common functional activities, developmental pathways, or genetic linkages and pleiotropy [9]. Over time this results in correlated evolution and thus covariation between traits. At the evolutionary level, deciphering how and why traits covary is critical for understanding the mechanisms that explain how morphological variation and covariation evolves [10,11].

Over the past several years an increasing number of studies have investigated patterns of morphological integration and covariation in numerous traits and across a wide variety of taxa [5,12–19]. Most have characterized morphological trait covariation within species, although some studies have also compared within-species patterns across multiple taxa [20–23] and others have examined interspecific patterns [24–28]. In such cases, understanding the evolution of morphological covariation requires an explicit macroevolutionary perspective, where patterns of covariation and integration are characterized across species. Here, the phylogenetic relationships among taxa must be considered, as species are not evolutionarily independent of one another [29–31]. To this end, the evolutionary associations between pairs of traits have been examined phylogenetically using elements in the evolutionary covariance matrix [32]. However, because hypotheses of morphological integration are typically interested in the degree of covariation between sets of traits, a pairwise trait approach is insufficient, as pairwise correlations do not capture the covariation between sets of traits treated as separate blocks of characters (much like Pearson's correlation coefficient does not fully capture the information described by R^2 in multiple regression). Thus, to evaluate patterns of morphological integration phylogenetically, alternative procedures must be used. Several studies have utilized phylogenetic regression to assess morphological covariation between two sets of traits of while accounting for phylogenetic non-independence (e.g., landmark configurations of face versus the braincase, or the ramus versus the alveolar region of the mandible) [23,25,26,33]. Alternatively, analyses may be performed on the set of phylogenetic independent contrast scores obtained for the two sets of traits [24,34,35].

One approach commonly used to characterize ahistorical patterns of morphological integration between blocks of traits is partial least squares (PLS). Partial least squares is a statistical procedure that quantifies the degree of covariation between sets of variables, based on the overall trait covariance matrix [see 5,36,37]. PLS can be used to assess patterns of covariation between sets of morphological data (as in studies of morphological integration) or between a set of morphological data and another dataset (such as diet or environmental data). One advantage of this approach is that neither set of variables is assumed to be dependent on the other, as in regression analyses [5,37]. This makes PLS a particularly useful tool for assessing the relationship between sets of traits that are hypothesized to covary but for which there is no *a priori* directional relationship posited between them.

Recently, patterns of morphological integration were evaluated in an explicit phylogenetic framework by evaluating covariation between sets of phylogenetic independent contrasts (PIC) for two sets of phenotypic traits [24,34] using PLS and other approaches. However, because there is a direct statistical relationship between PIC and phylogenetic generalized least squares methods (PGLS) [30,38,39], one may also consider using the evolutionary covariance matrix from PGLS directly as the basis for evaluating phylogenetic morphological integration. Indeed, several authors have mentioned this as a possibility [24,40], but to date none have demonstrated the equivalency (shown below). However, while PGLS- and PIC-based regression analyses provide identical statistical results under a Brownian motion evolutionary model [30,38,39], there are some advantages to implementing the PGLS-based approach for assessing morphological covariation in a phylogenetic context. First, PGLS approaches may be suitable for use under a variety of evolutionary models, including Brownian motion, Ornstein-Uhlenbeck, and models describing other evolutionary processes [41,42]. Additionally, PIC assumes a completely bifurcating tree whereas PGLS may be used with trees containing polytomies [38].

For these reasons, examining phylogenetic morphological integration using this PGLS-based algorithm should be more broadly applicable than the previously published PIC-based methods [24,34]. However, to date the full analytical procedure for implementing this approach has not been elucidated. Additionally, the statistical properties of methods that assess morphological integration in a phylogenetic context have not yet been explored. In this paper we outline an approach for estimating the strength of morphological integration in a phylogenetic context based on the evolutionary covariance matrix obtained from phylogenetic generalized least squares. Using computer simulations we show that statistical tests based on the approach display acceptable Type I error and appropriate statistical power for detecting phylogenetic morphological integration between blocks of traits. We then present a biological example assessing morphological integration in *Plethodon* salamanders demonstrating the utility of the method. Computer code written in R for implementing the procedure is also provided.

Materials and Methods

Estimating Phylogenetic Morphological Integration and Covariation Using Partial Least Squares

Partial least squares is used to assess covariation between subsets of data by utilizing a covariance matrix of the complete data. First, a $p \times p$ covariance matrix is constructed from an $N \times p$ data matrix ($\mathbf{Y}$) containing the p phenotypic values for each of N specimens. In this case, $\mathbf{Y}$ represents two subsets (blocks) of variables, $\mathbf{Y_1}$ and $\mathbf{Y_2}$ (for generalizations to three sets of variables see: [5]). Thus, the phenotypic covariance matrix may be expressed as the partitioned matrix:

$$\mathbf{S} = \begin{pmatrix} \mathbf{S}_{11} & \mathbf{S}_{12} \\ \mathbf{S}_{21} & \mathbf{S}_{22} \end{pmatrix} \qquad (1)$$

where $\mathbf{S_{11}}$ and $\mathbf{S_{22}}$ describe the covariance within blocks $\mathbf{Y_1}$ and $\mathbf{Y_2}$ respectively, and $\mathbf{S}_{12} = \mathbf{S}_{21}^t$ represents the covariation between blocks. Next, $\mathbf{S_{12}}$ is decomposed using singular-value decomposition: $\mathbf{S}_{12} = \mathbf{UDV}^t$ where the resulting $\mathbf{U}$ and $\mathbf{V}$ matrices contain linear combinations of variables for each block $\mathbf{Y_1}$ and $\mathbf{Y_2}$. $\mathbf{D}$ is a diagonal matrix where each diagonal element contains the covariance between corresponding pairs of vectors in $\mathbf{U}$ and $\mathbf{V}$ [36,37]. The largest value in $\mathbf{D}$ corresponds to the highest covariation between linear combinations of $\mathbf{Y_1}$ and $\mathbf{Y_2}$. This may be found by multiplying (projecting) the original data ($\mathbf{Y_1}$ and $\mathbf{Y_2}$) onto the vectors of linear combinations ($\mathbf{U}$ and $\mathbf{V}$ respectively) to obtain PLS scores, from which the maximal covariation between blocks may be calculated. One can also obtain the correlation between $\mathbf{Y_1}$ and $\mathbf{Y_2}$ using these PLS scores, which may then be used to statistically assess the association between $\mathbf{Y_1}$ and $\mathbf{Y_2}$ using permutation procedures [37,43].

At a macroevolutionary scale, patterns of morphological integration are assessed from phenotypic data obtained from set of N species related by a phylogeny. To statistically evaluate patterns of evolutionary covariation in a phylogenetic context, one first requires a model that describes the evolutionary process that generated the data. Like many approaches, we utilize a Brownian motion model [29,44–46]. Here, evolutionary changes in a single trait along the phylogeny are independent from time step to time step, with an expected displacement of zero and a variance among species (σ^2) that accumulates proportional to time. For multivariate data, evolutionary change corresponds to a shift in the position of a species in a multivariate trait space whose axes correspond to trait dimensions. This process is described by an evolutionary covariance matrix ($\mathbf{R}$) whose diagonal elements represent the expected variation for each trait dimension (σ^2), and whose off-diagonal elements express the evolutionary covariation in changes between trait dimensions [47,48]. Analytically this evolutionary covariance matrix may be estimated as:

$$\mathbf{R} = \frac{(\mathbf{Y} - E(\mathbf{Y}))^t \mathbf{C}^{-1} (\mathbf{Y} - E(\mathbf{Y}))}{N-1} \qquad (2)$$

where $\mathbf{Y}$ is a $N \times p$ matrix of phenotypic values for the N species, $E(\mathbf{Y})$ is a $N \times p$ matrix where each row contains the multivariate phylogenetic mean found at the root of the phylogeny $\hat{a} = (\mathbf{1}^t\mathbf{C}^{-1}\mathbf{1})^t\mathbf{C}^{-1}(\mathbf{1}^t\mathbf{C}^{-1}\mathbf{1})$, $\mathbf{1}$ is a $N \times 1$ column vector of ones, and $\mathbf{C}^{-1}$ represents the inverse of the $N \times N$ phylogenetic covariance matrix describing the evolutionary relationships among species [30,39,49,50]. Note that equation (2) describes the unbiased estimate of $\mathbf{R}$, as the denominator is N-1 (for discussion see [30]).

In phylogenetic comparative biology, numerous procedures estimate the evolutionary covariance matrices as the starting point for evaluating hypotheses of changes in evolutionary rates among traits [48,50–55] and shifts in trait covariances across the phylogeny [32,48]. For assessing patterns of phylogenetic morphological integration, the evolutionary covariance matrix is first obtained. Next, $\mathbf{R}$ is represented as a partitioned matrix describing the evolutionary covariation within and between two sets of variables $\mathbf{Y_1}$ and $\mathbf{Y_2}$:

$$\mathbf{R} = \begin{pmatrix} \mathbf{R}_{11} & \mathbf{R}_{12} \\ \mathbf{R}_{21} & \mathbf{R}_{22} \end{pmatrix}. \quad (3)$$

A singular-value decomposition of $\mathbf{R_{12}}$ is then performed: $\mathbf{R}_{12} = \mathbf{U_{R1}DV^t_{R2}}$ to obtain evolutionary linear combinations ($\mathbf{U_{R1}}$ and $\mathbf{V_{R2}}$) for the two sets of variables.

To estimate the evolutionary correlation between the blocks of phenotypic data, $\mathbf{Y_1}$ and $\mathbf{Y_2}$, projection scores on the first column of $\mathbf{U_{R1}}$ and $\mathbf{V_{R2}}$ are obtained. However, because $\mathbf{U_{R1}}$ and $\mathbf{V_{R2}}$ represent linear combinations from the evolutionary covariance matrix that takes phylogeny into account the original phenotypic data in $\mathbf{Y}$ must also be expressed phylogenetically prior to projection onto these axes. This is accomplished through phylogenetic transformation, as described in [30,56]. First, the transformation matrix $\mathbf{D}$ is found from an eigen-decomposition of the phylogenetic covariance matrix: $\mathbf{DCD}^t = \mathbf{I}$. Next, the original phenotypic data, $\mathbf{Y}$, are projected onto $\mathbf{D}$, thereby transforming them by the phylogeny:

$$\mathbf{Y}_{phy} = \mathbf{D}(\mathbf{Y} - \mathbf{1}\hat{a}^t) \quad (4)$$

To calculate evolutionary PLS scores, the two blocks of phylogenetically corrected data, $\mathbf{Y_{1phy}}$ and $\mathbf{Y_{2phy}}$, are multiplied by the first pair of vectors in $\mathbf{U_{R1}}$ and $\mathbf{V_{R2}}$. From these the evolutionary correlation between the two blocks of data, $\mathbf{r}_{PLS}$, is found by the correlation between the two vectors of PLS scores. Finally, the significance of this correlation, P_{rand}, may be assessed using phylogenetic permutation, where the phenotypic values for all species for one block (e.g., $\mathbf{Y_2}$) are permuted on the tips of the phylogeny, correlation scores are obtained from the permuted datasets, and these are compared to the observed value (for similar procedures see: [24,57,58]). Computer code written in R for implementing the approach is found in File S1.

Statistical Performance of the Approach

The procedure specified above provides a logical means of characterizing evolutionary patterns of morphological integration in a phylogenetic context (for alternative formulations see [24,26]). However, to date, none of the proposed tests of phylogenetic morphological integration have been evaluated in terms of their Type I error or statistical power (though the power of methods based on the RV coefficient are known to be sensitive to sample size: [59]). To alleviate this shortcoming, we examine the statistical performance of the hypothesis test proposed here, using a series of computer simulations. Simulations were conducted on a series of random phylogenies which differed in the number of taxa (N = 16, 32, 64, 128). For each simulation, the number of species was first selected and a random phylogeny was generated. Next, the total number of trait dimensions was selected (p = 6, 8, 10, 16, 20, 30), which were divided equally between two blocks: $\mathbf{Y_1}$ and $\mathbf{Y_2}$. To simulate phenotypic data, an initial $p \times p$ covariance matrix was constructed. Simulations assumed isotropic error ($\sigma^2 = 1.0$) for all trait dimensions, and these values were treated as the diagonal elements of the $p \times p$ covariance matrix. For the off-diagonal elements, covariation between trait dimensions was varied depending on simulation conditions. Simulations evaluating type I error rates used no initial association between trait dimensions ($\sigma_{12} = 0.0$), while simulations evaluating statistical power used increasing levels of covariation between trait dimensions ($\sigma_{12} = 0.2, 0.3, 0.4, 0.5, 0.7, 0.9$). Phenotypic data were then obtained by simulating multi-dimensional traits following a Brownian motion model of evolution. Thus, the following procedure was used to generate 1000 data sets: 1) generate a random phylogeny, 2) generate an initial covariance matrix, 3) simulate data. For each dataset, the degree of morphological integration was then estimated using the procedure described above, and was evaluated statistically using phylogenetic permutation. For all simulations, when the initial covariation between sets of traits was greater than zero (i.e., when $\sigma_{12} > 0.0$), the proportion of significant results (out of 1000) provides an estimate of the statistical power of the test for that level of input covariation. Likewise, when the initial covariation between sets of traits was equal to zero (when $\sigma_{12} = 0.0$), the proportion of significant results (out of 1000) provides an estimate of the Type I error rate of the test when no input covariation is provided.

Simulation results. For all simulations, hypothesis tests of phylogenetic morphological integration displayed appropriate Type I error rates near the nominal value of $\alpha = 0.05$. This pattern remained consistent across the range of trait dimensionality examined in this study, and was consistent across the range of species richness evaluated here (Fig. 1). This finding is important, as it implies that the Type I approach developed here is relatively robust to sample size and trait dimensionality. This is in contrast to alternative methods of assessing morphological integration based on the RV coefficient, which lose power as sample size decreases [59].

In addition, the statistical power of tests evaluating phylogenetic morphological integration increased as the degree of covariation between $\mathbf{Y_1}$ and $\mathbf{Y_2}$ increased. This implied that the approach is better capable of detecting patterns of morphological integration as the strength of that signal is large. As expected, for the same number of trait dimensions, statistical power increased as the number of taxa examined increased (viewed across panels in Fig. 1). For example, when $p = 10$ there is considerably higher power to detect integration between blocks of traits when examining trends across 32 species as compared to just 16 species; and still more power to detect trends with a higher number of taxa. Thus, as with other phylogenetic comparative approaches utilizing a larger number of species is desirable.

Surprisingly, statistical power also increased with increasing numbers of trait dimensions. Thus, for a given number of taxa, the approach is better able to detect trends for traits of higher dimensionality than for traits represented by fewer dimensions. Similar findings were recently shown for methods evaluating multivariate rates of evolution on phylogenies [52], where power increased with increasing dimensionality. This result implies that measuring more complex traits does not impinge on one's ability to evaluate morphological integration in high-dimensional multivariate data, as the procedure described here is capable of detecting these patterns. Similar results were obtained on phylogenies with different numbers of taxa (Fig. 1). Overall these simulations reveal that hypothesis tests of phylogenetic morphological integration and based on the evolutionary covariance matrix have appropriate Type I error and statistical power, and thus provide a useful means of detecting morphological integration and covariation in a phylogenetic context for high-dimensional phenotypic datasets.

A Biological Example

As a biological example of the approach described here we evaluate the degree of phylogenetic morphological integration in the skulls of *Plethodon* salamanders. Vertebrate skulls are perhaps the most intensively studied anatomical structure in terms of morphological integration, and considerable evidence suggests that morphological integration in vertebrate skulls is displayed in a wide variety of taxa [5,12,15,17,24,34,60,61]. The pervasive

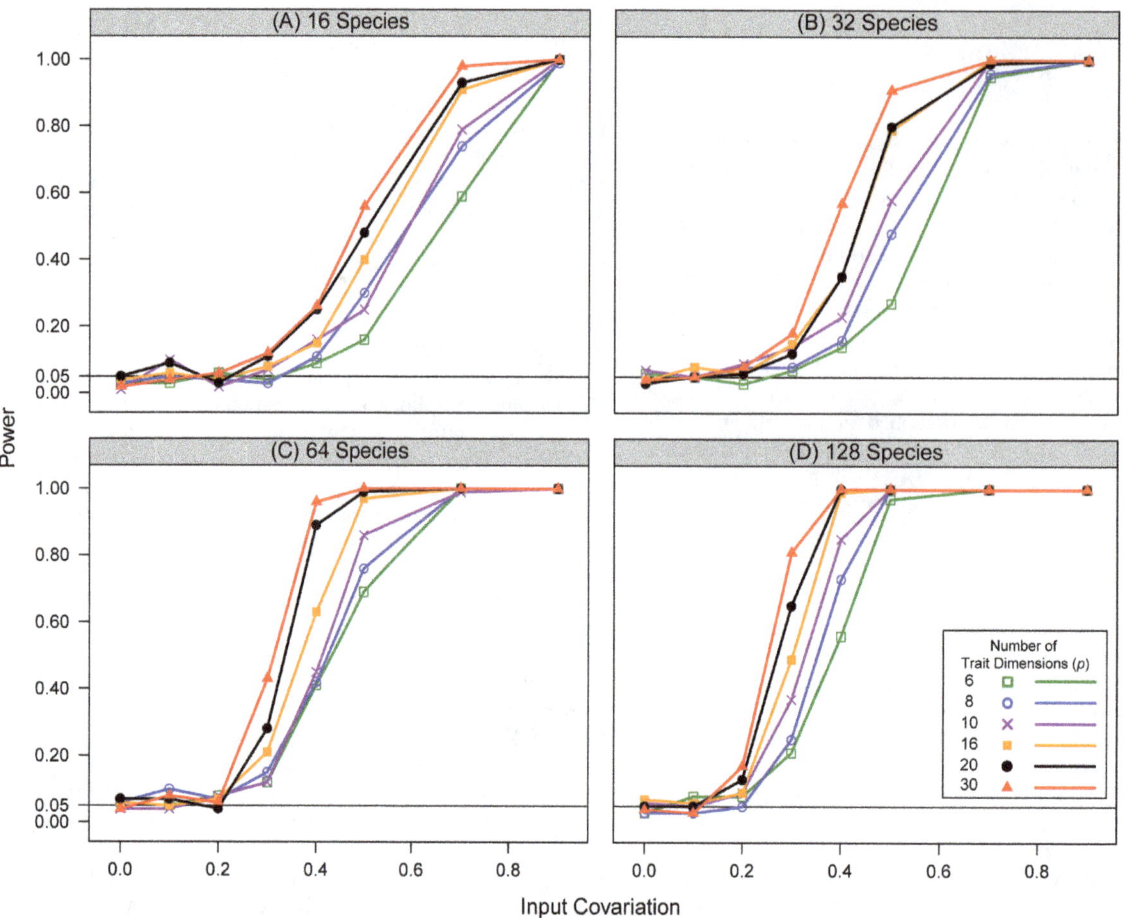

Figure 1. Type I error and statistical power of approach. Simulation results for type I error and statistical power of hypothesis testing procedures evaluating phylogenetic morphological integration. Type I error is found as the first point on each curve, where the input covariation between X and Y was 0.0 (see text). The horizontal line represents the expected type I error rate of 5%. Data were simulated under a Brownian motion model of evolution on randomly generated phylogenies containing: A) 16 species, B) 32 species, C) 64 species, and B) 128 species. Curves for increasing numbers of trait dimensions are shown.

degree of skull integration is due to the fact that the component parts of the vertebrate skull share common developmental pathways, are under common functional demands, and thus evolve jointly with one another [62,63]. In *Plethodon* salamanders, the head serves an important role during aggressive encounters, with the mandible and cranium operating together. Here, agonistic interactions begin with a series of visual displays between individuals [64], but these often escalate into direct attacks through biting [65–67]. Additionally, the head is also important for securing and immobilizing prey during foraging [43,68–69]. Thus, it may be expected that cranial shape and mandible shape will be highly integrated in *Plethodon,* although this hypothesis has never been empirically examined. Further, because head shape exhibits a strong genetic component [70] it is thus reasonable to evaluate whether selection has driven the evolution of integration at the interspecific level.

To test this hypothesis we quantified head shape from 691 adult salamanders from 18 species of *Plethodon* for which data were available (data from [71–76], see Table 1). Head shape was quantified using geometric morphometric methods [2,77]. First, 11 landmarks were digitized from images of the left-lateral side of each head (Fig. 2a). Next, the position of the jaw was standardized relative to the skull by rotating the jaw to a common articulation angle among specimens [78]. Specimens were then aligned using a Generalized Procrustes analysis [79], and a set of shape variables were obtained for each specimen (Procrustes tangent coordinates). Both the cranium and mandible were superimposed simultaneously to take into consideration the relative size differences of the two structures (for discussion see [62]). The mean head shape was then calculated for each of the 18 species. We recognize that the use of species means does not allow within-species variation to be evaluated [24]. However, unlike likelihood-based PGLS approaches [56], methods for evaluating the effects of intraspecific variation in the context of phylogenetic partial least squares have not yet been developed, and are outside of the scope of the present paper.

With these data, the degree of morphological integration between the mandible (landmarks 1–5) and the cranium (landmarks 6–11) was evaluated in a phylogenetic context. For this, a multi-gene time-calibrated molecular phylogeny for *Plethodon* was used as an estimate of the species-level relationships ([80]: Fig. 2b). Phylogenetic morphological integration was then mathematically characterized using the method described above, and statistical significance was determined using phylogenetic permutation. In addition, phylogenetic morphological integration was also evaluated using a procedure based on phylogenetic independent

Table 1. Number of adult specimens per species used in this study.

Species	*N*	*Species*	*N*
Plethodon albagula	23	*Plethodon montanus*	25
Plethodon cinereus	60	*Plethodon nettingi*	26
Plethodon electromorphus	73	*Plethodon richmondi*	106
Plethodon fourchensis	15	*Plethodon serratus*	11
Plethodon glutinosus	25	*Plethodon shenandoah*	30
Plethodon hoffmani	123	*Plethodon teyahalee*	25
Plethodon hubrichti	26	*Plethodon variolatus*	25
Plethodon jordani	25	*Plethodon Virginia*	23
Plethodon mississippi	25	*Plethodon yonahlossee*	25

Shape data were obtained using geometric morphometric methods, as described in the text.

contrasts [24], as implemented in MorphoJ [81]. Patterns of integration were then visualized using thin-plate spline deformation grids of both the cranium and mandible for exemplar specimens. All analyses were performed in R 3.0.1 (R Development Core Team [82]) using routines in the library geomorph [83,84], and new routines written by the authors (File S1).

Biological example results. Analyses indicated that there was significant phylogenetic morphological integration between the cranium and mandible across species of *Plethodon* ($r_{PLS} = 0.813$; $P_{rand} = 0.008$). Among species, mandible shapes varied from relatively slender and elongated to more robust and laterally compressed (Fig. 2c). Additionally, salamanders displayed crania that were relatively more slender laterally to relatively laterally compressed (Fig. 2c). Taken together, the morphological integration between crania and mandibles was best described as a shift from individuals exhibiting relatively elongated mandibles and crania (on the positive side of the PLS axes) to individuals displaying relatively more robust mandibles with relatively more compressed crania (towards the negative side of the PLS axes). These analyses reveal that there is a significant degree of evolutionary morphological integration in the skulls of *Plethodon* found when the phylogenetic relationships among taxa are taken

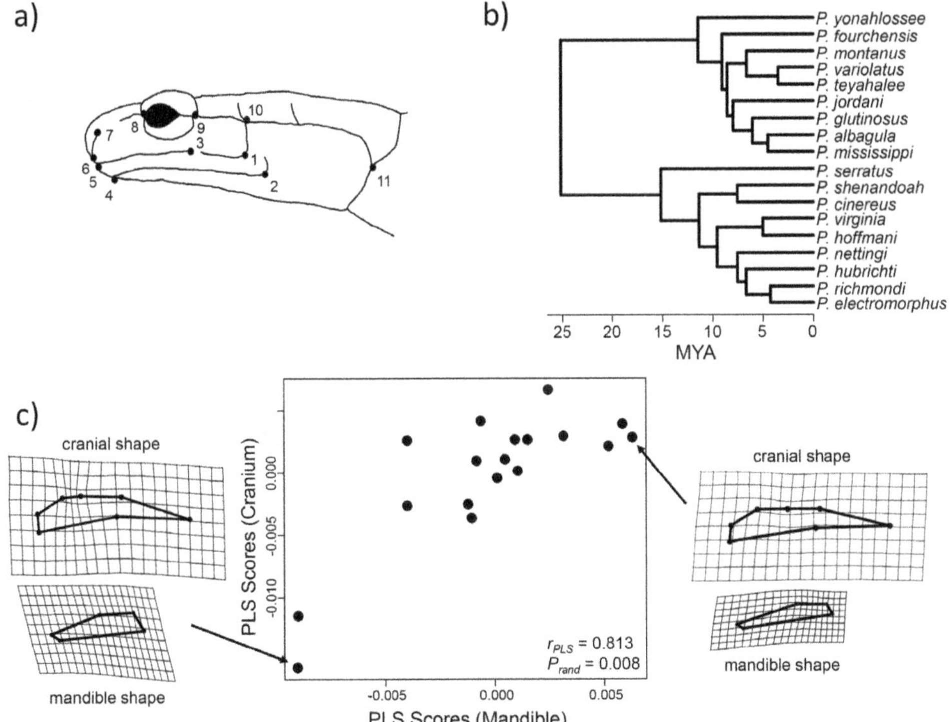

Figure 2. Data summary and analytical results evaluating morphological integration in *Plethodo* salamanders. a) Positions of 11 anatomical landmarks used to quantify head shape in *Plethodon* salamanders (image from [72]). b) Fossil-calibrated molecular phylogeny displaying the estimated phylogenetic relationships among the species of *Plethodon* examined here. c) Plot of scores along the first axis for mandible shape versus cranial shape found from a partial least squares analysis of the evolutionary covariance matrix (**R**). Thin-plate spline deformation grids for the representing the extreme forms along each axis are shown (magnified by 2X).

into consideration. Finally, a phylogenetic PLS analysis on the independent contrasts of shape confirmed that the approach developed here obtains identical results to those found using PIC-based methods under a Brownian motion model [24] for this biological example ($r_{PLS} = 0.813$; $P_{rand} = 0.008$).

Discussion

A common feature of organisms is that some traits covary with one another, a pattern termed morphological integration. Patterns of morphological integration may be evaluated at differing levels of biological organization (developmental, ontogenetic, evolutionary), but when characterized across species, the phylogenetic relationships among taxa must be taken into account [24]. In this paper we outlined how the evolutionary covariance matrix obtained from phylogenetic generalized least squares may be utilized to estimate the degree of phylogenetic morphological integration between two sets of variables. Using computer simulations under Brownian motion we found that the approach has appropriate Type I error and statistical power. This represents a necessary in-depth assessment of the statistical performance of phylogenetically-informed PLS methods. This approach is shown to be capable of evaluating patterns of morphological integration across differing numbers of species and for traits of different dimensionality. We then examined morphological integration between the cranium and mandible in *Plethodon* salamanders, and found significant evolutionary integration between the two structures across taxa. We also provide an implementation of this approach in the R programming language. This application of phylogenetically-informed PLS may be used to evaluate not only morphological integration (correlation between two sets of landmark coordinates) but also correlation between morphology and other multivariate data, such as ecological or behavioral variables.

The method evaluated here provides a useful complement to other approaches that evaluate morphological integration using phylogenetic independent contrasts [24]. Indeed, we have demonstrated that when implemented properly, both methods will yield identical results for the same dataset (assuming a Brownian motion model of evolution), though complications from polytomies are avoided with the PGLS approaches. Further, at least in theory, the approach proposed here may be used to estimate the degree of phylogenetic morphological integration under a broader set of evolutionary models that characterize the tempo and mode of evolution under distinct evolutionary processes. However, to date, the statistical framework for evaluating alternative evolutionary models such as Ornstein-Uhlenbeck using highly multi-dimensional data has not yet been developed. Nonetheless, evaluating evolutionary morphological integration using the PGLS-derived evolutionary covariation matrix is a logical and flexible generalization of the PIC-based approach.

Finally, this procedure provides a tool for assessing macroevolutionary hypotheses regarding morphological integration. Phenotypic integration can influence patterns of variability and evolvability [10,11,23,85,86]. Accordingly, this concept has been hypothesized to be a factor in the evolution of morphological and lineage diversity, such as in adaptive radiations [23,25,26,33]. To this end, the statistical framework herein represents a powerful method for estimating the degree of morphological integration and covariation in comparative datasets while taking in to account the non-independence of taxa.

Supporting Information

File S1 Computer Code for R. The function estimates the degree of phylogenetic morphological covariation between two sets of variables using partial least squares. The observed value is statistically assessed using phylogenetic permutation, where data for one block are permuted across the tips of the phylogeny, an estimate of the covariation between sets of variables, and compared to the observed value.

Acknowledgments

We thank E. Sherratt, C. Klingenberg, C. Fruciano and several anonymous reviewers for comments on earlier versions of this manuscript.

Author Contributions

Conceived and designed the experiments: DCA RF. Performed the experiments: DCA RF. Analyzed the data: DCA RF. Wrote the paper: DCA RF.

References

1. Adams DC, Rohlf FJ, Slice DE (2004) Geometric morphometrics: ten years of progress following the 'revolution'. Italian Journal of Zoology 71: 5–16.
2. Adams DC, Rohlf FJ, Slice DE (2013) A field comes of age: Geometric morphometrics in the 21st century. Hystrix 24: 7–14.
3. Olson EC, Miller RL (1951) A mathematical model applied to a study of the evolution of species. Evolution 5: 325–338.
4. Olson EC, Miller RL (1958) Morphological Integration. Chicago: University of Chicago Press.
5. Bookstein FL, Gunz P, Mitteroecker P, Prossinger H, Schaefer K, et al. (2003) Cranial integration in Homo: singular warps analysis of the midsagittal plane in ontogeny and evolution. Journal of Human Evolution 44: 167–187.
6. Klingenberg CP (2008) Morphological integration and developmental modularity. Annual Review of Ecology, Evolution, and Systematics 39: 115–132.
7. Magwene PM (2001) New tools for studying integration and modularity. Evolution 55: 1734–1745.
8. Mitteroecker P, Bookstein FL (2007) The conceptual and statistical relationship between modularity and morphological integration. Systematic Biology 56: 818–836.
9. Cheverud JM (1996) Developmental integration and the evolution of pleiotropy. American Zoologist 36: 44–50.
10. Chernof B, Magwene PM (1999) Morphological integration: forty years later. In: E.C O, Miller RL, editors. Morphological integration. Chicago: University of Chicago Press. 316–360.
11. Magwene PM (2009) Statistical methods for studying modularity: a reply to Mitteroecker and Bookstein. Systematic Biology 58: 146–149.
12. Goswami A (2006) Morphological integration in the carnivoran skull Evolution 60: 169–183.
13. Klingenberg CP, Badyaev AV, Sowry SM, Beckwith NJ (2001) Inferring developmental modularity from morphological integration: analysis of individual variation and asymmetry in bumblebee wings. American Naturalist 157: 11–23.
14. Laffont R, Renvoisé E, Navarro N, Alibert P, Montuire S (2009) Morphological modularity and assessment of developmental processes within the vole dental row (Microtus arvalis, Arvicolinae, Rodentia). Evolution and Development 11: 302–311.
15. Mitteroecker P, Bookstein FL (2008) The evolutionary role of modularity and integration in the hominid cranium. Evolution 62: 943–958.
16. Mitteroecker P, Gunz P, Meubauer S, Muller G (2012) How to explore morphological integration in human evoluion and development? Evolutionary Biology 39: 536–553.
17. Parsons KJ, Márquez E, Albertson RC (2012) Constraint and opportunity: the genetic basis and evolution of modularity in the cichlid mandible. American Naturalist 179: 64–78.
18. Young RL, Badyaev AV (2006) Evolutionary persistence of phenotypic integration: influence of developmental and functional relationships on complex trait evolution. Evolution 60: 1291–1299.
19. Zelditch ML, Wood AR, Bonett RM, Swiderski DL (2008) Modularity of the rodent mandible: integrating bones, muscles, and teeth. Evolution and Development 10: 756–768.
20. Gómez-Robles A, Polly PD (2012) Morphological integration in the hominin dentition: evolutionary, developmental, and functional factors. Evolution 66: 1024–1043.
21. Klingenberg CP, Duttke S, Whelan S, Kim M (2012) Developmental plasticity, morphological variation and evolvability: a multilevel analysis of morphometric

integration in the shape of compound leaves. Journal of Evolutionary Biology 25: 115–129.
22. Mitteroecker P, Gunz P, Bernhard M, Schaefer K, Bookstein FL (2004) Comparison of cranial ontogenetic trajectories among great apes and humans. Journal of Human Evolution 46: 679–698.
23. Monteiro LR, Nogueira MR (2010) Adaptive radiations, ecological specialization, and the evolutionary integration of complex morphological structures. Evolution 64: 724–744.
24. Klingenberg CP, Marugán-Lobón J (2013) Evolutionary covaration in geometric morphometric data: analyzing integration, modularity, and allometry in a phylogenetic context. Systematic Biology 62: 591–610.
25. Meloro C, Raia P, Carotenuto F, Cobb SN (2011) Phylogenetic signal, function and integration in the subunits of the carnivoran mandible. Evolutionary Biology 38: 465–475.
26. Nogueira MR, Peracchi AL, Monteiro LR (2009) Morphological correlates of bite force and diet in the skull and mandible of phyllostomid bats. Functional Ecology 23: 715–723.
27. Arnqvist G, Rowe L (2002) Correlated evolution of male and female morphologies in water striders. Evolution 56: 936–947.
28. Rüber L, Adams DC (2001) Evolutionary convergence of body shape and trophic morphology in cichlids from Lake Tanganyika. Journal of Evolutionary Biology 14: 325–332.
29. Felsenstein J (1985) Phylogenies and the comparative method. American Naturalist 125: 1–15.
30. Garland TJ, Ives AR (2000) Using the past to predict the present: confidence intervals for regression equations in phylogenetic comparative methods. American Naturalist 155: 346–364.
31. Harvey PH, Pagel MD (1991) The comparative method in evolutionary biology. Oxford: Oxford University Press.
32. Revell LJ, Collar DC (2009) Phylogenetic analysis of the evolutionary correlation using likelihood. Evolution 63: 1090–1100.
33. Dornburg A, Sidlauskas B, Santini F, Sorenson L, Near TJ, et al. (2011) The influence of an innovative strategy on the phenotypic diversification of triggerfish (family: Balistidae). Evolution 65: 1912–1926.
34. Santanta SE, Lofgren SE (2013) Does nasal echolocation influence the modularity of the mammal skull? Journal of Evolutionary Biology 26: 2520–2526.
35. Revell LJ, Harrison AS (2008) PCCA: a program for phylogenetic canonical correlation analysis. Bioinformatics 24: 1018–1020.
36. Bookstein FL, Streissguth A, Sampson P, Barr H (1996) Exploiting redundant measurement of dose and behavioral outcome: new methods from the teratology of alcohol. Developmental Psychology 32: 404–415.
37. Rohlf FJ, Corti M (2000) The use of partial least-squares to study covariation in shape. Systematic Biology 49: 740–753.
38. Blomberg SP, Lefevre JG, Wells JA, Waterhouse M (2012) Independent contrasts and PGLS regression estimators are equivalent. Systematic Biology 61: 382–391.
39. Rohlf FJ (2001) Comparative methods for the analysis of continuous variables: geometric interpretations. Evolution 55: 2143–2160.
40. Monteiro LR (2013) Morphometrics and the comparative method: studying the evolution of biological shape. Hystrix 24: 25–32.
41. Butler MA, King AA (2004) Phylogenetic comparative analysis: a modeling approach for adaptive evolution. American Naturalist 164: 683–695.
42. Hansen TF (1997) Stabilizing selection and the comparative analysis of adaptation. Evolution 51: 1341–1351.
43. Adams DC, Rohlf FJ (2000) Ecological character displacement in *Plethodon*: biomechanical differences found from a geometric morphometric study. Proceedings of the National Academy of Sciences, USA 97: 4106–4111.
44. Edwards AWF, Cavalli-Sforza LL (1964) Reconstruction of evolutionary trees. In: Heywood VH, McNeill J, editors. Phenetic and phylogenetic classification. London: Systematics Assocaition Publications. 67–76.
45. Felsenstein J (1973) Maximum-likelihood estimation of evolutionary trees from continuous characters. Americal Journal of Human Genetics 25: 471–492.
46. Felsenstein J (2004) Inferring phylogenies. Sunderland, MA.: Sinauer Associates.
47. Felsenstein J (1988) Phylogenies and quantitative characters. Annual Review of Ecology and Systematics 19: 445–471.
48. Revell LJ, Harmon LJ (2008) Testing quantitative genetic hypotheses about the evolutionary rate matrix for continuous characters. Evolutionary Ecology Research 10: 311–331.
49. Martins EP, Hansen TF (1997) Phylogenies and the comparative method: a general approach to incorporating phylogenetic information into the analysis of interspecific data. American Naturalist 149: 646–667.
50. O'Meara BC, C.Ane, Sanderson MJ, Wainwright PC (2006) Testing for different rates of continuous trait evolution using likelihood. Evolution 60: 922–933.
51. Adams DC (2013) Comparing evolutionary rates for different phenotypic traits on a phylogeny using likelihood. Systematic Biology 62: 181–192.
52. Adams DC (2014) Quantifying and comparing phylogenetic evolutionary rates for shape and other high-dimensional phenotypic data. Systematic Biology 63: 166–177.
53. Revell LJ (2008) On the analysis of evolutionary change along single branches in a phylogeny. American Naturalist 172: 140–147.
54. Thomas GH, Freckleton RP, Székely T (2006) Comparative analyses of the influence of developmental mode on phenotypic diversification rates in shorebirds. Proceedings of the Royal Society, B 273: 1619–1624.
55. Hohenlohe PA, Arnold SJ (2008) MIPoD: A hypothesis-testing framework for microevolutionary inference from patterns of divergence. American Naturalist 171: 366–385.
56. Blomberg SP, Garland T, Ives AR (2003) Testing for phylogenetic signal in comparative data: behavioral traits are more labile. Evolution 57: 717–745.
57. Adams DC (2014) A generalized Kappa statistic for estimating phylogenetic signal from shape and other high-dimensional multivariate data. Systematic Biology: In Review.
58. Klingenberg CP (2009) Morphometric integration and modularity in configurations of landmarks: tools for evaluating a priori hypotheses. Evolution and Development 11: 405–421.
59. Fruciano C, Franchini P, Meyer A (2013) Resampling-Based Approaches to Study Variation in Morphological Modularity. PLoS ONE 8: e69376.
60. Fischer-Rousseau L, Cloutier R, Zelditch ML (2009) Morphological integration and developmental progress during fish ontogeny in two contrasting habitats. Evolution and Development 11: 740–753.
61. Monteiro LR, Bonato V, Reis SFd (2005) Evolutionary integration and morphological diversification in complex morphological structures: Mandible shape divergence in spiny rats (Rodentia, Echimyidae). Evolution and Development 7: 429–439.
62. Klingenberg CP (2013) Cranial integration and modularity: insights into evolution and development from morphometric data. Hystrix 24: 43–58.
63. Lieberman DE, McBratney BM, Krovitz G (2002) The evolution and development of cranial form in *Homo sapiens*. Proceedings of the National Academy of Sciences, USA 99: 1134–1139.
64. Jaeger RG, Forester DC (1993) Social behavior of plethodontid salamanders. Herpetologica 49: 163–175.
65. Anthony CD, Wicknick JA, Jaeger RG (1997) Social interactions in two sympatric salamanders: Effectiveness of a highly aggressive strategy. Behaviour 134: 71–88.
66. Nishikawa KC (1985) Competition and the evolution of aggressive behavior in two species of terrestrial salamanders. Evolution 39: 1282–1294.
67. Jaeger RG, Kalvarskya ND, Shimizu N (1982) Territorial behaviour of the red-backed salamander: Expulsion of intruders. Animal Behavior 30: 490–496.
68. Maglia AM, Pyles RA (1995) Modulation of prey-capture behavior in *Plethodon cinereus* (Green) (Amphibia: Caudata). Journal of Experimental Zoology 272: 167–183.
69. Adams DC (2000) Divergence of trophic morphology and resource use among populations of *Plethodon cinereus* and *P. hoffmani* in Pennsylvania: a possible case of character displacement. In: Bruce RC, Jaeger RJ, Houck LD, editors. The biology of Plethodontid salamanders. New York: Klewer Academic/Plenum. 383–394.
70. Adams DC (2011) Quantitative genetics and evolution of head shape in *Plethodon* salamanders. Evolutionary Biology 38: 278–286.
71. Adams DC (2010) Parallel evolution of character displacement driven by competitive selection in terrestrial salamanders. BMC Evolutionary Biology 10: 1–10.
72. Adams DC, West ME, Collyer ML (2007) Location-specific sympatric morphological divergence as a possible response to species interactions in West Virginia *Plethodon* salamander communities. Journal of Animal Ecology 76: 289–295.
73. Arif S, Adams DC, Wicknick JA (2007) Bioclimatic modelling, morphology, and behaviour reveal alternative mechanisms regulating the distributions of two parapatric salamander species Evolutionary Ecology Research 9: 843–854.
74. Deitloff J, Petersen JD, Adams DC (2013) Complex species interactions lead to unpredictable outcomes in *Plethodon*. Herpetologica 62: 1–10.
75. Maerz JC, Myers EM, Adams DC (2006) Trophic polymorphism in a terrestrial salamander. Evolutionary Ecology Research 8: 23–35.
76. Myers EM, Adams DC (2008) Morphology is decoupled from interspecific competition in *Plethodon* salamanders in the Shenandoah Mountains. Herpetologica 64: 281–289.
77. Bookstein FL (1991) Morphometric tools for landmark data: geometry and biology. Cambridge: Cambridge University Press.
78. Adams DC (1999) Methods for shape analysis of landmark data from articulated structures. Evolutionary Ecology Research 1: 959–970.
79. Rohlf FJ, Slice DE (1990) Extensions of the Procrustes method for the optimal superimposition of landmarks. Systematic Zoology 39: 40–59.
80. Wiens JJ, Engstrom TN, Chippendale PT (2006) Rapid diversification, incomplete isolation, and the 'speciation clock' in North American salamanders (genus: Plethodon): testing the hybrid swarm hypothesis of rapid radiation. Evolution 60: 2585–2603.
81. Klingenberg CP (2011) MorphoJ: an integrated software package for geometric morphometrics. Molecular Ecology Resources 11: 353–357.
82. R Development Core Team (2013) R: a language and environment for statistical computing. Version 3.0.1. Available: http://cran.R-project.org R Foundation for Statistical Computing, Vienna.
83. Adams DC, Otárola-Castillo E (2012) geomorph: Software for geometric morphometric analyses. R package version 1.1–4. Available: http://CRAN.R-project.org/package = geomorph.

84. Adams DC, Otárola-Castillo E (2013) geomorph: an R package for the collection and analysis of geometric morphometric shape data. Methods in Ecology and Evolution 4: 393–399.
85. Clune J, Mouret JB, Lipson H (2013) The evolutionary origins of modularity. Proceedings of The Royal Society B Biological Sciences 280: 20122863.
86. Hallgrímsson B, Willmore K, Hall BK (2002) Canalization, developmental stability, and morphological integration in primate limbs. American Journal of Physical Anthropology 119: 131–158.

18

Evolution of Rapid Development in Spadefoot Toads Is Unrelated to Arid Environments

Cen Zeng[1], Ivan Gomez-Mestre[2], John J. Wiens[3]*

1 Department of Biology II, University of Munich, Munich, Germany, 2 Ecology, Evolution, and Development Group, Doñana Biological Station, Consejo Superior de Investigaciones Científicas, Seville, Spain, 3 Department of Ecology and Evolutionary Biology, University of Arizona, Tucson, Arizona, United States of America

Abstract

The extent to which species' life histories evolve to match climatic conditions is a critical question in evolutionary biology and ecology and as human activities rapidly modify global climate. GIS-based climatic data offer new opportunities to rigorously test this question. Superficially, the spadefoot toads of North America (Scaphiopodidae) seem to offer a classic example of adaptive life-history evolution: some species occur in extremely dry deserts and have evolved the shortest aquatic larval periods known among anurans. However, the relationships between the climatic conditions where spadefoots occur and the relevant life-history traits have not been explicitly tested. Here, we analyzed these relationships using GIS-based climatic data, published life-history data, and a time-calibrated phylogeny for pelobatoid frogs. Surprisingly, we find no significant relationships between life-history variables and precipitation or aridity levels where these species occur. Instead, rapid development in pelobatoids is strongly related to their small genome sizes and to phylogeny.

Editor: Hector Escriva, Laboratoire Arago, France

Funding: The authors thank the Elite Researcher Training Program of the Ministry of Education of China for funds allowing C. Zeng to visit J.J.W. and work on this project. The funders had no role in study design, data collection and analysis, decision to publish, or preparation of the manuscript.

Competing Interests: The authors have declared that no competing interests exist.

* E-mail: wiensj@email.arizona.edu

Introduction

Variation in climate over space and time may be an important factor driving evolutionary changes in life-history among and within species [1,2]. Studying this relationship between climate and life-history evolution has taken on new urgency as climate has begun to change rapidly and impact natural populations [3–6]. The combination of GIS-based climatic data and phylogenetic comparative methods now provides the opportunity to rigorously test hypotheses relating climatic variation to life-history variation among species (e.g. [7–9]). However, to our knowledge, no studies have used this approach to test for the environmental correlates of developmental rates.

The spadefoot toads of North America (Scaphiopodidae) seem to offer a classic example of adaptation in life-history variables to extreme climatic conditions. Although frog species richness is strongly correlated with mesic environments [10], scaphiopodid spadefoot toads occur in all the desert regions of North America, and some species occur in the driest regions within these deserts [11]. Seemingly in association with this environment, they have extremely short larval periods [12], and one species (*Scaphipus couchii*) that occurs in the driest regions of North America is thought to have the shortest aquatic larval period among all the > 6,000 species of frogs [13,14]. Spadefoot toads spend much of the year underground but are active on the surface during rainy periods (in summer for most species), when they emerge to forage and breed in temporary pools filled by rain [11]. These pools often dry quickly, and spadefoot toads appear to have evolved very rapid development to allow their eggs to develop and hatch, and the aquatic larvae to grow and metamorphose, before these pools dry [15–18]. It seems intuitive that regions with lower precipitation would tend to have smaller temporary pools that dry out more quickly (given their smaller size), which could lead to a strong relationship between macro-climatic precipitation levels where species occur and their rates of larval development. Many previous authors have noted that rapid pond drying leads to high tadpole mortality and that development is rapid in desert-dwelling tadpoles, and that rapid development may therefore be an adaptation allowing survival in these climates (e.g. [13–15,19]). However, the seven species of scaphiopodid spadefoot toads occur in a variety of habitats across North America, from arid deserts to mesic temperate forests [11,20]. No previous study has explicitly tested whether their rapid developmental rates are actually related to occurrence in more arid environments using explicit climatic data and phylogenetic comparative methods. For example, Buchholz & Hayes [18] suggested that developmental traits in pelobatoids were related to phylogeny rather than habitat, but without data on climate or use of phylogeny-based tests.

Here, we test the relationships between environmental conditions and life-history traits among species of pelobatoid frogs. Based on current classifications [21,22], the pelobatoid frogs include the scaphiopodids (North American spadefoot toads) and the pelobatids (Eurasian spadefoot toads) and two other families that interdigitate among the two clades of spadefoot toads (the Eurasian Pelodytidae and Asian Megophryidae). We synthesize existing data in the literature on relevant developmental traits in these species, specifically larval period and hatching time. We also include available data on genome size, given that small genome size is associated with rapid development in many organisms [23], although this has not been tested in a phylogenetic context in frogs (to our knowledge). We then test how these traits are related to environmental conditions where these species occur. We obtain GIS-based climatic data from georeferenced localities for these species, focusing on variables most likely to determine the water available from rainfall for larval development (annual precipitation, precipitation of the wettest quarter, precipitation seasonality, aridity). We then analyze these data in the context of a time-calibrated phylogeny. Specifically, we test whether pelobatoid species occurring in drier environments have shorter hatching times and larval periods, and whether shorter hatching times and larval periods are related to smaller genome sizes. We also test how life-history traits, genome size, and climatic distributions are related to the phylogeny. Our study also generates a well-supported, multi-locus, time-calibrated Bayesian phylogeny for scaphiopodids, providing a resource for comparative studies on this model system in evolution, ecology, development, and behaviour [14,17,24–29]. Our results show that evolution of rapid development in spadefoot toads is not related to occurrence in drier climates, but that there are significant relationships between developmental rates and genome size.

Materials and Methods

Life-history data

We searched the literature for data on relevant life-history variables for all pelobatoid species (Appendix S1), starting from the summary provided in Gomez-Mestre & Buchholz [24]. Whenever possible, we used only data measured in the field under natural conditions. For those few species lacking field-based data for a specific variable, data from the lab were used instead. We obtained data on larval period (from hatching of the eggs to approximately Gosner stage 42) and hatching time (from egg deposition to hatching). For each variable, when multiple records were available for one species, we obtained the maximum and minimum values from the available records and calculated the midpoint. Data were available for most scaphiopodid, pelobatid, and pelodytid species, but relevant data were available for only two megophryid species. We note that megophryids are geographically and climatically distinct from the other pelobatoid families, occurring primarily in mesic tropical and subtropical areas of Asia [21].

Data on genome sizes were obtained from T. R. Gregory's database (http://www.genomesize.com/). Data were available for only eight pelobatoid species, but these species included two from each genus for all the genera in the Pelobatidae, Pelodytidae, and Scaphiopodidae. Summary life-history data for each species for each variable are provided in Table S1.

Climatic data

To obtain climatic data for each species, we first obtained georeferenced locality data. We used species distribution data from HerpNET (www.herpnet.org) and the Global Biodiversity Information Facility (GBIF; www.gbif.org; Version 1.2.6). HerpNET and GBIF both provide a frequently updated database of museum specimen records. We searched each database for each species, and then combined the records into a set of unique localities for each species. Sample sizes ranged from 3 to 322 localities per species (mean = 89.1).

Localities for each species were visualized using DIVA-GIS version 7.5.0.0 and compared to species distribution maps from the IUCN Red List of Threatened Species, version 2012.2 [30]. The known distributions of these species are relatively stable and agreed upon by different sources [11,20,30]. Localities falling outside the IUCN distribution map were excluded. We also excluded localities with estimated elevations (from WorldClim, see below) that fell outside the range of reported elevations from IUCN.

Using this carefully vetted set of georeferenced localities, we then obtained climatic data from the WorldClim (version 1.3) database [31]. Data are based on averages from weather stations from the years ~1950–2000 (with spatial interpolation to localities between weather stations), with a spatial resolution of ~1 km^2. For each locality, we extracted data on annual precipitation (Bio12), precipitation seasonality (Bio15), and precipitation of the wettest quarter (Bio16). We expect these variables to have the strongest influence on rainfall available for filling temporary ponds for anuran reproduction. Maximum and minimum values across localites within the species range were obtained, as well as the midpoint of these two values, and the mean value for the species averaged across all localities.

We also used a measure of aridity per se, following Oufiero *et al.* [32]. For each species, this was calculated as

$$\log 10(Q), \text{where}: Q = P/((Tmax + Tmin)(Tmax - Tmin))x1000$$

where P is annual precipitation (mm; mean across localities across species range), Tmax is the maximum value of Bio5 (maximum temperature of warmest month) across the species range, and Tmin is the lowest value of Bio6 (minimum temperature of the coldest month). Arid environments have a lower Q [33]. We also used logQ2, in which Tmax is the mean value of Bio10 (mean temperature of warmest quarter) across localities and Tmin is the mean value of Bio11 (mean temperature of coldest quarter), but this gave similar results. A summary of the climatic data for each species is provided in Table S1.

Time-calibrated phylogeny

We estimated a time-calibrated phylogeny for pelobatoid frogs, since one including all relevant taxa was not available at the time we initiated our study (and one available now has some issues, see below). We first reduced the data matrix compiled by Pyron & Wiens [22] to include only the 16 pelobatoid species for which life-history data were available. However, these data were available for only two megophryid species, and life-history data were available for *Leptobrachium nigrops* but not sequence data. Therefore, rather than exclude this species and genus, we used sequence data from *Leptobrachium chapaense* to represent *L. nigrops* in the tree. We also excluded genes sampled for fewer than 4 included species. The resulting data matrix included 16 species and data from the mitochondrial ribosomal genes 12S and 16S, the mitochondrial protein-coding gene cytochrome *b*, and the nuclear protein-coding genes H3A, RAG1, RHO, SIA, and SLC8A3 (total length of combined alignments = 9,355 base pairs). Data were not available for all 8 genes for all 16 species, and some species were therefore missing data for some genes. However, both simulations and empirical studies suggest that including some missing data need not lead to inaccurate estimates of phylogeny, especially when a large number of characters is sampled overall (review in [34]).

The time-calibrated tree was estimated using the Bayesian uncorrelated lognormal approach in BEAST 1.5.4 [35,36]. We used the GTR + I + Γ model (following [22]) with 4 rate categories for Γ, and estimated base frequencies. We used a clock model with the uncorrelated, lognormal approach, and an estimated rate. The starting tree was based on a Yule speciation prior.

We initially used two fossil calibration points. For each point, we identified a fossil that represented the oldest taxon that could be confidently assigned to a clade of extant species. A fossil can be used to determine the minimum age of a clade, but the clade can be older than this oldest known fossil. We therefore used a lognormal prior distribution for each calibration point, with an offset equal to the minimum age of the oldest fossil (the youngest age of the oldest stratum in which it is found) and a mean of 5 and a standard deviation of 1 Myr. This combination of mean and standard deviation yields a 95% prior distribution that extends from just slightly older than the minimum age of the fossil to approximately 15 Myr older (an arbitrary but seemingly plausible range), with the highest probability slightly older than the age of the fossil.

We initially used the crown-group age of Pelobatoidea as being at least 50.3 Myr old, given a fossil scaphiopodid (*Scaphiopus guthriei*) from the Wind River formation (lower Eocene Wasatchian 50.3–55.4 Mya) following Rocek & Rage [37]. The 95% interval on the prior is 50.9–66.0 Myr.

We initially treated the crown-group age of Pelobatidae and Megophryidae as being at least 33.9 Myr old, given the fossil *Eopelobates grandis* which appears to be closely related to *Pelobates* [37,38], from the Chadron formation (33.9–38 Mya). The 95% interval on the prior for Pelobatidae+Megoprhyidae is 34.5–49.6 Myr. This clade may be older if the undescribed "Green River pelobatid" can be assigned to it (Wasatchian, 50.3–5.4 Mya; [37]). Further, the most recent common ancestor of Pelodytidae, Pelobatidae, and Megophryidae is also at least 33.9 Mya, given fossil *Pelodytes* from the late Eocene (37.2–33.9 Mya; [37]), but we did not use this fossil calibration (given that the pelobatid + megophryid calibration ensures that this clade is at least this old).

Initial analyses yielded family-level topologies that did not match those of Pyron & Wiens [22], most likely due to the lack of outgroups. Given that these family-level relationships are generally well supported when outgroups are included (e.g. [22,39–41]), we constrained these relationships. Specifically we constrained the clade: Pelodytidae+Pelobatidae+Megophryidae. Given that the pelobatid+megophryid clade is a fossil constraint, these two constraints enforce the Pyron & Wiens [22] topology for families. Importantly, the same set of family-level relationships is also found in other previous analyses of pelobatoid relationships, including those based on mitochondrial data only [39], nuclear data only [40], and combined nuclear and mitochondrial data [41].

We performed two independent runs each with 50,000,000 generations sampled every 1,000 generations. We used the maximum clade credibility trees with mean node heights. The first 10% of generations sampled were discarded as burn-in using TreeAnnotator version 1.5.4 and viewed using FigTree version 1.3.1 [35]. We confirmed that the two independent runs gave effective sample sizes (ESS) greater than 200 for the likelihood and selected clade ages, and that they converged on similar topologies and divergence dates. Trees from the two analyses were combined to yield a majority-rule consensus tree with mean branch lengths.

This initial analysis yielded estimated ages for Pelobatoidea and the family-level clades within it that were considerably younger than those estimated in previous studies (e.g. [40–42]). For example, the pelobatoid crown group was 53 Myr old in this tree, and ~130, 170 and 150 Myr old (respectively) in these previous studies. Therefore, we reran the analyses as above, but making two changes. First, for the crown age of Pelobatoidea, we used a normal prior distribution with a mean age of 150 Mya and a standard deviation of 10 Myr (95% prior interval: 133.6–166.4). This prior interval roughly corresponds to the range of estimated ages in previous studies. Second, we used the fossil calibration for *Scaphiopus guthrei* for the crown-group age of Scaphiopodidae, rather than the stem-group age (this choice is less conservative about the placement of this fossil but more in line with previous age estimates).

This second set of results gave an identical topology and similar relative branch lengths to the first analysis, but with absolute branch lengths (ages) similar to those estimated in previous studies. We used this second BEAST tree for our phylogenetic comparative analyses. The topology was very strongly supported, with only one node with a posterior probability <0.95 (Fig. 1). Therefore, we did not incorporate uncertainty in the phylogeny into our comparative analyses. This topology is available in nexus/newick format in Appendix S2.

Note that another study has recently estimated a large-scale time-calibrated tree for amphibians, including pelobatoids [43]. However, given the large number of taxa included in that study, the use of a somewhat suboptimal method for estimating divergence dates was necessary (penalized likelihood; [44]). Furthermore, that study [43] relied entirely on secondary calibration points (from [42]). Therefore, we prefer our estimate of divergence dates. Nevertheless, these estimates are actually quite similar (and similar to estimates in other recent studies [7,40–42]), and are based on nearly identical molecular datasets [22].

Phylogenetic comparative analysis

We tested the relationship between pairs of variables using phylogenetic generalized least squares (PGLS; [45]) as implemented in the R package *caper*, version 0.5 [46]. Prior to conducting these analyses, we found the best-fitting evolutionary model for each variable using the R packages *ape* [47] and *geiger* [48]. We compared the fit of the models using the estimated likelihood and Akaike information criterion (AIC), with an AIC difference of 4 or greater indicating support for alternative models [49]. We compared the Brownian motion (BM; perfect fit of a character to the phylogeny), Ornstein-Uhlenbeck (OU; equivalent to stabilizing selection around a single optimum), and estimated lambda (level of phylogenetic signal is estimated) models (see Table S2). We found that in most cases, model fit was similar between the OU and lambda models (AIC difference <4), with the exception of two climatic variables for which OU was strongly favored (annual precipitation, wettest quarter precipitation). We therefore used the lambda model in PGLS, given that this model was either favored and/or alternate models were not. We also performed a set of analyses using the OU model. Specifically, we repeated the PGLS analyses after using *geiger* to transform the tree based on the OU model and the estimated value of alpha. To estimate alpha, the selected variable was fitted to the OU model 10 times and the alpha with the minimum deviance (-2 * log-likelihood) was applied in the transformation [50]. PGLS results were generally similar using the lambda and OU models, and we present the results using the lambda model as our primary results. We present the OU results as supplementary information (Table S4).

We used PGLS to test the following specific hypotheses. (1) We predicted that overall hatching times and larval periods of species will be related to their mean values for climatic variables (annual precipitation, precipitation of the wettest quarter, aridity), assuming that species reduce their hatching times and larval period to allow them to metamorphose before temporary breeding ponds dry in more arid climates. For these analyses, we summarized variation in hatching times and larval periods within species based on midpoint values (midpoint between lowest and highest values within species). (2) We predicted that the minimum (shortest) hatching times and larval periods within species will be related to the lowest values for annual precipitation related variables across the range of each species. This second set of analyses was intended to address the possibility that overall species values might not reflect variation within species, and that we

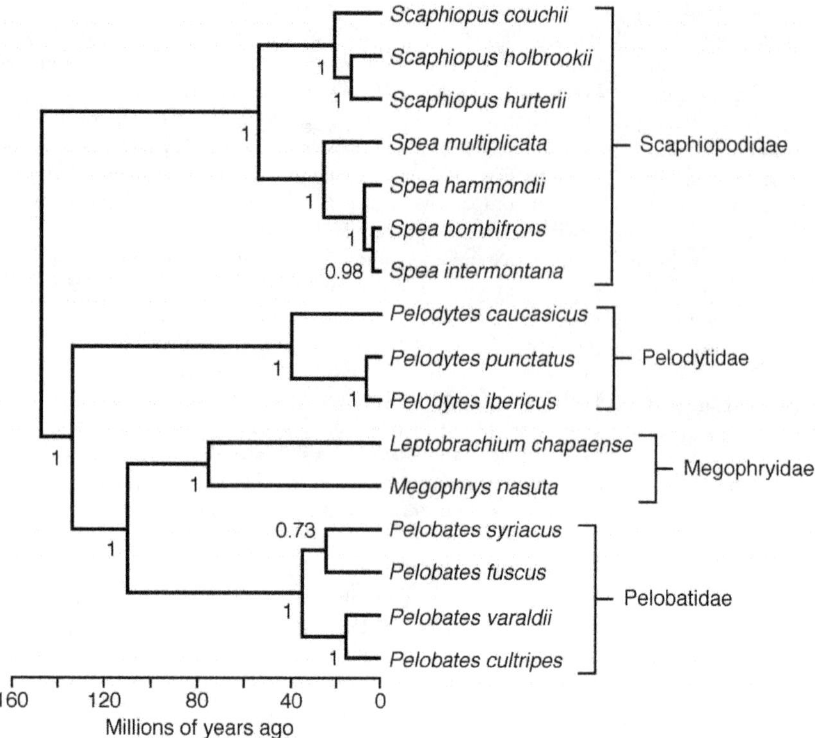

Figure 1. Time-calibrated phylogeny of pelobatoid frogs used in comparative analyses. Numbers at nodes indicate Bayesian posterior probabilities of clades.

expect shorter larval periods and hatching times in parts of the species range with lower rainfall. (3) We predicted that hatching times and larval period would be related, assuming that species occurring in drier environments will evolve to minimize both simultaneously. We examined both midpoint and minimum values for these variables within species. (4) We predicted that genome sizes would be smaller in species with more rapid development (shorter hatching times and larval periods).

We acknowledge that the methods described above could lead to a potential mismatch between developmental traits and climatic variables for specific localities (e.g. for a given species, the locality with the shortest recorded larval period may not correspond to the driest locality where the species occurs). Therefore, we performed an additional analysis in which both developmental and climatic values for each species were based on a single locality with the shortest recorded field-based larval period for that species (focusing on annual precipitation). However, there were some issues in this analysis. First, the data on larval period for four species could not be traced to specific localities (i.e. *Pelodytes caucasica, Pelobates syriacus, Pelobates varaldii, Megophrys nasuta*). For these species, we used data on the minimum recorded larval period (from the field) and the lowest annual precipitation across sampled localities. In several other species, it was not possible to trace the shortest recorded larval period known for that species to a specific locality. In these cases, we simply used the shortest larval period for a given species that could be traced to a specific locality. Although these larval periods were sometimes longer than the shortest larval periods recorded for that species, they may also be more reliable, and should provide a strong overall test of how climate and larval period are related. The data and references used are summarized in Appendix S3 and Table S3.

We also tested each variable for phylogenetic signal using lambda [51]. Given that the phylogenetic results were very strongly supported (Fig. 1) and similar to previous estimates (see above), we did not test the robustness of the results of the comparative analyses to alternative trees.

Results

Surprisingly, we find no relationship between larval period and the climatic variables nor between hatching time and the climatic variables (Table 1), using either mean/midpoint or minimum values. There is also no relationship when using data on larval period and climate from specific localities (Table 1). There is no significant relationship between midpoint hatching time and midpoint larval period, but there is a strong relationship using the minimum values within species for both variables (Fig 2a; Table 1). There are strong relationships between genome size and minimum hatching times and between genome size and minimum larval period (Fig. 2b,c), but not between genome size and mipdoint hatching times. Results are generally similar using the OU model (Table S4), especially in the non-significant relationships between climate and developmental rates. However, the relationships between minimum hatching time and minimum larval period and minimum hatching time and minimum genome size are no longer significant under the OU model (although the relationship between midpoint hatching time and larval period is), but OU is not the best fitting model for any of these variables, and so these results should not be preferred to our main results using the lambda model. Most variables (Table 2) show significant but not perfect phylogenetic signal (lambda $= 0.4$–0.6), except for precipitation seasonality (lamba <0.01) and genome size (lambda >0.95).

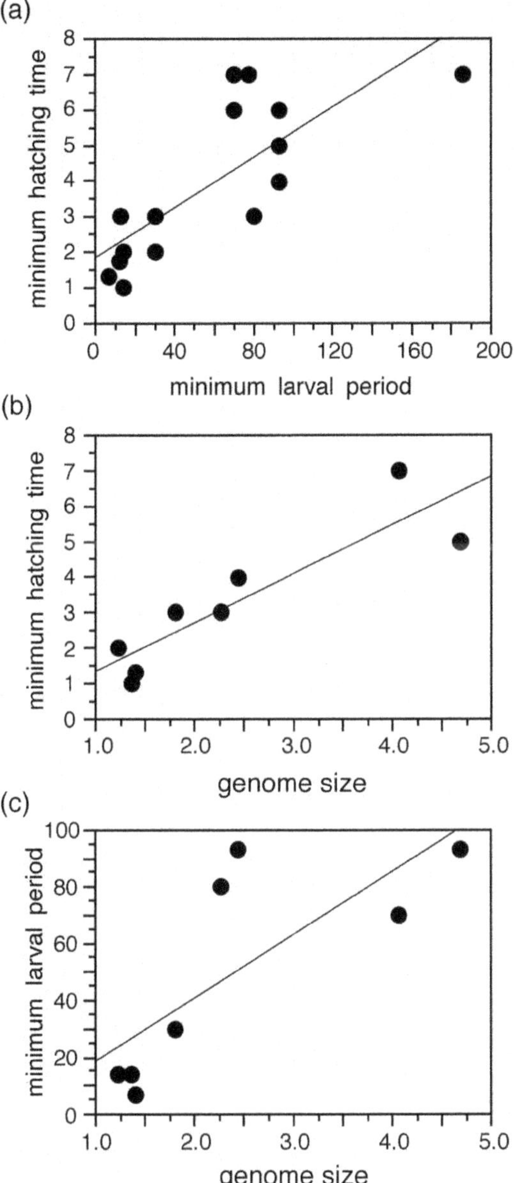

Figure 2. Relationships between selected life-history variables and genome size (from among the significant relationships in Table 1). For ease of visualization, we plot the raw data and standard regression lines (see Table 1 for PGLS results). Larval periods and hatching times are given in days; genome sizes are given as C-values in picograms.

Discussion

In this study, we use explicit GIS-based climatic data and phylogenetic comparative methods to test the hypothesis that short developmental times in spadefoot toads are associated with occurrence in more arid environments. Surprisingly, we find no relationship between developmental rates (larval period and hatching times) and climate in the geographic areas where these species occur. Instead, we find strong relationships between our two measures of developmental rates (Table 1), between developmental rates and genome sizes (Table 1), and between phylogeny and developmental rates, genome size, and most climatic variables (Table 2). Our conclusions about developmental rates and climate are largely consistent with those of Buchholz & Hayes [18], but are based on explicit statistical analyses of phylogeny and climate.

Why do we find no relationship between climate and developmental rates? The first question to address is whether the absence of this relationship is real or an artifact of our methods or data. We think that the most important source of error in our study is that our between-species analyses require reducing all variation among populations within a species to a single value for that species, and analyzing only values among species. Thus, it might be that within-species variability obscures between-species patterns. For example, a strong relationship between developmental rate and climate may only arise in dry parts of species ranges, and might be obscured by including life-history and climatic data from other parts of the species range. However, we still find no relationship when using minimum values for larval period and hatching time and the driest values for climatic variables across the species range, instead of midpoints and means (although we acknowledge that the localities for climate and life-history in these latter analyses are not precisely matched). In addition, when we do match data on larval period and climate for a specific locality for each of several species, we again find no relationship. Furthermore, we do find significant relationships between minimum larval periods and minimum hatching times, suggesting that significant relationships between life-history variables can be captured using our data and methods.

We argue that the lack of a strong relationship between climate and life-history instead reflects real patterns that are inconsistent with aridity and short development times being closely related. For example, inspecting the raw species data (Table S1), there are dramatic differences in life history between scaphiopodids (short development times) and pelobatids and pelodytids (longer developmental times), despite the overlapping distributions of climatic variables among these families. Furthermore, two of the three species of *Scaphiopus* (*S. holbrookii, S. hurterii*) occur in relatively mesic environments but nevertheless have relatively fast developmental times, as does the more arid-dwelling *S. couchii.* There are also relatively long development times in species that occur in relatively arid environments, such as *Spea intermontana* and *Spea hammondii.* In summary, these patterns help explain why no significant relationship between dry climates and rapid development was observed.

We note that these two main explanations for the lack of relationship between climate and life history are not mutually exclusive. Specifically, there are some patterns among pelobatoid species that are clearly inconsistent with a tight relationship between developmental times and climate. Nevertheless, there may still be important within-species variation in developmental times and climate that may reflect adaptive evolution, but that our between-species approach is relatively insensitive to. Similarly, phenotypic plasticity and local-scale conditions of temporary pools are also known to play an important role in determining developmental rates in spadefoot toads (e.g. [13–17,29]). However, it also appears that variation within species occurs within limited bounds, and that among species variation is much greater than variation within species (Table S1).

We also emphasize that our focus here is on the question of whether short development times are associated with occurrence in regions with dry climate. However, this is not the same as asking whether short larval periods are associated with use of temporary pools of short duration. In fact, species in mesic regions might select pools with short duration (e.g. *Scaphiopus holbrookii*), whereas species in more arid regions with longer development times may utilize more permanent aquatic larval sites (e.g. *Spea hammondii,*

Table 1. Relationships between climatic and life-history variables in pelobatoid frogs using phylogenetic generalized least squares (PGLS).

Variables	R^2	P-value
midpoint larval period ~ mean annual precipitation	0.0180	0.7768
midpoint larval period ~ mean precip. wettest quarter	0.0275	0.6802
midpoint larval period ~ mean precip. seasonality	0.0002	0.9964
midpoint larval period ~ aridity (logQ)	0.0147	0.8141
midpoint hatching time ~ mean annual precipitation	0.1016	0.2659
midpoint hatching time ~ mean precip. wettest quarter	0.0559	0.4833
midpoint hatching time ~ mean precip. seasonality	0.1564	0.1288
midpoint hatching time ~ aridity (logQ)	0.0949	0.2903
minimum larval period ~ min. annual precipitation	0.0026	0.9640
minimum hatching time ~ min. annual precipitation	0.0373	0.6158
larval period ~ annual precipitation (specific localities)	0.0063	0.9160
midpoint hatching time ~ midpoint larval period	0.0987	0.2760
minimum hatching time ~ minimum larval period	**0.5981**	**0.0001**
midpoint larval period ~ genome size	**0.6245**	**0.0124**
minimum larval period ~ genome size	**0.6566**	**0.0089**
midpoint hatching time ~ genome size	0.3159	0.1405
minimum hatching time ~ genome size	**0.7860**	**0.0017**

Significant relationships ($P<0.05$) are boldfaced. Median larval periods and median hatching times refers to the midpoint between the highest and lowest values reported for a species (Appendix S1). Minimum refers to the lowest value. For climatic variables, mean refers to the mean among localities for a species, and min. the lowest value among localities within a species.

Spea intermontana; [17]). Such patterns may help explain the weak relationship between large-scale climate and development time.

In contrast to the comparisons with climate, our results show a strong relationship between genome sizes and developmental rates in pelobatoids (Table 1; Fig. 2b,c). This result is supported despite the limited number of pelobatoid species with available data on genome sizes (8 species vs. 16 species for most other variables). To our knowledge, no previous studies have tested for a relationship between developmental rates and genome size in anurans using phylogenetic comparative methods. However, various non-phylogenetic studies have been performed that suggested a relationship between DNA content and larval period (e.g. [52–56]). Further, phylogenetic comparative analyses in salamanders also suggested a relationship between embryonic period (equivalent to hatching time here) and DNA content [57,58].

The exact causal relationships between genome size and developmental rate in pelobatoids are unclear. One hypothesis is that rapid development is difficult with larger genomes, leading to evolution of smaller genome sizes in rapidly developing species. Gregory [23] suggested that large genome size acts as a constraint on rapid development, but that other factors drive the evolution of developmental rate besides genome size.

Intriguingly, we find that many climatic variables and life-history traits show relatively strong relationships with the phylogeny (based on values of Pagel's [51] lambda; Table 2), but genome size shows the strongest relationship of all (lambda > 0.95). We speculate that genome size may act to constrain

Table 2. Estimation of phylogenetic signal in the traits analyzed here, based on fit to a Brownian motion model of trait evolution using Pagel's [51] lambda.

Variable	Lambda
Mean annual precipitation	0.6426
Mean precip. wettest quarter	0.6631
Mean precip. seasonality	3.88E-07
Aridity (logQ)	0.7264
Minimum annual precipitation	0.5061
Midpoint larval period	0.7435
Midpoint hatching time	0.4605
Genome size	0.9627

Lambda varies from 0 to 1, with higher values indicating stronger phylogenetic signal.

evolutionary changes in developmental rates among species, and might help underlie the relationship between trait variation and phylogeny seen in traits relating to developmental rate.

In this study, we show that developmental rates in spadefoot toads are not significantly related to occupation of relatively arid environments, despite the observation that some pelobatoid species with very fast rates occur in very dry environments. We show instead that these measures of developmental rates are significantly related to each other, to phylogenetic history, and to genome size. We note that our results do not rule out the possibility of strong relationships between climate and life history among populations within species nor an important role for phenotypic plasticity and local-scale conditions in determining developmental rates within populations.

Supporting Information

Table S1 Summary of life-history and climatic data for each species.

Table S2 Comparison of three evolutionary models for each climatic and life-history variable.

Table S3 Summary of life-history and climatic data for each species using matched localities for climatic and developmental data for 12 of the 16 species (see Appendix S3).

Table S4 Relationships between climatic and life-history variables in pelobatoid frogs using phylogenetic generalized least squares (PGLS), utilizing the OU model.

Appendix S1 Data on life history variables in pelobatoid frogs and associated literature references.

Appendix S2 Time-calibrated phylogeny used in the comparative analyses in nexus/newick format.

Appendix S3 Data on larval period matched to specific localities.

Acknowledgments

We thank the Elite Researcher Training Program of the Ministry of Education of China for funds allowing C. Zeng to visit J.J.W. and work on this project. We thank D. Buckley and an anonymous referee for helpful feedback on the manuscript.

Author Contributions

Conceived and designed the experiments: CZ IGM JJW. Performed the experiments: CZ JJW. Analyzed the data: CZ JJW. Wrote the paper: CZ IGM JJW.

References

1. Stearns SC (1992) The evolution of life histories. Oxford: Oxford University Press.
2. Roff DA (2002) Life history evolution. Sunderland: Sinauer Associates, Sunderland, Massachusetts.
3. Visser ME, Van Noordwijk AJ, Tinbergen JM, Lessells CM (1998) Warmer springs lead to mistimed reproduction in great tits (*Parus major*). Proc. R. Soc. Lond. B 265: 1867–1870.
4. Husby A, Kruuk LEB, Visser ME (2009) Decline in the frequency and benefits of multiple brooding in great tits as a consequence of a changing environment. Proc. R. Soc. Lond. B 276: 1845–1854.
5. Sinervo B (2010) Erosion of lizard diversity by climate change and altered thermal niches. Science 328: 894–899.
6. Lane JE, Kruuk LEB, Charmantier A, Murie JO, Dobson FS (2012) Delayed phenology and reduced fitness associated with climate change in a wild hibernator. Nature 489: 554–556.
7. Gomez-Mestre I, Pyron RA, Wiens JJ (2012) Phylogenetic analyses reveal unexpected patterns in the evolution of reproductive modes in frogs. Evolution 66: 3687–1700.
8. Lambert SM, Wiens JJ (2013) Evolution of viviparity: a phylogenetic test of the cold-climate hypothesis in phrynosomatid lizards. Evolution 67: 1715–1728.
9. Bonett R., Steffen M, Lambert SM, Wiens JJ, Chippindale PT (2014) Evolution of paedomorphosis in plethodontid salamanders: ecological correlates and re-evolution of metamorphosis. Evolution 68: 466–482.
10. Buckley LB, Jetz W (2007) Environmental and historical constraints on global patterns of amphibian richness. Proc. R. Soc. Lond. B 274: 1167–1173.
11. Stebbins RC (2003) A field guide to western reptiles and amphibans. Third edition. New York: Houghton-Mifflin Company.
12. Wright AH, Wright AA (1949) Handbook of frogs and toads of the United States and Canada. Ithaca: Cornell Univ. Press.
13. Newman RA (1987) Effects of density and predation on *Scaphiopus couchii* tadpoles in desert ponds. Oecologia 71: 301–307.
14. Kulkarni S, Gomez-Mestre I, Moskalik C., Storz B, Buchholz D (2011) Evolutionary reduction of developmental plasticity in desert spadefoot toads. J. Evol. Biol. 24: 2445–2455.
15. Newman RA (1988) Adaptive plasticity in development of *Scaphiopus couchii* tadpoles in desert ponds. Evolution 42: 774–783.
16. Morey S, Reznick D (2000) A comparative analysis of plasticity in larval development in three species of spadefoot toads. Ecology 81: 1736–1749.
17. Morey SR, Reznick DN (2004) The relationship between habitat permanence and larval development in California spadefoot toads: field and laboratory comparisons of developmental plasticity. Oikos 104: 172–190.
18. Buchholz DR, Hayes TB (2002) Evolutionary patterns of diversity in spadefoot toad metamorphosis (Anura: Pelobatidae). Copeia 2002: 180–189.
19. Mayhew WW (1965) Adaptations of the amphibian, *Scaphiopus couchi*, to desert conditions. Amer. Midl. Natur. 55: 95–109.
20. Conant RC, Collins JT (1998) Reptiles and amphibians of eastern and central North America, 2nd edn. Boston: Houghton-Mifflin.
21. Vitt LJ, Caldwel JP (2009) Herpetology: an introductory biology of amphibians and reptiles. Third edition. San Diego: Academic Press.
22. Pyron RA, Wiens JJ (2011) A large-scale phylogeny of Amphibia including over 2,800 species, and a revised classification of extant frogs, salamanders, and caecilians. Mol. Phylogenet. Evol. 61: 543–583.
23. Gregory TR (2002) Genome size and developmental complexity. Genetica 115: 131–146.
24. Gomez-Mestre I, Buchholz DR (2006) Developmental plasticity mirrors differences among taxa in spadefoot toads linking plasticity and diversity. Proc. Natl. Acad. Sci.U.S.A 103: 19021–19026.
25. Pfennig KS (2007) Facultative mate choice drives adaptive hybridization. Science 318: 965–967.
26. Pfennig DW, Martin RA (2010) Evolution of character displacement in spadefoot toads: different proximate mechanisms in different species. Evolution 64: 2331–2341.
27. Ledón-Rettig C, Pfennig DW (2011) Emerging model systems in eco-evo-devo: the environmentally responsive spadefoot toad. Evol. Devol. 13: 391–400.
28. Bazazi S, Pfennig KS, Handegard NO, Couzin ID (2012) Vortex formation and foraging in polyphenic spadefoot toad tadpoles. Behav. Ecol. Sociobiol. 66: 879–889.
29. Gomez-Mestre I, Kulkarni S, Buchholz DR (2013) Mechanisms and consequences of developmental acceleration in tadpoles responding to pond drying. PLOS ONE 8: e84266.
30. IUCN (2012) The IUCN Red List of Threatened Species Version 2012.2. [Database]. URL: www.iucnredlist.org.
31. Hijmans RJ, Cameron SE, Parra JL, Jones PG, Jarvis A (2005) Very high resolution interpolated climate surfaces for global land areas. Int. J. Climatol. 25: 1965–1978.
32. Oufiero CE, Adolph SC, Gartner GEA, Garland T, Jr. (2011) Latitudinal and climatic variation in body size and dorsal scale rows in *Sceloporus* lizards: a phylogenetic perspective. Evolution 65: 3590–3607.
33. Tieleman BI, Williams JB, Bloomer P (2002) Adaptation of metabolism and evaporative water loss along an aridity gradient. Proc. R. Soc. Lond. B 270: 207–214.
34. Wiens JJ, Morrill MC (2011) Missing data in phylogenetic analysis: reconciling results from simulations and empirical data. Syst. Biol. 60: 719–731.
35. Drummond A, Rambaut A (2007) BEAST: Bayesian evolutionary analysis by sampling trees. BMC Evol. Biol. 7: 214–221.

36. Drummond AJ, Ho SYW, Phillips MJ, Rambaut A (2006) Relaxed phylogenetics and dating with confidence. PLoS Biol. 4: e88.
37. Rocek Z, Rage J-C (2000) Tertiary Anura of Europe, Africa, Asia, North America, and Australia. In: H Heatwole, RL Carroll, eds. Amphibian Biology. Vol. 4. Chipping Norton, Australia: Surrey Beatty. pp. 1333–1387.
38. Yuan W, Keqin G, Xing X (2000) Early evolution of discoglossid frogs: new evidence from the Mesozoic of China. Naturwissenschaften 87: 417–420.
39. Garcia-Paris M, Buchholz DR, Parra-Olea G (2003) Phylogenetic relationships of Pelobatoidea re-examined using mtDNA. Mol. Phylogenet. Evol. 28: 12–23.
40. Wiens JJ (2007) Global patterns of diversification and species richness in amphibians. Am. Nat. 170: S86–S106.
41. Roelants K, Gower DJ, Wilkinson M, Loader SP, Biju SD, et al. (2007) Global patterns of diversification in the history of modern amphibians. Proc. Nat. Acad. Sci. U.S.A. 104: 887–892.
42. Wiens JJ (2011) Re-evolution of lost mandibular teeth in frogs after more than 200 million years, and re-evaluating Dollo's law. Evolution 65: 1283–1296.
43. Pyron RA, Wiens JJ (2013) Large-scale phylogenetic analyses reveal the causes of high tropical amphibian diversity. Proc. R. Soc. Lond. B280: 20131622
44. Sanderson MJ (2002) Estimating absolute rates of molecular evolution and divergence times: a penalized likelihood approach. Mol. Biol. Evol. 19: 101–109
45. Martins EP, Hansen TF (1997) Phylogenies and the comparative method: a general approach to incorporating phylogenetic information into the analysis of interspecific data. Am. Nat. 149: 646–667.
46. Orme D, Freckleton R, Thomas G, Petzoldt T, Fritz S, et al. (2012) CAPER: Comparative Analyses of Phylogenetics and Evolution in R. R package version 0.5. http,//CRAN.R-project.org/package = caper.
47. Paradis E, Claude J, Strimmer K (2004) APE: analyses of phylogenetics and evolution in R language. Bioinformatics 20: 289–290.
48. Harmon LJ, Weir J, Brock C, Glor RE, Challenger W (2008) GEIGER: a statistical package for investigating evolutionary radiation in a comparative context. Bioinformatics 24: 129–131.
49. Burnham KP, Anderson DR (2002) Model selection and multimodel inference: a practical information-theoretic approach, 2nd ed. New York: Springer-Verlag.
50. Hansen TF (1997) Stabilizing selection and the comparative analysis of adaptation. Evolution 51: 1341–1351.
51. Pagel M (1999) Inferring the historical patterns of biological evolution. Nature 401: 877–884.
52. Goin OB, Goin CJ, Bachmann K (1968) DNA and amphibian life history. Copeia 1968: 532–540.
53. Bachmann K (1972) Nuclear DNA and developmental rate in frogs. Quart. J. Florida Acad. Sci. 35: 225–231.
54. Oeldorf E, Nishioka M, Bachmann K (1978) Nuclear DNA amounts and developmental rate in Holarctic Anura. Z. Zool. Syst. Evolutionsforsch. 16: 216–224.
55. Horner HA, Macgregor HC (1983) C-value and cell volume: their significance in the evolution and development of amphibians. J. Cell Sci. 63: 135–146.
56. Camper JD, Ruedas LA, Bickham JW, Dixon JR (1993) The relationship of genome size with developmental rates and reproductive strategies in five families of neotropical bufonoid frogs. Genet. (Life Sci. Adv.) 12: 79–87.
57. Pagel M, Johnstone RA (1992) Variation across species in the size of the nuclear genome supports the junk-DNA explanantion for the C-value paradox. Proc. R. Soc. Lond. B 249: 119–124.
58. Jockusch EL (1997) An evolutionary correlate of genome size change in plethodontid salamanders. Proc. R. Soc. Lond. B 264: 597–604.

19

A Tri-Component Conservation Strategy Reveals Highly Confident MicroRNA-mRNA Interactions and Evolution of MicroRNA Regulatory Networks

Chen-Ching Lin[1], **Ramkrishna Mitra**[1], **Zhongming Zhao**[1,2,3,4]*

1 Department of Biomedical Informatics, Vanderbilt University School of Medicine, Nashville, Tennessee, United States of America, **2** Department of Psychiatry, Vanderbilt University School of Medicine, Nashville, Tennessee, United States of America, **3** Department of Cancer Biology, Vanderbilt University School of Medicine, Nashville, Tennessee, United States of America, **4** Center for Quantitative Sciences, Vanderbilt University School of Medicine, Nashville, Tennessee, United States of America

Abstract

MicroRNAs are small non-coding RNAs that can regulate expressions of their target genes at the post-transcriptional level. In this study, we propose a tri-component strategy that combines the conservation of microRNAs, homology of mRNA coding regions, and conserved microRNA binding sites in the 3′ untranslated regions to discover conserved microRNA-mRNA interactions. To validate the performance of our conservation strategy, we collected the experimentally validated microRNA-mRNA interactions from three databases as the golden standard. We found that the proposed strategy can improve the performance of existing target prediction algorithms by approximately 2–4 fold. In addition, we demonstrated that the proposed strategy could efficiently retain highly confident interactions from the intersection results of the existing algorithms and filter out the possible false positive predictions in the union one. Furthermore, this strategy can facilitate our ability to trace the homologues in different species that are targeted by the same miRNA family because it combines these three features to identify the conserved miRNA-mRNA interactions during evolution. Through an extensive application of the proposed conservation strategy to a study of the miR-1/206 regulatory network, we demonstrate that the target mRNA recruiting process could be associated with expansion of miRNA family during its evolution. We also uncovered the functional evolution of the miR-1/206 regulatory network. In this network, the early targeted genes tend to participate in more general and development-related functions. In summary, the conservation strategy is capable of helping to highlight the highly confident miRNA-mRNA interactions and can be further applied to reveal the evolutionary features of miRNA regulatory network and functions.

Editor: Zhang Zhang, Beijing Institute of Genomics, Chinese Academy of Sciences, China

Funding: This work was partially supported by National Institutes of Health grants (R01LM011177, R03CA167695, P30CA68485, P50CA095103, and P50CA090949) and Ingram Professorship Funds (to ZZ). The funders had no role in study design, data collection and analysis, decision to publish, or preparation of the manuscript.

* Email: zhongming.zhao@vanderbilt.edu

Introduction

MicroRNAs (miRNAs) are small, highly conserved non-coding RNA molecules that are ~22 nucleotides in length and are involved in numerous biological processes, such as development, differentiation, and growth [1–4]. By complementarily binding to target mRNA transcripts, miRNAs can trigger gene down-regulation or translational repression [5,6]. So far, multiple algorithms have been developed for miRNA target prediction, and these algorithms vary from each other in their uses of additional refining strategies [4,7–9]. For example, miRanda measures the thermodynamic stability between miRNAs and their putative target mRNAs [7,8,10], TargetScan searches the conserved seed pairing regions in the 3′ untranslated regions (UTRs) of genes using whole genome alignment [4,11], and mimiRNA incorporates the expression profiles of miRNAs and mRNAs [9]. Among these algorithms, TargetScan has been reported to possess more robust prediction performance in various cellular systems [12]. One major reason for TargetScan's superior execution is its utilization of conservation information across species, which can efficiently reduce the number of false positive predictions [13].

Recently, the evolution of miRNAs has been studied extensively [14–18]. A miRNA is rarely lost during evolution once it has been established in a species [14–18]. The low secondary loss rate of miRNAs during evolution has been successfully applied to investigate the phylogeny of eukaryotic organisms [16,18,19]. Collectively, these studies indicated that the majority of miRNAs are highly conserved. Therefore, the conservation of miRNAs should be included in the identification of conserved miRNA-mRNA interactions. After reviewing several target prediction strategies, it became apparent that sequence conservation criteria in miRNA binding regions could increase overall precision and achieve better performance [12]. TargetScan reportedly possesses superior target prediction performance because of its utilization of conservation information; however, a high false positive miRNA-target prediction rate was also observed [20,21]. Hence, an advanced conservation-based strategy that can accomplish improved target prediction performance is necessary. During miRNA evolution, the conserved miRNA-mRNA interactions may derive from the conservation traits of (1) miRNA, (2) coding region of target mRNA, and (3) miRNA binding sites in the 3′ UTR of the target mRNA. Therefore, an appropriate strategy to identify highly conserved miRNA-mRNA interactions should incorporate all three components into its algorithm to fully take into account the miRNA regulatory mechanisms. In this study, we proposed a conservation strategy to incorporate these three components into existing algorithms. This strategy combined miRNA conservation, mRNA coding region homology, and conserved miRNA binding sites in the 3′ UTRs into miRNA target predictions (Fig. 1). This conservation-based strategy was then used to discover the conserved miRNA-mRNA interactions at a large scale and investigate the evolution of the miRNA regulatory network and functions. Using the experimentally validated miRNA-mRNA interactions as the gold standard, we found that our strategy could improve the performance of the existing miRNA target prediction algorithms, including TargetScan. Finally, through an extensive application of our strategy to study the evolution of the miR-1/206 family, we demonstrated the evolutionary connections between this miRNA family and its regulatory network. Intriguingly, an evolutionary development (evo-devo) characteristic was observed in this network.

Methods

The sequences of mature miRNAs and 3′ UTR of mRNAs

The mature miRNA sequences from eight species, Caenorhabditis elegans, Drosophila melanogaster, Danio rerio, Xenopus tropicalis, Ornithorhynchus anatinus, Bos taurus, Mus musculus, and Homo sapiens, were obtained from miRBase Release 19 [22]. The 3′ UTR sequences of mRNAs in the above eight species and the homologous genes across species were obtained from Ensembl BioMart [23].

MicroRNA-mRNA interactions

In this study, three algorithms, TargetScan [24–26], miRanda [8], and MultiMiTar [27], were used to predict the possible miRNA-mRNA interactions in eight species independently. These three algorithms use different information on miRNA target prediction. TargetScan focuses on seed complementary [22–24]; miRanda considers the thermodynamic properties between miRNA mature sequence and binding sites on target mRNA 3′ UTR [8]; MultiMiTar is a machine-learning based method that utilizes important miRNA-targeting specificity features from both the seed and out of seed interacting regions [27]. Besides using these three algorithms separately, we also built other three combinations of putative target gene pools from the above three algorithms. The first combination is the intersection of predicted miRNA-mRNA interactions from these three algorithms. However, the intersection would be biased by the minimum putative target set. Thus, the union, which collected all predicted results of these three algorithms, was considered to be the second combination data set. The intersection and union are believed to have reduced false positive and false negative prediction results respectively. To better utilize these two data sets, we created a combined miRNA-target mRNA set from the intersection and union as the third combination. This combination is termed as "IntSec(hsa)," which combines the intersection interactions in humans with the union of those interactions in the other seven species. In other words, IntSec(hsa) possessed the most strict predicted results in the species of interest, i.e. human, and the largest interaction set as evolutionary references in other species. In addition, IntSec(hsa) can be used to test if the proposed strategy is capable of filtering out the false positive reference interactions in other species while retaining the highly confident miRNA-mRNA interactions in the studied species.

Experimentally validated miRNA-mRNA interactions

To assess the performance of the proposed conservation strategy, we compiled an experimentally validated miRNA-mRNA interaction dataset from the union of three databases, TarBase V5.0 [28], miRecords [29], and miRTarBase V4.4 [30], as the gold standard. Finally, 21,849 experimentally validated miRNA-mRNA interactions in humans were collected and used.

The tri-component conservation strategy

In this study, we proposed a tri-component conservation strategy to discover conserved miRNA-mRNA interactions, to improve the performance of existing miRNA target prediction algorithms, and to investigate the evolution of miRNA regulatory networks (Fig. 1). This strategy combined the conservations of miRNAs, coding regions, and miRNA binding sites in the 3′ UTR. First, the evolutionarily conserved miRNA families were obtained from miRBase [22,31]. For one miRNA family conserved across n species, there would be at least n member miRNAs. This step groups evolutionarily conserved miRNAs into families. In one species, one target mRNA set can be predicted for and assigned to a mature miRNA by an existing algorithm. Therefore, for one miRNA family conserved across n species, up to n number of target sets in n species can be predicted by one algorithm. Then, one target mRNA of the miRNA family i and its orthologues, which were predicted as target mRNAs of the miRNA family i in other species, were considered to be conserved target mRNAs of the miRNA family i. This step selects target mRNAs with conserved coding regions. Accordingly, our strategy required the conserved miRNA binding sites located in the homologue genes' 3′ UTR and targeted by the members of one miRNA family. Consequently, the conserved miRNA-mRNA interactions with conserved miRNAs, target mRNA coding regions, and miRNA binding sites in the 3′ UTR can be identified by the tri-component conservation strategy.

Additionally, we further defined the conservation level of one conserved miRNA-mRNA interaction by the number of species in which this miRNA-mRNA interaction could be detected. For example, for a miRNA family with n species, a target mRNA that meets the criteria of the strategy in k species would be assigned a conservation level of k. To have further restriction, we required the conserved miRNA-mRNA interactions to be detected in both the oldest and youngest species; thus, k is from 2 to n.

Results

Improving miRNA target prediction using the tri-component conservation strategy

In this study, we developed a tri-component conservation strategy that combined the conservations of miRNAs, mRNA coding region, and miRNA binding sites in the 3′ UTR to predict highly conserved and confident miRNA-mRNA interactions (Fig. 1). This strategy was applied to three target prediction algorithms (TargetScan [24–26], miRanda [8], and MultiMiTar [27]) and three combination datasets (intersection, union, and

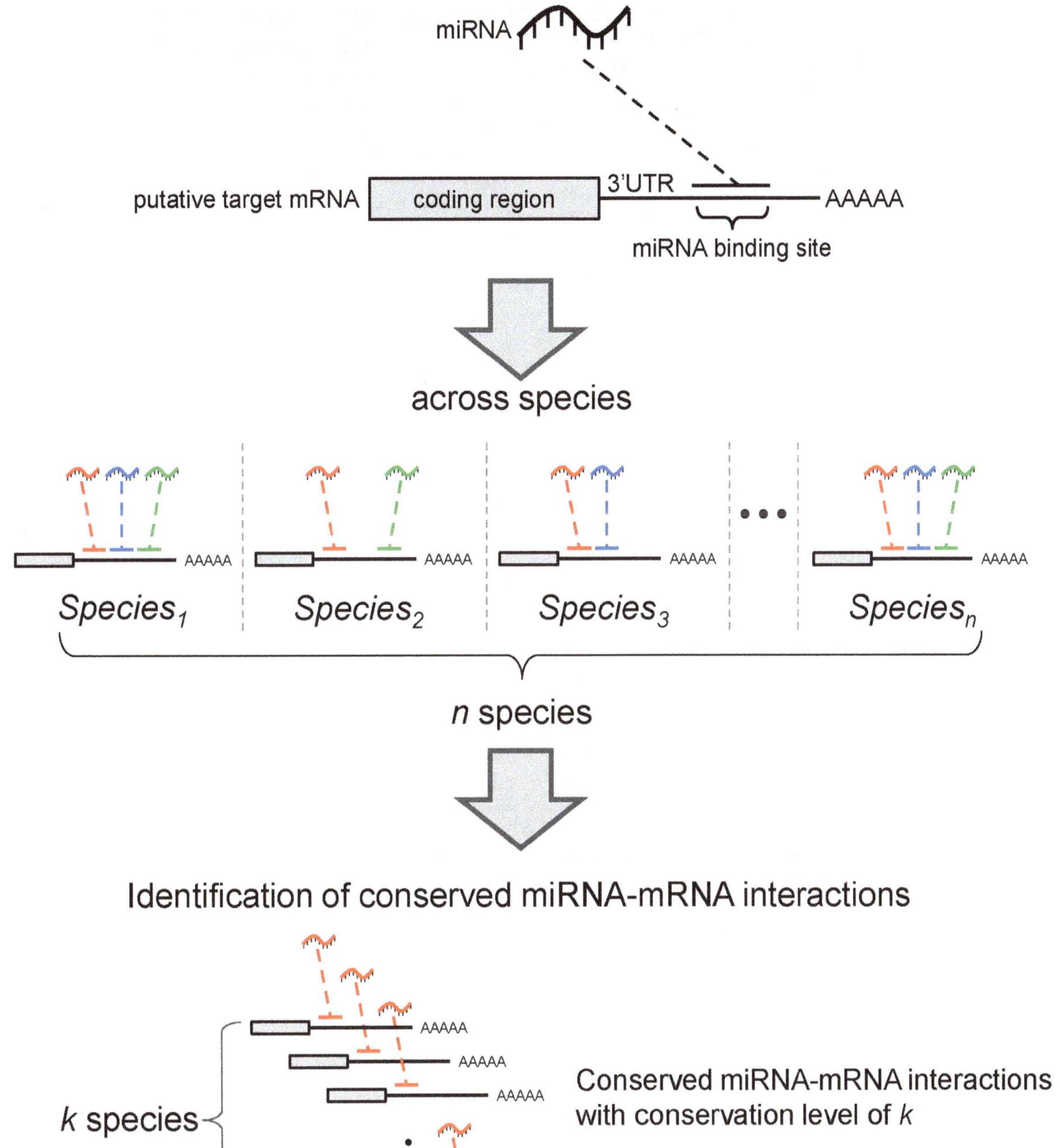

Figure 1. The tri-component conservation strategy scheme. The scheme of the proposed conservation strategy to identify the conserved miRNA-mRNA interactions is shown. The upper section shows the three major components of miRNA: the regulation-miRNA, mRNA coding region, and 3′ UTR of target mRNA. In the middle section, each color represents a member of one miRNA family. The putative target mRNAs are from homologues in each species. The lower section shows a miRNA-mRNA interaction conserved across *k* species. We further restricted the conserved miRNA-mRNA interactions that must be detected in both the oldest and youngest species; thus, *k* is from 2 to *n*.

IntSec(hsa)) across eight species (see Methods). Notably, the third combination dataset, IntSec(hsa), combines the intersection of three algorithms in humans and the union of three algorithms in the other seven species. Furthermore, a total of 21,849 experimentally validated miRNA-mRNA interactions in humans collected from three databases (TarBase V5.0 [28], miRecords [29], and miRTarBase V4.4 [30]) were used as the gold standard to evaluate the target prediction performance. The work-flow of our strategy was described in Figure S1.

After applying our conservation strategy, the precision and F-measure values substantially increased by 2–4 fold compared to the original algorithms and combination data sets, i.e. intersection, union, and IntSec(hsa) (Fig. 2). F-measure, which is the harmonic mean of precision and recall, was used to assess the overall prediction performance in this study. This improvement indicated that our conservation strategy could efficiently identify highly confident (experimentally validated) miRNA-mRNA interactions from the original algorithms. Importantly, the conservation strategy in our study could improve the performance of TargetScan, which also incorporated conservation information into its own algorithm. However, TargetScan used the UTRs of the reference species based on orthology; that is, it used the aligned genomic regions between the reference species genome and the studied species based on whole genome alignment [24,32]. In other words, the UTRs used by TargetScan in the reference species might not be a 3′ UTR of a gene. Different from TargetScan, our conservation strategy simultaneously considered the miRNA conservation, coding region homology, and conserved binding sites in the 3′ UTR of the (homologues) target mRNA. Accordingly, the overall improved performance elucidates that the underlying conservation strategy is useful to gain more confident results.

The conservation level of miRNA-mRNA interactions also affects the precision (Fig. 2). The conservation level is defined by the number of species in which this conserved miRNA-mRNA interaction can be detected. In most of the used data sets, as the conservation level decreased, the precision decreased and then became convergent after the conservation level of 6 (Fig. 2A). However, overall precision remained stronger than that of applying original algorithms only. This observation shows that the conservation strategy is very stable on predicting highly confident miRNA-mRNA interactions. In addition, this result also suggests that a higher conservation level could lead to a more precise prediction of the true miRNA-mRNA interactions. Notably, the F-measure of the intersection (IntSec) was dramatically decreased in the highest conservation level (Fig. 2B). This observation could be caused by overly stringent limitations on the intersection. However, the best F-measure was reached by applying our conservation strategy to IntSec(hsa). This observation also demonstrates that the conservation strategy can efficiently retain highly confident miRNA-mRNA interactions of the intersection in the studied species and filter out possible false positive predictions of the unions in other reference species.

The evolution of the miR-1/206 family regulatory network: an extensive application of the tri-component conservation strategy

In contrast to the other target prediction algorithms, our proposed conservation strategy combined three major components (1) conservation of miRNAs (2) orthologues of target genes, and (3) conserved miRNA binding sites in the 3′ UTR. We combined these three features to identify the conserved miRNA-mRNA interactions during evolution. This strategy facilitated our ability to trace the homologues in different species that are targeted by the same miRNA family. Due to this intrinsic advantage, the conservation strategy can be further applied to study the evolution of miRNA regulatory networks. In this study, we used the miR-1/206 family to demonstrate this application. The combination putative target gene dataset of IntSec(hsa) was used to perform this analysis.

MiR-1/206 is a highly conserved miRNA family from non-vertebrates to mammals (Fig. 3A) [18]. During its evolution, miR-1/206 branched into two subfamilies, miR-1 and miR-206 [33]. This observation of the highly similar mature sequences within each subfamily but relative dissimilarity between these two subfamilies (Fig. S2) warranted further investigation on their regulations, such as the gene networks regulated by these two subfamilies. Notably, the member miRNAs in the miR-1/206 family possess completely identical seed regions but different mature sequences (Fig. S2). Therefore, the miRNA-mRNA interactions predicted by the seed-based target prediction algorithms would be all the same between these two subfamilies. As a result, the evolution of miRNA-regulated networks between these two subfamilies can't be observed. Accordingly, the combination target mRNA dataset is very proper to be used to discover the evolution of networks regulated by miRNAs in the same family. Therefore, we studied the miR-1/206 regulatory network in humans to discover the connections between the evolutions of the miR-1/206 family and its regulatory network. The human target genes identified by the conservation strategy were further grouped by the most distant (targeted) species in which their homologues were targeted by miR-1/206 (Fig. 3B). In general, the most distant species with homologues of a human gene is considered the species with the farthest evolutionary distance. In this study, with the intrinsic advantage of the conservation strategy, we extensively define the most distant (targeted) species of a human miRNA-

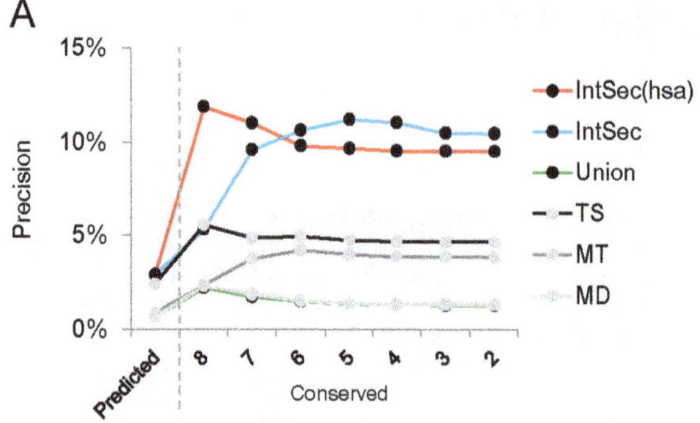

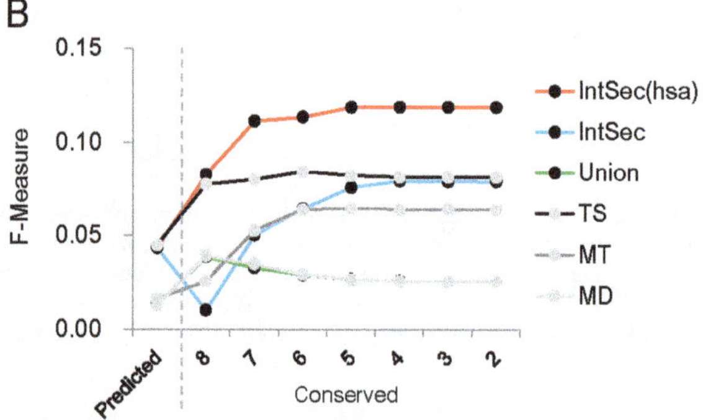

Figure 2. The improved performance of the conservation strategy. The performances of the conservation strategy and miRNA target prediction algorithms were evaluated by (A) precision and (B) F-measure. There are three algorithms (TS: TargetScan, MD: miRanda, and MT: MultiMiTar) and three combinations (IntSec: intersection, Union: union, and IntSec(hsa): intersection in humans with unions in other reference species). The results from the original algorithms/combinations were labeled "Predicted" (the left side of the dashed line). The results of the conserved miRNA-mRNA interactions identified by our strategy were labeled "Conserved" (the right side of the dashed line). The numbers along the X-axis indicate the conservation level of the conserved miRNA-mRNA interactions. Both the precision and F-measure are improved after applying the proposed conservation strategy. In two plots (2A and 2B), MD and union nearly overlap.

target gene as the species with the most distant homologues targeted by the same miRNA family. Notably, the number of target gene was dramatically increased from *D. melanogaster* to *D. rerio* (increased by 2.9 fold, Fig. 3B and Fig. S3) when the miR-1/206 family branched to two subfamilies. This observation suggested that the variety of mature miRNA in one miRNA family could be reflected by the changes in its regulatory network during evolution. A previous study also reported that the size of miRNA family could affect the accumulation of their conserved target genes [34].

The functional evolution of the miR-1/206 regulatory network was investigated as well. Functions of genes were annotated with their biological process category in Gene Ontology (GO) [35]. For each gene group, the involved functions that have $P \leq 0.05$ as derived from the hypergeometric test were defined as significantly enriched. In addition, significantly enriched functions were ranked by the number of annotated genes, and the top 10 significantly enriched functions were listed in Table S1. The representative functions were summarized from the top 10 enriched biological processes in each gene group (Table S1) and labeled according to the corresponding gene group (Fig. 3B). We observed a series of variations in miR-1/206 regulatory functions during its evolution. The development-related functions first evolved in *C. elegans*, *D. melanogaster*, and *D. rerio*, and the functions involved in stimulus response also evolved in *D. rerio*. The cellular transport/localization-related functions then evolved in *X. tropicalis*. In *O. anatinus* and *B. taurus*, the miR-1/206 family evolved to regulate metabolic processes in cells. Additionally, signaling pathway-related biological processes and two more specific functions, DNA replication proofreading and muscle organ development, evolved in *M. musculus*. Finally, in *H. sapiens*, miR-1/206 regulatory functions evolved into positive regulations of transcription/gene expression. More importantly, we observed an association between the evolution of miR-1/206 regulatory network and its regulatory development-related functions. During the evolution of the miR-1/206 regulatory network, "multicellular organismal development" first evolved in *C. elegans*. This biological process participates in the developmental progression of a multicellular organism from its initial stage to late stage. "System process," the function involved in the development of an organ system during a multicellular organismal process, evolved in *D. melanogaster*. Then, "organ development" and a more specific biological process, "muscle organ development," evolved in *D. rerio* and *M. musculus*, respectively. Interestingly, the miR-1/206 family had been found to play a key role in the development of muscle organs [36–38]. These observations suggested that, from older to younger species, miR-1/206 regulatory developmental functions have evolved from a drastic to a mild level, i.e., from organismal level to organ-specific (i.e., muscle). In other words, an evo-devo feature of miR-1/206 regulatory functions was revealed by applying the proposed conservation strategy. Of note, the association between the evolution of miRNAs and the organismal complexity had been recently reported [39,40]. Furthermore, investigating the GO level of enriched functions revealed that older target genes tend to be enriched in functions with a lower GO level (Fig. 3C). The evolutionary distances relative to *H. sapiens* were calculated by TimeTree [41] and represented with million years ago (Mya). In addition to using the top 10 enriched functions to perform the GO level analysis, analyses using the top 30, 20, and 10% enriched functions were also conducted and showed consistent conclusions (Fig. S4). In other words, early targeted genes of miR-1/206 family tend to participate in more general functions, and late ones tend to participate in more specific functions. However, genes with higher GO level might reflect more studies than those with lower GO level. To confirm this potential bias, we retrieved the number of publications for each gene from NCBI PubMed, which roughly reflects the extent of studies of the genes. We did not find the older targeted genes had more publications (Fig. S5). This preliminary analysis indicated no substantial bias on the extent of studies of each gene. In summary, this observation reconfirmed the evo-devo characteristic of miR-1/206 regulatory developmental functions from invertebrates to vertebrates and mammals.

Discussion

In this study, we proposed a tri-component conservation strategy to identify the conserved miRNA-mRNA interactions and demonstrated its ability to improve the performance of existing target prediction algorithms. The improved performance of the proposed conservation strategy implies that conserved miRNA-mRNA interactions might be highly confident [12]. Even though the conservation strategy improved the performance of the three miRNA target prediction algorithms, its precision and F-measure are still relatively low. The highest precision is about 12% as reached by IntSec(hsa) at the most stringent conservation level of 8 (Fig. 2A), and the best F-measure is 0.12, also reached by IntSec(hsa), at a moderate conservation level of 5 (Fig. 2B). The low F-measure might indicate a relatively higher false negative rate in our strategy. The inadequate performance may result from the small and incomplete experimentally validated miRNA-mRNA interaction dataset. To confirm this, we removed those miRNAs with <200 experimentally validated targets and re-calculated the precision. We found that the highest precision achieved 37% by IntSec(hsa) with the most stringent conservation level of 8. Using a pooled miRNA data set (miR-1, miR-30, miR-155, miR-16, and let-7b), Selbach *et al.* [12] reported the precision of their miRNA target prediction approach, pSILAC, was approximately 30–60%. Interestingly, the precision of IntSec(hsa) was 56% when using the same miRNA data set. These observations further confirmed our explanations and pointed out that the proposed strategy might be capable of obtaining highly confident miRNA-mRNA interactions from the existing prediction algorithms. The best performance was observed for IntSec(hsa). IntSec(hsa) combines the intersection target gene set in humans and the union in other species. The intersection dataset was the smallest with expectation to possess a high precision rate, while the union created the largest dataset with a high recall rate. In other words, IntSec(hsa) integrated the smallest but highly confident target set in humans with the target sets as large as possible in other species as reference. This combination achieved the best performance on predicting experimentally validated miRNA-mRNA interactions. Thus, this observation indicated that the conservation strategy had a robust trade-off between precision and recall. Moreover, through our strategy, the IntSec(hsa) could take advantages of both the intersection and union. The results of MD and the union datasets were almost the same (Fig. 2A and 2B), suggesting that the union dataset was dominated by the prediction results of MD. Additionally, species-specific miRNA-mRNA interactions might be omitted by the innate manipulation of the conservation strategy. This shortcoming could be improved by using a group of closely related species as a reference (e.g., using mammals or primates as the references to predict miRNA-mRNA interactions in humans). Briefly, our conservation strategy improves the performance of predicting highly confident miRNA-mRNA interactions. In addition, we applied the conservation strategy to study the evolution of the miR-1/206 family. This extensive application further revealed the evolutionary connections between

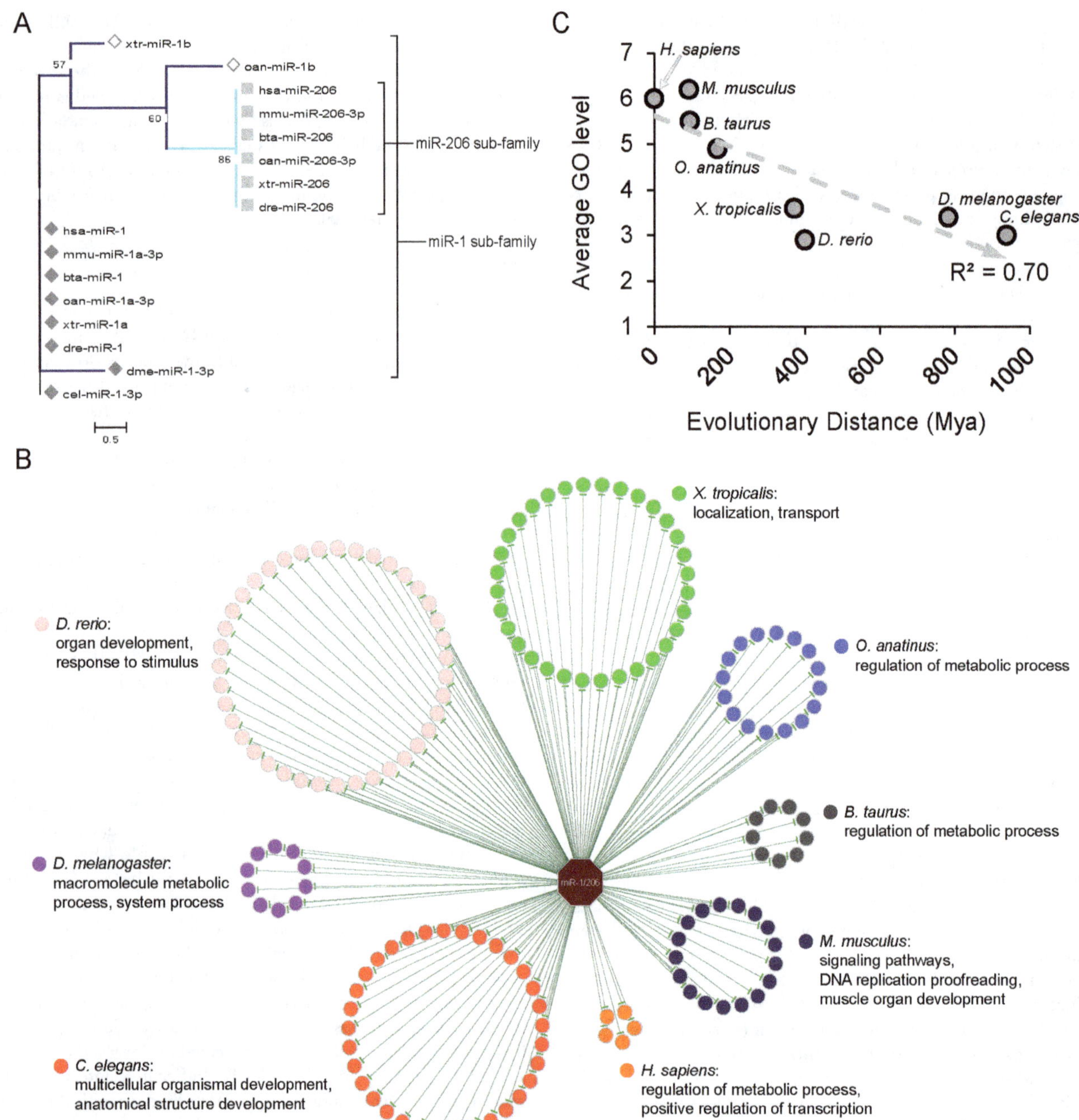

Figure 3. Evolutionary analyses of the miR-1/206 family regulatory network. (A) The phylogenetic tree of miR-1/206 family. This tree was drawn by MEGA 5.2.2 (Neighbor-Joining algorithm, 500 bootstrap replications) [42]. Blue: the branch of miR-1 subfamily; light blue: miR-206 subfamily. This tree shows that miR-1 subfamily existed before *C. elegans* and miR-206 subfamily before *D. rerio*. (B) The regulatory network of the miR-1/206 family in humans. The miR-1/206 family is represented by an octagon in the center of the network. Circles denote target genes of miR-1/206 in humans. Circle colors denote the most distant species in which the gene was targeted by the miR-1/206 family. The representative enriched functions specific to each species are listed under each species name. (C) The correlation between the Gene Ontology (GO) level of the top 10 enriched functions in miR-1/206 human target genes and the evolutionary distance. Target genes of older species tend to be enriched with more general biological functions, represented by lower levels of GO terms. (Mya: Million Years Ago).

the miR-1/206 family and its regulatory network and demonstrated the functional evolution of the miR-1/206 regulatory network.

Supporting Information

Figure S1 The work-flow of the tri-component conservation strategy. First, we obtained the mature miRNA sequences from miRBase 19 and 3′ UTR sequences from Ensembl BioMart for eight studied species. With the above two datasets, we run three existing target prediction algorithms [8,24–27] to produce putative miRNA-mRNA interactions (MMIs) for one studied species. In this study, human is the studied species. Consequently, for each species, we obtained three putative MMI sets from three existing algorithms. Furthermore, two combinational MMI sets, i.e., intersection and union, have been obtained. Next, we executed this target prediction process on eight studied species. After this step, there would be eight putative MMI sets for each algorithm or each combinational dataset. Next, we created IntSec(hsa) that was consisted of the intersection MMIs in humans and the union ones in the other seven species. We denoted these six MMI sets, i.e. TargetScan, miRanda, MultiMiTar, intersection, union, and IntSec(hsa), as combinations. Until here, we obtained eight putative MMI sets for each combination. Furthermore, for eight species, we obtained miRNA family from miRBase [22] and homologues information from Ensembl BioMart [23], respectively. The member miRNAs in one miRNA family are evolutionary conserved. Then, for each combination, we grouped putative target genes into homologues target gene sets across eight species. The MMIs, formed by genes in homologues target gene set and the member miRNAs of one miRNA family in different species, have been identified as the conserved MMIs of the corresponding miRNA family. The strategy was depicted in Fig. 1. Furthermore, the number of species in which the conserved MMI was formed has been denoted as its conservation level of the observed conserved target genes. To have further restriction, we required the conserved MMIs to be detected in both the oldest and youngest species of the homologues target gene set. Finally, we compiled an experimentally validated MMI set from the union of three databases, TarBase V5.0 [28], miRecords [29], and miRTarBase V4.4 [30]. Using this MMI set as the gold standard, we can evaluate the performance of each MMI combination.

Figure S2 The mature sequences of miR-1/206 family. This figure shows the mature sequences of miR-1/206 family. The background colors represented the different types of nucleotides. The RNAs in seed regions were colored in white.

Figure S3 The size variety of miR-1/206 regulatory network during evolution. The human target gene sizes in the most distant species were shown at y-axis. There is a dramatic increasing of target gene size in between *D. melanogaster* and *D. rerio*.

Figure S4 The correlation between the evolutionary distance and GO level. The correlation that older target genes tend to be enriched in lower level GO functions was further confirmed by other three criteria, top 20, 30, and 10%. (Mya: Million Years Ago).

Figure S5 The correlation between the evolutionary distance and the number of literatures. The correlation that older target genes don't tend to be studied more was further confirmed. (Mya: Million Years Ago).

Table S1 The top 10 enriched functions in the most distant species.

Acknowledgments

The authors thank Rebecca Hiller Posey for proofreading an earlier draft of the manuscript, and the two reviewers whose comments helped improve the quality of this work.

Author Contributions

Conceived and designed the experiments: ZZ CCL. Performed the experiments: CCL RM. Analyzed the data: CCL RM. Contributed reagents/materials/analysis tools: CCL RM. Wrote the paper: ZZ CCL RM.

References

1. Bartel DP (2004) MicroRNAs: genomics, biogenesis, mechanism, and function. Cell 116: 281–297.
2. Filipowicz W, Bhattacharyya SN, Sonenberg N (2008) Mechanisms of post-transcriptional regulation by microRNAs: are the answers in sight? Nat Rev Genet 9: 102–114.
3. Flynt AS, Lai EC (2008) Biological principles of microRNA-mediated regulation: shared themes amid diversity. Nat Rev Genet 9: 831–842.
4. Kim VN, Nam JW (2006) Genomics of microRNA. Trends Genet 22: 165–173.
5. Doench JG, Sharp PA (2004) Specificity of microRNA target selection in translational repression. Genes Dev 18: 504–511.
6. Guo H, Ingolia NT, Weissman JS, Bartel DP (2010) Mammalian microRNAs predominantly act to decrease target mRNA levels. Nature 466: 835–840.
7. John B, Enright AJ, Aravin A, Tuschl T, Sander C, et al. (2004) Human MicroRNA targets. PLoS Biol 2: e363.
8. Enright AJ, John B, Gaul U, Tuschl T, Sander C, et al. (2003) MicroRNA targets in Drosophila. Genome Biol 5: R1.
9. Ritchie W, Flamant S, Rasko JE (2010) mimiRNA: a microRNA expression profiler and classification resource designed to identify functional correlations between microRNAs and their targets. Bioinformatics 26: 223–227.
10. Betel D, Koppal A, Agius P, Sander C, Leslie C (2010) Comprehensive modeling of microRNA targets predicts functional non-conserved and non-canonical sites. Genome Biol 11: R90.
11. Friedman RC, Farh KK, Burge CB, Bartel DP (2009) Most mammalian mRNAs are conserved targets of microRNAs. Genome Res 19: 92–105.
12. Selbach M, Schwanhausser B, Thierfelder N, Fang Z, Khanin R, et al. (2008) Widespread changes in protein synthesis induced by microRNAs. Nature 455: 58–63.
13. Ritchie W, Rasko JE, Flamant S (2013) MicroRNA target prediction and validation. Adv Exp Med Biol 774: 39–53.
14. Peterson KJ, Dietrich MR, McPeek MA (2009) MicroRNAs and metazoan macroevolution: insights into canalization, complexity, and the Cambrian explosion. Bioessays 31: 736–747.
15. Heimberg AM, Sempere LF, Moy VN, Donoghue PC, Peterson KJ (2008) MicroRNAs and the advent of vertebrate morphological complexity. Proc Natl Acad Sci U S A 105: 2946–2950.
16. Heimberg AM, Cowper-Sal-lari R, Semon M, Donoghue PC, Peterson KJ (2010) microRNAs reveal the interrelationships of hagfish, lampreys, and gnathostomes and the nature of the ancestral vertebrate. Proc Natl Acad Sci U S A 107: 19379–19383.
17. Sempere LF, Martinez P, Cole C, Baguna J, Peterson KJ (2007) Phylogenetic distribution of microRNAs supports the basal position of acoel flatworms and the polyphyly of Platyhelminthes. Evol Dev 9: 409–415.
18. Sempere LF, Cole CN, McPeek MA, Peterson KJ (2006) The phylogenetic distribution of metazoan microRNAs: insights into evolutionary complexity and constraint. J Exp Zool B Mol Dev Evol 306: 575–588.
19. Lee CT, Risom T, Strauss WM (2007) Evolutionary conservation of microRNA regulatory circuits: an examination of microRNA gene complexity and conserved microRNA-target interactions through metazoan phylogeny. DNA Cell Biol 26: 209–218.
20. Baek D, Villen J, Shin C, Camargo FD, Gygi SP, et al. (2008) The impact of microRNAs on protein output. Nature 455: 64–71.
21. Bartel DP (2009) MicroRNAs: target recognition and regulatory functions. Cell 136: 215–233.
22. Kozomara A, Griffiths-Jones S (2011) miRBase: integrating microRNA annotation and deep-sequencing data. Nucleic Acids Res 39: D152–157.
23. Flicek P, Ahmed I, Amode MR, Barrell D, Beal K, et al. (2013) Ensembl 2013. Nucleic Acids Res 41: D48–55.
24. Lewis BP, Burge CB, Bartel DP (2005) Conserved seed pairing, often flanked by adenosines, indicates that thousands of human genes are microRNA targets. Cell 120: 15–20.
25. Ruby JG, Stark A, Johnston WK, Kellis M, Bartel DP, et al. (2007) Evolution, biogenesis, expression, and target predictions of a substantially expanded set of Drosophila microRNAs. Genome Res 17: 1850–1864.
26. Jan CH, Friedman RC, Ruby JG, Bartel DP (2011) Formation, regulation and evolution of Caenorhabditis elegans 3′ UTRs. Nature 469: 97–101.
27. Mitra R, Bandyopadhyay S (2011) MultiMiTar: a novel multi objective optimization based miRNA-target prediction method. PLoS One 6: e24583.
28. Papadopoulos GL, Reczko M, Simossis VA, Sethupathy P, Hatzigeorgiou AG (2009) The database of experimentally supported targets: a functional update of TarBase. Nucleic Acids Res 37: D155–158.
29. Xiao F, Zuo Z, Cai G, Kang S, Gao X, et al. (2009) miRecords: an integrated resource for microRNA-target interactions. Nucleic Acids Res 37: D105–110.

30. Hsu SD, Lin FM, Wu WY, Liang C, Huang WC, et al. (2011) miRTarBase: a database curates experimentally validated microRNA-target interactions. Nucleic Acids Res 39: D163–169.
31. Burge SW, Daub J, Eberhardt R, Tate J, Barquist L, et al. (2013) Rfam 11.0: 10 years of RNA families. Nucleic Acids Res 41: D226–232.
32. Grimson A, Farh KK, Johnston WK, Garrett-Engele P, Lim LP, et al. (2007) MicroRNA targeting specificity in mammals: determinants beyond seed pairing. Mol Cell 27: 91–105.
33. Tani S, Kuraku S, Sakamoto H, Inoue K, Kusakabe R (2013) Developmental expression and evolution of muscle-specific microRNAs conserved in vertebrates. Evol Dev 15: 293–304.
34. Fahlgren N, Jogdeo S, Kasschau KD, Sullivan CM, Chapman EJ, et al. (2010) MicroRNA gene evolution in Arabidopsis lyrata and Arabidopsis thaliana. Plant Cell 22: 1074–1089.
35. Ashburner M, Ball CA, Blake JA, Botstein D, Butler H, et al. (2000) Gene ontology: tool for the unification of biology. The Gene Ontology Consortium. Nat Genet 25: 25–29.
36. Chen Y, Gelfond J, McManus LM, Shireman PK (2011) Temporal microRNA expression during in vitro myogenic progenitor cell proliferation and differentiation: regulation of proliferation by miR-682. Physiol Genomics 43: 621–630.
37. Chen JF, Tao Y, Li J, Deng Z, Yan Z, et al. (2010) microRNA-1 and microRNA-206 regulate skeletal muscle satellite cell proliferation and differentiation by repressing Pax7. J Cell Biol 190: 867–879.
38. Townley-Tilson WH, Callis TE, Wang D (2010) MicroRNAs 1, 133, and 206: critical factors of skeletal and cardiac muscle development, function, and disease. Int J Biochem Cell Biol 42: 1252–1255.
39. Berezikov E (2011) Evolution of microRNA diversity and regulation in animals. Nat Rev Genet 12: 846–860.
40. Xu J, Zhang R, Shen Y, Liu G, Lu X, et al. (2013) The evolution of evolvability in microRNA target sites in vertebrates. Genome Res 23: 1810–1816.
41. Hedges SB, Dudley J, Kumar S (2006) TimeTree: a public knowledge-base of divergence times among organisms. Bioinformatics 22: 2971–2972.
42. Tamura K, Peterson D, Peterson N, Stecher G, Nei M, et al. (2011) MEGA5: molecular evolutionary genetics analysis using maximum likelihood, evolutionary distance, and maximum parsimony methods. Mol Biol Evol 28: 2731–2739.

20

Developmental and Evolutionary History Affect Survival in Stressful Environments

Gareth R. Hopkins*, Edmund D. Brodie Jr., Susannah S. French

Department of Biology and the Ecology Center, Utah State University, Logan, Utah, United States of America

Abstract

The world is increasingly impacted by a variety of stressors that have the potential to differentially influence life history stages of organisms. Organisms have evolved to cope with some stressors, while with others they have little capacity. It is thus important to understand the effects of both developmental and evolutionary history on survival in stressful environments. We present evidence of the effects of both developmental and evolutionary history on survival of a freshwater vertebrate, the rough-skinned newt (*Taricha granulosa*) in an osmotically stressful environment. We compared the survival of larvae in either NaCl or $MgCl_2$ that were exposed to salinity either as larvae only or as embryos as well. Embryonic exposure to salinity led to greater mortality of newt larvae than larval exposure alone, and this reduced survival probability was strongly linked to the carry-over effect of stunted embryonic growth in salts. Larval survival was also dependent on the type of salt (NaCl or $MgCl_2$) the larvae were exposed to, and was lowest in $MgCl_2$, a widely-used chemical deicer that, unlike NaCl, amphibian larvae do not have an evolutionary history of regulating at high levels. Both developmental and evolutionary history are critical factors in determining survival in this stressful environment, a pattern that may have widespread implications for the survival of animals increasingly impacted by substances with which they have little evolutionary history.

Editor: Daniele Canestrelli, Tuscia University, Italy

Funding: This research was supported financially by the Utah State University (USU) Department of Biology and the Ecology Center. The USU Merril-Cazier Library's Open Access Funding Initiative and the USU Ecology Center kindly assisted with publication costs. The Natural Sciences and Engineering Research Council of Canada (NSERC) provided additional financial support to GRH. The funders had no role in study design, data collection and analysis, decision to publish, or preparation of the manuscript.

Competing Interests: The authors have declared that no competing interests exist.

* E-mail: gareth.r.hopkins@gmail.com

Introduction

Natural and anthropogenic stressors are commonplace throughout the environment. The ways in which stressors impact organisms, and their ability to successfully respond to these stressors is of paramount importance to our understanding of biological systems. For organisms with complex life cycles, the ability to respond to a given stressor may vary depending on life history stage, and there may be carry-over effects from one stage to the next [1] (see Table S1 in Supporting Information). However, organisms may or may not have an evolutionary history of regulating the stressor in question, and this may also affect their ability to effectively respond [2,3]. We propose that both an organism's developmental history of exposure to a stressor (developmental history hypothesis) and its evolutionary history of regulating that stressor (evolutionary history hypothesis) play critical roles in the survival of organisms in stressful environments.

It has been suggested that the earlier in an organism's life history environmental stressors are experienced, the more severe the lasting consequences will be [4–6], and there is strong empirical evidence across animal taxa for this assertion (Table S1). This forms the basis of our developmental history hypothesis. In humans, for example, the environment of the womb can significantly affect an individual's chances of cardiac and other diseases later in life [5,7,8]. In birds, the temperature at which eggs are incubated can affect hatchling body composition, growth, immunocompetence and thermoregulatory ability [4,9]. Developmental temperature also affects survival, growth and behavior of juvenile reptiles (e.g., [10]) (Table S1). Elevated CO_2 as embryos results in decreased larval settlement success of sea urchins [11], and the ability of bryozoans to produce large, successful colonies is dependent on their embryonic experience and growth [12]. Thus, embryonic exposure to stressors can be critical to an animal's future fitness (Table S1).

Parsing critical life history stages, however, is not trivial, and many studies have given contradictory evidence for the developmental history hypothesis. For example, while multiple studies have shown that embryonic environment can significantly affect an individual's chances of success in later life (Table S1), others have shown that it is the larval or juvenile environment that has the greatest influence on survival, growth, or reproduction (e.g., [13,14]). Still others have shown that while the embryonic environment has a significant role to play in later life, its effect may be dependent on the environment animals experience later in life (e.g., [15,16,17]). Experiments are often not designed to isolate the effects of environment on a specific life history stage from those of another (e.g., [18,19–21]), and thus, consistent knowledge of the environmental and carry-over effects across multiple life history stages is lacking (but see [14,17,22,23]).

While there is a strong empirical basis for the developmental history hypothesis (even with the conflicting evidence and limitations identified above), there is much less known regarding the evolutionary history hypothesis. Organisms in most habitats today face both natural stressors with which they have an evolutionary history, and thus evolved physiological mechanisms of regulating (e.g., CO_2, temperature, NaCl), and novel stressors with which they do not have this same evolutionary history (e.g., pesticides, flame retardants, commercial non-NaCl-based deicing salts), and thus lack the physiological mechanisms to regulate. The effects of developmental history must therefore be placed in this environmental and evolutionary context. While many studies have documented the significant effects of unfamiliar substances such as pollutants on evolutionarily-naïve organisms (e.g., reviewed by [24] for amphibians), these cannot be directly compared to stressors with which the organism has an evolutionary history, and thus a means of regulating, as the nature of the two stressors is usually very different (i.e., comparing the effect of a herbicide with the effect of NaCl). At this point, we do not know how the potentially important effects of an organism's evolutionary history with a stressor may interact with its developmental history of exposure with the stressor.

To address these concerns, we tested the effects of developmental and evolutionary history on survival in stressful environments. We chose the rough-skinned newt (*Taricha granulosa* Skelton; Caudata: Salamandridae) as our model, an osmotically sensitive organism, and salinity as its stressor. Specifically, we tested the effects of both NaCl and $MgCl_2$ on the post-hatching survival of newt larvae that had either been exposed to salt as both embryos and larvae or just as larvae. Salinity is an excellent stressor to use to test our two hypotheses, as it is a naturally occurring abiotic component of aquatic habitats, and is known to have significant carry-over effects from the embryonic to post-hatching life-stages in a variety of organisms, [15,23,25–27] (Table S1). We used salt concentrations that were within environmentally relevant limits of freshwater aquatic systems impacted by either natural (i.e., estuaries) or anthropogenic (i.e., road deicing salts) sources of salts [29,30]. The two most common sources of salinity in North America today are two different salts, NaCl and $MgCl_2$, only one of which most organisms have an evolutionary history of regulating. Sodium chloride (NaCl) is one of the most common osmolytes, and organisms have an evolutionary history of regulating this in a variety of habitats, whereas $MgCl_2$ has not been identified as a common vertebrate osmolyte [31], and Mg^{2+} is not found in substantial concentrations in most freshwater habitats, nor the precipitation that feeds them (including in the newts' range) [32]. Therefore, animals do not have the same evolutionary history of physiological regulation of this ion. Nevertheless, $MgCl_2$ is now the second most commonly used road deicer in North America (behind NaCl), and is used exclusively in some areas of the continent [33]. Thus, there is the potential that organisms will encounter $MgCl_2$ in substantial quantities in their environment. We found that both salts caused significant developmental carry-over effects from the embryonic environment on larval survival, but that the salts differed in their effects on larval survival, according to the differential evolutionary history that amphibians have with regulating the two stressors. As more and more freshwater animals, mostly maladapted to salt, will be forced to cope with increasing salinization of their habitats due to the application of road deicing salts [34,35–36], landscape modification and agricultural waste [37–40], and rising sea-levels [41–43], understanding the effects of both developmental and evolutionary history of salinity exposure will have important implications for both life history and evolutionary theory, as well as conservation efforts.

Materials and Methods

Ethics Statement

Adult rough-skinned newts (*Taricha granulosa*) (not an endangered or protected species) were collected by dip-net and hand from Soap Creek ponds (44°40′13.22″N, 123°16′39.65″W) under Oregon Department of Fish and Wildlife Scientific Taking Permit #062-11. Access to these ponds was granted by Joe Beatty, Oregon State University. The Utah State University Institutional Animal Care and Use Committee (IACUC) approved the collection and use of animals in this research, and all experimental protocols (approved protocol #1524). Animals were euthanized at the completion of experiments with MS-222, in accordance with the approved IACUC protocol (#1524).

Experimental Procedure

As reported in a previous study ([44] for detailed methods on habitat, field collection, rearing eggs and preparing salt solutions), we reared eggs from 16 different gravid wild-caught female rough-skinned newts (*Taricha granulosa*) from a single, salt-naïve population from Benton County, Oregon, in a laboratory environmental control chamber at 7°C. This population is truly salt-naïve [44], being highly philopatric to freshwater ponds that are separated by hundreds of meters from small county roads that are not salted (Kendal Weeks, Oregon Department of Transportation Road Maintenance, personal communication; Kent Mahler, Benton County Road Maintenance, personally communication). While $MgCl_2$ is widely used in Oregon as its exclusive deicer, it is also not applied to the nearest stretch of highway to these ponds, located over 4 km away (Kendal Weeks, Oregon Department of Transportation Road Maintenance, personal communication). See [44] for additional details on this habitat. Eggs from wild-caught females were randomized to one of six different salt treatments, made with laboratory grade NaCl (Thermo Fisher Scientific, Fair Lawn, NJ, USA), $MgCl_2$ (Acros Organics, Fair Lawn, NJ, USA) and distilled water (Low NaCl, Low MgCl = 1.0 g/l Cl^-; Medium NaCl, Medium $MgCl_2$ = 1.5 g/l Cl^-; High NaCl, High $MgCl_2$ = 2.0 g/l Cl^-) and a control (20% Holtfreter's Solution = 0.7 g/l Cl^- [45]). Those eggs that survived these treatments were used in the present experiment. At hatching, the size (total length) and developmental stage [46] of hatchlings were recorded (see [44] for full methods and results).

Eggs that were reared in a salt treatment remained in that salt treatment as larvae (Fig. 1). Approximately 7 times more control eggs were reared than salt treatment eggs, so that control eggs could be randomized to new larval treatments in the present experiment (similarly to [26]) (Fig. 1). Eggs were monitored daily and all larvae were transferred to their new treatment solution within 12 hours of hatching. This direct transfer, following a similar protocol of Petranka and Doyle [47], was meant to mimic the sharp spike in Cl^- concentrations found in road-side environments that immediately occurs within hours of a deicing event or snowmelt [48–50], where minimal to no time is allowed for acclimation. While gradual acclimation of low salinity levels have led to increased tolerance in some amphibians (e.g., [51]) it has also led to increased susceptibility in others [52], and is less environmentally relevant to examining the sudden spikes of salinity seen in habitats due to road deicing salt application. In addition, while the salt concentrations used were typical for those immediately resulting from deicing events [48,53], they were also well below recorded NaCl and $MgCl_2$ LD-50 values for other amphibian larvae [37,54,55].

Larvae were housed in sibling groups of up to 5 individuals (keeping offspring from different female and treatment combina-

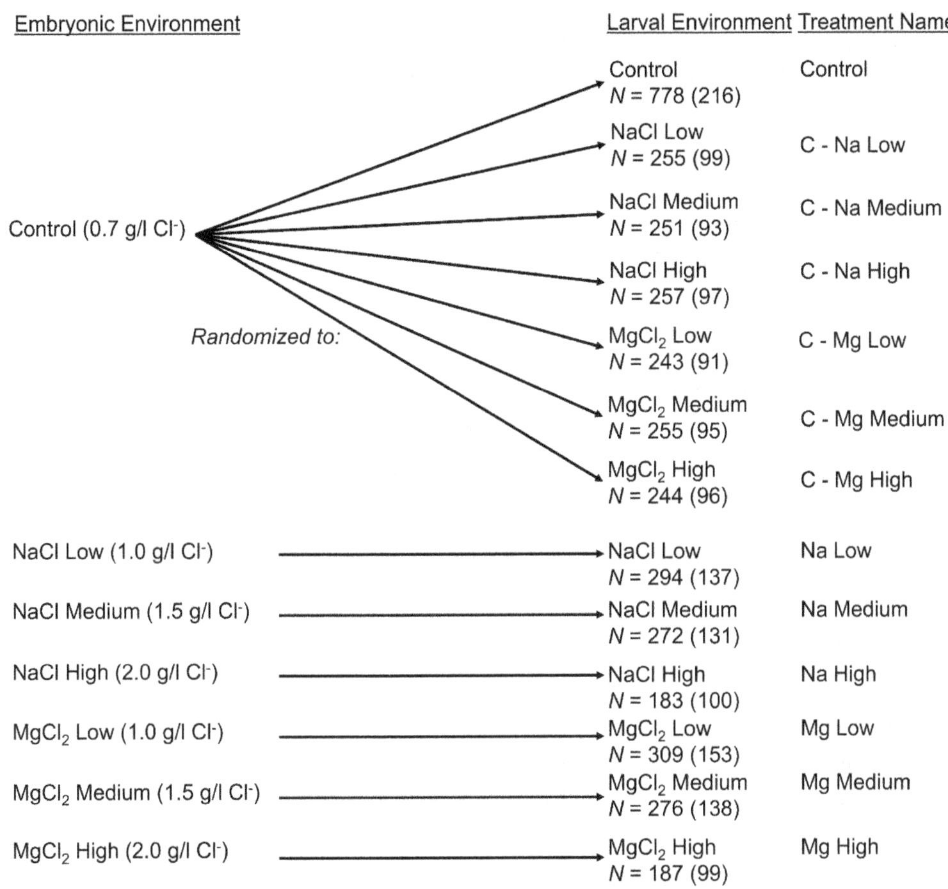

Figure 1. Outline of experimental design. Embryonic and larval environments, salinity concentrations, treatment names, and sample sizes are shown. Newt eggs were reared in either a freshwater control, or one of six salt treatments. Upon hatching, embryos that were reared in salt stayed in that salt, whereas embryos reared in control either stayed in control or were randomized to one of the six salt treatments for the larval environment. The name of each treatment combination is listed, and sample sizes are given under each larval environment (numbers outside of parentheses indicate total number of individuals in the treatment, whereas numbers inside parentheses indicate number of containers in the treatment (up to five sibling larvae were reared in the same container, and individuals within containers were treated as nested subsamples. See Methods for more details).

tions separate) in 12.5 cm diameter, 10.5 cm deep round plastic containers, filled with 400 ml of solution. Each container was randomized to a location in a growth chamber set at 7°C, with a 12 h light: dark photoperiod. Containers were checked daily for larval survival, and dehydration. Dead larvae were noted and removed, and a small amount of distilled water was added to each container if necessary, to compensate for evaporation. *Taricha granulosa* larvae retain some embryonic yolk for up to approximately two weeks after hatching, and do not engage in feeding on prey before then. As we did not want to confound our survival results with possible negative effects of the salt treatments on larval prey, we only conducted this experiment for 15 days post-hatching; if a larva was alive at day fifteen, it was recorded as alive for the purposes of the analysis. A similar endpoint has also been used in a previous study on post-hatching survival of frog larvae in road deicing salt [47].

Statistical Analysis

For survival analyses, individual larvae were treated as subsamples within containers, which were treated as subsamples nested within individual female. Larval survival was analyzed using a binomial distribution, with a generalized linear mixed model blocking on individual female as a random effect. We first compared the survival of control newts (i.e., those reared in control as eggs and larvae) to newts in all other treatments for each salt, and then ran separate models to compare survival among salt treatments (minus control) for both larvae that were reared in control and those reared in salt as eggs, with Tukey-adjusted multiple comparisons among individual treatment levels, when an overall significant effect of treatment was found. We were, however, primarily interested in comparing and contrasting the effects of embryonic and larval environment on larval survival. As we did not have a complete factorial design in this study (e.g. embryonic low $MgCl_2$ + larval high NaCl treatment combination), for this analysis, we analyzed the effects of the two different salt types separately, using embryonic and larval treatments as fixed effect factors in our models. We then analyzed the effect of embryonic versus larval environment on larval survival for each salt [56]. In these analyses, larval treatment had three levels, low, medium and high, and embryonic treatment had two levels, control and salt. This enabled a direct statistical comparison to be made of larval survival between animals that were reared as eggs in control or, for example, low $MgCl_2$, for larvae that were reared in low $MgCl_2$. We conducted Tukey-adjusted multiple comparisons, specifically comparing larval survival in each salt treatment level between eggs that were reared in either that salt treatment or control, for cases in which an overall significant effect of either embryonic treatment, larval treatment, or their interaction was found. Analyses were conducted using PROC GLIMMIX in SAS software version 9.3, with significance set at $\alpha = 0.05$.

As embryonic exposure to salt affected the size and developmental stage at hatching of newts, as did differences among individual mothers (females) [44], we wanted to further assess the potential contribution of these variables, as well as embryonic and larval treatments in general, in explaining any overall effects of salt treatment in either embryonic or larval environments on larval survival. To do this, we conducted multivariate classification analyses, which measure variable importance in a model's ability to correctly classify larvae as having died or survived. As only one out of 778 newt larvae died after being reared in control as both an embryo and larvae (see Results), we restricted our analyses to larvae reared in salt post-hatching. We used three validated classification procedures [57], logistic regression, Classification Trees [58] and Random Forests [57,59], and in each case assessed variable importance by examining the relative classification performance of models incorporating or not incorporating key variables.

Specifically, we assessed the ability of the models to correctly classify larvae as having died (sensitivity). For the full model, we included all larval and embryonic variables of potential interest, including: larval treatment, embryonic treatment, length at

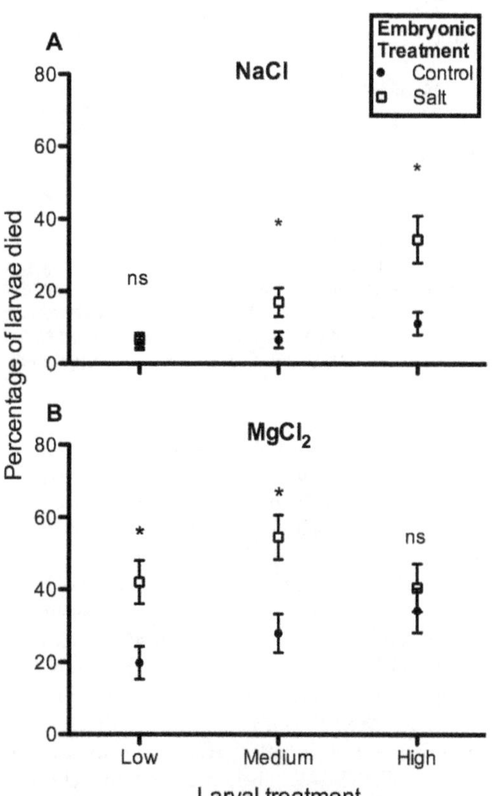

Figure 2. Percentage (mean ± SE) of larvae that died in each salt treatment. (A) NaCl, (B) $MgCl_2$. Only 1 out of 778 larvae in Control died, and thus only results for mortality in salt treatments are shown. Direct comparisons are made between the mortality of larvae reared as embryos in salt (open squares) or control (closed circles). Asterisks indicate significant differences (Tukey-adjusted multiple comparisons) between the percentages of larvae died in each of these treatments (i.e., for the larval treatment Medium NaCl, significantly more larvae died when reared as eggs in that salt, than did larvae reared as eggs in control). "ns" = no significant difference between treatments.

Table 1. Effects of embryonic environment, larval environment, and their interaction on larval survival in NaCl and $MgCl_2$.

Salt type	Embryonic environment			Larval environment			Embryonic x Larval environments		
	F	df (n,d)	*p*	*F*	df (n,d)	*p*	*F*	df (n,d)	*p*
NaCl	11.19	1,74	**0.0013**	9.61	2,74	**0.0002**	2.28	2,74	0.1095
$MgCl_2$	18.34	1,73	**<0.0001**	2.22	2,73	0.1162	2.00	2,73	0.1429

Significant effects are listed in bold.

hatching, developmental stage at hatching, and female identity. We then withdrew the larval treatment variable, and reassessed the model's sensitivity, withdrew all embryonic variables (leaving only larval treatment and individual female) and again reassessed the model's sensitivity, to assess the potential relative contribution of larval environment in predicting larval mortality. As well as assessing variable importance in this manner, all three classification methods also provide separate indicators of variable importance [57]. This is achieved through a variable importance plot in Random Forests, a classification plot in Classification Trees, and the variable with the largest Wald Chi-Square value in logistic regression. We chose the most important variable identified in each of these methods from the original full model, and reinserted it back into the model including only larval treatment and female identity, and assessed whether the inclusion of this identified variable increased model performance. Classification analyses were completed in SAS (logistic regression) and R (R Development Core Team, 2008, www.R-project.org) (Classification Trees and Random Forests). Finally, as length at hatching was identified as a key variable of importance in predicting larval mortality (see Results), we compared the mean length at hatching of larvae that died versus survived in each treatment using t-tests in SAS software version 9.3, with significance set at $\alpha = 0.05$.

Results

After 14 days, only one out of 778 larvae reared in control as both egg and larva ("control" treatment) died in this treatment, which was significantly fewer than in any other treatment (all $p < 0.001$). The survival of the remaining larvae, all experiencing salts in their larval environment, was then compared. There was a significant effect of larval salt treatment on larval survival for both newts that were reared embryonically in salt ($F_{5,74} = 16.54$, $p < 0.0001$) and control ($F_{5,73} = 7.81$, $p < 0.0001$). For animals that were reared as eggs in salt and stayed in that salt as larvae, significantly more larvae died in low and medium $MgCl_2$ than in those corresponding concentrations of NaCl (all Tukey adjusted multiple comparisons $p < 0.0001$), with a similar percentage of larvae dying in high $MgCl_2$ as high NaCl (Tukey adjusted $p = 0.98$). For animals that were reared as eggs in control and then transferred to salt as larvae, marginally more larvae died in low $MgCl_2$ than low NaCl (Tukey adjusted $p = 0.0698$), and significantly more larvae died in medium and high $MgCl_2$ than the corresponding concentrations of NaCl (Tukey adjusted $p < 0.02$).

Increased salt concentration, in both the embryonic and larval environments, generally resulted in increased larval mortality (with the exception of high $MgCl_2$) (Fig. 2). For both salts, larval survival was significantly affected by embryonic environment (Table 1). For the majority of treatment levels, larvae that were reared as eggs in control solution survived significantly better than larvae that were reared as eggs in salt treatments (Fig. 2). For NaCl, both embryonic and larval treatments significantly affected survival of larvae in this salt, but for $MgCl_2$, only embryonic treatment significantly explained larval survival (Table 1). There were no significant interacting effects of embryonic and larval environments on larval survival (Table 1).

Eggs that were reared in salt water resulted in smaller larvae at hatching than those reared in control [44]. Classification analyses with three different methods all revealed length at hatching as the consistently most important variable in determining larval survival (Table 2), further strengthening the evidence of the importance of embryonic environment on survival post-hatching. Although Classification Trees and Random Forests had better sensitivity than logistic regression (as was expected [57]), the ability of models, using any of the classification methods, to correctly classify larvae as having died declined dramatically with the exclusion of embryonic variables (i.e., larval treatment and female identity alone was a very poor classifier of larval survival), but recovered substantially with the re-inclusion of length at hatching as a predictor variable (Table 2), further identifying it as a critical

Table 2. Classification analyses for predicting whether or not newt larvae died ("sensitivity"), for data excluding control data (i.e., only newts in salt as larvae).

Model	Classification method	Model sensitivity (%) (percent larvae correctly classified as having died)	Change in model sensitivity from full model sensitivity (%)	Most important variable identified
Full (Larval Treatment, Egg Treatment, Length & Stage at Hatching, Female)	Logistic Regression	47.70	.	Length at Hatching
	Classification Trees	69.65	.	Length at Hatching
	Random Forests	64.78	.	Length at Hatching
Just Embryonic Variables (Egg Treatment, Length & Stage at Hatching, Female) (not Larval Treatment)	Logistic Regression	39.82	−7.88	Egg Treatment
	Classification Trees	64.39	−5.26	Length at Hatching
	Random Forests	59.26	−5.52	Length at Hatching
Just Larval Treatment and Female	Logistic Regression	13.67	−34.03	Larval Treatment
	Classification Trees	19.84	−49.81	Larval Treatment
	Random Forests	14.32	−50.46	Larval Treatment
Just Larval Treatment, Female, & Length at Hatching	Logistic Regression	44.42	−3.28	Length at Hatching
	Classification Trees	61.76	−7.89	Length at Hatching
	Random Forests	52.56	−12.22	Length at Hatching

Three multivariate classification methods were utilized (logistic regression, Classification Trees, and Random Forests) to determine the most important variables predicting larval survival in salt. See Methods and Results for more details regarding these analyses and their interpretation.

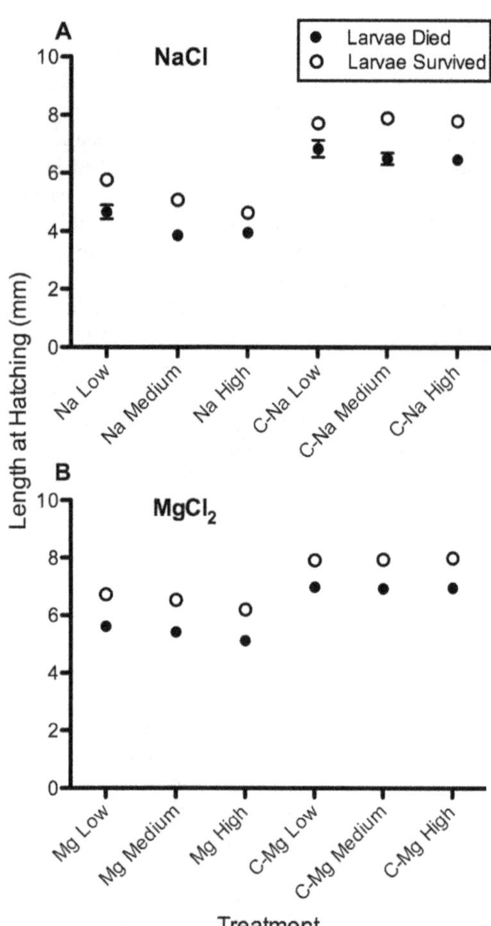

Figure 3. Mean (±SE) lengths at hatching (mm) of larvae that died (closed circles) or survived (open circles) in each salt treatment. (A) NaCl, (B) $MgCl_2$. In all treatments, larvae that survived averaged larger at hatching than those that died (all t-tests, $p<0.01$). See [44] for full results on length at hatching.

variable for predicting larval survival. Larvae that survived, in each of the treatments, were significantly larger at hatching, on average, than larvae that died (Fig. 3; for all t-tests, $p<0.01$).

Discussion

Developmental and evolutionary history each significantly affected the survival of newt larvae in salts, and thus the importance of both hypotheses was supported. Eggs appear to be a critical life history stage for this amphibian in osmotically stressful environments. Animals that were exposed to salt as embryos and survived hatched at a smaller size than animals that did not experience embryonic salinity. Stunting of embryonic growth put amphibian larvae at greater risk for salt-induced mortality (Table 2). However, our results show that it is also important to understand the evolutionary history an organism has with a stressor. Even though there was no difference in egg mortality between embryos reared in NaCl or $MgCl_2$ [44], more larvae died in $MgCl_2$ than in NaCl (Fig. 4). While newt larvae have evolved with natural sources of NaCl in their environment, which they can osmoregulate, such common regulation of $MgCl_2$ does not appear to have evolved. Understanding this evolutionary history, as well as parsing critical life history stages is imperative to understand the effects of stressors on the life history of an organism.

The majority of organisms have complex life cycles, and the experiences of one life stage can have profound impacts on those in subsequent stages [1] (Table S1). Embryonic salinity is known to affect the post-hatching survival, growth and development of marine and estuarine invertebrates, such as barnacles [23], crabs [15,25,27], horseshoe crabs [26] and tunicates [28]. While all life history stages of amphibians have, individually, repeatedly been found to be extremely sensitive to salt [21,47,54,55,60–68], with a few notable exceptions such as *Fejervarya cancrivora* [69,70], the relative sensitivity of each life history stage, and potential downstream effects of salinity from one stage to the next, have been less studied. In one of the only other studies on amphibians to examine embryonic carry-over effects of salinity, frog larvae (*Lithobates sylvaticus*) reared in salt water (NaCl-based) as eggs had reduced survival in salt compared to larvae that were reared in freshwater as eggs [47]. This study also found that growth and development of larvae that survived was depressed in those animals reared embryonically in salt, also suggesting carry-over effects of embryonic exposure to salt [47]. Snodgrass et al [21] also found that *Bufo americanus* toadlets exposed to stormwater pond sediment (which had an increased conductivity mainly due to road deicing salt) as embryos were smaller at metamorphosis than embryos and larvae exposed to freshwater (although the relative effects of embryonic vs. larval exposure were not separated). Other studies have also shown potential carry-over effects of larval salinity exposure on metamorphic traits important for adult fitness [37,60]. These results all clearly show that Qiu and Qian's [23] statement regarding marine invertebrates, that "osmotic stress experienced in one life-stage can be passed over to the next life-stage", can apply to freshwater vertebrates as well.

This pattern of decreased post-hatching survival as a consequence of embryonic exposure has also been found in amphibians

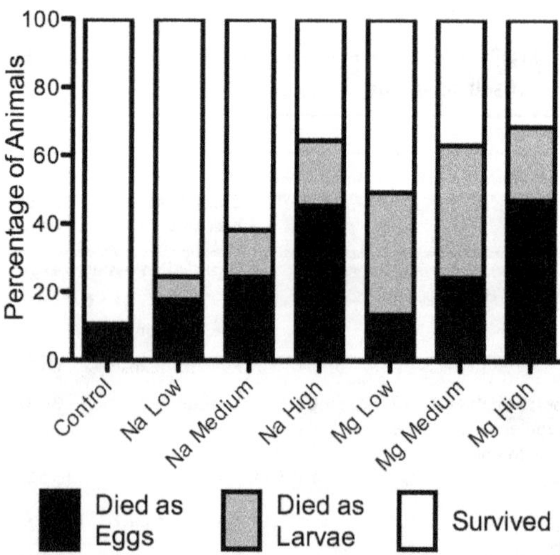

Figure 4. Mortality of newt eggs (black bars) and larvae (grey bars) in each salt treatment. This figure shows only larvae that were reared as both eggs and larvae in salt. The percentage of individuals that survived in each treatment is indicated in white. All percentages are calculated based on the total number of eggs that started in each treatment (Control = 2577, low NaCl = 363, medium NaCl = 366, high NaCl = 345, low $MgCl_2$ = 369, medium $MgCl_2$ = 369, high $MgCl_2$ = 354; [44]), some of which either died (black bars), or survived to hatching and were reared in salt, where they either died (grey bars) or survived (white bars).

in response to other stressors, such as nitrite [71] and pesticides [22]. Thus, studies that do not examine effects at each life history stage and do not consider the potential for cascading effects across stages may seriously underestimate the cumulative effects of exposure to stressors [19,71–73].

One of the primary ways that osmotic stress affects the embryonic stage to influence post-hatching survival is through the retardation of growth and development. Newt eggs that were reared in salt water hatched sooner, smaller and less developed than newts reared in a freshwater control [44], and this resulting reduced length at hatching appears to be the single most important variable in predicting next-stage (larval) survival in salt water (Table 2). Size at hatching/birth is well known to have important implications on larval, juvenile, and adult health and survival in a wide variety of taxa, ranging from sea snails [74] and bryozoans [12], to birds [75] and humans [7]. Furthermore, this link between size and fitness has been identified as key to life history theory [12]. Among amphibians, hatching early, smaller and less developed, is known to affect larval survival, the onset of feeding competence, competitive and predatory interactions, and larval growth rate and timing of metamorphosis [6,76–80]. Similar to our findings, small, less developed amphibian larvae are more susceptible to pollutants than are large larvae [81,82]. Smaller larval rough-skinned newts are also more vulnerable to be injured and die in predatory encounters with dragonfly nymphs [83]. Thus, even if smaller hatchlings are able to survive short-term in osmotically stressful environments (which seems unlikely from our results (Fig. 3)), or even if compensatory growth occurred later in development, a host of other fitness consequences of this initial stunted embryonic growth and development are still likely later in life [84], further emphasizing the importance of the embryonic environment for life-time fitness.

While the effects of the two salt types were not significantly different on embryonic survival [44], there were differences in the larval stage, whereby $MgCl_2$ had relatively greater effects on survival (Fig. 4). This is in spite of the fact that embryos actually hatched slightly larger at $MgCl_2$ than at NaCl [44]. Although most amphibian eggs, like those of many other aquatic organisms [85], have little means of osmoregulating at the salt concentrations used in this study [37,63,86], and thus the effects of NaCl and $MgCl_2$ at this life history stage are equally destructive (any affect of evolutionary history is minimized in the absence of regulatory ability), amphibian larvae have evolved to osmoregulate Na^+ and Cl^- ion concentrations in their body through the use of integumental and gill Na^+ pumps [31,87–90]. Larvae have not evolved this same ability to regulate Mg^{2+} ions, however, and thus larvae in NaCl were able to attempt osmoregulation to survive in this solution whereas larvae in $MgCl_2$ were not. In addition to lacking this evolutionary history of osmoregulation, Mg^{2+} has also been shown to be inhibitory to important osmoregulatory skin ion pump functioning in other amphibian larvae [91]. Whereas the effects of NaCl on larvae act in a typical dose-response fashion (Fig. 2a), it appears that any concentration of $MgCl_2$ is detrimental to larvae (Fig. 2b), as they have less means to regulate it. This may explain why larval salinity concentration significantly influences larval survival for animals in NaCl, but not those in $MgCl_2$ (Table 1). The fact that Mg High had lower mortality than Mg Low or Medium (Fig. 2b) may be due to a number of possible reasons, including hormesis [92]. In the only other studies on the effects of $MgCl_2$ on amphibian larvae to date, both Dougherty and Smith [62] and Harless et al [54] also found that this emerging deicing salt was more toxic to frog tadpoles than NaCl. Magnesium chloride may in fact, be more toxic than NaCl to life in general, as studies have found that otherwise salt-tolerant plants [93,94] and archaea [95] are often intolerant of $MgCl_2$, and the threshold for biological processes in $MgCl_2$ is lower than other salts, including NaCl [95]. These results make sense from an evolutionary perspective, given the small quantities of Mg^{2+} generally found in most aquatic ecosystems, relative to the higher quantities of Na^+ found in precipitation and the ocean [32], and thus many organisms may not have an evolutionary history of regulating Mg^{2+} in high concentrations in their environment.

Vulnerability of a particular life history stage can be described as the ability of that life history stage to regulate the stressor in question. Using this criterion, it appears that eggs are the most vulnerable life history stage to salts overall in amphibians (this study; [61,65,66,96]) and effects on embryonic development at this stage have profound survival consequences in later life history stages, even possibly affecting population viability indirectly through influencing post-embryonic (larval) mortality [97]. Similarly, amphibian larvae cannot successfully osmoregulate in $MgCl_2$, and thus all life history stages are particularly vulnerable to this evolutionarily novel but emerging deicing agent, which is now the second most commonly used road deicer in North America [33].

Conclusions

Understanding the evolutionary history of an organism with its stressor, and the differential sensitivity of life history stages to that stressor are critical in assessing the vulnerability of organisms to stressful environments. It is now apparent that embryonic exposure to a stressor can have profound implications on the post-hatching survival and fitness of organisms in practically all animal taxa (Table S1), through influencing growth and development in this critical life history stage. In post-hatching individuals, however, even the largest, best-developed organism can only successfully deal with stressors that they have evolved to regulate. As the world of these organisms becomes increasingly impacted by anthropogenic factors, understanding this evolutionary history and its survival implications at and across different life history stages will be critical for the future conservation of animals in increasingly stressful environments.

Supporting Information

Table S1 Animal phyla where components of the embryonic environment have been demonstrated to have significant carry-over effects post-hatching. This list is not exhaustive, but is representative of the diversity and breadth of this phenomenon throughout the animal kingdom.

Acknowledgments

We thank B. Parrish for substantial help in the laboratory, as well as L. Neuman-Lee, N. Kiriazis, B. Gall, A. Stokes, T. Stokes, and B. Rowland for assistance. Thanks also to S. Durham and Z. Stopher for valuable help with data analysis, and J. Beatty and Oregon State University for access to their research ponds. B. and E. Gall kindly collected newts under Oregon Department of Fish and Wildlife permit # 062-11. We also thank G. Smith, Z. Stopher, and M. Baker for constructive feedback on an earlier version of the manuscript.

Author Contributions

Conceived and designed the experiments: GRH EDB. Performed the experiments: GRH. Analyzed the data: GRH SSF. Contributed reagents/materials/analysis tools: EDB SSF. Wrote the paper: GRH EDB SSF. Designed the analysis: SSF.

References

1. Pechenik JA, Wendt DE, Jarrett JN (1998) Metamorphosis is not a new beginning. BioScience 48: 901–910.
2. Calow P (1991) Physiological costs of combating chemical toxicants: ecological implications. Comparative and Biochemical Physiology 100C: 3–6.
3. Badyaev AV (2005) Stress-induced variation in evolution: from behavioural plasticity to genetic assimilation. Proceedings of the Royal Society B 272: 877–886.
4. DuRant SE, Hopkins WA, Wilson AF, Hepp GR (2011) Incubation temperature affects the metabolic cost of thermoregulation in a young precocial bird. Functional Ecology 26: 416–422.
5. Lindström J (1999) Early development and fitness in birds and mammals. Trends in Ecology and Evolution 14: 343–348.
6. Orizaola G, Dahl E, Laurila A (2010) Compensating for delayed hatching across consecutive life-history stages in an amphibian. Oikos 119: 980–987.
7. Barker DJP (2006) Adult consequences of fetal growth restriction. Clinical obstetrics and gynecology 49: 270–283.
8. Gluckman PD, Hanson MA, Beedle AS (2007) Early life events and their consequences for later disease: a life history and evolutionary perspective. American Journal of Human Biology 19: 1–19.
9. DuRant SE, Hopkins WA, Hawley DM, Hepp GR (2011) Incubation temperature affects multiple measures of immunocompetence in young wood ducks (*Aix sponsa*). Biology Letters 8: 108–111.
10. Webb GJW, Cooper-Preston H (1989) Effects of incubation temperature on crocodiles and the evolution of reptilian oviparity. American Zoologist 29: 953–971.
11. Dupont S, Dorey N, Stumpp M, Melzner F, Thorndyke M (2013) Long-term and trans-life-cycle effects of exposure to ocean acidification in the green sea urchin *Strongylocentrotus droebachiensis*. Marine Biology 160: 1835–1843.
12. Marshall DJ, Bolton TF, Keough MJ (2003) Offspring size affects the post-metamorphic performance of a colonial marine invertebrate. Ecology 84: 3131–3137.
13. Mandrillon A-L, Saglio P (2007) Effects of embryonic exposure to conspecific chemical cues on hatching and larval traits in the common frog (*Rana temporaria*). Chemoecology 17: 169–175.
14. Räsänen K, Laurila A, Merilä J (2002) Carry-over effects of embryonic acid conditions on development and growth of *Rana temporaria* tadpoles. Freshwater Biology 47: 19–30.
15. Giménez L, Anger K (2003) Larval performance in an estuarine crab, *Chasmagnathus granulata*, is a consequence of both larval and embryonic experience. Marine Ecology Progress Series 249: 251–264.
16. McKnight CM, Gutzke WHN (1993) Effects of the embryonic environment and of hatchling housing conditions on growth of young snapping turtles. Copeia 1993: 475–482.
17. Watkins TB, Vraspir J (2006) Both incubation temperature and posthatching temperature affect swimming performance and morphology of wood frog tadpoles (*Rana sylvatica*). Physiological and Biochemical Zoology 79: 140–149.
18. Kusch RC, Krone PH, Chivers DP (2007) Chronic exposure to low concentartions of waterborne cadmium during embryonic and larval development results in the long-term hindrance of antipredator behavior in zebrafish. Environmental Toxicology and Chemistry 27: 705–710.
19. Rohr JR, Palmer BD (2005) Aquatic herbicide exposure increases salamander desiccation risk eight months later in a terrestrial environment. Environmental Toxicology and Chemistry 24: 1253–1258.
20. Rohr JR, Sager T, Sesterhenn TM, Palmer BD (2006) Exposure, postexposure, and density-mediated effects of atrazine on amphibians: breaking down net effects into their parts. Environmental Health Perspectives 114: 46–50.
21. Snodgrass JW, Casey RE, Joseph D, Simon JA (2008) Microcosm investigations of stormwater pond sediment toxicity to embryonic and larval amphibians: variation in sensitivity among species. Environmental Pollution 154: 291–297.
22. Greulich K, Pflugmacher S (2003) Differences in susceptibility of various life stages of amphibians to pesticide exposure. Aquatic Toxicology 65: 329–336.
23. Qiu J-W, Qian P-Y (1999) Tolerance of the barnacle *Balanus amphitrite amphitrite* to salinity and temperature stress: effects of previous experience. Marine Ecology Progress Series 188: 123–132.
24. Egea-Serrano A, Relyea RA, Tejedo M, Torralva M (2012) Understanding of the impact of chemicals on amphibians: a meta-analytic review. Ecology and Evolution 2: 1382–1397.
25. Charmantier G, Giménez L, Charmantier-Daures M, Anger K (2002) Ontogeny of osmoregulation, physiological plasticity and larval export strategy in the grapsid crab *Chasmagnathus granulata* (Crustacea, Decapoda). Marine Ecology Progress Series 229: 185–194.
26. Ehlinger GS, Tankersley RA (2004) Survival and development of horsehoe crab (*Limulus polyphemus*) embryos and larvae in hypersaline conditions. Biological Bulletin 206: 87–94.
27. Giménez L (2002) Effects of prehatching salinity and initial larval biomass on survival and duration of development in the zoea 1 of the estuarine crab, *Chasmagnathus granulata*, under nutritional stress. Journal of Experimental Marine Biology and Ecology 270: 93–110.
28. Thiyagarajan V, Qian P-Y (2003) Effect of temperature, salinity and delayed attachment on development of the solitary ascidian *Styela plicata* (Lesueur). Journal of Experimental Marine Biology and Ecology 290: 133–146.
29. Environment Canada (2001) Priority Substances List Assessment Report: Road Salts. Environment Canada, Ottawa pp. 283.
30. Vernberg FJ, Vernberg WB (2001) The coastal zone: past, present, and future. Columbia, South Carolina: University of South Carolina Press.
31. Shoemaker VH, Nagy KA (1977) Osmoregulation in amphibians and reptiles. Annual Reviews of Physiology 39: 449–471.
32. Drever JI (1997) The geochemistry of natural waters: surface and groundwater environments. Upper Saddle River, NJ: Prentice Hall.
33. National Transportation Research Board (2007) National Cooperative Highway Research Program Report 577: Guidelines for the Selection of Snow and Ice Control Materials to Mitigate Environmental Impacts. Transportation Research Board of the National Academies, Washington, D.C.: National Cooperative Highway Research Program.
34. Kaushal SS, Groffman PM, Likens GE, Belt KT, Stack WP, et al. (2005) Increased salinization of fresh water in the northeastern United States. Proceedings of the National Academy of Sciences of the United States of Amerca 102: 13517–13520.
35. Thunqvist E (2004) Regional increase of mean chloride concentration in water due to the application of deicing salt. Science of the Total Environment 325: 29–37.
36. Cañedo-Argülles M, Kefford BJ, Piscart C, Prat N, Schäfer RB, et al. (2013) Salinization of rivers: an urgent ecological issue. Environmental Pollution 173: 157–167.
37. Chinathamby K, Reina RD, Bailey PCE, Lees BK (2006) Effects of salinity on the survival, growth and development of tadpoles of the brown tree frog, *Litoria ewingii*. Australian Journal of Zoology 54: 97–105.
38. Christy MT, Dickman CR (2002) Effects of salinity on tadpoles of the green and golden bell frog (*Litoria aurea*). Amphibia-Reptilia 23: 1–11.
39. Kearney BD, Byrne PG, Reina RD (2012) Larval tolerance to salinity in three species of Australian anuran: an indication of saline specialisation in *Litoria aurea*. PLoS One 7: e43427.
40. Williams WD (2001) Anthropogenic salinisation of inland waters. Hydrobiologia 466: 329–337.
41. Gornitz V (1995) Sea-level rise: a review of recent past and near-future trends. Earth Surface Processes and Landforms 20: 7–20.
42. Nicholls RJ, Hoozemans FMJ, Marchand M (1999) Increasing flood risk and wetland losses due to global sea-level rise: regional and global analyses. Global Environmental Change 9: S69–S87.
43. Purcell KM, Hitch AT, Klerks PL, Leberg PL (2008) Adaptation as a potential response to sea-level rise: a genetic basis for salinity tolerance in populations of a coastal marsh fish. Evolutionary Applications 1: 155–160.
44. Hopkins GR, French SS, Brodie Jr ED (2013) Potential for local adaptation in response to an anthropogenic agent of selection: effects of road deicing salts on amphibian embryonic survival and development. Evolutionary Applications 6: 384–392.
45. Armstrong JB, Duhon ST, Matacinski GM (1989) Raising the axolotl in captivity. In: Armstrong JB, Malacinski GM, editors. Developmental Biology of the Axolotl. New York: Oxford University Press. pp. 220–227.
46. Harrison RG (1969) Harrison stages and description of the normal development of the spotted salamander, *Amblystoma punctatum* (Linn.). In: Harrison RG, editor. Organization and Development of the Embryo. New Haven, CT.: Yale University Press. pp. 44–66.
47. Petranka JW, Doyle EJ (2010) Effects of road salts on the composition of seasonal pond communities: can the use of road salts enhance mosquito recruitment? Aquatic Ecology 44: 155–166.
48. Borst M, Brown RA (2013) Chloride released from three permeable pavement surfaces after winter salt application. Journal of the American Water Resources Association 50: 29–41.
49. Whitfield PH, Wade NL (1992) Monitoring transient water quality events electronically. Water Resources Bulletin 28: 703–711.
50. Whitfield PH, Wade NL (1996) Transient water quality events in British Columbia coastal streams. Water Science and Technology 33: 151–161.
51. Wu C-S, Yang W-K, Lee T-H, Gomez-Mestre I, Kam Y-C (2014) Salinity acclimation enhances salinity tolerance in tadpoles living in brackish water through increased Na+, K+-ATPase expression. Journal of Experimental Zoology 321A: 57–64.
52. Hua J, Pierce BA (2013) Lethal and sublethal effects of salinity on three common Texas amphibians. Copeia 2013: 562–566.
53. Hoffman RW, Goldman CR, Paulson S, Winters GR (1981) Aquatic impacts of deicing salts in the central Sierra Nevada mountains, California. Journal of the American Water Resources Association 17: 280–285.
54. Harless ML, Huckins CJ, Grant JB, Pypker TG (2011) Effects of six chemical deicers on larval wood frogs (*Rana sylvatica*). Environmental Toxicology and Chemistry 30: 1637–1641.
55. Langhans M, Peterson B, Walker A, Smith GR, Rettig JE (2009) Effects of Salinity on survivorship of wood frog (*Rana sylvatica*) tadpoles. Journal of Freshwater Ecology 24: 335–336.
56. Able KW, Palmer RE (1988) Salinity effects on fertilization success and larval mortality of *Fundulus heteroclitus*. Copeia 1988: 345–350.
57. Cutler DR, Edwards TC, Beard KH, Cutler A, Hess KT, et al. (2007) Random forests for classification in ecology. Ecology 88: 2783–2792.

58. Breiman L, Friedman JH, Olshen RA, Stone CJ (1984) Classification and regression trees. Monterey, California: Wadsworth and Brooks/Cole.
59. Breiman L (2001) Random forests. Machine Learning 45: 15–32.
60. Alexander LG, Lailvaux SP, Pechmann JHK, De Vries PJ (2012) Effects of salinity on early life stages of the Gulf Coast toad, *Incilius nebulifer* (Anura: Bufonidae). Copeia 2012: 106–114.
61. Brand A, Snodgrass JW, Gallagher MT, Casey RE, Van Meter R (2010) Lethal and sublethal effects of embryonic and larval exposure of *Hyla versicolor* to stormwater pond sediments. Archives of Environmental Contamination and Toxicology 58: 325–331.
62. Dougherty CK, Smith GR (2006) Acute effects of road de-icers on the tadpoles of three anurans. Applied Herpetology 3: 87–93.
63. Karraker NE, Gibbs JP (2011) Road deicing salt irreversibly disrupts osmoregulation of salamander egg clutches. Environmental Pollution 159: 833–835.
64. Karraker NE, Gibbs JP, Vonesh JR (2008) Impacts of road deicing salt on the demography of vernal pool-breeding amphibians. Ecological Applications 18: 724–734.
65. Karraker NE, Ruthig GR (2009) Effect of road deicing salt on the susceptibility of amphibian embryos to infection by water molds. Environmental Research 109: 40–45.
66. Padhye AD, Ghate HV (1992) Sodium chloride and potassium chloride tolerance of different stages of the frog, *Microhyla ornata*. Herpetological Journal 2: 18–23.
67. Viertel B (1999) Salt tolerance of *Rana temporaria*: spawning site selection and survival during embryonic development (Amphibia, Anura). Amphibia-Reptilia 20: 161–171.
68. Collins SJ, Russell RW (2009) Toxicity of road salt to Nova Scotia amphibians. Environmental Pollution 157: 320–324.
69. Dunson WA (1977) Tolerance to high temperature and salinity by tadpoles of the Philippine Frog, *Rana cancrivora*. Copeia 1977: 375–378.
70. Gordon MS, Schmidt-Nielsen K, Kelly HM (1961) Osmotic regulation in the crab-eating frog (*Rana cancrivora*). Journal of Experimental Biology 38: 659–678.
71. Griffis-Kyle KL (2005) Ontogenetic delays in effects of nitrite exposure on tiger salamanders (*Ambystoma tigrinum*) and wood frogs (*Rana sylvatica*). Environmental Toxicology and Chemistry 24: 1523–1527.
72. Pahkala M, Laurila A, Merilä J (2001) Carry-over effects of ultraviolet-B radiation on larval fitness in *Rana temporaria*. Proceedings of the Royal Society B 268: 1699–1725.
73. Smith GR, Waters MA, Rettig JE (2000) Consequences of embryonic UV-B exposure for embryos and tadpoles of the plains leopard frog. Conservation Biology 14: 1903–1907.
74. Moran AL, Emlet RB (2001) Offspring size and performance in variable environments: field studies on a marine snail. Ecology 82: 1597–1612.
75. Gorman HE, Nager RG (2004) Prenatal developmental conditions have long-term effects on offpsring fecundity. Proceedings of the Royal Society B 271: 1923–1928.
76. Boone MD, Scott DE, Niewiarowski PH (2002) Effects of hatching time for larval Ambystomatid salamanders. Copeia 2002: 511–517.
77. Touchon JC, Warkentin KM (2010) Short- and long-term effects of the abiotic egg environment on viability, development and vulnerability to predators of a Neotropical anuran. Functional Ecology 24: 566–575.
78. Touchon JC, Warkentin KM (2013) Effects of plastic hatching timing carry over through metamorphosis in red-eyed treefrogs. Ecology 94: 850–860.
79. Warkentin KM (1995) Adaptive plasticity in hatching age: a response to predation risk trade-offs. Proceedings of the National Academy of Sciences of the United States of Amerca 92: 3507–3510.
80. Warkentin KM (1999) Effects of hatching age on development and hatchling morphology in the red-eyed treefrog, *Agalychnis callidryas*. Biological Journal of the Linnean Society 68: 443–470.
81. Beebee TJC (1986) Acid tolerance of natterjack toad (*Bufo calamita*) development. Herpetological Journal 1: 78–81.
82. Cooke AS (1979) The incluence of rearing density on the subsequent response to DDT dosing for tadpoles of the frog *Rana temporaria*. Bulletin of Environmental Contamination and Toxicology 21: 837–841.
83. Gall BG, Stokes AN, French SS, Schlepphorst EA, Brodie III ED, et al. (2011) Tetrodotoxin levels in larval and metamorphosed newts (*Taricha granulosa*) and palatability to predatory dragonflies. Toxicon 57: 978–983.
84. Metcalfe NB, Monaghan P (2001) Compensation for a bad start: grow now, pay later? Trends in Ecology and Evolution 16: 254–260.
85. Charmantier G (2010) Ontogeny of osmoregulation in crustaceans: a review. Invertebrate Reproduction and Development 33: 177–190.
86. Gosner KL, Black IH (1957) The effects of acidity on the development and hatching of New Jersey frogs. Ecology 38: 256–262.
87. Alvarado RH, Dietz TH (1970) Effect of salt depletion on hydrominearal balance in larval *Ambystoma gracile* - I. Ionic composition. Comparative and Biochemical Physiology 33: 85–92.
88. Alvarado RH, Dietz TH (1970) Effect of salt depletion on hydromineral balance in larval *Ambystoma gracile* - II. Kinetics of ion exchange. Comparative and Biochemical Physiology 33: 93–110.
89. Alvarado RH, Moody A (1970) Sodium and chloride transport in tadpoles of the bullfrog *Rana catesbeiana*. American Journal of Physiology 218: 1510–1516.
90. Bernabò I, Bonacci A, Coscarelli F, Tripepi M, Brunelli E (2013) Effects of salinity stress on *Bufo balearicus* and *Bufo bufo* tadpoles: tolerance, morphological gill alterations and Na+/K-ATPase localization. Aquatic Toxicology 132–133: 119–133.
91. Alvarado RH, Cox TC (1985) Action of polyvalent cations on sodium transport across skin of larval and adult *Rana catesbeiana*. Journal of Experimental Zoology 236: 127–136.
92. Calabrese EJ (2003) Hormesis: the dose-response revolution. Annual Reviews of Pharmacology and Toxicology 43: 175–197.
93. Ashraf M, McNeilly T, Bradshaw AD (1989) The potential for evolution of tolerance to sodium chloride, calcium chloride, magnesium chloride and seawater in four grass species. New Phytologist 112: 245–254.
94. Wu L (1981) The potential for evolution of salinity tolerance in *Agrostis stolonifera* L. and *Agrostis tenuis* Sibth. New Phytologist 89: 471–486.
95. Hallsworth JE, Yakimov MM, Golyshin PN, Gillion JLM, D'Auria G, et al. (2007) Limits of life in MgCl2-containing environments: chaotropicity defines the window. Environmental Microbiology 9: 801–813.
96. Beebee TJC (1985) Salt tolerances of natterjack toad (*Bufo calamita*) eggs and larvae from coastal and inland populations in Britain. Herpetological Journal 1: 14–16.
97. Vonesh JR, De la Cruz O (2002) Complex life cycles and density dependence: assessing the contribution of egg mortality to amphibian declines. Oecologia 133: 325–333.

21

Patterns of MiRNA Expression in Arctic Charr Development

Kalina H. Kapralova, Sigrídur Rut Franzdóttir, Hákon Jónsson, Sigurður S. Snorrason, Zophonías O. Jónsson*

Institute of Life- and Environmental Sciences, University of Iceland, Reykjavík, Iceland

Abstract

Micro-RNAs (miRNAs) are now recognized as a major class of developmental regulators. Sequences of many miRNAs are highly conserved, yet they often exhibit temporal and spatial heterogeneity in expression among species and have been proposed as an important reservoir for adaptive evolution and divergence. With this in mind we studied miRNA expression during embryonic development of offspring from two contrasting morphs of the highly polymorphic salmonid Arctic charr (*Salvelinus alpinus*), a small benthic morph from Lake Thingvallavatn (SB) and an aquaculture stock (AC). These morphs differ extensively in morphology and adult body size. We established offspring groups of the two morphs and sampled at several time points during development. Four time points (3 embryonic and one just before first feeding) were selected for high-throughput small-RNA sequencing. We identified a total of 326 conserved and 427 novel miRNA candidates in Arctic charr, of which 51 conserved and 6 novel miRNA candidates were differentially expressed among developmental stages. Furthermore, 53 known and 19 novel miRNAs showed significantly different levels of expression in the two contrasting morphs. Hierarchical clustering of the 53 conserved miRNAs revealed that the expression differences are confined to the embryonic stages, where miRNAs such as sal-miR-130, 30, 451, 133, 26 and 199a were highly expressed in AC, whereas sal-miR-146, 183, 206 and 196a were highly expressed in SB embryos. The majority of these miRNAs have previously been found to be involved in key developmental processes in other species such as development of brain and sensory epithelia, skeletogenesis and myogenesis. Four of the novel miRNA candidates were only detected in either AC or SB. miRNA candidates identified in this study will be combined with available mRNA expression data to identify potential targets and involvement in developmental regulation.

Editor: Christophe Antoniewski, CNRS UMR7622 & University Paris 6 Pierre-et-Marie-Curie, France

Funding: This project was supported by grants from The Icelandic Centre for Research (RANNIS, http://http://www.rannis.is/), grants #100204011, #110285-0061 and #130756-051, and the University of Iceland Research Fund. The funders had no role in study design, data collection and analysis, decision to publish, or preparation of the manuscript.

Competing Interests: The authors have declared that no competing interests exist.

* Email: zjons@hi.is

Introduction

Since the initial discoveries of lin-4 and let-7 miRNAs have emerged as key regulators of animal development (reviewed in [1,2]). These small (~22 nt) non coding RNAs, regulate gene expression by inducing mRNA degradation or translational repression, making for a specific and "fine-tunable" response (reviewed in [3]). miRNAs originate from different parts of the genome (intergenic regions, exons or intronic sequences [4] and are transcribed either as independent transcriptional units or as clusters of several miRNAs (reviewed in [5]). A common feature of all miRNA genes, regardless of their genome location, is the folding of their primary transcript into a stem-loop structure. This hairpin structure is recognized and converted into a miRNA-miRNA* duplex by the miRNA processing machinery (see [5]). One of the strands, dubbed the "mature" miRNA, is then loaded into the miRISC complex while the complementary "star" sequence is often degraded [6]. In most cases of miRNA mediated gene regulation the target repertoire is determined by the "seed" region (nt 2-8 located at the 5' end of the mature miRNA) of the miRNA [7].

In general miRNAs are highly conserved among taxa [5]. Comparative studies show how new miRNAs have continuously been emerging during the evolution of metazoan genomes [8,9] through various mechanisms including gene duplications of preexisting miRNAs followed by changes in their sequences, or *de novo* appearance from random hairpins [10]. However, once they become integrated into the regulatory network, their primary sequence and particularly their seed region, becomes subject to strict selective constraints [5,11]. Variation in timing and expression patterns among species suggests that these molecules may play an important role in shaping physiological differences. For example comparison of two fish species (medaka and zebrafish) showed that heterochrony in miRNAs expression is associated with neuromast and craniofacial development [12]. This was suggested to reflect the differences in formation of the head and sensory epithelia observed between medaka and zebrafish. Morphological differences arising in development can potentially drive evolutionary change, adaptive divergence and speciation (discussed in [1]). More specifically, it has been suggested that miRNAs may generally cover more restricted regulatory niches than transcription factors and thus frequently be more important in terminal differentiation programs [13]. It has also been proposed that miRNAs are involved in enhancing species evolvability by stabilizing gene expression and signaling cascades leading to the increased distinctness of developmental phenotypes, thereby increasing heritability of traits and facilitating natural selection ([14] and discussion in [15]).

Arctic charr as a model species to study adaptive divergence

The high level of phenotypic polymorphism present in Northern freshwater systems offers an excellent opportunity to study adaptive divergence [16]. These watersheds, with their rivers and lakes, were formed after the last glacial epoch 10 000–16 000 years ago. The short evolutionary history characterized by physical variability and topographic dynamics sets a stage where the early steps of divergence may be playing out in multiple locations and species. Studies of whitefish (*Coregonus clupeaformis*), threespine stickleback (*Gasterostreus aculeatus*) and Arctic charr (*Salivelunus alpinus*) have shown that fish inhabiting these systems exhibit an extremely high level of inter-population variation in phenotype with many populations diversifying along a benthic to limnetic habitat axis [17–20]. Although Arctic charr in Iceland originates from a single Atlantic lineage [21], this species shows an extremely high level of variation in phenotype between populations and many examples of polymorphism (i.e. sympatric morphs) have been documented [17,22–24]. The Arctic charr morphs of Lake Thingvallavatn constitute an extreme example of local phenotypic diversity. Four morphs grouped into two morphotypes have been described in the lake: a limnetic morphotype represented by planktivorous (PL) and piscivorous (PI) charr, with pointed snout and evenly protruding jaws, and a derived, benthic morphotype represented by small (SB) and large benthivorous (LB) charr, blunt snout, short lower jaw and relatively large pectoral fins [25]. These morphs also differ extensively in life history characteristics (size and age at maturity) and embryology [26–29]. The morphs also exhibit strikingly clear differentiation in ecology as reflected in different habitat use, diet and endoparasite fauna [27,28,30]. Several common garden experiments have shown that some key morph specific traits have a definite genetic basis [31,32]. A recent study, using neutral microsatellite markers, revealed significant but subtle genetic differentiation between the three most common morphs in Lake Thingvallavatn, which is consistent with a scenario of early evolution of reproductive isolation, followed by slow divergence by drift with restrictive gene flow [33]. Notably, a study of immune system genes revealed more pronounced genetic differentiation among the morphs in the lake, consistent with a scenario where parts of the immune systems have diverged substantially among Arctic charr morphs from Lake Thingvallavatn [34]. The adaptive nature of the trophic morphology and feeding behavior of the Thingvallavatn morphs has been demonstrated in a series of laboratory rearing experiments [29,31,35]. Moreover the role of developmental heterochrony in the evolution of the Thingvallavatn Arctic charr morphs was demonstrated in a study showing that some skeletal elements of the head start ossifying earlier and/or faster in small benthivorous embryos than in embryos derived from the planktivorous morph [35].

Some of the key differences in functional traits that define the charr morphs are without doubt rooted in differences in the expression of developmental genes. We hypothesize that miRNAs may, through their potentially stabilizing effect of phenotypes [14], play a fundamental role in the divergence of developmental processes that induce differential cranial morphologies in Arctic charr morphs. As a first step of addressing this hypothesis we utilized high-throughput sequencing techniques to identify and annotate Arctic charr miRNAs and to study their expression during the development of two contrasting Arctic charr morphologies. To this end we used a common garden set up to generate embryonic series of two contrasting Arctic charr morphotypes, a benthic morphotype, represented by the SB-charr from Thingvallavatn and a limnetic morphotype represented by fish from the Hólar aquaculture stock (AC). These two morphs differ greatly in adult size, color and head morphology (Figure 1): SB are small, dark and have a sub-terminal mouth and rounded snout whereas AC are large, silvery and have a pointed snout and a longer lower jaw. We sampled AC and SB embryos at four developmental timepoints reflecting important events in Arctic charr craniofacial development and used high-throughput sequencing to quantify differences in miRNA expression between the morphs. More specifically we identified and annotated Arctic charr miRNAs using homology to known miRNAs in other species. Furthermore, we identified a large set of novel miRNA candidates by aligning reads to genomic sequences from the closely related Atlantic salmon, *Salmo salar*. Expression levels for both known and novel miRNAs were compared between AC and SB.

Materials and Methods

Sampling and methodology

All sampling from the wild and rearing in aquaculture was performed according to Icelandic law and with proper permissions. Fish from Lake Thingvallavatn were caught by the authors for the purpose of this study with fishing permissions obtained from the Thingvellir National Park Commission and the owner of the Mjóanes farm. SSS and ZOJ hold special permits for sampling fish from nature for scientific purposes according to Icelandic law (clause 26 of law 61/2006 on salmonid fishing). Control fish from Hólar aquaculture stock were obtained from a national breeding programme, and were not specifically bred for the purpose of this project. These fish are held at the arctic charr breeding station, a quarantined rearing and holding facility, at Hólar University College. After stripping for gametes, parent fish were killed by a sharp blow to the head and checked for absence of breathing when placed in water. Setting up crosses and the subsequent killing of parents was performed by the authors. Ethics committee approval is not needed for regular or scientific fishing in Iceland (The Icelandic law on animal protection, Law 15/1994, last updated with Law 157/2012). The rearing of embryos was performed according to Icelandic regulations (licence granted to Hólar University College aquaculture and experimental facilities) in Verið, Sauðárkrókur, Iceland. Sampling of embryos was performed by University College Aquaculture Research Station (HUC-ARC) personnel. HUC-ARC has an operational license according to Icelandic law on aquaculture (Law 71/2008), that includes clauses of best practices for animal care and experiments. For this study the last gestation age at which embryos were sacrificed was 434 (τ_s) units. For RNA extraction, samples were flash frozen in RNA later. Prior to freezing eggs were permeabilized by puncture with a needle. Samples for staining (not described in this study) were treated with an overdose of phenoxyethoanol before fixing.

For this study we used developmental time-series from pure crosses of two Arctic charr morphs, Hólar aquaculture charr (AC) and small benthic charr (SB) from Lake Thingvallavatn. These strains were selected mainly for their pronounced differences in body size, coloration and head morphology (Figure 1). As stated above, the AC crosses were made with parents from the Hólar breeding programme [36]. Fish from the small benthic morph (SB)

Figure 1. Two contrasting Arctic charr morphs differing in size, coloration and head morphology. Top: Arctic charr from aquaculture stock (AC) is large, silvery and has a pointed snout and long lower jaw, Bottom: Small benthic charr from Thingvallavatn (SB) is small, dark and has a sub-terminal mouth and rounded snout.

were caught in Lake Thingvallavatn using gill-nets. Eggs from several females were pooled and fertilized using milt from several males from the same group. Eggs were reared at approximately 5°C in a hatching tray (EWOS, Norway) under constant water flow and in complete darkness at the Holar University College experimental facilities in Verið, Sauðárkrókur. Exact water temperature was recorded twice daily to estimate the relative age of the embryos using tau-somite (τ_s) units defined as the time it takes for one somite pair to form at a given temperature [37]. Embryos were collected throughout development and either fixed in 4% PFA or stored in RNAlater (Ambion) at −80°C. Based on embryos sampled at different developmental stages and stained with alcian blue (cartilage) and alizarin red (bone), four time-points (141, 161, 200 and 434 τ_s) were selected to represent important stages of bone and cartilage development. Stages 141, 161 and 200 are embryonic whereas stage 434 is a fry stage and for simplicity these stages will be referred to as stages 1, 2, 3 and 4, respectively. Two independent samplings were performed: one was used for high-throughput small-RNA-sequencing (miRNA-seq) and the other one for qRT-PCR.

Small RNA sequencing

Total RNA from each stage of each morph was isolated from a pool of 6 whole embryos and enriched for small RNAs using the mirVana kit (Ambion). The purity and amount of small RNA was verified on a BioAnalyzer (Agilent Technologies). The samples were prepared for sequencing following the small RNA v1.5 sample preparation protocol from Illumina. Briefly, 3' and 5' RNA adapters were ligated to small RNAs, which were subsequently, reverse transcribed into DNA and PCR amplified. The samples were then run on polyacrylamide gels and the DNA eluted from bands corresponding to 20–30 nucleotide RNA fragments. miRNA and transcriptome sequencing (mRNA-seq) was performed at deCODE Genetics (Reykjavik, Iceland) using the TruSeq smallRNA (v1.5) kit (Illumina) on an Illumina $GAII_X$ instrument. Raw reads were submitted to NCBI Sequence Read Archive (SRA) under accession number SRP039492.

miRNA-seq data processing

Raw reads were processed with cutadapt [38] as follows: First, adaptor sequences were removed and only reads with adaptors were kept. Next, we used the FastX toolkit (http://hannonlab.cshl.edu/fastx_toolkit/index.html) (script available on request) and the quality scores associated with the reads to remove bases with a Phred based quality score [39] of less than 20 from read ends. Sequences retaining less than 15 nucleotides after filtering were discarded. Reads where 10% or more of the bases had a Phred quality score lower than 20 were also discarded. Finally, identical reads were reduced to one copy with the redundancy noted in the read name. The sequence filtering and collapsing was repeated for each sample.

Annotation of ncRNAs

To annotate sequences using known RNAs we used Rfam version 10.1 [40] and miRBase [41–44] version 20 databases. The Rfam database was searched with HMMER (version 3.0; http://hmmer.janelia.org/) with an e-value cutoff of 0.01. For the miRBase the ssearch command from the fasta package version 36.3.6d [45] was used to detect homology between the mature miRNAs and the collapsed sequences (e-value cutoff 0.01).

On a quest for novel miRNAs

To identify novel miRNAs we used a probabilistic model of miRNA biosynthesis implemented in miRDeep 2 [46]. As a sequenced Arctic charr genome is not currently available, we used the genome sequences from the closely related Atlantic salmon [47] for reference. Collapsed reads were aligned to the Atlantic salmon genome with the mirDeep 2 mapping program (10 minimum reads per miRNA) for both morphs with the time-points combined. To facilitate mapping, collapsed sequences with strictly lower read count than 4 were omitted from the detection. The miRDeep2 algorithm then uses the reference regions bracketing the aligned reads to compute a hairpin structure and estimates the probability of each sequence being a true miRNA precursor based on the position of the reads, their frequency, the energetic stability of their secondary structure and conservation of the 5′ ends.

Sequences with log score greater or equal to 2 were considered as potential miRNAs, the predicted hairpins were searched against hairpins from mirBase (version 20) with blastall (version 2.2.26, -W 7) [48]. Candidates were annotated as known miRNAs if the alignment length was greater or equal to 60 nucleotides and expected value for the match was lower than 0.01 (-e 0.01) otherwise the hairpins were classified as novel.

PCR amplification and sequencing of miRNA clusters

To assess the degree of sequence conservation for genomic clusters containing known and novel miRNAs between Arctic charr and Atlantic salmon we selected 4 clusters containing known miRNAs (miR-19c, 18b* and 20b; miR-133a and miR-133b and miR-143-3p and miR-143-5p; miR-219-3p and miR-219-5p) and 3 clusters containing novel miRNA candidates (sal-nov-235, sal-nov-242 and sal-nov-343) and PCR amplified their genomic regions from the Arctic charr genome. Primers were designed with Primer3 (http://primer3.wi.mit.edu/) (Table S1). The same PCR program was used for all primer pairs: an initial denaturation at 95°C for 5 min; 35 cycles of 95°C for 45 seconds; 45 seconds at a 53°C; 1 min at 72°C, then a final step of 10 min at 72°C. PCR products were treated with ExoSap and sequenced on an Applied Biosystems 3500xL Genetic Analyzer using BigDye chemistry. Raw sequencing data was base-called by Sequencing Analysis Software v5.4 with KBTMBasecaller v1.41 (Applied Biosystems), and run through Phred and Phrap, prior to trimming primer sequences, visual editing of ambiguous bases and putative polymorphisms in Consed [49]. Fasta files were exported and aligned with ClustalW (http://www.ebi.ac.uk/Tools/msa/clustalw2/, and manually inspected for alignment errors in Genedoc (www.psc.edu/biomed/genedoc). All sequences were deposited in Genebank under the accession numbers [KJ573796-KJ573802]. These sequences were then searched using blast against the salmon database. The conservation between Arctic charr and salmon ranged between 91–94% for the known miRNAs and 92–98% for the novel miRNA candidates. Mismatches were always located outside of the miRNA mature-star sequence.

Differential expression analysis

The R package edgeR [50] was used to study the differential expression of conserved and novel miRNA candidates between morphs and among developmental time-points in a generalized linear model, where the additive covariates (no-interaction) corresponded to developmental time-point and different morphs. The normalization factors were calculated for each sample using the function calcNormFactors. As there are no replicates in any of the experimental conditions, the options method = "deviance", robust = TRUE and subset = NULL were used for estimating the common dispersion (function estimateGLMCommonDisp) parameters as recommended by the edgeR user manual. The trended and tagwise dispersion were also estimated (function estimateGLMTrendedDisp and estimateGLMTagwiseDisp) with default options. The statistical significance of the terms was assessed by comparing likelihood difference to a reduced model without the time or the morph terms, with the function glmLRT in edgeR. The first 20 bases of each annotated sequence (novel and previously described miRNAs) were used as an identifier and the counts were aggregated for sequences that share the first 20 bases. This allowed us to work at the sequence level without lumping together isoforms (isomiRs). An entry (first 20 bases) was only considered for the statistical testing if the counts per million reads were strictly greater than 3 in at least two experimental points resulting in 1862 tags. We adjusted for multiple testing using the Benjamini-Hochberg false discovery rate [51]. The R script used for this analysis is available in File S1.

Descriptive analysis

Cluster analysis was performed using the heatmap function and plotted using the gplots package in R (http://www.r-project.org/). Prior to the clustering analysis expression levels for each miRNA were normalized across samples using a Variance Stabilizing Transformation.

Real-time quantitative PCR analysis

In order to verify the observed differential expression between morphs in our miRNA-seq data, we selected 9 miRNAs (sal-miR-17, sal-miR-26a, sal-miR-30b, sal-miR-122, sal-miR-140, sal-miR-181a*, sal-miR-196a, sal-miR-199a and sal-miR-206) for qPCR analysis. We concentrated on the 3 embryonic stages, as in both our analyses (for morph or developmental effect) the expression profiles between the samples of the last stage appeared to be very similar (see results). For the qPCR analysis two separate RNA extractions (biological replicates) were used for each data point. RNA was extracted from pools of 6 whole embryos/fry using a standard TRI Reagent (Sigma) protocol and treated with DNaseI (New England Biolabs) in order to limit genomic DNA contamination. All samples were from the same sampling effort and were extracted and processed simultaneously. cDNA was synthesized using the Exiqon universal cDNA Synthesis Kit II. The consistency of the cDNA synthesis among samples was verified using a spike in template along with a Control primer set (Exiqon). For the qPCRs we used SYBR Green master mix (Exiqon) and LNA primers (Exiqon). All qPCRs were done in duplicates (technical replicates) in a 10 µl reaction volume in 96 well-PCR plates on an ABI 7500 real-time PCR System (Applied Biosystems) following manufacturer instructions (Exiqon). The same PCR program was used for all miRNA primer pairs: starting with a 2 min hold at 50°C followed by a 10 min initial denaturation at 95°C and 45 cycles of 10 sec denaturation at 95°C and 1 min annealing/extension at 60°C. A melting curve analysis was performed at the end of each PCR to verify the specificity of the amplification. U2 spliceosomal snRNA (Primer sequence: GGTACTGCAATACCGGGG) was initially selected as a reference gene. However, the use of non-miRNA genes as reference has been shown to be problematic and the use of mean expression is often more appropriate [52]. We therefore opted to use the geometric mean for the expression values of the miRNAs under study as a reference. Relative expression (fold change) for each miRNA compared to stage 1 in AC was calculated in R using a script provided in File S2.

Results

Small RNA sequencing descriptive statistics

In order to identify miRNAs involved in Arctic charr development and morph differences, we made 8 small RNA libraries from four developmental time-points of two contrasting morphs of Arctic charr. The sequencing depth ranged from 29.1 to 33.9 million reads with a mean depth of 32.4 million reads per sample. After removing the adapters using cutadapt [38] and filtering out low quality reads using the FastX toolkit (http://hannonlab.cshl.edu/fastx_toolkit/), we obtained on average 24.8 million reads per sample (Table 1). The size distribution of the collapsed reads of all 8 libraries accounting for redundancy is shown in Figure S1. The majority of the reads were 21–23 nucleotides, corresponding to the typical miRNA size range. Details of the size distribution for both unique and collapsed reads for all 8 libraries are shown in Figure S2. All 8 libraries showed similar distribution with a peak at 21–23 nt. Furthermore, annotation of the collapsed reads using the Rfam database confirmed that our small RNA libraries were highly enriched with miRNAs (Table 2).

A total of 326 conserved and 427 novel miRNA candidates were found in the data

All collapsed reads were compared to the mature miRNA sequences available in miRBase (release 19) using ssearch [45] and 326 candidates (Table S2) were identified with high confidence (e-value <0.001). The 10 most abundant miRNAs account for 65% of the total conserved miRNAs (Figure 2) with sal-miR-206 and sal-miR-1 alone accounting for 36% of the total miRNAs. We identified 427 novel miRNA candidates (Table S3) of which 37% were represented by the 10 highest expressed putative miRNAs. We sequenced the genomic regions of three novel miRNAs (sal-nov-235, sal-nov-242 and sal-nov-334). They were all highly conserved between Arctic charr and Salmon (Table S1). Furthermore their mature and star sequences are located in highly conserved blocks in medaka, fugu, tetraodon and stickleback. Several of the conserved miRNAs were present in two or three isoforms (isomiRs). For example sal-miR-451 exists in 3 isoforms (Table S2). Two of these (sal-miR-451_1 and sal-miR-451_3) are highly conserved among vertebrates, whereas the third (sal-miR-451_2) has not previously been described in other species. This derived isoform differs in one base (G->U substitution) located at the 3' end of the mature sequence and is the predominant isoform of sal-miR-451 in our data (Table S2). Another interesting example is sal-miR-152, where 4 isoforms are found in our data (Table S2) with the most abundant being the ancestor sequence. The three other isoforms are one mutation away from the ancestral form. Interestingly, these mutations (T->A, C, or G) are located at the same site (position 5) for all 3 derived isoforms (Table S2).

51 known miRNAs and 6 novel miRNA candidates are differentially expressed among developmental stages

We found 51 known miRNAs to be differentially expressed among developmental time-points. Hierarchical clustering (Figure 3, see also Table S4 for background data) of miRNA expression showed that the 8 samples grouped into four clusters according to developmental time: one major division separates the three embryonic stages (1, 2 and 3) from the last post-hatching stage (4) and three divisions for each of the three embryonic stages (Figure 3). This major division between embryonic and post hatching stages indicates a clear shift in miRNAs expression between these developmental phases. The second division separates stage 3 from stages 1 and 2 and the third division separates stages 1 and 2 (Figure 3). Interestingly, there are two major divisions in the miRNA expression pattern clustering: node one depicts miRNAs that are highly expressed during the embryonic stages and their expression decreases in the last stage while the second node includes miRNAs with high expression in stage 4 and low expression in the embryonic stages. For example members of the 430 family (miR-430 a, b, c and d) are highly expressed in the embryonic stages and their expression decreases markedly in late development. In addition other miRNAs, such as sal-miR-219 a and b and miR-181c, show higher expression in the embryonic stages. On the other hand, miRNAs such as sal-miR-22a, 140, 182, 183, 192, 215 and different members of the let-7 family show increasing expression over time. Of the novel candidates, 6 putative miRNAs were found to be differentially expressed among developmental points (Table 3). Three of them sal-nov-1, sal-nov-5 and sal-nov-18 are also differentially expressed between morphs.

53 known miRNAs and 19 novel miRNA candidates are differentially expressed between AC and SB embryos

We tested for differential expression between morphs using a Generalised Linear Model and adjusted for multiple testing using the Benjamin-Hochberg false discovery rate as decribed in methods. We found 53 miRNAs to be differentially expressed between AC and SB. These miRNAs cluster by morph during the embryonic stages (stages 1-3) (Figure 4, see also Table S4 for background data). During these 3 stages miRNAs such as sal-miR-130, 133, 153, 17, 30, 451, 219, 26, 199a and 145 are highly expressed in AC, whereas sal-miR- 206, 133, 122, 181a, 192, 196a and 223 are highly expressed in SB. The expression of some of these "morph specific" miRNAs for example sal-miR-130, 153, 17, 30b and 30c in AC and sal-miR-196a, 206, 192 and 122 in SB observed in the embryonic stages decreases markedly in the last stage. During the last stage the observed miRNA expression differences between the two morphs disappear (Figure 4). Of the novel miRNA candidates, 19 putative miRNAs were found to be differentially expressed between AC and SB (Table 3). Two of them, sal-nov-4 and 5 were only expressed in AC and at most/all stages whereas expression of another putative novel miRNA, sal-

Table 1. Summary of read numbers from small RNA sequencing.

Sample	Number of reads (NR)	NR after adapter trimming	NR after collapsing
SB 1	33.4 M	23.1 M	2.4 M
SB 2	32.8 M	28.2 M	0.9 M
SB 3	33.9 M	30.5 M	0.7 M
SB 4	29.1 M	21.6 M	0.7 M
AC 1	32.8 M	24.2 M	1.7 M
AC 2	32.5 M	21.1 M	2.0 M
AC 3	31.3 M	23.2 M	0.7 M
AC 4	33.6 M	26.2 M	1.0 M

Number of reads (NR, in millions of reads) in high-throughput data from small RNA libraries of four developmental points and two morphs of Arctic charr. AC = Aquaculture charr, SB = Small Benthic charr.

Table 2. High-throughput reads annotated using the Rfam database.

snc RNA	Number of reads
miRNA	50841311
rRNA	49033
mRNA	5755
tRNA	24451
SNORD	151004
U	208259
sno	25857

nov-3, was only detected in SB offspring and at all four developmental points. With three exceptions (sal-nov-4, 5 and 16) all of the differentially expressed putative miRNAs showed very low expression levels (Table 3). We selected 9 miRNAs and further studied their expression by qPCR using independent biological replicates. The selection was based on the dynamics and degree of differential expression between morphs and/or among developmental points seen in the sequencing data. We concentrated on the three embryonic stages, as in both our analyses (for morph or developmental effect), the expression profiles between the samples of stage as the expression profiles between the samples of stage 4 appeared to be very similar. Eight of these miRNAs (miR-17, miR-26a, miR-30b, miR-140, miR-181a*, miR-196a, miR-199a and miR-206) amplified well (Figure 5), whereas miR-122 showed double peaks in melting curve analysis and was discarded from further analysis. Five (miR 17, 26a, 30b, 140 and 206) out of eight miRNAs tested with qPCR showed similar expression patterns to what was expected from the high-throughput sequencing (Figure 5, A–E). Three miRNAs (miR-196a and miR-199a and miR-181a) exhibited similar expression patterns in one or two of the three stages under study, (Figure 5, F–H).

Discussion

The molecular mechanisms underlying the development of Arctic charr morphologies are likely to have parallels in other vertebrate species and studying them is of interest in both developmental and evolutionary contexts. Given the recent divergence in northern populations of Arctic charr, it is likely that the observed phenotypic polymorphism rooted in development in this species arose mostly by differences in gene regulation as opposed to changes in protein coding sequences. In a recent study [53] two genes involved in matrix remodeling in bone formation (sparc and mmp2) showed consistent differences in expression during the development of embryos derived from benthic and limnetic morphs of Arctic charr, suggesting that these genes might be involved in the development of these distinct Arctic charr morphologies. Little is known about what controls such differences in expression. While numerous studies demonstrate the involvement of transcription factors and other regulatory elements in the phenotypic evolution of birds [54] and fish [55–60], there are few known examples of miRNAs playing a role in morphological variation. In a recent study Arif *et al.* [61] experimentally demonstrated that differences in the "naked valley" phenotype

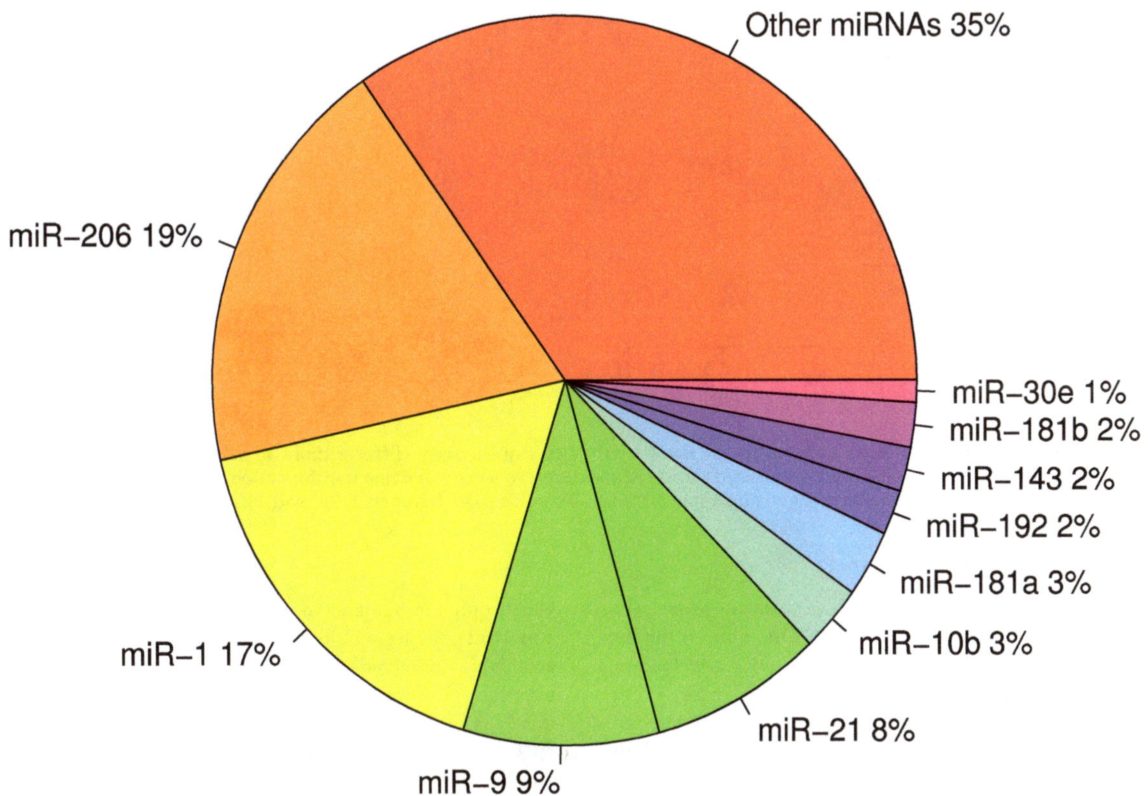

Figure 2. Relative abundance of known miRNAs in all samples combined. Together the 10 most abundant known miRNAs constitute 65% of all known miRNAs.

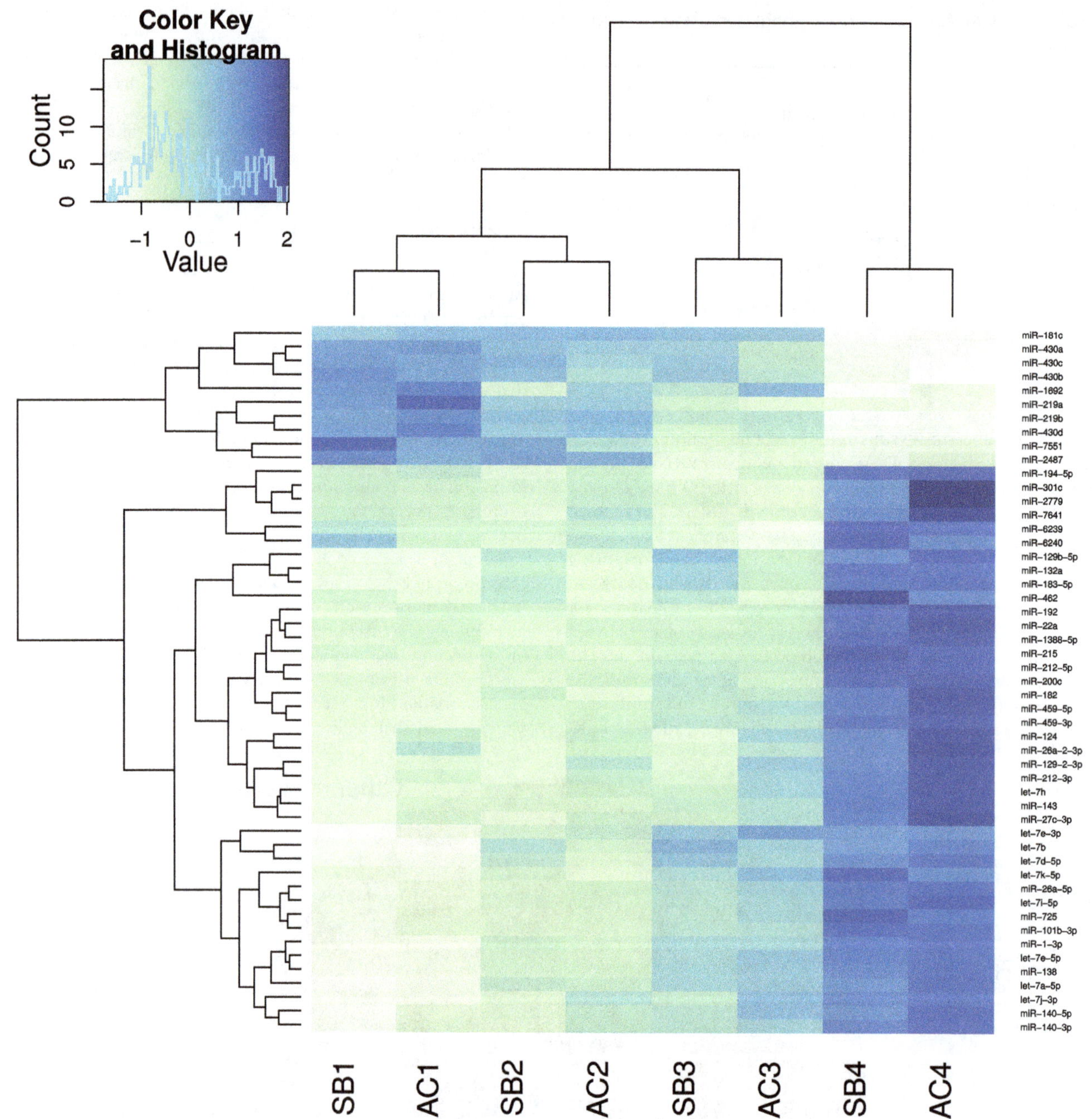

Figure 3. Heat-map showing relative expression of the 51 miRNAs significantly differentially expressed among developmental stages. Expression levels for each miRNA were normalized across samples using variance stabilizing transformation. Blue denotes high and white low relative expression. AC stands for Aquaculture charr and SB stands for Small benthic charr. Numbers 1, 2, 3 and 4 depict the four developmental time-points.

observed among natural populations of *D. melanogaster* were caused by variation of miR-92a expression. Our study is the first phase of assessing the involvement of miRNAs in the development of key morphological traits and their potential role in the morphological evolution of the highly polymorphic Arctic charr. In so doing we also hope to shed light on some of the developmental circuitry operating at these levels of development.

Using small RNA-seq we found 326 known and 427 novel miRNA candidates in Arctic charr. A few of the candidates, termed novel (Tables 3 and S2) are absent from miRbase but have been previously identified in other salmonid species [62–64]. When only the 326 known miRNAs are considered, the 10 most abundant ones account for 65% of reads (Figure 2). These miRNA are highly conserved among taxa and have been shown to have important functions during development. The two most abundant miRNAs in our data, miR-206 and miR-1, together account for 36% of the total known miRNAs (Figure 2). Their role in skeletogenesis and myogenesis has been studied in some detail, for example miR-206 has been found to induce myogenic differentiation [65–67] while inhibiting osteoblast differentiation [68], and miR-1 has been found to regulate skeletal muscle and cardiac development [69,70]. These miRNAs are highly conserved in

Table 3. Novel miRNA candidates differentially represented between morphs and/or developmental stages in the small-RNA-seq data.

Name	Expressed	SB1	SB2	SB3	SB4	AC1	AC2	AC3	AC4
sal-nov_1	Morph/Time	7	4	1	1	17	31	25	10
sal-nov_2	Morph	8	13	16	9	104	89	151	88
sal-nov_3	Morph	34	78	63	17	0	0	0	0
sal-nov_4	Morph	0	0	0	0	13076	5397	5755	560
sal-nov_5	Morph/Time	0	0	0	0	1800	563	343	0
sal-nov_6	Morph	47	9	14	18	73	58	23	28
sal-nov_7	Morph	25	5	1	9	48	45	14	5
sal-nov_8	Morph	207	168	20	7	256	117	47	71
sal-nov_9	Morph	130	59	13	54	268	361	173	70
sal-nov_10	Morph	16	6	12	10	45	53	52	25
sal-nov_11	Morph	8	13	16	9	104	89	151	88
sal-nov_12	Morph	8	12	13	6	14	17	15	6
sal-nov_13	Morph	21	44	112	133	113	110	236	308
sal-nov_14	Morph	2	32	31	53	27	25	96	103
sal-nov_15	Morph	3	22	32	31	6	7	8	10
sal-nov_16	Morph	644	1472	1725	983	966	648	944	1011
sal-nov_17	Morph	24	30	27	2	65	41	69	50
sal-nov_18	Morph/Time	13	89	121	372	3	11	17	219
sal-nov_19	Morph	8	4	9	0	207	63	107	15
sal-nov_20	Time	70	274	967	5358	128	213	988	6078
sal-nov_21	Time	3472	7090	12625	26369	5005	4127	10188	20766
sal-nov_22	Time	121	107	56	1	206	71	26	0

Included are miRNA names, differential expression and number of raw reads per stage per morph.

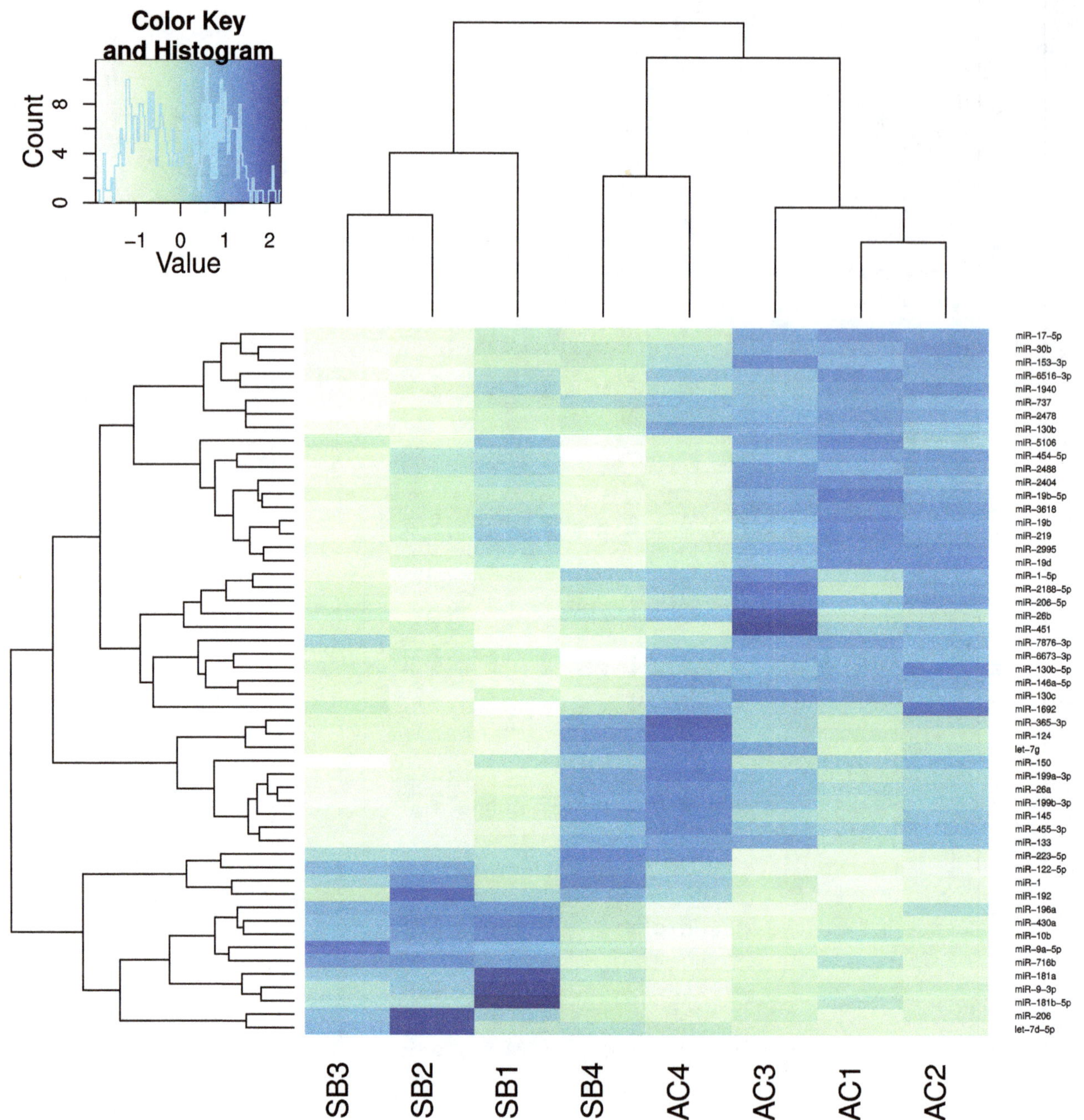

Figure 4. Heat-map showing relative expression of the 53 miRNAs significantly differentially expressed between AC and SB morphs. Expression levels for each miRNA were normalized across samples using variance stabilizing transformation. Blue denotes high and white low relative expression. AC stands for Aquaculture charr and SB stands for Small benthic charr. Numbers 1, 2, 3 and 4 depict the four developmental points.

animal evolution [8] with miR-1 retaining its muscle-specific expression from *C. elegans* to human [71]. Other highly expressed miRNAs are involved in cardiogenesis (miR-21), neurogenesis (miR-9), gut and gall bladder development (miR-143 and miR-192) [72]. One of the oldest miRNAs in the animal kingdom (miR-10) [8] is also among the 10 most highly expressed miRNAs in our data. Encoded in the intron of hoxB4, this miRNA is suggested to play a role in anterior posterior patterning.

A few miRNAs exist in multiple forms in our data. Among these the most interesting examples are miR-152 and miR-451. In the case of miR-152 the polymorphism is located in the seed region, which suggests functional divergence. In the case of miR-451 the ancestral and derived forms differ in one base (G->U substitution) located in the 3' end of the mature sequence. Interestingly the derived form of miR-451 is also the most abundant one. Although not as essential as the seed region, 3' miRNA-target pairing has a role in defining target specificity within miRNA families [73]. The derived form of miR-451 might have evolved following the whole genome duplication Salmonids underwent 25–100 million years ago or as a result of gene duplication. Other possibilities for the presence of this miRNA in our data include post-transcriptional editing of the ancestral form. However, without a sequenced

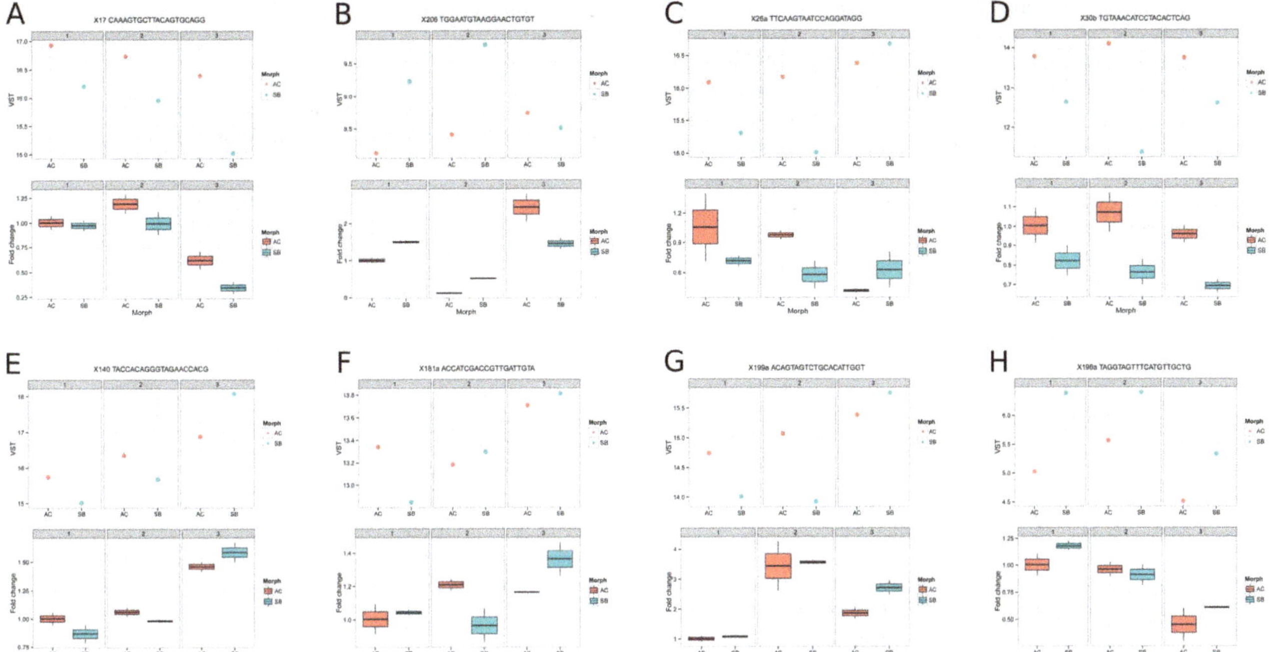

Figure 5. Comparison of expression of 8 selected miRNAs (miR-17, 26a, 30b, 140, 181a, 196a and 199a) at three developmental time-points for two contrasting Arctic charr morphs (AC and SB) quantified by small RNA-seq (upper panel) or qPCR (lower panel).

Arctic charr genome, distinguishing between evolutionary scenarios represents a challenge. The derived form of mir-451 might be specific to Arctic charr as it is not present in the salmon genome and has not been reported in rainbow trout. As miRNAs often co-evolve with their targets [5] further phylogenetic analysis will help shedding light on the evolution of miR-152 and miR-451 and their targets in Arctic charr.

51 previously annotated miRNAs were found to be differentially expressed among developmental points. The cluster analysis of these miRNAs showed a clear shift in miRNA expression between the embryonic stages and the post-hatching stage, visible from both dendrograms. Although samples grouped by developmental stage, the major division was between stages 1, 2, 3 and stage 4. These findings were further confirmed by the existence of two major clades of miRNA expression: one containing miRNAs highly expressed during the embryonic stages and one the last stage. Among the miRNAs showing high expression in early development are the members of the 430 family (miR-430 a, b, c and d). In zebrafish these miRNAs are involved in the maternal to zygotic transition by deadenylation and clearing of maternal transcripts [74]. The majority of the miRNAs showed higher expression in the last developmental stage. Examples include miR-1, members of the let-7 family, miR-22a, miR-140, miR-182, miR-183, miR-192 and miR-215. The evolutionarily ancient and highly conserved let-7 family is involved in the regulation of the timing of developmental events in *C. elegans*, in particular the transition from larval stage 4 (L4) to adult [75]. In vertebrate development let-7 is temporally regulated and it is thought to play a role in late temporal transitions during development [75]. Other miRNAs found to be highly expressed in the last developmental stage are also involved in major developmental processes such as muscle differentiation (miR-1), endochondrial bone development (miR-140) and neuromast differentiation (miR-182 and 183) [12,82]. Overall 72 miRNAs (19 novel miRNA candidates and 53 conserved miRNAs) were found to be differentially expressed between AC and SB at the developmental points under study. Of those sal-miR-196a, sal-miR-206, 122, 192, 196a, 223 and 181a were more highly expressed in SB whereas sal-miR-26a, 30b, 17-5p, 153-3p, 130b and c, 199a were more highly expressed in AC (Figure 4). All of the conserved miRNAs showing variation in expression between the Arctic charr morphs have been found to play an important role in development. For example, miR-196a, which is encoded in a Hox cluster, has been found to be involved in axial and appendicular patterning in chicken [76] and zebrafish [77]. Another muscle specific miRNA, miR-206, shows large expression differences between SB and AC especially at stage 3, where there is a 2.5 fold difference between the two morphs. This miRNA is involved in muscle differentiation and its expression is up-regulated by MyoD in differentiating muscle fibers. Loss of function of MyoD leads to down-regulation of miR-206 and severe deformities in the craniofacial elements [78]. miR-206 has also been shown to directly affect osteoblast differentiation and its overexpression in the osteoblasts of transgenic mice leads to bone abnormalities [68].

Other conserved miRNAs, miR-130b and c, miR-133, miR-199a-3p, miR-26a and miR-451 were highly expressed in AC throughout early development compared to SB. Some of these miRNAs are involved in myogenesis and skeletogenesis, for example, miR-199a-3p is important for normal skeletal development. In mouse a knockdown of Dnm3os (the primary precursor of a miR-214-miR-199a cluster) leads to skeletal abnormalities such as craniofacial hypoplasia [79]. miR-26 contributes to neurogenesis and myogenesis [80] and is involved in rainbow trout embryonic development [62] whereas miR-451 has been found to be involved in erythroid maturation in zebrafish [81]. Nineteen novel miRNA candidates were found to be differentially expressed between the two Arctic charr morphs in this study. Of those, two were only expressed in AC while one putative miRNA was only expressed in SB. These novel miRNA candidates were detected in most/all stages in one morph and not detectable in any of the stages in the other morph (Table 3), therefore it is unlikely that they represent a sequencing or technical error. As we used the Salmon genome to detect novel miRNAs, none of the novel miRNAs is likely to be morph or even Arctic charr specific, although expression differences can be expected. Several scenarios exist as to why these miRNAs are not expressed in both morphs, for example they might have been lost, their sequence might have been modified leading to the instability of the miRNA secondary structure or their expression repressed.

miRNAs are understudied in fishes and at present represent just a fraction of the miRNAs described in mammals. In the latest release of mirBase (version 20) from June 2013 there are only 255 mature miRNA sequences available for zebrafish, whereas 2578 mature sequences have been described in humans. Here we find 427 novel miRNA candidates, which are not Arctic charr specific. Some of these miRNAs are in highly conserved blocks (sal-nov235, 242, 334) among fishes, indicating that these miRNAs might have arisen early in the evolutionary history of fishes.

Concluding Remarks and Future Directions

The theoretical underpinnings of our study are based on the general proposition that differences in the level, timing and pattern of miRNA expression or acquisition of new miRNAs can influence variation in developmental circuits, so as to generate diverse and possibly discrete morphological phenotypes, thereby creating substrate for natural selection to act upon. We use a system of two contrasting morphs of Arctic charr and as a first step we surveyed miRNA expression at four developmental stages thereby homing in on the miRNA genes that may have a bearing on the morphological and functional differences between the morphs. Differences in expression levels were detected in 72 miRNAs. Interestingly, the majority of these miRNAs (53/72) are evolutionarily stable and have been previously described as part of important developmental processes such as neurogenesis, erythropoiesis, skeleto- and myogenesis, specifically in craniofacial elements. Some miRNAs (e.g. the let-7 and miR430 families) show indications of differences in timing of expression. Other miRNAs (sal-miR-152 and sal-miR-451) exhibit sequence divergence. We are currently working on follow up experiments e.g. looking for the putative targets of the interesting miRNA candidates found in this study and defining their expression pattern using *in situ* hybridization in embryos derived from additional morphs and populations.

Supporting Information

Figure S1 Length distribution of reads in the small-RNA-seq data for all samples combined. A major peak is observed at 22 nt, corresponding to the typical miRNA size.

Figure S2 Length distribution of reads in individual miRNA-seq samples. A–D: Small benthic (SB) stages 1–4; E–H: Aquaculture (AC) stages 1–4. Left panel: redundant reads, Right panel: unique reads.

Table S1 Background information and primers used for amplification of selected miRNA clusters. The names of *S. salar* contigs and miRNAs in clusters used for primer design, sequence identity (%) between Arctic charr and Salmon, and forward and reverse primer sequences are shown.

Table S2 Conserved Arctic charr miRNAs. 326 known miRNAs were identified in the small-RNA-seq data. Included are miRNA names, miRNA sequences, number of raw reads per sample and names of miRNA orthologs.

Table S3 Putative novel arctic charr miRNAs. 427 novel miRNA candidates. Included are novel miRNA candidate IDs, miRDeep2 score (the log-odds score assigned to the hairpin by miRDeep2), significant ($p<0.005$) randfold p-value of the putative miRNA to form a hairpin structure, consensus mature, consensus star, consensus precursor sequences and genomic coordinates in Salmon.

Table S4 Background data for figures 3 and 4. Normalized number of reads in all 8 samples for miRNAs that showed differential expression between morphs and/or developmental time-points.

File S1 R code used for differential expression analysis of miRNA-seq data.

File S2 R code used to calculate relative expression (fold change) for each miRNA compared to stage 1 in AC.

Acknowledgments

We acknowledge Bjarni K. Kristjánsson, Einar Svavarsson and Soizic Le Deuff for assisting with the sampling of parents, generation and maintenance of and sampling from embryo groups. We thank Jóhannes Guðbrandsson and Arnar Pálsson for discussions and advice on statistics. We also thank Valerie H. Maier and Ehsan Pashay Ahi for discussing the project at various stages. We thank Guðbjörg þ. Örlygsdóttir, Steinunn Snorradóttir and Ólafur þ. Magnússon at deCODE Genetics for technical support.

Author Contributions

Conceived and designed the experiments: SRF ZOJ SSS KHK. Performed the experiments: KHK SRF ZOJ. Analyzed the data: HJ KHK SRF. Contributed reagents/materials/analysis tools: HJ KHK SRF SSS ZOJ. Wrote the paper: KHK ZOJ SRF HJ SSS.

References

1. Plasterk RHA (2006) Micro RNAs in animal development. Cell 124: 877–881.
2. Mishima Y (2012) Widespread roles of microRNAs during zebrafish development and beyond. Dev Growth Differ 54: 55–65. doi:10.1111/j.1440-169X.2011.01306.x
3. Li J, Zhang Z (2012) miRNA regulatory variation in human evolution. Trends Genet: 1–9. doi:10.1016/j.tig.2012.10.008
4. Zhao Y, Srivastava D (2007) A developmental view of microRNA function. Trends Biochem Sci 32: 189–197. doi:10.1016/j.tibs.2007.02.006
5. Berezikov E (2011) Evolution of microRNA diversity and regulation in animals. Nat Rev Genet 12: 846–860. doi:10.1038/nrg3079
6. Bartel DP, Lee R, Feinbaum R (2004) MicroRNAs: Genomics, Biogenesis, Mechanism, and Function. Genomics: The miRNA Genes. 116: 281–297.
7. Lewis BP, Shih I, Jones-Rhoades MW, Bartel DP, Burge CB (2003) Prediction of Mammalian MicroRNA Targets. Cell 115: 787–798.
8. Heimberg AM, Sempere LF, Moy VN, Donoghue PCJ, Peterson KJ (2008) MicroRNAs and the advent of vertebrate morphological complexity. Proc Natl Acad Sci U S A 105: 2946–2950. doi:10.1073/pnas.0712259105
9. Tarver JE, Sperling EA, Nailor A, Heimberg AM, Robinson JM, et al. (2013) miRNAs: Small Genes with Big Potential in Metazoan Phylogenetics. Mol Biol Evol 30: 2369–2382.

10. Liu N, Okamura K, Tyler DM, Phillips MD, Chung W-J, et al. (2008) The evolution and functional diversification of animal microRNA genes. Cell Res 18: 985–996. doi:10.1038/cr.2008.278
11. Saunders MA, Liang H, Li W-H (2007) Human polymorphism at microRNAs and microRNA target sites. Proc Natl Acad Sci U S A 104: 3300–3305.
12. Ason B, Darnell DK, Wittbrodt B, Berezikov E, Kloosterman WP, et al. (2006) Differences in vertebrate microRNA expression. Proc Natl Acad Sci U S A 103: 14385–14389. doi:10.1073/pnas.0603529103
13. Hobert O (2008) Gene regulation by transcription factors and microRNAs. Science (80-) 319: 1785–1786. doi:10.1126/science.1151651
14. Peterson KJ, Dietrich MR, McPeek MA (2009) MicroRNAs and metazoan macroevolution: insights into canalization, complexity, and the Cambrian explosion. Bioessays 31: 736–747.
15. Ebert MS, Sharp P (2012) Roles for microRNAs in conferring robustness to biological processes. Cell 149: 515–524. doi:10.1016/j.cell.2012.04.005
16. Wilson AJ, Gíslason D, Skúlason S, Snorrason SS, Adams CE, et al. (2004) Population genetic structure of Arctic charr, Salvelinus alpinus from northwest Europe on large and small spatial scales. Mol Ecol 13: 1129–1142. doi:10.1111/j.1365-294X.2004.02149.x
17. Snorrason SS, Skúlason S (2004) Adaptive speciation in northern fresh water fishes — patterns and processes. In: Dieckmann U, Metz H, Doebeli M, Tautz D, editors. Adaptive speciation. Cambridge University Press, Cambridge. pp.210–228.
18. Robinson BW, Schluter D (2000) Natural selection and the evolution of adaptive genetic variation in northern freshwater fishes. Adaptive genetic variation in the wild. New York: Oxford University Press. pp.65–94.
19. Schluter D (1993) Adaptive Radiation in Sticklebacks: Size, Shape, and Habitat Use Efficiency. Ecology 3: 699–709.
20. Schluter D, McPhail JD (1993) Character displacement and replicate adaptive radiation. Trends Ecol Evol 8: 197–200.
21. Brunner PC, Douglas MR, Osinov A, Wilson CC, Bernatchez L (2001) Holarctic phylogeography of Arctic charr (Salvelinus alpinus L.) inferred from mitochondrial DNA sequences. Evolution 55: 573–586.
22. Gíslason D, M Ferguson M, Skúlason S, Snorrason S, Ferguson MM, et al. (1999) Rapid and coupled phenotypic and genetic divergence in Icelandic Arctic char (Salvelinus alpinus). Can J Fish Aquat Sci 56: 2229–2234.
23. Cresko WA, Amores A, Wilson C, Murphy J, Currey M, et al. (2004) Parallel genetic basis for repeated evolution of armor loss in Alaskan threespine stickleback populations. Proc Natl Acad Sci U S A 101: 6050–6055. doi:10.1073/pnas.0308479101
24. Woods PJ, Skulason S, Snorrason SS, Kristjansson BK, Malmquist HJ, et al. (2012) Intraspecific diversity in Arctic charr, Salvelinus alpinus, in Iceland: I. Detection using mixture models.
25. Snorrason S, Skúlason S, Sandlund O, Malmquist H, Jonsson B, et al. (1989) Shape polymorphism in Arctic charr, Salvelinus alpinus, in Thingvallavatn, Iceland. Pysiological Ecol Japan 1: 393–404.
26. Jonsson B, Skúlason S, Snorrason SS, Sandlund OT, Malmquist HJ, et al. (1988) Life History Variation of Polymorphic Arctic Charr (Salvelinus alpinus) in Thingvallavatn, Iceland. Can J Fish Aquat Sci 45: 1537–1547.
27. Malmquist HJ, Snorrason SS, Skulason S, Jonsson B, Sandlund OT, et al. (1992) Diet differentiation in polymorphic Arctic charr in Thingvallavatn, Iceland. J Anim Ecol 61: 21–35. doi:10.2307/5505
28. Sandlund OT, Gunnarson K, Jonasson PM, Jonsson B, Lindem T, et al. (1992) The Arctic charr Salvelinus alpinus in Thingvallavatn. Oikos 64: 305–351.
29. Skulason S, Noakes DLG, Snorrason SS (1989) Ontogeny of trophic morphology in four sympatric morphs of arctic charr Salvelinus alpinus in Thingvallavatn, Iceland. Biol J Linn Soc 38: 281–301. doi:10.1111/j.1095-8312.1989.tb01579.x
30. Frandsen F, Malmquist HJ, Snorrason SS (1989) Ecological parasitology of polymorphic Arctic charr, Salvelinus alpinus (L.), in Thingvallavatn, Iceland. J Fish Biol 34: 281–297.
31. Skúlason S, Snorrason SS, Ota D, Noakes DLG (1993) Genetically based differences in foraging behaviour among sympatric morphs of arctic charr (Pisces: Salmonidae). Anim Behav 45: 1179–1192.
32. Skúlason S, Snorrason SS, Noakes DLG, Ferguson MM (1996) Genetic basis of life history variations among sympatric morphs of Arctic char, Salvelinus alpinus. Can J Fish Aquat Sci 53: 1807–1813.
33. Kapralova KH, Morrissey MB, Kristjánsson BK, Olafsdóttir GÁ, Snorrason SS, et al. (2011) Evolution of adaptive diversity and genetic connectivity in Arctic charr (Salvelinus alpinus) in Iceland. Heredity (Edinb) 106: 472–487.
34. Kapralova KH, Gudbrandsson J, Reynisdottir S, Santos CB, Baltanás VC, et al. (2013) Differentiation at the MHCIIα and Cath2 Loci in Sympatric Salvelinus alpinus Resource Morphs in Lake Thingvallavatn. PLoS One 8: e69402.
35. Eiriksson GM, Skulason S, Snorrason SS (1999) Heterochrony in skeletal development and body size in progeny of two morphs of Arctic charr from Thingvallavatn, Iceland. J Fish Biol 55: 175–185.
36. Svavarsson E (2007) Árangur í kynbótum á bleikju og næstu skref [reference in icelandic]. Fræðaþing landbúnaðarins (conference proceedings) 4: 121–125.
37. Gorodilov YN (1996) Description of the early ontogeny of the Atlantic salmon, Salmo salar, with a novel system of interval (state) identification. Environ Biol Fishes 47: 109–127. doi:10.1007/BF00005034
38. Martin M (2011) Cutadapt removes adapter sequences from high-throughput sequencing reads. EMBnet.jounal 17: 10–12.
39. Ewing B, Green P (1998) Base-Calling of Automated Sequencer Traces Using Phred. II. Error Probabilities. Genome Res 8: 186–194.
40. Burge SW, Daub J, Eberhardt R, Tate J, Barquist L, et al. (2013) Rfam 11.0: 10 years of RNA families. Nucleic Acids Res 41: D226–32.
41. Griffiths-Jones S (2004) The microRNA Registry. Nucleic Acids Res 32: D109–11. doi:10.1093/nar/gkh023
42. Griffiths-Jones S, Grocock RJ, van Dongen S, Bateman A, Enright AJ (2006) miRBase: microRNA sequences, targets and gene nomenclature. Nucleic Acids Res 34: D140–4.
43. Griffiths-Jones S, Saini HK, van Dongen S, Enright AJ (2008) miRBase: tools for microRNA genomics. Nucleic Acids Res 36: D154–8.
44. Kozomara A, Griffiths-Jones S (2011) miRBase: integrating microRNA annotation and deep-sequencing data. Nucleic Acids Res 39: D152–7.
45. Pearson WR, Lipman DJ (1988) Improved tools for biological sequence comparison. Proc Natl Acad Sci 85: 2444–2448.
46. Friedländer MR, Chen W, Adamidi C, Maaskola J, Einspanier R, et al. (2008) Discovering microRNAs from deep sequencing data using miRDeep. Nat Biotechnol 26: 407–415. doi:10.1038/nbt1394
47. Di Génova A, Aravena A, Zapata L, González M, Maass A, et al. (2011) SalmonDB: a bioinformatics resource for Salmo salar and Oncorhynchus mykiss. Database (Oxford) 2011: bar050. doi:10.1093/database/bar050
48. Altschul SF, Gish W, Miller W, Myers EW, Lipman DJ (1990) Basic local alignment search tool. J Mol Biol 215: 403–410.
49. Gordon D, Abajian C, Green P (1998) Consed: A Graphical Tool for Sequence Finishing. Genome Res 8: 195–202. doi:10.1101/gr.8.3.195
50. Robinson MD, McCarthy DJ, Smyth GK (2010) edgeR: a Bioconductor package for differential expression analysis of digital gene expression data. Bioinformatics 26: 139–140.
51. Benjamini Y, Hochberg Y (1995) Controlling the False Discovery Rate: A Practical and Powerful Approach to Multiple Testing. J R Stat Soc Ser B 57: 289–300.
52. Mestdagh P, Van Vlierberghe P, De Weer A, Muth D, Westermann F, et al. (2009) A novel and universal method for microRNA RT-qPCR data normalization. Genome Biol 10: R64.
53. Ahi EP, Guðbrandsson J, Kapralova KH, Franzdóttir SR, Snorrason SS, et al. (2013) Validation of reference genes for expression studies during craniofacial development in arctic charr. PLoS One 8: e66389.
54. Mallarino R, Campàs O, Fritz JA, Burns KJ, Weeks OG, et al. (2012) Closely related bird species demonstrate flexibility between beak morphology and underlying developmental programs. Proc Natl Acad Sci U S A 109: 16222–16227.
55. Sylvester JB, Rich CA, Loh Y-HE, van Staaden MJ, Fraser GJ, et al. (2010) Brain diversity evolves via differences in patterning. Proc Natl Acad Sci U S A 107: 9718–9723.
56. Roberts RB, Hu Y, Albertson RC, Kocher TD (2011) Craniofacial divergence and ongoing adaptation via the hedgehog pathway. Proc Natl Acad Sci U S A 108: 13194–13199.
57. Manousaki T, Hull PM, Kusche H, Machado-Schiaffino G, Franchini P, et al. (2013) Parsing parallel evolution: ecological divergence and differential gene expression in the adaptive radiations of thick-lipped Midas cichlid fishes from Nicaragua. Mol Ecol 22: 650–669. doi:10.1111/mec.12034
58. Henning F, Jones JC, Franchini P, Meyer A (2013) Transcriptomics of morphological color change in polychromatic Midas cichlids. BMC Genomics 14: 171.
59. Filteau M, Pavey SA, St-Cyr J, Bernatchez L (2013) Gene coexpression networks reveal key drivers of phenotypic divergence in lake whitefish. Mol Biol Evol 30: 1384–1396.
60. Yamamoto Y, Byerly MS, Jackman WR, Jeffery WR (2009) Pleiotropic functions of embryonic sonic hedgehog expression link jaw and taste bud amplification with eye loss during cavefish evolution. Dev Biol 330: 200–211.
61. Arif S, Murat S, Almudi I, Nunes MDS, Bortolamiol-Becet D, et al. (2013) Evolution of mir-92a Underlies Natural Morphological Variation in Drosophila melanogaster.
62. Ramachandra RK, Salem M, Gahr S, Rexroad CE, Yao J (2008) Cloning and characterization of microRNAs from rainbow trout (Oncorhynchus mykiss): their expression during early embryonic development. BMC Dev Biol 8: 41. doi:10.1186/1471-213X-8-41
63. Ma H, Hostuttler M, Wei H, Rexroad CE, Yao J (2012) Characterization of the rainbow trout egg microRNA transcriptome. PLoS One 7: e39649.
64. Bekaert M, Lowe NR, Bishop SC, Bron JE, Taggart JB, et al. (2013) Sequencing and Characterisation of an Extensive Atlantic Salmon (Salmo salar L.) MicroRNA Repertoire. PLoS One 8: e70136.
65. Kim HK, Lee YS, Sivaprasad U, Malhotra A, Dutta A (2006) Muscle-specific microRNA miR-206 promotes muscle differentiation. J Cell Biol 174: 677–687.
66. Sweetman D, Goljanek K, Rathjen T, Oustanina S, Braun T, et al. (2008) Specific requirements of MRFs for the expression of muscle specific microRNAs, miR-1, miR-206 and miR-133. Dev Biol 321: 491–499.
67. Dey BK, Gagan J, Dutta A (2011) miR-206 and -486 induce myoblast differentiation by downregulating Pax7. Mol Cell Biol 31: 203–214. doi:10.1128/MCB.01009-10
68. Inose H, Ochi H, Kimura A, Fujita K, Xu R, et al. (2009) A microRNA regulatory mechanism of osteoblast differentiation. Proc Natl Acad Sci U S A 106: 20794–20799. doi:10.1073/pnas.0909311106

69. Chen J-F, Mandel EM, Thomson JM, Wu Q, Callis TE, et al. (2006) The role of microRNA-1 and microRNA-133 in skeletal muscle proliferation and differentiation. Nat Genet 38: 228–233.
70. Liu N, Williams AH, Kim Y, McAnally J, Bezprozvannaya S, et al. (2007) An intragenic MEF2-dependent enhancer directs muscle-specific expression of microRNAs 1 and 133. Proc Natl Acad Sci U S A 104: 20844–20849.
71. Niwa R, Slack FJ (2007) The evolution of animal microRNA function. Curr Opin Genet Dev 17: 145–150. doi:10.1016/j.gde.2007.02.004
72. Wienholds E, Kloosterman WP, Miska E, Alvarez-Saavedra E, Berezikov E, et al. (2005) MicroRNA expression in zebrafish embryonic development. Science 309: 310–311.
73. Brennecke J, Stark A, Russell RB, Cohen SM (2005) Principles of microRNA-target recognition. PLoS Biol 3: e85.
74. Giraldez AJ, Mishima Y, Rihel J, Grocock RJ, Van Dongen S, et al. (2006) Zebrafish MiR-430 promotes deadenylation and clearance of maternal mRNAs. Science 312: 75–79. doi:10.1126/science.1122689
75. Pasquinelli AE, Reinhart BJ, Slack F, Martindale MQ, Kuroda MI, et al. (2000) Conservation of the sequence and temporal expression of let-7 heterochronic regulatory RNA. Nature 408: 86–89.
76. McGlinn E, Yekta S, Mansfield JH, Soutschek J, Bartel DP, et al. (2009) In ovo application of antagomiRs indicates a role for miR-196 in patterning the chick axial skeleton through Hox gene regulation. Proc Natl Acad Sci U S A 106: 18610–18615.
77. He X, Yan Y-L, Eberhart JK, Herpin A, Wagner TU, et al. (2011) miR-196 regulates axial patterning and pectoral appendage initiation. Dev Biol 357: 463–477. doi:10.1016/j.ydbio.2011.07.014
78. Hinits Y, Williams VC, Sweetman D, Donn TM, Ma TP, et al. (2011) Defective cranial skeletal development, larval lethality and haploinsufficiency in Myod mutant zebrafish. Dev Biol 358: 102–112.
79. Watanabe T, Sato T, Amano T, Kawamura Y, Kawamura N, et al. (2008) Dnm3os, a non-coding RNA, is required for normal growth and skeletal development in mice. Dev Dyn 237: 3738–3748. doi:10.1002/dvdy.21787
80. Wong CF, Tellam RL (2008) MicroRNA-26a targets the histone methyltransferase Enhancer of Zeste homolog 2 during myogenesis. J Biol Chem 283: 9836–9843.
81. Pase L, Layton JE, Kloosterman WP, Carradice D, Waterhouse PM, et al. (2008) miR-451 regulates zebrafish erythroid maturation in vivo via its target gata2. Blood 113: 1794–1804. doi:10.1182/blood-2008-05-155812
82. Papaioannou G, Inloes JB, Nakamura Y, Paltrinieri E, Kobayashi T (2013) let-7 and miR-140 microRNAs coordinately regulate skeletal development. Proc Natl Acad Sci U S A 110: E3291–300.

22

A New Flow-Regulating Cell Type in the Demosponge *Tethya wilhelma* – Functional Cellular Anatomy of a Leuconoid Canal System

Jörg U. Hammel*[¤a], Michael Nickel[¤b]

Institut für Spezielle Zoologie und Evolutionsbiologie mit Phyletischem Museum, Friedrich-Schiller-Universität Jena, Erbertstr. 1, 07743, Jena, Germany

Abstract

Demosponges possess a leucon-type canal system which is characterized by a highly complex network of canal segments and choanocyte chambers. As sponges are sessile filter feeders, their aquiferous system plays an essential role in various fundamental physiological processes. Due to the morphological and architectural complexity of the canal system and the strong interdependence between flow conditions and anatomy, our understanding of fluid dynamics throughout leuconoid systems is patchy. This paper provides comprehensive morphometric data on the general architecture of the canal system, flow measurements and detailed cellular anatomical information to help fill in the gaps. We focus on the functional cellular anatomy of the aquiferous system and discuss all relevant cell types in the context of hydrodynamic and evolutionary constraints. Our analysis is based on the canal system of the tropical demosponge *Tethya wilhelma*, which we studied using scanning electron microscopy. We found a hitherto undescribed cell type, the reticuloapopylocyte, which is involved in flow regulation in the choanocyte chambers. It has a highly fenestrated, grid-like morphology and covers the apopylar opening. The minute opening of the reticuloapopylocyte occurs in an opened, intermediate and closed state. These states permit a gradual regulation of the total apopylar opening area. In this paper the three states are included in a theoretical study into flow conditions which aims to draw a link between functional cellular anatomy, the hydrodynamic situation and the regular body contractions seen in *T. wilhelma*. This provides a basis for new hypotheses regarding the function of bypass elements and the role of hydrostatic pressure in body contractions. Our study provides insights into the local and global flow conditions in the sponge canal system and thus enhances current understanding of related physiological processes.

Editor: David J. Schulz, University of Missouri, United States of America

Funding: Funding for this research came from Deutsche Forschungs Gemeinschaft (www.dfg.de) research grant HA 6405/1-1 to JUH. The funder had no role in study design, data collection and analysis, decision to publish, or preparation of the manuscript.

Competing Interests: The authors have declared that no competing interests exist.

* Email: joerg.hammel@uni-jena.de

¤a Current address: Center for Materials and Coastal Research, Helmholtz-Zentrum Geesthacht, Max-Planck-Straße 1, 21502, Geesthacht, Germany
¤b Current address: Bionic consulting, Bruckenäcker 4, 70565, Stuttgart, Germany

Introduction

Sponges are sessile filter-feeding animals. Accordingly, the canal or aquiferous system is their most distinct anatomical feature. Functionally speaking it can be considered the most important organizational unit besides the skeletal elements which give the sponge its structure. In accordance with their feeding habits, all physiological processes in sponges rely on the ability to process high volumes of water through the body. Only in this way are they able to obtain the required nutrients and oxygen and get rid of metabolic waste products.

Research into the biomechanics and fluid dynamics of filter-feeding and into biological fluid transport systems in general has revealed a close interdependence between hydrodynamic constraints, the micro- and macro-morphology of the cellular elements involved and, indeed, the structure of the anatomy in its entirety [1–6]. A number of hydrodynamic constraints and optimality principles have been suggested to play a role in shaping the general architecture of the canal system [3], but the key features appear to be flow resistance and pressure drop [2]. Pressure drop can be understood as the resistance which fluid encounters when it passes through a filter. In the incurrent canal system in sponges, small apertures in the form of ostia and prosopyles contribute significantly to the pressure drop within the system (Figure 1). Further on, the apopylar apertures and the microvilli collar of the choanocyte chambers are also thought to play a significant role (Figure 1). While the effect of pressure drop in sponges has been considered to varying extents in general models of flow on an organismal scale, almost nothing is known about the influence of cell morphologies on local flow conditions or their implication for hydrodynamics on an organismal scale. Local flow regimes are of the utmost importance, however, especially when it comes to functional considerations such as nutrient uptake and gas exchange.

From a biological perspective resistance has a significant influence on two central aspects of filter feeding. On the one hand it determines the power required to move the fluid through the system. On the other hand it determines, in the context of morphological constraints and anatomy, the flow velocity of the

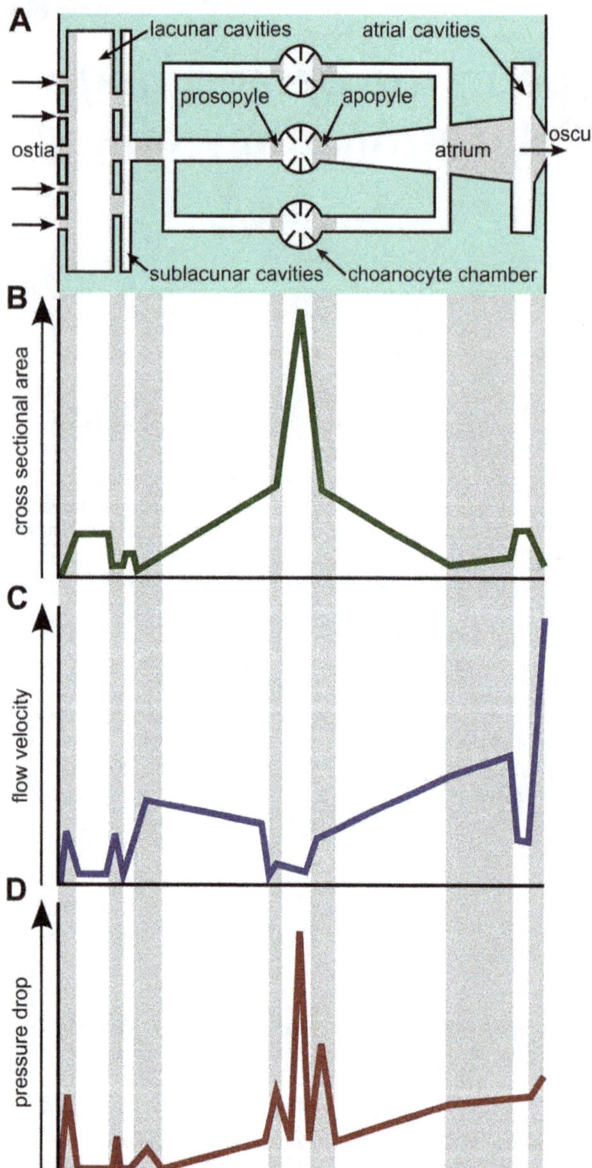

Figure 1. Scheme of hydrodynamic conditions in different sections of the leuconoid canal system based on morphometric and anatomical data on the sponge canal system as well as on fundamental physical laws in hydrodynamics [3,4,6,9–11,46]. (A) Structural representation of the main canal system elements in the direction of flow. (B) Schematic diagram of the change of available total cross sectional area along the flow path. (C) Schematic diagram of flow velocities in the canal system. (D) Schematic diagram of the change of pressure drop along the flow path.

fluid in the canal system. Particle capture rates are greatly influenced by the prevailing Reynolds number and are therefore related to flow velocity and anatomy [7,8]. We are consequently faced with complex interdependencies between cellular morphology and anatomy, energy expenditure and filter-feeding. In order to understand these complex relationships in sponges we need detailed information regarding the hierarchy and three-dimensional architecture of the canal system, quantitative morphometric data pertaining to individual canal segments, flow velocity measurements and detailed morphological data regarding the cellular entities involved in the canal system. The morphometric and anatomical data pertaining to the architecture of the canal system and the cell types involved then needs to be integrated into basic fluid dynamic theory in order to gain a deeper and more detailed understanding of the hydrodynamic situation as a whole in sponge canal systems. Current understanding is based on general information regarding leucon-type canal systems [3,4,9,10] and recent specific morphometric and hierarchical data pertaining to the aquiferous system [6]. Flow velocity within the canal system is affected most prominently by the total available cross-sectional area of every functional unit in it (Figure 1A–C) [3,4,11]. Slower flow velocities are caused by an increase in total available cross-sectional area on any given hierarchical level [4,9]. However, the cross-sectional area of single segments on a hierarchical level is usually small. Overall increases in cross-sectional area are related to increases in the number of small sized segments on the respective level [6,12]. As the lower cross-sectional area of small sized canals is a consequence of their smaller diameter we can draw from the following two equations a direct relationship between pressure drop and resistance:

$$R = \frac{8 \cdot \eta \cdot l}{\pi \cdot r^4} \tag{1}$$

$$\Delta P = Q \cdot R \tag{2}$$

Where R is resistance, η the viscosity of the fluid, l the length of a canal segment, r the canal diameter, ΔP is pressure drop and Q is flow. According to equation (1), radius has the greatest influence on resistance, which allows us to conclude that numerous small sized canals will lead to high resistance and therefore necessitate a high level of pumping power. Equation (2) describes the relationship between pressure drop within the system and flow, viz. resistance. Sites with high local resistance in the system contribute significantly to pressure drop, especially when small sized elements are involved (Figure 1D). All considerations so far have remained on a local scale, however, focusing on single canal system elements. In order to come up with a comprehensive functional morphological interpretation, the complete architecture of the canal system and the specific sub-elements defined in the context of hydrodynamics as functional units need to be taken into account on both the local and the organismal scale. In order to do this, two fundamental principles of resistance theory have to be considered. (1) Total resistance for serial segments is the sum of all the segments included. (2) For segments arranged in parallel, total resistance is given by the following equation.

$$R_{tot} = \frac{1}{\sum_{k=1}^{n} \frac{1}{R_k}} \tag{3}$$

As a consequence, the high resistance of numerous small sized canal segments - on any hierarchical level - turns out to make a much smaller contribution to total resistance on the organismal scale than indicated by the high individual values.

At present, the model for flow regimes in sponges [9] considers some of the physical and hydrodynamic constraints mentioned above [3,4,11], but with regard to morphological and architectural information is restricted to statistical morphometric data [3]. Modern imaging and analysis techniques have made detailed and even complete morphometric data available for biophysical considerations of general canal system anatomy [6,12–14]. The studies in question have revealed that the architectural complexity of the canal system anatomy in leucon-type sponges is much higher than previously thought, featuring phenomena such as bypass elements or highly asymmetric branching which need to be included in an updated flow model in order to explain canal system hydrodynamics on a local scale as well as an organismal one. However, in order to obtain a sufficiently detailed picture of the hydrodynamics of the canal system to put together a new biophysical model of flow, data from a single species needs to be available for all the prerequisites mentioned above. Flow inside the canal system of sponges is influenced not only by the system's gross morphological architecture but subject too to constraints imposed by cellular elements. Most studies into sponge aquiferous systems have focused either on the architecture and morphology of the canal system in general or on the way in which choanocytes work. The present study aims to provide an overview, from a functional morphological and hydrodynamic perspective, of all relevant cellular structures within the leucon-type poriferan aquiferous system of one exemplary species.

The tropical demosponge *Tethya wilhelma* SARÀ, SARÀ, NICKEL & BRÜMMER 2001 was chosen as a model on which to assess the way in which the morphology of cellular elements of the canal system relates to functional morphological aspects derived from hydrodynamic constraints. The general architecture of the canal system had already been examined for this species on an organismic scale [6,15]. Being one of the rare sponge species continuously cultivable under laboratory conditions [16–19] and even exhibiting regular asexual reproduction by budding [20], *T. wilhelma* is an emerging model demosponge for various types of functional investigation including physiological, genetic and morphological studies.

Morphologically speaking, the following series of elements are considered the functional modules of the aquiferous system [21]: Ostia>(sub dermal lacunae)>incurrent canals>prosopyls>choanocyte chambers>apopyles>excurrent canals>oscule(s). Ostia are the microscopic incurrent openings into the system, while the oscule or oscules are the excurrent openings. The choanocyte chambers act as displacement pumps and generate the pressure differential which drives the water through the system [10]. Their in- and excurrent openings are called the proso- and apopyle.

There are large discrepancies in our current morphological understanding of the various elements of and cell types involved in the aquiferous system. Although some cell types (e.g. endopinacocytes and choanocytes in particular) have been studied in detail, thorough comparative cytological studies based on broad taxon sampling are scarce. The most comprehensive review is to be found in Simpson's compendium of sponge biology [21], though the information in it is unfortunately fairly general. A more recent and detailed study into cell types in demosponges focuses on systematic and evolutionary aspects of aquiferous system characters [22]. Detailed morphological studies of cell types which contribute to functionally important elements of the aquiferous system help us, when they consider the hydrodynamic environment in which such cells are found, to assess their functional role [23,24]. This applies to apopylar cells (cone cells), central cells and any other cell type located in hydrodynamically pivotal sites in choanocyte chambers.

Theoretical and experimental investigations into choanocyte chambers have shown on the basis of choanocyte arrangement and orientation that the chambers can be understood as positive displacement pumps or, in technical terms, as peristaltic pumps [10,11]. Experimentally and theoretically consistent models for filter feeding in sponges do exist, though definitive experimental evidence is still lacking since science currently lacks the technical observation methodologies for *in vivo* studies [1,25]. However, in order to complement our understanding of functional morphology, the present study is intended to provide a detailed analysis of cell types within the canal system of *T. wilhelma* with respect to their impact on local flow and consequences for hydrodynamics on an organismic level.

Results

Canal system compartments and anatomical details

The canal system architecture in *T. wilhelma* is of the leucon type with some striking manifestations of specific canal system elements. The incurrent canal system features voluminous cortical lacunar sub-dermal cavities. This cortical lacunar network is connected to an underlying network of sub-lacunar cavities located at the choanosome/cortex boundary. Both lacunar systems consist of an extensive network of anastomosing oval-shaped/flat canals. Branching off from the lacunar- and sub-lacunar cavities, high numbers of ramifying canals lead into the choanosome. Due to the roughly globular shape of the body, the canals of the incurrent and excurrent canal systems are significantly intertwined in the choanosome region. Within the excurrent canal system the atrium region stands out by virtue of its volume and can be characterized as a larger sized canal resembling a vestibule which opens directly into the outflow opening (oscule) (Figure 2). Depending on the state of morphological (re-)organization and environmental flow conditions, varying numbers of oscules are present, from one in the majority of cases to several in more rare cases.

Ostia

Specimens of *T. wilhelma* exhibit ostia of varying sizes, with no direct correlation with body size discernible - at least not in the specimens investigated here (Figure 3). The diameters of single ostia in all the specimens studied (N = 10) ranged from fully closed to a typical maximum of <15 µm. Ostia greater than this in diameter were present only in very low numbers. Depending on environmental flow conditions, ostia appear as single openings, in small groups or as ostia fields (Figure 3A). Smaller sized ostia are formed by intracellular pores (Figure 3B, Figure S1), whereas larger ones are made up like intercellular ostia by groups of several cells (Figure 3C). In both cases the exopincocytes involved in the formation of ostia are in direct contact with adjacent exopincocytes and endopinacocytes. Where specimens of *T. wilhelma* had been cultured under steady flow conditions over a long period of time, ostia fields covering the topmost portion of the surface of the sponge were observable. In this case, ostia were generally larger (up to 43 µm). Tissue bridges between the ostia usually varied in length between 5 µm and 20 µm.

Choanocyte chambers

Choanocyte chambers are almost globular in *T. wilhelma* and possess one apopylar and one to several prosopylar openings (Figure 4A). The number of choanocytes within a choanocyte chamber is dependent on chamber size and body size (~50–90 choanocytes/chamber, 70±13 choanocytes/chamber (N = 15 taken from 4 specimens)). The choanocytic prosopyle is formed by an interstice between adjacent choanocytes which lack filopodial extensions, which means that the prosopyle itself lacks any kind of specialized choanocytic prosopylar structure (Figure 4A).

Prosopyles

Prosendopinacocytes form internal, single-cell pores known as pinacocytic prosopyles (Figure 4F). The mean diameter of these

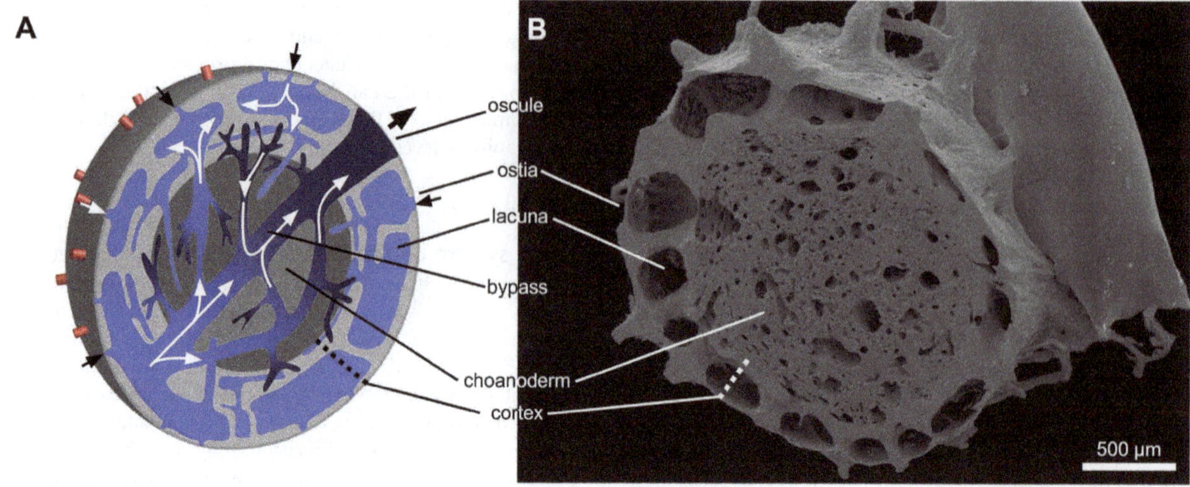

Figure 2. Schematic organization (A) and habitus (B) of *T. wilhelma* aquiferous system. (A) Potential flow directions in the canal system are indicated with arrows (after [15]). A color gradient from light to dark blue in the canals indicates the allocation of the corresponding elements to the incurrent and excurrent system. Due to the presence of bypasses in the canal system flow directions cannot be assigned with certainty to all sections. This might even cause backflows from the excurrent to the incurrent system. Main features/structures of the canal system are labeled in the scanning electron micrograph (B) as well as in the schematic drawing (A).

pore-based openings into the choanocyte chambers is about 7.4 μm. The prosendopinacocytes which form the pinacocytic prosopyle come into direct contact with the basal part of choanocyte cell bodies (Figure 4G).

Apopyles

The choanocytic apopyle is formed by apopylar cells (Figure 4B–D), two to three of which (depending on the size of the choanocyte chamber) form a ring-like structure (Figure 4B). Each apopylar cell bears a single cilium 3.9 μm in length (Figure 4D). In a cross-sectional view the ring formed by apopylar cells around the apopylar opening displays a characteristic double cone shape [26] (Figure 4C). On the choanocytic face the apopylar cells come into contact with choanocytes by way of a thin velum which forms the edge of the inner part of the ring/pore structure. This velum comes into direct contact with the choanocyte microvilli collar. The single cilium of the apopylar cells projects into the apopylar opening (Figure 4B–D). Facing the apopyle the cells connect to an apopylar pore-forming apendopinacocyte, which in turn touches a hitherto undescribed cell type spanning the apopylar opening (Figure 4B, Figure 5).

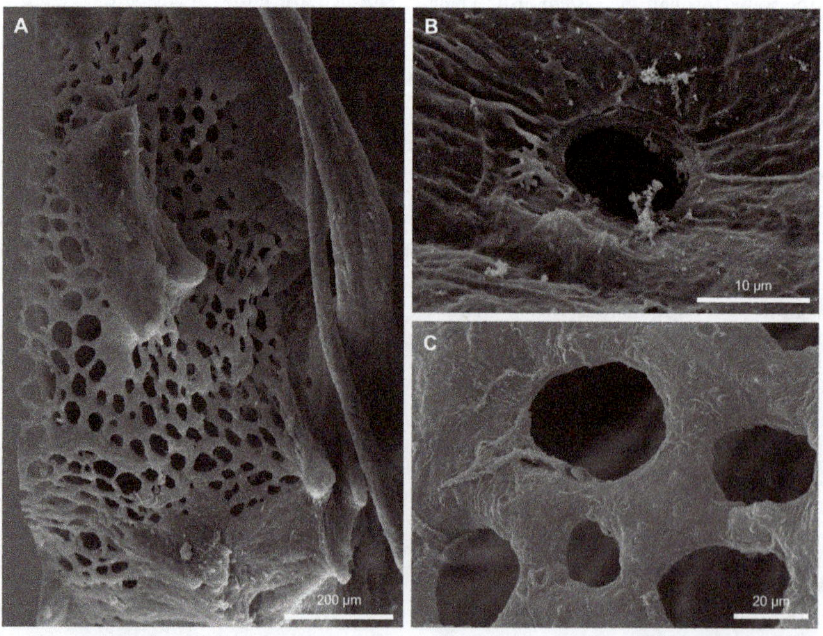

Figure 3. Scanning electron micrograph of an ostia pore field (A), a single ostium (B) and details of ostia in an ostia pore field (C).

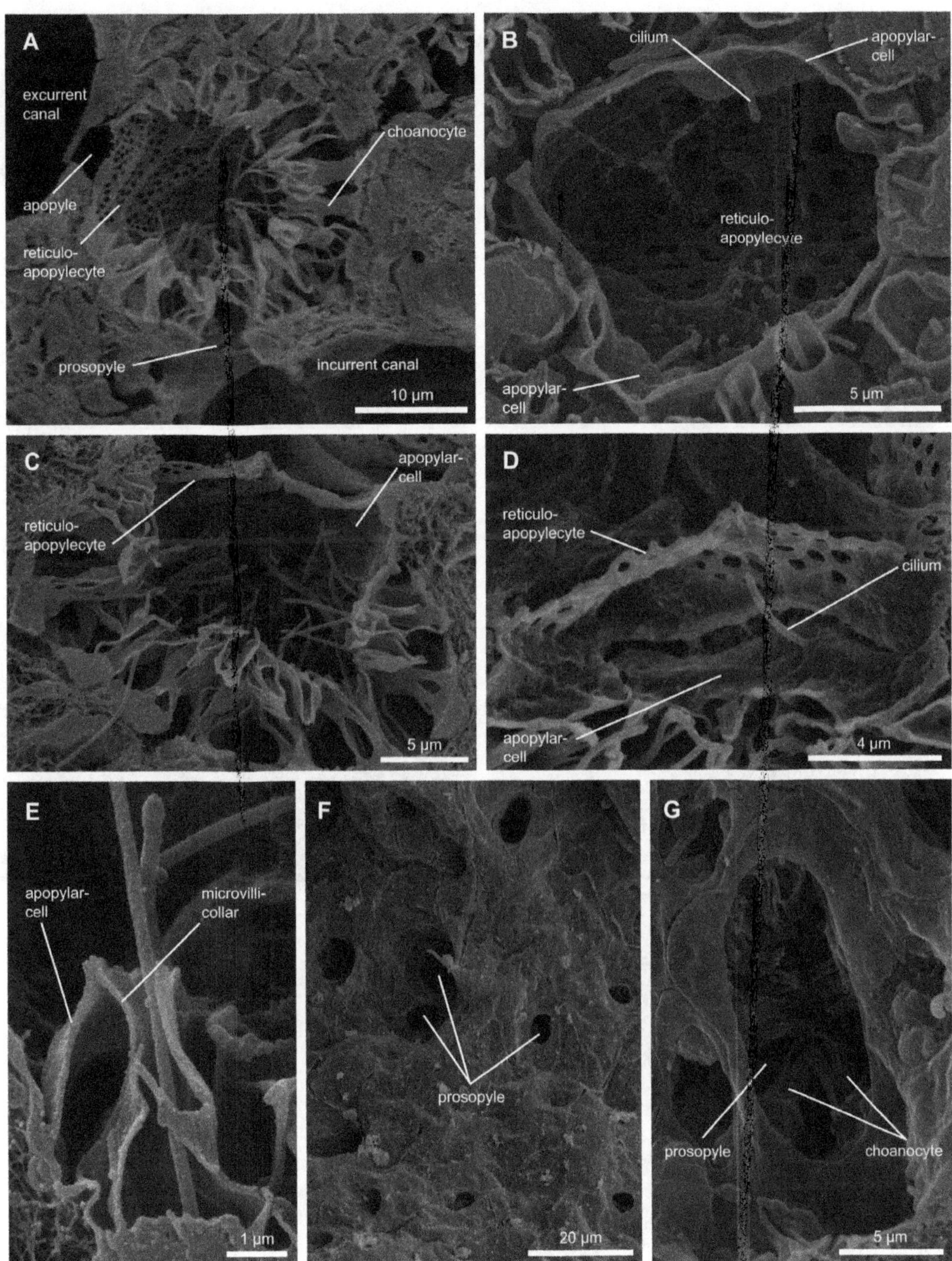

Figure 4. Scanning electron micrograph of cellular structures in the choanocyte chamber. (A) Overview of a choanocyte chamber connected to an incurrent- and excurrent canal with the relevant cellular prosopylar and apopylar elements and the location of the new cell type: reticuloapopylocyte. (B) Circular arrangement of apopylar cells and the position adjacent to reticuloapopylocyte. Hydrodynamic sealing of apopylar velum and microvilli collar. (C) Arrangement of cilium bearing apopylar cells, choanocytes and reticuloapopylocytes in the choanocytic apopyle. (D) Detailed view of an apopylar cell with its cilium directing into the flow at the apopyle. (E) Detailed view of the apopylar velum and microvilli collar contact side which results in a hydrodynamic sealing. (F) Overview of prosopylar openings in the incurrent canal system. (G) Pore cell forming a prosopylar opening. In the background microvilli collars of choanocytes are visible.

A new mesh-forming cell type within the apopyle - Reticuloapopylocyte

Reticuloapopylocytes – a previously unknown type of cell - have a high number of small intracellular pores which give them a mesh or grid-like morphology (Figure 5). These pores have openings of about 0.53 µm±0.07 µm (N=82, taken from 1 specimen) (Figure 5E–F) and are found in an opened and closed state (Figure 5D). Reticuloapopylocytes, then, are able to adopt a gradient of opening states from totally open and highly fenestrated to partially or almost completely closed. When all reticuloapopylocyte pores are open, the functional cross-sectional area of the

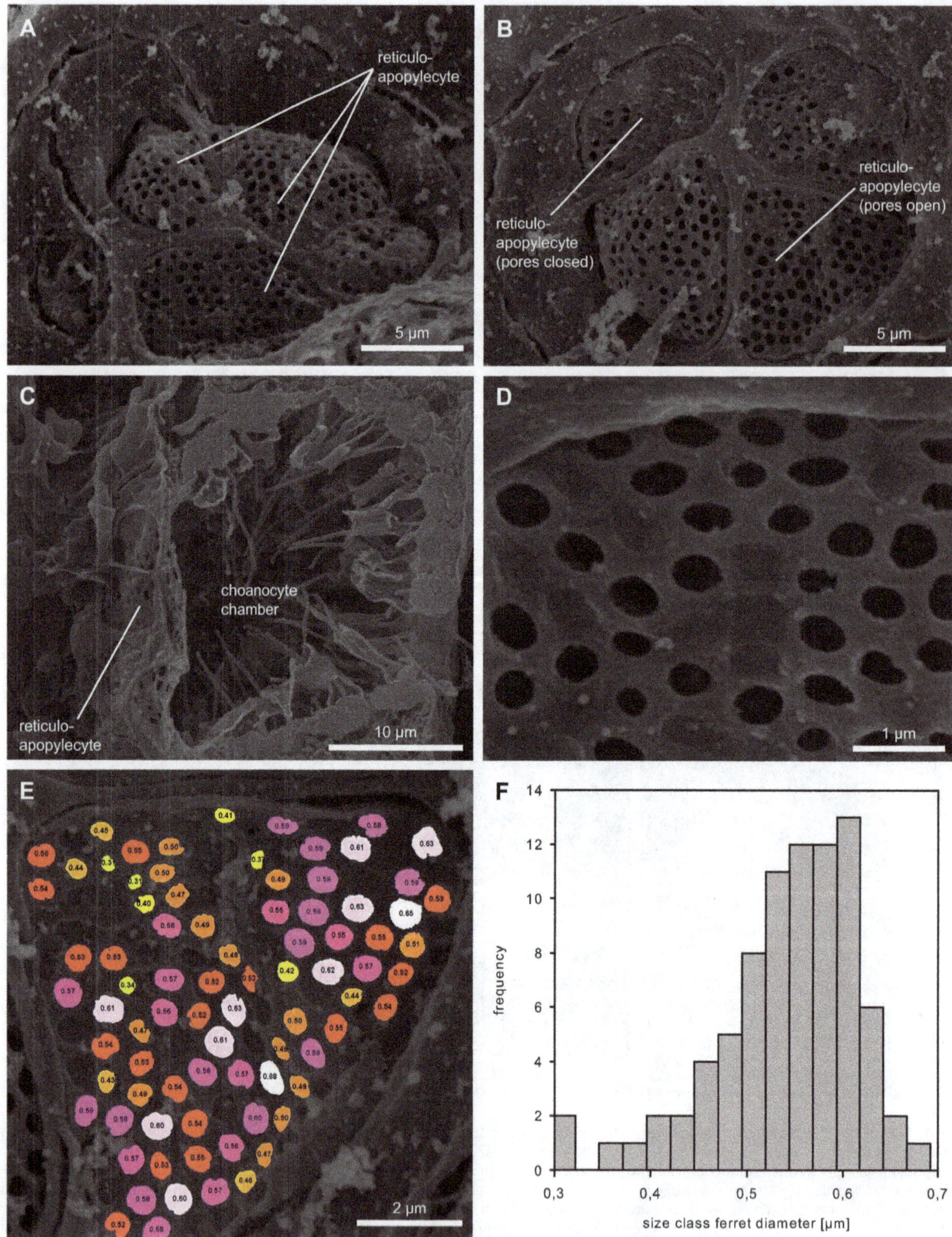

Figure 5. Scanning electron micrographs of reticuloapopylocytes. (A) View on reticuloapopylocytes from the excurrent canal with adjacent endopinacocytes and most of the pores open. (B) View on reticuloapopylocytes from the excurrent canal with one cell having most of the pores closed. (c) Overview of the position of reticuloapopylocytes in the apopyle (cross section through a choanocyte chamber). (D) Detailed view on pores of reticuloapopylocytes in an open and closed state. (E) Color coded and labeled ferret pore diameter of reticuloapopylocyte. (F) Distribution of ferret pore diameters in reticuloapopylocytes.

apopyle equals approximately 50% of the total area which would be present if the reticuloapopylocyte was absent. Typically, the cross-sectional area available to flow is much lower. The cell itself is very thin, usually below 0.5 µm, which is why the high level of fenestration leads to a grid-like morphology. Where a single reticuloapopylocyte spans the apopylar opening, it is almost circular in shape. In the case of larger apopylar openings, two or more reticuloapopylocytes form a mesh-like covering (Figure 5A–C).

Using the pore measurements presented in figure 5E–F, we calculated how reticuloapopylocytes contribute to the resistance of flow. Taking as a basis the cross-sectional area of pores and entire cells, we calculated the radius of pores and the radius of the apopylar opening. For the sake of simplification, we assumed that both were circular. By putting the measurements presented into equations 1 and 3, we calculated reticuloapopylocyte resistance to be $4.12 \cdot 10^{-3}$ Pa s μm^{-3}. In order to compare this value, we then calculated the resistance of the same apopyle opening without the reticuloapopylocyte and found it to be $3.13 \cdot 10^{-3}$ Pa s μm^{-3}. An apopylar opening with the same available cross-sectional area as the reticuloapopylocyte (12.87 μm^2) would give rise to a single apopyle with a radius of 2.03 µm and a resistance of $5.45 \cdot 10^{-3}$ Pa s μm^{-3}. The resistance of an apopyle with a reticuloapopylocyte is therefore 1316 times greater than that of the same apopylar opening unaltered. A smaller apopyle with the same available cross-sectional area as observed in the reticuloapopylocyte would lead to a 17-fold increase in resistance compared to the reference apopyle.

Pinacocytes

The prosendopinacocytes lining the walls of the lacunar and sublacunar cavities and the incurrent canal walls are less than 0.5 µm thick except for a small swelling incorporating the nucleus. Their overall shape is irregular and adopted to the local canal geometry (Figure 6A–D). The prosendopinacocytes in our study never displayed the T-shaped or umbrella-like morphology characteristic of exopincocytes (Figure 6F).

T. wilhelma possesses two types of apendopinacocytes which line the walls of excurrent canals and the atrium region, respectively. The type present in and around the atrium region bears a single 5.5 µm±0.79 µm (N = 16, taken from 4 specimens) long cilium (Figure 6C,E). Monociliated apendopinacocytes exhibit a fusiform cell morphology and appear to be arranged in a highly ordered fashion within the atrium region (Figure 6A,C). As in the case of prosendopinacocytes, the main cell body is very thin, usually below 0.5 µm, with the exception of the part holding the nucleus. Away from the atrium, monociliated apendopinacocytes become less frequent and non-ciliated apendopinacocytes start to dominate in lining the canal walls. Non-ciliated apendopinacocytes are no different on the micro morphological level to non-ciliated prosendopinacocytes.

Discussion

1. Morphology

Ostia. The diameters displayed by ostia in *T. wilhelma* were highly variable, ranging from total closure to more than 40 µm when open. The ability to open and close ostia within a relatively short period of time for flow-regulating purposes has been documented in a number of different sponge species (e.g. [27,28]). For this reason ostia diameters and numbers within specimens appear highly variable at any given time.

Pinacocytes. Biophysically, pinacocytes encounter a number of mechanical forces including shear stress and drag which are generated by flow along the canal system. Some of these forces result from direct interactions between the fluid and the pinacocyte surface which in turn contribute to general flow resistance and the resulting velocity profile. The boundary layer of the flow profile is particularly important in the context of particle feeding as it is involved in the slowdown and sedimentation of particles for phagocytosis along the canal walls [29].

The morphologies of apendopinacocytes, and most likely endopinacocytes in general, might reflect local hydrodynamics [30]. For the purposes of comparison, arterial endothelial cells have been shown under pulsatile but unidirectional laminar flow to align in the direction of flow [5]. In areas of flow separation and/or flow reversal (e.g. branching), they adopt an unaligned polygonal-shaped organization [5]. However, since our knowledge of local flow regimes in canals is very limited, it cannot yet be claimed with certainty that there is a direct correlation between endopinacocyte morphology and flow. Nevertheless, the fact that apendopinacocytes in *T. wilhelma* are aligned in an ordered way in the atrium region in particular is of great interest, for it is theoretically possible, taking fluid dynamics and morphometric data into account [6], that flow there might develop a pronounced unidirectional laminar profile.

T. wilhelma apendopinacocytes in and around the atrium region are monociliated. A morphologically similar cell type is characteristic of all Homoscleromorpha [22,31,32]. However, the monociliated endopinacocytes of Homoscleromorpha bear a much longer cilium and have been proposed to be actively involved in flow generation, something which is highly unlikely in *T. wilhelma* where the short cilium would make flow generation by apendopinacocytes relatively inefficient compared to that by choanocytes [10]. We propose as an alternative that the short apendopinacocyte cilium in *T. wilhelma* functions as a stereocilium and is involved in local flow sensing. The fact that the monociliated apendopinacocytes of the freshwater sponge *Ephydatia muelleri* (Lieberkühn, 1856), which are located in exactly the same position as in *T. wilhelma*, have recently been demonstrated to have a sensory function backs up this claim [30]. The nonmotile primary cilium in *Ephydatia muelleri* consists of 9 circularly arranged microtubule doublets ("9+0" fashion), but lacks the central ones ("9+1" fashion) characteristic of motile cilia and flagellae [30,33].

Choanocyte chambers. The choanocyte chambers in *T. wilhelma* exhibit two specializations which are presumed to have a substantial impact on local and global fluid dynamics: (1) monociliated apopylar cells and (2) reticuloapopylocytes. In *T. wilhelma* apopylar cells form a ring-shaped reduction of the choanocytic apopylar opening which is double cone-shaped in cross-section. A functional morphological interpretation of the location of this cell type in a hydrodynamically pivotal site is discussed below. Apart from their role in preventing back flow, the function of apopylar cells is currently unclear, especially with regard to the cilium. However, since the cilium projects freely into the apopylar opening we propose that it is involved in flow sensing. Verifying this experimentally, however, will be technically challenging. As in the case of monociliated apendopinacocytes, ultrastructural data pertaining to microtubule arrangement might help to answer this question.

2. Functional Anatomy

Hydrodynamic situation in sections of the canal system and implications for the function of cell types. The development of ostia pore fields (see Figure 3), as observed in T. wilhelma under steady state flow conditions, can be explained as a result of fundamental fluid dynamic principles. As explained by

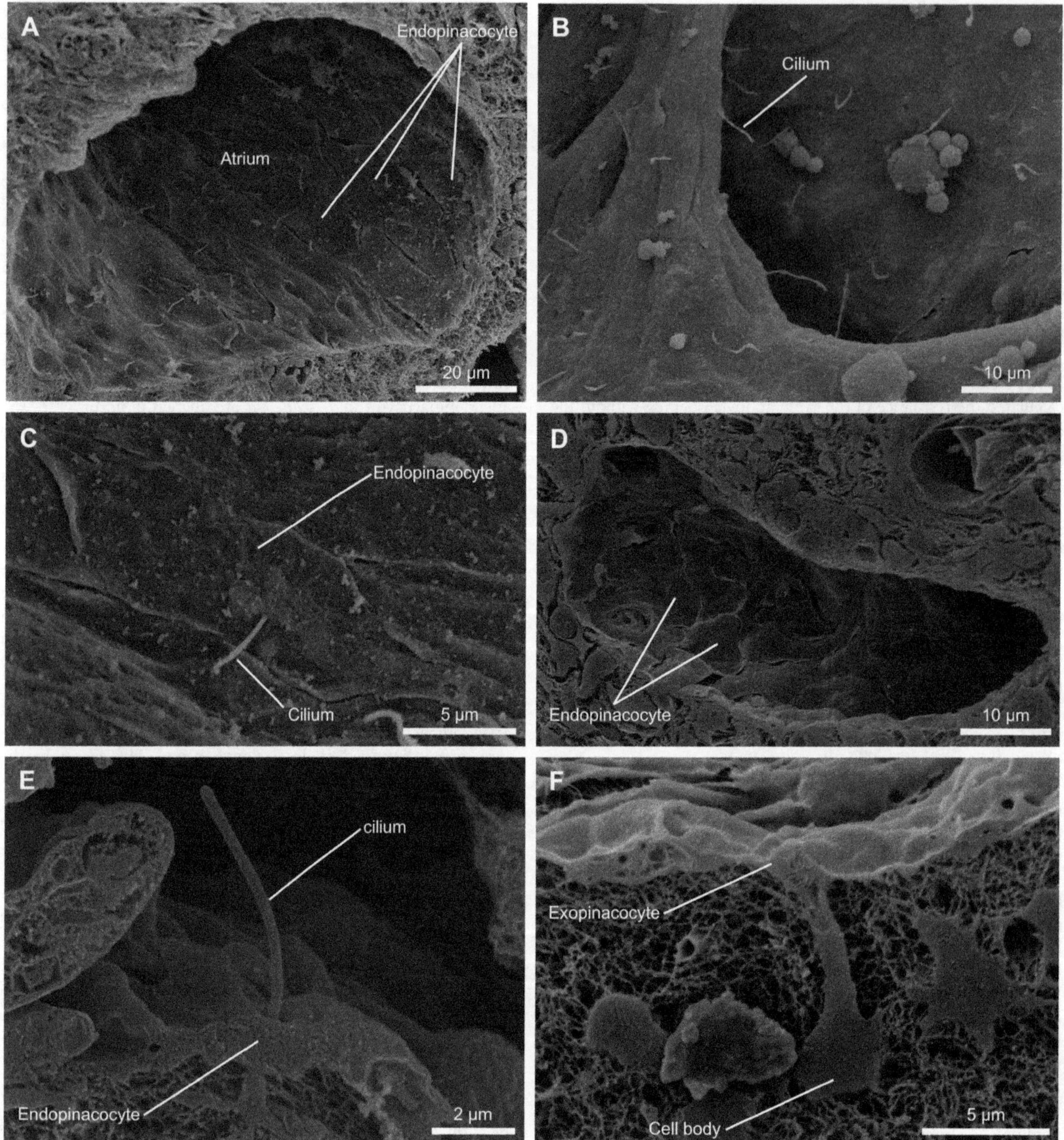

Figure 6. Scanning electron micrographs of pinacocytes. (A) Highly ordered apendopinacocytes in the atrium region. (B) Monociliated apendopinacocytes in the excurrent canal system. (C) Detailed view of a monociliated apendopinacocyte. (D) Prosendopinacocytes lining the walls of the incurrent canal system. (E) Detail of the cilium of an apendopinacocyte. (F) Cross section of an exopinacocyte lining the outer surface of *T. wilhelma*. Note the T-shaped umbrella like cross sectional morphology with the cell body of the pinacocyte sunk into the extra cellular matrix.

equations 1 and 3 in the introduction, pore fields drastically reduce the total resistance of the global influx and therefore reduce global pumping energy costs. Even though the parallel arrangement of small sized elements in the canal system reduces resistance on an organismic scale, resistance in each single element remains high. Therefore, the systemic resistance of individual canal segments influences the amount of water passing through certain areas of the sponge body. This can be quantified by the term perfusion, the amount of water passing through a defined volume of the sponge body over a given time interval. Consequently, resistance is a factor which can be used directly to control the perfusion of certain areas of the sponge body and to adjust local flow. No studies to our knowledge have yet addressed this aspect of local flow regulation from a detailed theoretical and experimental perspective. However, it seems on the basis of all the available data and fluid dynamics models that a local regulation of perfusion is possible within specific areas of the sponge, and that this is most efficient in regions which (1) have a significant impact on flow resistance and (2) can be mechanically modified by the sponge. Both requirements are fulfilled when it comes to ostia and the oscule, and in principle in the case of small sized canal segments too. As T. wilhelma normally only possesses one oscule, flow theory and the continuity of flow would suggest that oscule contraction would only cause very slight variation in local flow. This is supported by studies into ostia contraction in T. wilhelma (unpublished data) and other species, which have demonstrated that single ostia can be contracted individually [34]. Unless new methodologies become available, however, it will only be possible to demonstrate this quantitatively and experimentally in a transparent sponge species which permits in situ high resolution flow measurements to be taken within the canal system. The question of whether and how small sized canal segments influence perfusion is closely related to the pronounced regular body contractions observed in T. wilhelma. Predicting the effects of canal contractions on local flow during a contraction and expansion cycle is difficult, as information on the exact dynamics of canal contractions can only be obtained indirectly from the overall kinetics inferable from time-lapse sequences [13,35]. However, local body contractions and contraction waves across the body have been reported both for T. wilhelma and other sponges [36] and are presumed to be related to local changes in canal diameter and to result in changes in perfusion (see equation 1).

In terms of local hydrodynamics, the most complex functional unit within the canal system is the choanocyte chamber. From experimental and theoretical studies into sponges and choanoflagellates, we know a good deal about particle filtering at the level of choanocytes (e.g. [2,10,25,37]). However, we still lack detailed knowledge of flow fields in choanocyte chambers. A schematic drawing of simulated flow fields is given in figure 7. Hydrodynamically pivotal sites within the choanocyte chamber are marked with stars (Figure 7) and refer to structures with a significant impact on flow resistance. These include the prosopylar openings (Figure 1D), where resistance is determined by the diameter of the opening. It is presumed that the small size of these openings causes flow to accelerate compared to its velocity in adjacent canal segments. Predicting the situation for choanocytes is difficult as we lack information on how flow in the near field surrounding the choanocytes is affected by neighboring cells. In choanoflagellates, which are morphologically and functionally very similar to choanocytes microvilli collar height, density, spacing, angle and flagella length have been demonstrated to be interdependent [2]. The choanocytes in T. wilhelma have a smaller number of almost erect microvilli which are oriented parallel to each other and can be expected to reduce resistance to flow. This in turn can be expected to reduce pressure drop at the level of choanocyte chambers, if velocity is the fixed parameter or a slower flow an pumping capacity compared to choanoflagellates if pressure drop is the reference constant determining flow. Downstream in the direction of flow apopylar openings form the next anatomical structure crucial to pressure drop. In T. wilhelma, as in some other sponges [23,26,28,31,38–43], apopylar cells directly adjacent to the apopylar opening form a cone-shaped ring structure which makes contact with the neighboring choanocytes. The exact function of this structure is hard to pinpoint. Comparative experimental studies into flow fields around sessile and free swimming choanoflagellates might serve as a starting point. The studies in question have demonstrated that the boundary layer (e.g. the height above the substratum in a sessile choanoflagellate) has a significant influence on far and near field flow in terms of the development of eddies [16]. Applying these observations to

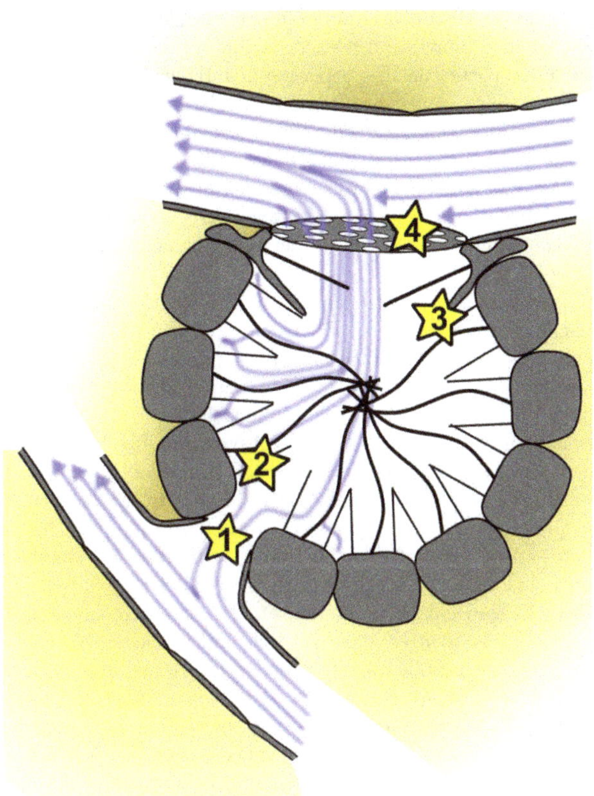

Figure 7. Schematic drawing of a choanocyte chamber with indicated flow directions and hydrodynamically pivotal sites (stars): 1. prosopyle, 2. microvilli collar, 3. contact side between apopylar velum of monociliated apopylar cells and microvilli collar of choanocytes at the apopylar opening, 4. reticuloapopylocyte.

choanocyte chambers may suggest that if no additional structures were present, eddies would develop between choanocytes and the apopylar opening. The direction of flow of eddies in this location would be opposite to the direction of outflow and would result in a significant disturbance of flow at the apopylar opening. In order to prevent the development of eddies in this location an additional boundary structure is needed. In T. wilhelma, the cone-shaped ring of apopylar cells around the opening fulfils this requirement by forming a ceiling seal with the microvilli collar tips of adjacent choanocytes, thus seeming to prevent backflow through eddies, which would significantly reduce local and global pumping efficiencies.

3. Functional aspects of the new cell type

From a hydrodynamic point of view, reticuloapopylocytes are the second functional morphological extravagance to be found in connection with *T. wilhelma* choanocyte chambers. Their location in the canal system and their morphology give rise to a number of hypotheses regarding their function. Reticuloapopylocytes might (1) serve as filtering devices, (2) be related to passive flow, and (3) serve as local flow-regulating devices.

A role in particle filtration, suggested by their sieve-like nature, can very likely be ruled out. We have never observed particles stuck on reticuloapopylocytes, nor witnessed any phagocytosic events. Considering the size of the pore(s) (~0.5 μm) and the size of a typical food particle (2 µm–5 µm), we would expect the pores to be clogged by retained particles within a very short period of time. From particle feeding experiments and our understanding of hydrodynamic constraints, we know that the majority of particles are restrained with great efficiency by the microvilli collar of choanocytes at the latest [25,44]. In other words, in terms of efficiency, an additional downstream filtering element in the form of reticuloapopylocytes is simply not necessary, which renders this potential function obsolete under parsimonious evolutionary principles.

Experimental and theoretical studies into filter feeding animals, including several sponge species, have demonstrated using a Venturi tube principle how even actively pumping species benefit from and make use of ambient flow-induced passive ventilation [11,45,46]. A recent work on hexactinellids provides detailed calculations of the dimensions of canal system elements (especially canal segments, choanocyte chambers and their openings) in relation to their role in fostering passive flows [46]. In this context the presence of large bypass elements [15] and the highly asymmetric nature of branching in *T. wilhelma* [6] could be interpreted as factors which promote passive flows. However, this hypothesis is speculative as the impact of bypass elements on flow patterns inside sponges is not yet well understood on either the local or the organismic scale. It is therefore currently impossible to prove or reject this hypothesis for *T. wilhelma*. What is more, a closer look at the morphology and dimensions of apopylar openings in *T. wilhelma* in the context of resistance theory does not support the hypothesis of passive ventilation by ambient flow. This is underlined by the resistance values we calculated for reticuloapopylocyte-bearing apopyles, which are about 1300 time greater than in unchanged apopyles and 8000 times greater than in the hexactinellid *Aphrocallistes vastus* [46], where ambient current-induced passive flow has been demonstrated. We would expect the much greater pressure drop/resistance generated at fenestrated apopyles in comparison to non-specialized apopylar openings to prevent the induction of passive flow through choanocyte chambers in *T. wilhelma*.

The third hypothesis regarding local flow regulation is related to the fact that individual reticuloapopylocyte intracellular pores have been observed in both an open and a closed state, and to the detection of a specific myosin-heavy chain expression pattern in this new cell type [47] which indicates its ability to actively modify its state of opening. In this it is strikingly reminiscent of intracellular ostia, which possess the ability to open and close relatively rapidly in order to regulate flow [48–50]. Altering the available cross-sectional area of the apopyle by entirely or partially closing individual pores changes the resistance of the apopyle. Closing pores leads to (has the capacity to lead to?) a reduction in the volume of flow and possibly even to a complete shutdown of individual choanocyte chambers in distinct areas of the sponge body. A reduction in the volume of flow at an apopyle will result in a change in the perfusion of the portion of the sponge body in question. The ability to alter flow rates on a local scale with consequences on the regional and even organismal levels qualifies the reticuloapopylocyte as a simple and highly precise fine-tuning device. Theoretically, reticuloapopylocytes permit a gradual adjustment of resistance at the apopyle by closing increasing numbers of pores to create an almost continuous decrease in flow. However, as these cells are to be found deep in the sponge body and are thus not accessible to *in vivo* light microscopy, direct experimental evidence to back up or refute this hypothesis will be difficult to obtain.

4. Functional constraints in the evolution of apopylar elements

Body contraction-expansion cycles have been demonstrated in representatives of all four major lineages of sponges ([36] and Nickel unpublished data). Of all the species studied so far, the amplitude and frequency of body contractions have been highest in *T. wilhelma* [13,36]. The primary effectors of body contraction are endopinacocytes [35]. In the course of a body contraction cycle the canal lumen disappears almost entirely. The change in canal diameter leads to an increase in resistance in the canal system. This change in the hydrodynamic situation in the canal system during a body contraction cycle gives rise to three different functional constraints with regard to the evolution of apopylar elements: (1) Risk of damage to canal system elements caused by increasing pressure in the contraction phase. (2) A need to modify the perfusion of body parts, something which can be influenced by contraction and expansion phases (3) A need to generate increased Gauge pressure during the inflation of the canal system in the second kinetic phase (see [13,35]) of the expansion cycle.

An increase in Gauge pressure within the canal system during the relatively rapid contraction phase is the result of cumulative resistance caused by the reduction in canal diameter and the presence of just a single oscule through which all residual water has to be expelled. The increased Gauge pressure leads to constraint (1), which primarily affects all delicate structures in the canal system (e.g. choanocytes). From a technical point of view the solution would be a pressure regulator. In a very simple way in *T. wilhelma*, the reticuloapopylocytes constitute just such pressure regulators. A comparable role has been demonstrated for the morphologically highly similar sieve plates in the phloem of plants [51].

The exact role of body contractions in sponges is unclear. One hypothesis proposes a physiological need to flush the canal system by exchanging all the water in the aquiferous system in the course of a body contraction cycle. Experimental studies into body contraction cycles in different sponge species have demonstrated the presence of contraction waves which travel over the sponge body ([35] and own unpublished data) Over the course of a body contraction cycle, canal diameters undergo alterations which result in changes in resistance. These changes affect perfusion rates, as formulated by constraint (2) on the principle described in section 2 above.

An analysis of body contraction kinetics in sponges has revealed four different sub-phases [35]. The contraction and expansion stages exhibit two distinct kinetic phases each. Endopinacocytes have been identified as effectors of contraction [35]. The two different kinetic phases of the *T. wilhelma* expansion cycle are thought to have two effectors. In the early and more rapid expansion phase elastic energy loaded into a distinct higher ordered sub-volume of the extracellular matrix is released [52]. This results in a partial inflation of the aquiferous system which enables the choanocyte chambers to start working again. In the second, much slower kinetic phase, we propose that Gauge pressure plays a role in fully inflating the canal system. Fulfilling this functional constraint (3) basically requires the presence of two specific components of the sponge aquiferous system - reticuloapopylocytes and bypass elements. Reticuloapopylocytes increase Gauge pressure by increasing resistance, while bypass elements form direct connections between the incurrent and excurrent canal system [6,12,14,15]. Their function and impact on flow in sponges is still under debate, but hydrodynamics and resistance theory might shed light on their functional role in the context of body contraction cycles in *T. wilhelma*. The increased back pressure in the incurrent canal system generated by the presence of reticuloapopylocytes in pumping choanocyte chambers is coupled to the excurrent canal system via bypass elements. This increases Gauge pressure throughout the system, helping it to inflate.

A large number of the hypotheses and interpretations discussed above are based on theoretical considerations and fundamental physical rules, the morphology of specific cell types and the morphometric information available on the canal system. Again, experimental verification *in vivo* is not currently possible due to the lack of optical live imaging techniques for structures deep inside the sponge body. Non-destructive approaches, e.g. x-ray videography and tomography or magnetic resonance imaging, do not provide the required spatial and/or temporal resolution needed to simultaneously analyze morphology, flow and the kinetics of contraction. Furthermore, we are faced with highly complex interdependencies between the phenomena in question - e.g. pressure drop and gauge pressure being caused by bypasses and reticuloapopylocytes. A solution to this dilemma might be computational fluid dynamic modelling approaches based on exact canal system geometries obtained from biological entities. Depending on the effect to be studied, modeling approaches might enable us to reject and formulate new hypotheses, or even test the influence of specific structural elements by modifying the geometries used (e.g. including/excluding bypass elements). However, this would require detailed information on the morphology of the canal system, volume flow and temporal analysis data pertaining to the kinetics of body contractions.

Conclusions

Reticuloapopylocytes, described here in *Tethya wilhelma*, represent a new and functionally distinct type of cell. On the basis of related functional morphological and hydrodynamic constraints, we evaluated a range of hypotheses pertaining to the function of this new cell and its effect on local and organismic flow conditions. Compared to our understanding of the functional morphology and influence on fluid dynamics of the other cell types discussed in the present study, our knowledge of the apopyle in leuconoid canal systems is patchy, especially when it comes to understanding its role in flow conditions on a local and organismic scale and its relationship to particle filtration in general. All the studies concerned with flow in sponges so far have focused mainly on the relationship between flow conditions and the architecture of the canal system in general, or concentrated on ecological aspects. However, if we break groups of cells in the aquiferous system down into functional units, the most interesting one is constituted by choanocytes and apopyle-related cells. The fact that a putative flow-regulating cell type is able to cut off every single choanocyte chamber and connected canal system elements from a highly parallelized canal system configuration raises the question of whether the apopyle is in fact a general regulative element in all sponges. Further research needs to focus on morphological changes in apopyles which reflect functional plasticity, e.g. during contraction events or pumping arrests. This will require a highly differentiated fixation scheme for functional states which will have to be characterized, analysed and understood in detail.

Materials and Methods

Sponge material

Individuals of *T. wilhelma* were sampled from the type location in the aquarium of the zoological-botanical garden 'Wilhelma' (Stuttgart). As *T. wilhelma* is not considered an endangered or protected species, no special sampling permits were required to retrieve material for scientific experiments from the aquarium section of the zoological-botanical garden. A continuous culture of sponges was maintained in a 180 l aquarium at 26°C using artificial seawater under a light/dark cycle of 12:12 h. The sponges were fed regularly with commercial invertebrate food (Artifical Plancton, Aquakultur Genzel) [13].

Scanning electron microscopy

Specimens of *T. wilhelma* were fixed overnight in a precooled iso-osmolar solution of 1.25% glutaraldehyde, followed by a contrasting step in iso-osmolar 1% OsO_4 solution for 1.5 h. They were desilified in 5% hydrofluoric acid for 1 h and then embedded in styrenemethacrylate [53]. After semi-thin sectioning, we dissolved the plastic around the remaining sponge using xylene-treatment and dehydrated the samples in increasing concentrations of acetone. Specimens were critically point dried in an Emitech K850 CPD system and sputter coated in an Emitech K500 SC system. SEM images were taken on a Philips XL30ESEM instrument.

Morphometric measurements

Morphometric measurements of reticuloapopylocytes and other cells were performed using ImageJ [54]. For the analysis of reticuloapopylocyte pore sizes pores were semi-automatically segmented using the level sets algorithm in Fiji [55]. The ferret diameters (min and max) and area of reticuloapopylocytes and all segmented pores were measured using functions in ImageJ.

Acknowledgments

We are grateful to Martin S. Fischer (Jena) for infrastructure and financial support, Katja Felbel and Benjamin Weiss (Jena) for excellent technical assistance, Isabel Koch and Alex Mendosa (Wilhelma Stuttgart) for additional supply of sponges complementing our own cultures, Isabel Heim (Neubulach) for aquaristic knowledge, Christopher Arnold, Florian Wolf, Henry Jahn and Josefine Gaede for aquarium maintenance. David J. Schulz (Missouri) and an anonymous reviewer provided valuable comments to this manuscript.

Author Contributions

Conceived and designed the experiments: JUH MN. Performed the experiments: JUH MN. Analyzed the data: JUH MN. Contributed reagents/materials/analysis tools: JUH MN. Wrote the paper: JUH MN.

References

1. Riisgård HU, Larsen PS (2010) Particle capture mechanisms in suspension-feeding invertebrates. Marine Ecology Progress Series 418: 255–293.
2. Pettitt ME, Orme BAA, Blake JR, Leadbeater BSC (2002) The hydrodynamics of filter feeding in choanoflagellates. European Journal of Protistology 38: 313–332.
3. Murray CD (1926) The physiological principle of minimum work. I. The vascular system and the cost of flood volume. Proceedings of the National Academy of Sciences 12: 207–214.
4. LaBarbera M (1990) Principles of Design of Fluid Transport Systems in Zoology. Science 249: 992–1000.

5. Waters SL, Alastruey J, Beard DA, Bovendeerd PH, Davies PF, et al. (2011) Theoretical models for coronary vascular biomechanics: progress & challenges. Prog Biophys Mol Biol 104: 49–76.
6. Hammel JU, Filatov MV, Herzen J, Beckmann F, Kaandorp JA, et al. (2012) The non-hierarchical, non-uniformly branching topology of a leuconoid sponge aquiferous system revealed by 3D reconstruction and morphometrics using corrosion casting and X-ray microtomography. Acta Zoologica 93: 160–170.
7. Humphries S (2009) Filter feeders and plankton increase particle encounter rates through flow regime control. Proceedings of the National Academy of Sciences of the United States of America 106: 7882–7887.
8. Jorgensen CB (1983) Fluid Mechanical Aspects of Suspension Feeding. Marine Ecology Progress Series 11: 89–103.
9. Reiswig HM (1975) The aquiferous systems of three marine demospongiae. Journal of Morphology 145: 493–502.
10. Larsen PS, Riisgard HU (1994) The Sponge Pump. Journal of Theoretical Biology 168: 53–63.
11. Vogel S (1983) Life in moving fluids. The physical biology of flow. Princeton: Princeton University Press.
12. Bavestrello G, Burlando B, Sarà M (1988) The architecture of the canal systems of *Petrosia ficiformis* and *Chondrosia reniformis* studied by corrosion casts (Porifera, Demospongiae). Zoomorphology 108: 161–166.
13. Nickel M (2004) Kinetics and rhythm of body contractions in the sponge *Tethya wilhelma* (Porifera: Demospongiae). Journal of Experimental Biology 207: 4515–4524.
14. Burlando B, Bavestrello G, Sarà M (1990) The aquiferous systems of *Spongia officinalis* and *Cliona viridis* (Porifera) based on corrosion cast analysis. Bollettino di Zoologia 57: 233–239.
15. Nickel M, Donath T, Schweikert M, Beckmann F (2006) Functional morphology of *Tethya* species (Porifera): 1. Quantitative 3D-analysis of *Tethya wilhelma* by synchrotron radiation based X-ray microtomography. Zoomorphology 125: 209–223.
16. Schippers KJ, Sipkema D, Osinga R, Smidt H, Pomponi SA, et al. (2012) Cultivation of Sponges, Sponge Cells and Symbionts: Achievements and Future Prospects. Advances in Sponge Science: Physiology, Chemical and Microbial Diversity, Biotechnology 62: 273–337.
17. Fosså SA, Nilsen AJ (1996) Kapitel 3: Schwämme. Korallenriff-Aquarium, Band 5 Einzellige Organismen, Schwämme, marine Würmer und Weichtiere im Korallenriff und für das Korallenriff-Aquarium. Bornheim: Birgit Schmettkamp Verlag. pp. 35–65.
18. Arndt W (1933) Haltung und Aufzucht von Meeresschwämmen. In: Abderhalden E, editor. Handbuch der Biologischen Arbeitsmethoden, Vol I: Methoden der Meeresbiologie. Berlin: Urban & Schwarzenberg. pp. 443–464.
19. Kinne O (1977) Cultivation of animals - research cultivation, 3: Porifera. In: Kinne O, editor. Marine Ecology, Vol III (Cultivation). London: Wiley Interscience. pp. 627–664.
20. Hammel JU, Herzen J, Beckmann F, Nickel M (2009) Sponge budding is a spatiotemporal morphological patterning process: Insights from synchrotron radiation-based x-ray microtomography into the asexual reproduction of *Tethya wilhelma*. Frontiers in Zoology 6: 19.
21. Simpson TL (1984) The cell biology of sponges. Berlin Heidelberg New York: Springer.
22. Boury-Esnault N (2006) Systematics and evolution of Demospongiae. Canadian Journal of Zoology 84: 205–224.
23. De Vos C, Boury-Esnault N. (1990) The apopylar cell of sponges. In: Rützler K, editor; Woods Hole, MA. Smithsonian Institution Press, Washington, D.C. pp. 153–158.
24. Reiswig HM, Brown MJ (1977) The central cells of sponges. Zoomorphology 88: 81–94.
25. Leys SP, Eerkes-Medrano DI (2006) Feeding in a Calcareous Sponge: Particle Uptake by Pseudopodia. The Biological Bulletin 211: 157–171.
26. Langenbruch PF (1988) Body Structure of Marine Sponges: V. Structure of Choanocyte Chambers in Some Mediterranean and Caribbean Haplosclerid Sponges Porifera. Zoomorphology 108: 13–22.
27. Harrison FW (1972) Phase Contrast Photo Micrography of Cellular Behavior in Spongillid Porocytes Porifera Spongillidae. Hydrobiologia 40: 513–517.
28. Weissenfels N (1980) Structure and Function of the Fresh Water Sponge *Ephydatia fluviatilis* Porifera: 7. The Porocytes. Zoomorphologie 95: 27–40.
29. Shimeta J, Jumars PA (1991) Physical Mechanisms and rates of particle capture by suspensionfeeders. Oceanography and Marine Biology - An Annual Review 29: 191–257.
30. Ludeman D, Farrar N, Riesgo A, Paps J, Leys S (2014) Evolutionary origins of sensation in metazoans: functional evidence for a new sensory organ in sponges. BMC Evolutionary Biology 14: 3.
31. Vacelet J, Boury-Esnault N, De Vos C, Donadey C (1989) Comparative study of the choanosome of Porifera: II The Keratose sponges. Journal of Morphology 201: 119–129.
32. Boury-Esnault N, De Vos C, Donadey C, Vacelet J (1984) Comparative study of the choanosome of Porifera: I The Homoscleromorpha. Journal of Morphology 180: 3–17.
33. Sorokin S (1962) Centrioles and the formation of rudimentary cilia by fibroblasts and smooth muscle cells. The Journal of Cell Biology 15: 363–377.
34. Leys SP, Meech RW (2006) Physiology of coordination in sponges. Canadian Journal of Zoology 84: 288–306.
35. Nickel M, Scheer C, Hammel JU, Herzen J, Beckmann F (2011) The contractile sponge epithelium sensu lato–body contraction of the demosponge *Tethya wilhelma* is mediated by the pinacoderm. Journal of Experimental Biology 214: 1692–1698.
36. Nickel M (2010) Evolutionary emergence of synaptic nervous systems: what can we learn from the non-synaptic, nerveless Porifera? Invertebrate Biology 129: 1–16.
37. Leys SP, Hill A (2012) The physiology and molecular biology of sponge tissues. Adv Mar Biol 62: 1–56.
38. Weissenfels N (1982) Structure and Function of the Fresh Water Sponge *Ephydatia fluviatilis* Porifera: 9. Scanning Electron Microscope Histology and Cytology. Zoomorphology 100: 75–88.
39. Weissenfels N (1981) Structure and Function of the Fresh Water Sponge *Ephydatia fluviatilis* Porifera: 8. The Origin and Development of the Flagellated Chambers and Their Junction with the Excurrent Canal System. Zoomorphology 98: 35–46.
40. Langenbruch PF, Simpson TL, Scalera Liaci L (1985) Body structure of marine sponges: III The structure of choanocyte chambers in *Petrosia ficiformis* (Porifera, Demospongiae). Zoomorphology 105: 383–387.
41. Langenbruch PF, Jones WC (1990) Body structure of marine sponges: VI. Choanocyte chamber structure in the Haplosclerida (Porifera, Demospongiae) and its relevance to the phylogenesis of the group. Journal of Morphology 204: 1–8.
42. Langenbruch PF, Scalera Liaci L (1990) Structure of choanocyte chambers in Haplosclerid sponges. In: Rützler K, editor; Woods Hole, MA. Smithsonian Institution Press, Washington, D.C. pp. 245–251.
43. Saller U (1990) Formation and construction of asexual buds of the freshwater sponge *Radiospongilla cerebellata* (Porifera, Spongillidae). Zoomorphology 109: 295–301.
44. Riisgard HU, Larsen PS (2001) Minireview: Ciliary Filter Feeding and Bio-Fluid Mechanics – Present Understanding and Unsolved Problems. Limnology and Oceanography 46: 882–891.
45. Vogel S (1977) Current-induced flow through living sponges in nature. Proc Natl Acad Sci U S A 74: 2069–2071.
46. Leys SP, Yahel G, Reidenbach MA, Tunnicliffe V, Shavit U, et al. (2011) The sponge pump: the role of current induced flow in the design of the sponge body plan. PLoS One 6: e27787.
47. Steinmetz PRH, Kraus JEM, Larroux C, Hammel JU, Amon-Hassenzahl A, et al. (2012) Independent evolution of striated muscles in cnidarians and bilaterians. Nature 487: 231–U1508.
48. Jones WC (1962) Is there a nervous system in sponges? Biological Reviews 37: 1–50.
49. Emson RH (1966) The reactions of the sponge Cliona celata to applied stimuli. Comparative Biochemistry and Physiology 18: 805–827.
50. Elliott GR, Leys SP (2007) Coordinated contractions effectively expel water from the aquiferous system of a freshwater sponge. Journal of Experimental Biology 210: 3736–3748.
51. Jensen KH, Mullendore DL, Holbrook NM, Bohr T, Knoblauch M, et al. (2012) Modeling the hydrodynamics of Phloem sieve plates. Front Plant Sci 3: 151.
52. Nickel M, Bullinger E, Beckmann F (2006) Functional morphology of *Tethya* species (Porifera): 2. Three-dimensional morphometrics on spicules and skeleton superstructures of *T. minuta*. Zoomorphology 125: 225–239.
53. Weissenfels N (1982) Scanning electron microscope histology of spongy *Ephydatia fluviatilis* material. Microscopica Acta 85: 345–350.
54. Rasband WS (1997–2014) ImageJ. Bethesda, Maryland, USA; Available: http://rsb.info.nih.gov/ij/: National Institutes of Health.
55. Schindelin J, Arganda-Carreras I, Frise E, Kaynig V, Longair M, et al. (2012) Fiji: an open-source platform for biological-image analysis. Nat Meth 9: 676–682.

23

Functional Conservation of the *Drosophila gooseberry* Gene and Its Evolutionary Alleles

Wei Liu[1,3], Lei Xue[2,3]*

1 College of Veterinary Medicine, Northwest Agriculture & Forest University, Yangling, Shaanxi, China, 2 School of Life Science and Technology, Tongji University, Shanghai, China, 3 Institute for Molecular Biology, University of Zürich, Zurich, Switzerland

Abstract

The *Drosophila Pax* gene *gooseberry* (*gsb*) is required for development of the larval cuticle and CNS, survival to adulthood, and male fertility. These functions can be rescued in *gsb* mutants by two *gsb* evolutionary alleles, *gsb*-Prd and *gsb*-Pax3, which express the *Drosophila* Paired and mouse Pax3 proteins under the control of *gooseberry cis*-regulatory region. Therefore, both Paired and Pax3 proteins have conserved all the Gsb functions that are required for survival of embryos to fertile adults, despite the divergent primary sequences in their C-terminal halves. As *gsb*-Prd and *gsb*-Pax3 uncover a *gsb* function involved in male fertility, construction of evolutionary alleles may provide a powerful strategy to dissect hitherto unknown gene functions. Our results provide further evidence for the essential role of *cis*-regulatory regions in the functional diversification of duplicated genes during evolution.

Editor: Shree Ram Singh, National Cancer Institute, United States of America

Funding: This work has been supported by the following three funds: 1. Swiss National Science Foundation, Grant No. 31-40874.94; 2. National Natural Science Foundation of China, Grant No. 30971681; 3. Fund from NWAF (No. Z11021005). The funders had no role in study design, data collection and analysis, decision to publish, or preparation of the manuscript.

Competing Interests: The authors have declared that no competing interests exist.

* E-mail: lei.xue@tongji.edu.cn

Introduction

During early *Drosophila* embryogenesis, the antero-posterior axis is progressively defined by the activities of four classes of segmentation genes: maternal coordinate genes, zygotic gap genes, pair-rule genes, and segment-polarity genes [1–4]. In addition to their roles in patterning the embryonic epidermis, many segmentation genes participate in other developmental programs like neurogenesis [5], myogenesis [6], and development of imaginal discs [7].

The *Drosophila gooseberry* (*gsb*) gene, initially identified as a member of the segment-polarity gene class [1], is required after germ band extension to maintain the ventral epidermal expression of *wingless* (*wg*), which suppresses ubiquitous denticle formation, through a *wg-gsb* autoregulatory loop [8]. In the central nervous system (CNS), *gsb* is essential for the activation of *gooseberry neuro* (*gsbn*) in a segmentally repeated pattern [9], for the differentiation of certain neuroblasts, and for the formation of the posterior commissure in each segment [10–14]. Since all known *gsb* mutant alleles are embryonic lethal [11], possible postembryonic functions of *gsb* remain largely unknown. Recently, *gsb* has been found to sustained expression of synaptic homeostasis, indicating the existence of postembryonic functions [15].

gsb encodes a transcription factor including two DNA binding domains in its N-terminal moiety, a paired-domain and a *prd*-type homeodomain [16–18]. Both domains are highly conserved in the N-terminal halves of the *Drosophila* Paired (Prd) and mouse Pax3 proteins, whose C-terminal halves, however, seem unrelated in their primary sequences to the C-terminal portion of Gsb [17,19]. *prd* is a member of the pair-rule gene class, specifying position along the antero-posterior axis with a double-segment periodicity and regulating the expression of segment-polarity genes [20]. The *Pax3* gene, a mutation in which is responsible for the *Splotch* phenotype in mice [21] and Waardenburg's syndrome I in humans [22,23], plays a pivotal role in myogenesis [24].

Despite their divergent developmental functions, Gsb and Pax3 proteins are able to substitute for most functions of Prd when expressed under the control of the complete *prd cis*-regulatory region in *prd*-Gsb and *prd*-Pax3 transgenes [25]. While *prd*-Pax3 is able to rescue the cuticular phenotype of *prd* mutants, *prd*-Gsb can further rescue *prd* mutants to adulthood [25], though the rescued males show reduced accessory glands and are sterile [26]. Taken together, these results indicate that Gsb, and Pax3 proteins have retained most functions of Prd despite their highly diverged C-terminal halves and further point to the *cis*-regulatory region as an important determinant for the functional diversification of these three genes. However, these experiments left unanswered the question of whether Prd and Pax3 proteins could substitute for the normal functions of Gsb.

To address this question, we produced two "evolutionary alleles" [25] of *gsb*, namely *gsb*-Prd and *gsb*-Pax3, which express Prd or Pax3 proteins under the control of the complete *gsb cis*-regulatory region. We show that both transgenes are able to rescue *gsb* mutants to fertile adults, albeit at reduced efficiencies, which suggests that both Prd and Pax3 proteins have conserved all normal functions of Gsb. We conclude that the divergent functions of *gsb*, *prd*, and *Pax3* genes are predominantly determined by their different *cis*-regulatory regions and are further modified by their protein coding regions. These results provide additional evidence to our previous model that the acquisition of different *cis*-

regulatory elements is the primary mechanism in the evolution of new functions [25]. Since some of the rescued males are sterile, *gsb* is important for male fertility. This discovery of a male fertility function of *gsb* suggests that the construction of "evolutionary alleles" may serve as a powerful tool to reveal the hitherto unknown functions of a gene.

Results

Characterization of two hypomorphic *gsb* alleles

The *gsb* gene was initially uncovered by two large deficiencies, *Df(2R)IIX62* and *Df(2R)Kr*^{SB1}, obtained in a screen for embryonic segmentation mutants [1]. Transheterozygotes of the two deficiencies have lost at least two genes in addition to *gsb* (**Figure 1A**). Their cuticle shows a strong segment-polarity phenotype (Fig. 2C), which is indistinguishable from that of homozygous *Df(2R)IIX62* embryos (**Figure 2B**) [1].

Two alleles affecting only the *gsb* gene were identified late, including a point mutation, *gsb*^{525}, and a P-element insertion, *gsb*^{P1155} [11]. In *gsb*^{525}, the codon of the first amino acid of the homeodomain is mutated to a TAA stop codon. In *gsb*^{525} embryos, the *gsb* mRNA level is much reduced by stage 11, presumably because *gsb* activity depends on the *wg-gsb* autoregulatory loop [8], and no Gsb protein is detected by immunostaining, while the protein product of the Gsb target, Gsbn, is barely detectable. The fact that *gsb*^{525}*/Df(2R)IIX62* embryos exhibit only a weak cuticular phenotype (**Figure 2D**), while that of *gsb*^{525} embryos (**Figure 2E**) is nearly wild-type (**Figure 2A**) implies that *gsb*^{525} is not a null allele [11]. Its hypomorphic nature might be explained in two not mutually exclusive ways: the cuticular function of *gsb* is provided either by a Gsb^{525} protein truncated before the homeodomain but including the entire paired domain, or by undetectable levels of wild-type Gsb protein generated by a low probability of read-through at the *ochre* nonsense mutation. To elucidate this question, we prepared two rescue constructs. *gsb0*-525 contains the same mutation as *gsb*^{525}, whereas *gsb0*-ΔHC encodes only the truncated Gsb^{525} protein (**Figure 1B**). Both of these two constructs are under the control of the *gsb* upstream region including the *gsb* cuticle enhancers GEE and GLE [26]. Evidently, only *gsb0*-525 can rescue the cuticle phenotype (**Figure 2G**), whereas *gsb0*-ΔHC cannot (**Figure 2H**). This demonstrates that in *gsb*^{525} embryos an undetectable level of wild-type Gsb protein is produced that is nearly enough to rescue the cuticular function of *gsb*.

gsb^{P1155} is also an interesting allele. It is an insertion of a P element located only 54 bp upstream of the *gsb* transcription start site (**Figure 1C**). This P-element insertion leads to largely reduced *gsb* mRNA and protein levels in homozygous embryos. While these mutants show a wild-type cuticular phenotype (**Figure 2F**) and

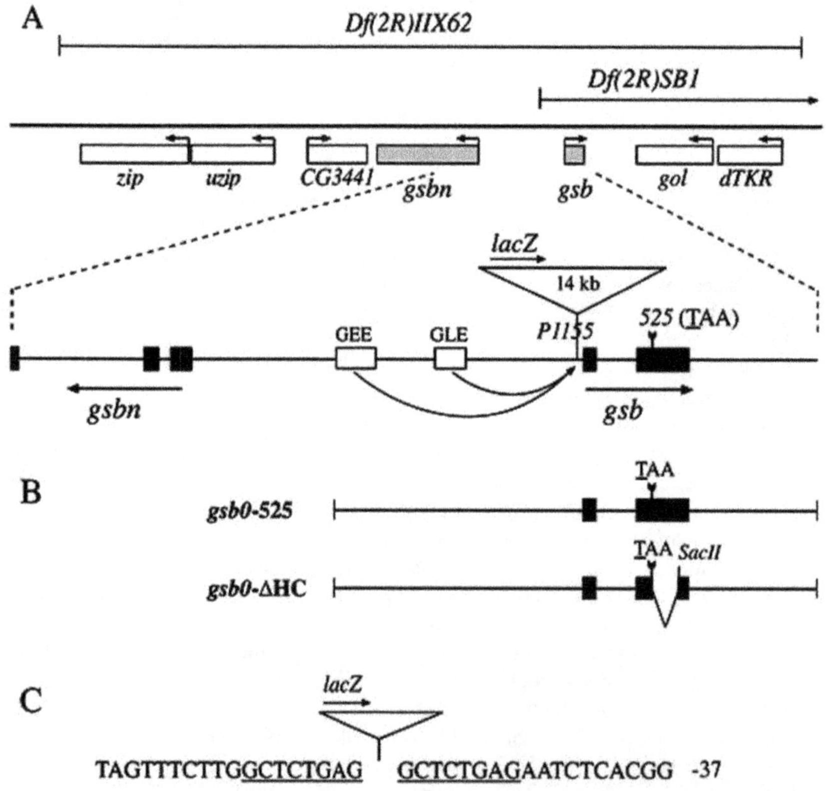

Figure 1. Locus of the *gsb* gene. (**A**) *gsb* mutant alleles. The two deficiencies, *Df(2R)IIX62* and *Df(2R)SB1*, as well as the two hypomorphic alleles, *gsb525* and *gsbP1155*, are depicted. Neighboring genes uncovered by *Df(2R)IIX62, zip, uzip, CG3441*, and *gsbn* upstream of *gsb*, *gol* and *dTKR* downstream of *gsb*, and their direction of transcription are indicated (the rigth telomere of the second chromosome is to the right). Exons are marked by black boxes in the enlarged portion of (**A**) and also in (**B**). (**B**) Map of *gsb0*-525 abd *gsb0*-ΔHC transgenes. Both transgenes contain the upstream epidermis enhancers of *gsb*, GEE and GLE (**Fig. 1A**; Li et al., 1993), the *gsb* promoter, and the entire 3′ UTR of *gsb*. In *gsb0*-ΔHC, 519 bp of coding region between the *gsb525* mutation and a *SacII* site are deleted, resulting in a shift of the open reading frame after the *gsb525* nonsense mutation. (**C**) Sequence surrounding the *gsbP1155* insertion site. The negative numbers refer to nucleotides upstream of the transcription start site. The eight nucleotides, duplicated during insertion of the P-element, are underlined.

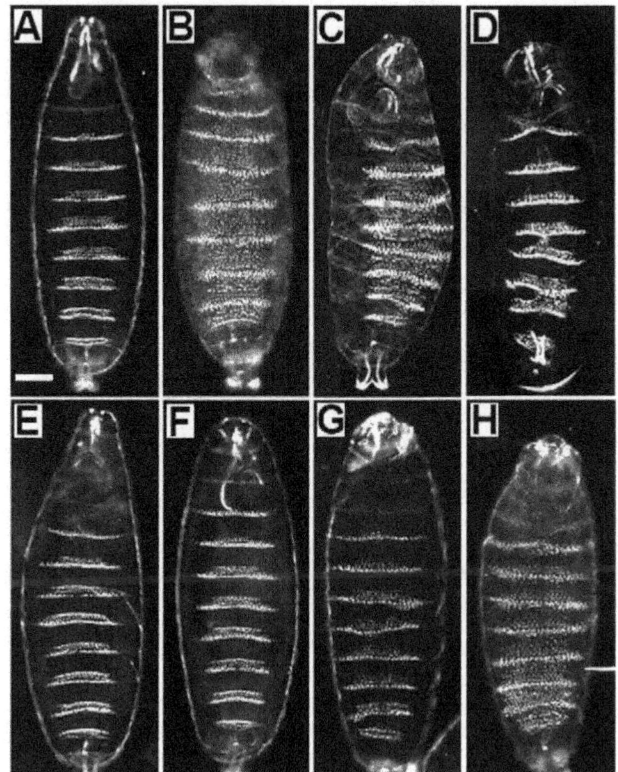

Figure 2. Cuticular phenotypes of *gsb* mutants. (**A**) *Df(2R)IIX62/CyO*, (**B**) *Df(2R)IIX62*, (**C**) *Df(2R)IIX62/Df(2R)KrSB1*, (**D**) *Df(2R)IIX62/gsb525*, (**E**) *gsb525*, (**F**) *gsbP1155*, (**G**) *Df(2R)IIX62; gsb0-525* (**H**) *Df(2R)IIX62; gsb0-ΔHC*. Note in strong *gsb* mutants (**B**, **C**), the ventral naked cuticle region of each segment is transformed into denticle belt, generating an overall denticle pattern, which is in contrast to wild-type (**A**). Scale bar: 50 um.

only mild CNS defects [11], we observed a strongly reduced expression of *gsbn* (data not shown). It follows that gsb^{P1155} is a weaker allele than gsb^{525}.

Generation of *gsb*-Prd and *gsb*-Pax3 transgenic flies

Previous work demonstrated that a *gsb* rescue construct, *gsb-res*, was able to perform all the known *gsb* functions and rescue *gsb* mutants to adulthood [9,11], which suggests that all essential *gsb* enhancer elements are included in this *gsb* transgene (**Figure 3**). To examine whether and to what extent the Prd and Pax3 proteins are able to substitute for the normal functions of Gsb, two rescue constructs, namely *gsb*-Prd and *gsb*-Pax3, were obtained by replacing the *gsb* coding region in *gsb-res* by that of *prd* and *Pax3*, respectively (**Figure 3**). Transgenic files were generated by P-element-mediated transformation in the *Drosophila* germlines [31]. Several independent lines were obtained for each construct. Only transgenic lines that were homozygous viable were selected for further investigation.

It has been previously shown that in wild-type embryos, Gsb protein is initially expressed during blastoderm at the end of cellularization in eight stripes in every other segment, which correspond to the odd-numbered Gsb stripes [9]. At gastrulation, the even-numbered Gsb stripes emerge between the odd-numbered stripes to generate a segmentally repeated expression pattern. Toward the end of germ band extension, Gsb protein reaches its highest levels in the ectoderm and becomes laterally restricted to the neuroectodermal region (**Figure 4A**). As expected, in *gsb*-Prd and *gsb*-Pax3 embryos, the Prd protein (**Figure 4B**) and *Pax3* mRNA (**Figure 4C**) are expressed in patterns that are indistinguishable from that of endogenous Gsb protein (**Figure 4A**). At this time of development the endogenous Prd protein is barely detectable in the epidermis [27].

Rescue of *gsb* target gene expression by *gsb*-Prd and *gsb*-Pax3

Previous work has shown that Gsb is required to maintain late *wg* expression in the ventral epidermis through a *wg-gsb* autoregulatory loop [8]. In homozygous *Df(2R)IIX62* embryos, Wg starts to decay in the ventral epidermis after 6 hours [8] and is no longer detectable at stage 13 (**Figure 5B**), while it remains expressed in wild-type embryos (**Figure 5A**). By introducing *gsb*-Prd or *gsb*-Pax3 transgenes into such *gsb* mutant embryos, the Wg expression pattern is fully rescued by one copy of either transgene (**Figure 5C, D**).

Beginning with stage 9, Gsb is expressed in delaminating neuroblasts, where it is required for the activation of *gsbn* [9]. This is apparent from a complete loss of Gsbn expression in *Df(2R)IIX62/gsb*^{525} embryos at the extended germ band stage (**Figure 5F**), while Gsbn expression is strongly expressed in the CNS of wild-type embryos at this stage (**Figure 5E**). Gsbn

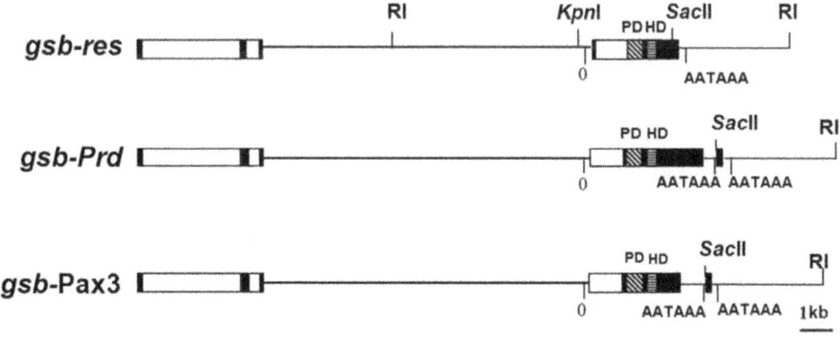

Figure 3. Map of *gsb-res*, *gsb*-Prd and *gsb*-Pax3 transgenes. The *gsb-res* transgene corresponds to the enlarged 20-kb genomic fragment in **Fig. 1A**, which includes the *gsb* transcribed region as well as adjacent 14-kb upstream and 3-kb downstream sequences [9]. The upstream sequence also contains the 5′ portion of the *gsbn* up to part of the third exon. In *gsb*-Prd and *gsb*-Pax3 transgenes, the *gsb* coding region (except of a small region encoding the C-terminus) is replaced by *prd* and *Pax3* cDNAs, while upstream and downstream regions are retained. The *gsb* intron is also retained by inserting it between sequences of the *gsb* and *prd* or *Pax3* leaders. Coding regions are indicated as black boxes except for the paired-domain (PD) and the *prd*-type homeodomain (HD) which are hatched. The *gsb* and *gsbn* introns are indicated as open boxes. The transcription start of *gsb* is marked by 0, and poly(A) addition signals AATAAA are indicated.

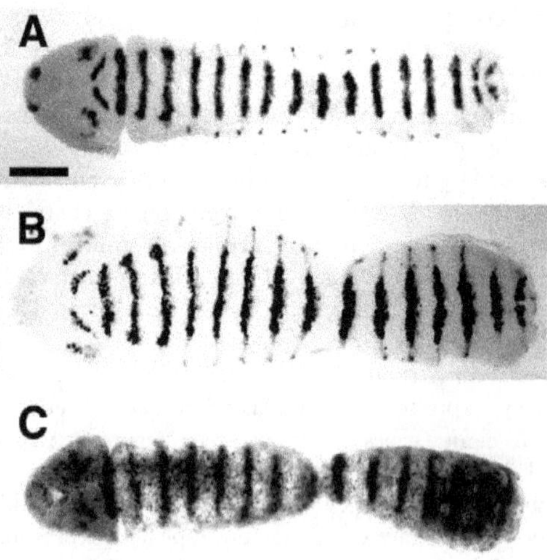

Figure 4. Expression of Gsb and Prd proteins and *Pax3* mRNA under control of the *gsb cis*-regulatory region. Expression of Gsb protein in wild-type embryos (ry^{506}; **A**), of Prd protein in transgenic *gsb*-Prd embryos (**B**), and of *Pax3* mRNA in transgenic *gsb*-Pax3 embryos (**C**) at the extended germ band stage. Wild-type embryos were stained with anti-Gsb antiserum and transgenic embryos, collected from homozygous *gsb*-Prd or *gsb*-Pax3 stocks, were stained with anti-Prd antiserum or hybridized *in situ* with digoxigenin-labeled *Pax3* cDNA. Unfolded embryos are shown and oriented with their anterior to the left. Scale bar: 100 um.

expression in such mutants are rescued by *gsb*-Prd or *gsb*-Pax3 transgene, respectively (**Figure 5G, H**). Taken together, these results demonstrate that Prd and Pax3 proteins can substitute for Gsb function in the transcriptional activation of two essential target genes.

Rescue of *gsb*⁻ cuticular phenotype by *gsb*-Prd and *gsb*-Pax3

One conspicuous feature of the *Drosophila* larva is the metameric ventral cuticular pattern, which crucially depends in each segment on the products of the segment-polarity genes. Embryos lacking *gsb* function exhibit a segment-polarity cuticle defect [1], which consists of mirror image duplications of denticle belts into the posterior portions of each segment where naked cuticle would develop in wild-type embryos (compare **Figure 6A, B**). This phenotype is caused by the loss of late Wg expression, which is required to repress the ubiquitous denticle formation in the ventral epidermis [8]. Consistent with the result that both *gsb*-Prd and *gsb*-Pax3 can rescue the late Wg expression in *gsb* mutants, both transgenes are able to fully rescue the cuticular phenotype of homozygous *Df(2R)IIX62* embryos when present as a single copy (**Figure 6C, D**). It follows that Prd and Pax3 proteins are able to perform the cuticular function of Gsb.

Rescue of *gsb*⁻ CNS phenotype by *gsb*-Prd and *gsb*-Pax3

In addition to its function in patterning the epidermis, *gsb* plays an important role in the development of the embryonic CNS [9–12]. Most prominently, posterior commissures (**Figure 7A**) are missing or reduced in each segment of $Df(2R)IIX62/gsb^{525}$ embryos (**Figure 7B**). This CNS phenotype can be fully rescued by one copy of the *gsb*-Prd (**Figure 7C**) or *gsb*-Pax3 transgene (**Figure 7D**), which indicates that Prd and Pax3 proteins are able to replace the Gsb function in the CNS.

Rescue of *gsb* mutants to adulthood by *gsb*-Prd and *gsb*-Pax3

To test if Prd and Pax3 proteins are able to substitute for all Gsb functions, we tested the ability of *gsb*-Prd and *gsb*-Pax3 transgenes to rescue *gsb* mutants to adulthood. For this purpose, a deficiency, *Df(2R)IIX62*, and two strong alleles of *gsb*, gsb^{525} and gsb^{P1155}, were used. Homozygous or heterozygous combinations of these three alleles are lethal during embryogenesis, which shows that *gsb* is required for postembryonic viability. Although rescue efficiencies are less than half of that of *gsb-res*, one copy of *gsb*-Prd or *gsb*-Pax3 is able to rescue about a quarter of $Df(2R)IIX62/gsb^{525}$ embryos to adulthood (**Table 1**). For all three transgenes, two copies result in 50% higher rescue efficiencies than one copy (**Table 1**), which suggests that at least one *gsb* function required for the viability is dosage dependent. Consistent with this interpretation, one copy of the transgenes is able to rescue a much higher proportion of $Df(2R)IIX62/gsb^{P1155}$ or gsb^{525}/gsb^{P1155} embryos to adulthood (**Table 1**). Therefore, both Prd and Pax3 proteins are able to

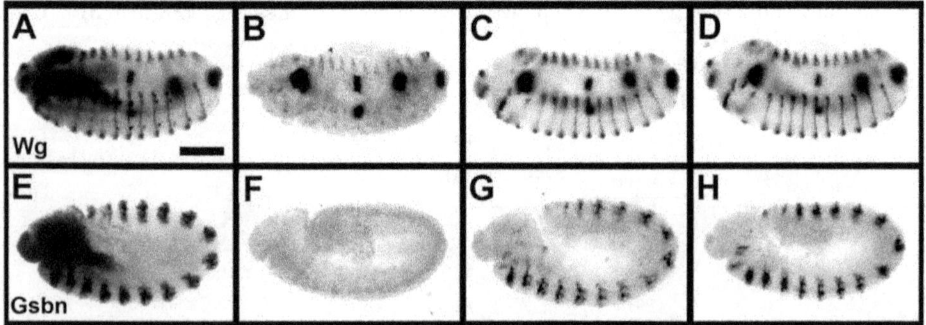

Figure 5. Rescue of Wg and Gsbn expression in *gsb* mutant embryos by *gsb*-Prd and *gsb*-Pax3 transgenes. Expression of Wg (**A–D**) and Gsbn (**E–H**) proteins in wild-type (**A, E**), homozygous *Df(2R)IIX62* (**B–D**) or transheterozygous *Df(2R)IIX62/gsb525* (**F–H**) *gsb* mutant embryos carrying no (**B, F**), one copy of the *gsb*-Prd (**C, G**) or *gsb*-Pax3 (**D, H**) transgene. Embryos at stage 13 (**A–D**) or stage 10 (**E–H**) are oriented with their anterior to the left and dorsal side up. Embryos were collected from crosses between *Df(2R)IIX62/*CyO, *hb*-LacZ; *gsb*-Prd/+ or *Df(2R)IIX62/CyO, hb*-LacZ; *gsb*-Pax3/+ males and *Df(2R)IIX62/CyO, hb*-LacZ (**A–D**) or *gsb525/CyO, hb*-LacZ females (**E–H**), and double stained for ß-galactosidase and or Gsbn protein with rabbit antiserum against ß-galactosidase and anti-Wg monoclonal antibodies or rabbit anti-Gsbn antiserum. Embryos stained with ß-galactosidase have at least one copy of wild-type *gsb* allele and were used as control (**A, E**). One quarter of the embryos did not stain for ß-galactosidase. Half of these embryos did not express Wg in the ventral epidermis and Gsbn in the CNS as expected for *gsb* mutants. The other half displayed rescued expression patterns, which suggested the presence of the transgenes. Scale bar: 100 um.

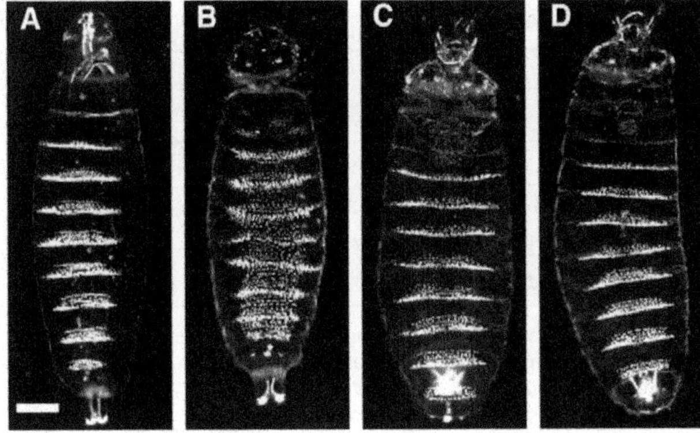

Figure 6. Rescue of the cuticular phenotype of gsb mutant embryos by gsb-Prd and gsb-Pax3 transgenes. Ventral view of cuticle preparations of wild-type (*ry*^{506}; **A**) and homozygous *Df(2R)IIX62* embryos without (**B**) and with one copy of the gsb-Prd (**C**) or gsb-Pax3 transgene (**D**) are shown under dark-field illumination (anterior is up). Wild-type and *gsb* mutant embryos were collected from the *Df(2R)IIX62/SM1* stock, while gsb mutant embryos carrying one copy of the transgenes were collected from crosses between *Df(2R)IIX62/SM1; gsb-Prd* or *Df(2R)IIX62/SM1; gsb-Pax3* males and *Df(2R)IIX62/SM1* females. *gsb* mutants were distinguished from wild type by the presence of the zip phenotype, a deformed head structure resulting from the deletion of the zip gene, which is uncovered by *Df(2R)IIX62* [30]. Scale bar: 50 um.

substitute for all Gsb functions required for survival to adulthood, albeit at lower efficiencies.

gsb is required for male fertility

Since all known *gsb* mutant alleles are lethal during embryogenesis [1,9,11], the adult functions of *gsb* remain unknown. Interestingly, most of the *Df(2R)IIX62/gsb*^{525} males rescued by one copy of *gsb*-Prd or *gsb*-Pax3 are sterile (**Table 2**), while females are fully fertile (data not shown). Therefore, *gsb* is endowed with a function that is essential for male fertility. Two copies of *gsb*-Prd or *gsb*-Pax3 result in significantly enhanced fertilities of *Df(2R)IIX62/gsb*^{525} males (**Table 2**), which suggests that this male fertility function is also dosage dependent. Consistent with this explanation, one copy of *gsb-res* rescues fertility in 39% of the *Df(2R)IIX62/gsb*^{525} males, while two copies rescue male fertility almost completely (**Table 2**). In addition, one copy of *gsb*-Prd or *gsb*-Pax3 is able to rescue fertility in about half of the *Df(2R)IIX62/gsb*^{P1155} males and in three quarters of the *gsb*^{525}/*gsb*^{P1155} males (**Table 2**), whereas one copy of *gsb-res* suffices to fully rescue male fertility in these two mutant combinations (**Table 2**). We conclude that *gsb* is required for male fertility, a function for which both Prd and Pax3 proteins are able to substitute.

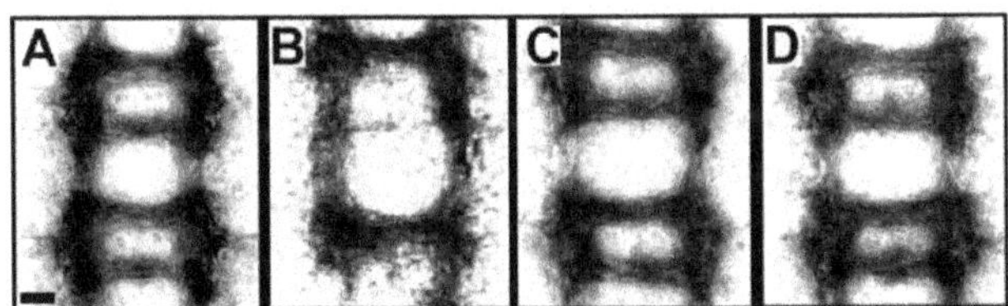

Figure 7. Rescue of the CNS phenotype of *gsb* mutant embryos by *gsb*-Prd and *gsb*-Pax3 transgenes. Patterns of longitudinal and commissural axons in the CNS of wild-type (*ry*^{506}; **A**) and *Df(2R)IIX62/gsb525* embryos without (**B**) and with one copy of the *gsb*-Prd (**C**) or *gsb*-Pax3 transgene (**D**). Embryos at stage 15 were collected from crosses between *Df(2R)IIX62/CyO, hb*-LacZ; *gsb*-Prd/+ or *Df(2R)IIX62/CyO, hb*-LacZ; *gsb*-Pax3/+ males and *gsb525/CyO, hb*-LacZ females, and double stained with rabbit antiserum against ß-galactosidase and monoclonal antibody BP102. One quarter of the embryos did not stain for ß-galactosidase as expected. Half of these embryos have missing or reduced posterior commissures as expected for *gsb* mutants, the other half displays fully rescued commissural patterns as in wild-type embryos. Scale bar: 10 um.

Discussion

Evolutionary alleles of *gsb*

The *Drosophila gsb* and *prd* and mouse *Pax3* genes encode transcription factors that share in their N-terminal moieties two DNA binding domains, a paired-domain and a *prd*-type homeodomain [16–19]. The homology between the N-terminal parts of the three proteins suggests that they were derived from a common ancestor, and thus might have retained some same abilities, despite their divergent C-terminal sequences and apparently distinct developmental functions [32]. Indeed, *gsb*-Prd and *gsb*-Pax3, which express Prd or Pax3 protein under the control of the *gsb cis*-regulatory region, are able to execute all *in vivo* functions of *gsb*, though less efficiently. Hence, both Prd and Pax3 may be considered as leaky mutant proteins of Gsb, whereas *gsb*-Prd and *gsb*-Pax3 are hypomorphic or 'evolutionary' alleles of *gsb*, as the coding regions of the three genes have been derived from a common ancestral gene during the course of evolution. These two 'evolutionary' alleles are weaker than the weakest previously known *gsb* allele, *gsb*^{P1155}, which generates a normal cuticular pattern but displays a weak CNS phenotype and is homozygous lethal during embryogenesis [11]. As these two new alleles have uncovered the previously unknown function of *gsb* required for male fertility, construction of evolutionary alleles may serve as an additional approach to discover unknown functions of a gene [25].

Although the N-terminal portions of the three proteins are rather conserved, their C-terminal parts have diverged to an extent that no obvious similarity in the primary sequences could be perceived [17,19]. Thus, it is particularly interesting that both Prd and Pax3 proteins have retained the potential to perform all the normal functions of Gsb, which suggests that all the important functional motives in the C-terminal part of Gsb have been conserved in the C-termini of Prd and Pax3, presumably in the 3-D structures. It follows that the functional diversification of *gsb*, *prd*, and *Pax3* reside in their *cis*-regulatory rather than their divergent C-terminal coding regions. Therefore, our results are consistent with, and add further weight to, the hypothesis that the acquisition of new enhancer elements by a gene plays a dominant role in evolution [25,26].

Evolutionary relationship between Gsb, Prd and Pax3 proteins

Our previous work has shown that Pax3 can perform only the cuticle function, but not the viability and male fertility functions of Prd [25,33]. Here we report that Pax3 is able to substitute for all Gsb functions in promoting embryonic CNS and cuticle development, postembryonic viability, and male fertility. Thus, in terms of functional conservation, Pax3 seems to be more closely related to Gsb than to Prd. It follows that Gsb and Pax3 are functionally also closer to the common ancestor than Prd. As an independent test of this conclusion, it would be interesting to see if Gsb is a better substitute for Pax3 functions than Prd.

In support of this hypothesis, Pax3 resembles Gsb better than Prd in primary sequences. For Gsb and Pax3, but not Prd, share an octapeptide that is located between the paired-domain and the *prd*-type homeodomain [16,19,32]. In addition, Prd possesses near

Table 1. Rescue of *gsb* mutant embryos to viable adults by *gsb*-Prd and *gsb*-Pax3 transgenes.

	gsb-res (%)		*gsb*-Prd (%)		*gsb*-Pax3 (%)	
	1 copy	2 copies	1 copy	2 copies	1 copy	2 copies
Df(2R)IIX62/gsb^{525}	62 (238/385)	96 (194/203)	21 (90/429)	31 (104/339)	27 (88/326)	**41 (96/234)**
Df(2R)IIX62/gsb^{P1155}	86 (607/707)	nd	51 (144/284)	nd	74 (192/260)	**nd**
gsb^{525}***/gsb***^{P1155}	**99 (344/346)**	**nd**	**61 (230/376)**	**nd**	**77 (226/293)**	**nd**

Percentage of rescued *gsb*$^-$ flies harboring one or two copies of *gsb-res*, *gsb*-Prd or *gsb-Pax3* transgenes (actual numbers of rescued flies per total number of expected *gsb* mutants are given in parentheses). *Df(2R)IIX62/gsb*^{525} flies carrying one or two copies of the transgenes were obtained as offspring from the crosses between *Df(2R)IIX62/SM1; P/P* (P stands for the transgenes) males and *gsb*^{525}*/SM1* or *gsb*^{525}*/SM1; P/P* females. *Df(2R)IIX62/gsb*^{P1155} and *gsb*^{525}*/gsb*^{P1155} flies carrying one copy of the transgenes were obtained from the crosses between *Df(2R)IIX62/SM1; P/P* or *gsb*^{525}*/SM1; P/P* males and *gsb*^{P1155}*/SM1* females. nd, not determined.

its C-terminal end a PRD repeat [34], which is also found in the products of several other genes that are important for early development [34,35], but not in Gsb and Pax3. Therefore, the common ancestor of Gsb, Prd, and Pax3 probably included, in addition to the paired-domain and the *prd*-type homeodomain, the octapeptide in between. After duplication and separation during the course of evolution, Gsb and Pax3 retained these three motives while Prd lost the octapeptide, but instead, obtained the PRD repeat.

In addition to its embryonic functions, *gsb* is also required for male fertility. This function appears to be dosage dependent, as better rescue efficiencies were achieved by either increasing the copy number of the transgenes or using weaker *gsb* mutant alleles (**Table 2**). Interestingly, *prd* is also required for male fertility, in particular for the development of accessory glands [33,34]. Since Gsb is able to substitute for all Prd functions that are required for survival to adulthood [25], but not its male fertility function [28], the male fertility function of Prd might have evolved after its separation from Gsb or have been subjected to strong selection during the course of evolution.

Dosage effect of *Pax* genes

Pax genes encode transcription regulators characterized by the presence of the paired-domain [32]. In vertebrates, *Pax* genes exhibit strong dosage effects, as most *Pax* genes are haploinsufficient [36], and overexpression of *Pax6* in mice leads to severe eye abnormalities [37]. In *Drosophila*, *prd* shows haploinsufficiency in an adult segmentation phenotype, and the *prd* evolutionary allele *prd*-Gsb displays strong dosage effects for all *prd* functions required for survival to adulthood [25]. In addition, overexpression of *eyeless*, the *Drosophila* homolog of *Pax6*, results in a small eye phenotype [38]. Here we show that one copy of the *gsb* rescue construct, *gsb-res*, is able to rescue only 62% of the *Df(2R)IIX62/gsb*^{525} mutants to adulthood (**Table 1**), of which only 39% of the males are fertile (**Table 2**). However, higher rescue efficiencies were scored in both cases by two copies of the transgene (**Table 1**, **2**), which indicates a dosage dependence of *gsb* functions in promoting viability and male fertility. This interpretation was confirmed by the use of two different combinations of *gsb* mutants, and by two *gsb* evolutionary alleles, *gsb*-Prd and *gsb*-Pax*3* (**Table 1**, **2**). A dosage effect was also reported for *gsb* functions in embryonic cuticle and CNS development, as reflected by differences in penetrance of the cuticle and CNS phenotypes in various combinations of different *gsb* mutant alleles [11]. Since the hypomorphic *gsb* mutants, *gsb*^{525} and *gsb*^{P1155}, display a normal cuticle but defects in the CNS [11], and one copy of *gsb-res* is able to fully rescue the CNS phenotype but to rescue the viability and male fertility functions only partially (**Table 1, 2**) in *Df(2R)IIX62/gsb*^{525} mutants, the cuticle function is least sensitive while the viability and male fertility functions are most sensitive to a decrease in the level of Gsb activity. The incomplete rescue of the viability and male fertility functions in *Df(2R)IIX62/gsb*^{525} mutants by one copy of *gsb-res* may result from two effects. First, the deficiency *Df(2R)IIX62*, which deletes, in addition to *gsb*, several other genes including *gsbn* [17,30], which is downstream of *gsb*, might affect the viability and male fertility. Second, *gsb-res* expresses Gsb protein at a subnormal level [9,11], which may result from a position effect of the P-element insertion or from the absence of additional *gsb* enhancer element(s) from the transgene.

The male fertility function of *gsb*

In addition to its embryonic functions, *gsb* is also required for the male fertility. This function appears dosage dependent, for better rescue efficiencies were achieved by either increasing the copy number of the transgenes or using weaker *gsb* mutant alleles (**Table 2**). *gsb* may get involved in male fertility via several means.

Table 2. Rescue of fertility of *gsb* mutant males by *gsb*-Prd and *gsb*-Pax3 transgenes.

	gsb-res (%)		*gsb*-Prd (%)		*gsb*-Pax3 (%)	
	1 copy	2 copies	1 copy	2 copies	1 copy	2 copies
Df(2R)IIX62/gsb^{525}	39 (36/92)	91 (20/22)	9 (2/23)	30 (6/20)	15 (3/20)	**35 (6/17)**
Df(2R)IIX62/gsb^{P1155}	92 (90/98)	nd	43 (16/37)	nd	48 (16/33)	**nd**
gsb^{525}***/gsb***^{P1155}	**95 (74/78)**	**nd**	**77 (23/30)**	**nd**	**75 (21/28)**	**nd**

Percentage of fertile males among *gsb* mutant males that were rescued by one or two copies of *gsb-res*, *gsb*-Prd or *gsb*-Pax3 transgenes (actual numbers of fertile males per total number of rescued males are given in parentheses). Rescued males were obtained from crosses described in legend of table 1 and were placed individually with at least three wild-type virgin females in fresh vials to score fertility. nd, not determined.

First, Gsb plays pivotal role in the development of ejaculatory duct that is required for the transfer of accessory gland secretions and sperm to females during copulation. Ejaculatory duct also secretes components of seminal fluid that might be essential for sperm fertility [39]. Second, Gsb is expressed in the secondary cells of adult accessory glands, suggesting a role of Gsb in the regulation of accessory gland secretions that are crucial for the male fertility (33). Third, males heterozygous for *Df(2R)IIX62*, which deletes *gsb* and its downstream gene *gsbn*, behave less aggressive in copulation (data not shown). This phenotype can be rescued by adding one copy of *gsb-res* (data not shown), implying the impaired Gsb-Gsbn pathway is responsible for this behavioral defect. In support of this interpretation, both Gsb and Gsbn are expressed in the leg and antenna imaginal discs (W.L., L.X. and M.N., unpublished observation), suggesting a role of *gsb* and *gsbn* in the development of leg and antenna, both of which have been shown to be important for eliciting proper male sexual behavior [40].

Interestingly, *prd* is also required for male fertility, for *prd* mutant males rescued by two differently modified *prd* transgenes, *prd*-Gsb [25] and *prd*Res [41], are sterile, despite their capabilities to copulate and transfer sperm to females [33]. These males have severely reduced or no accessory glands [33,41], suggesting *prd* is essential for accessory gland development. Hence, *prd* and *gsb*, though both are required for male fertiltiy, are involved in distinct developmental programs during metamorphosis. Since Gsb is able to substitute for all Prd functions that are required for survival to adulthood [25], but not its male fertility function [33], the male fertility function of Prd might have evolved after its separation from Gsb or have been subjected to strong selection during the course of evolution.

Materials and Methods

Plasmid constructions and generation of transgenic flies

Mutations were introduced into *gsb*0-525 and *gsb*0-ΔHC by PCR mutagenesis. Taking pKSpL5-Gsb [27] as template, the following primers were used: gsb-8 (5′-GTC GTC CGG GCT AGC CTT TAT TTC CT-3′), gsb-11 (5′-GGA AAT AAA GGC GAT CGC GGA CG -3′, gsb-12 (5′-CGT CCG CGA TCG CCT TTA TTT CC-3′), T3 primer, and T7 primer. Fragments containing the mutations were cloned into *gsb*-0 [27], the *gsb* complete leader region and intron were also recovered.

The *gsb*-Prd and *gsb*-Pax3 constructs were derived from *gsb-res* [9] in three steps. First, the 1-kb *gsb* intron was obtained as a PCR product with the primer gint1 (5′-GTC TAG AGT AAG CAC CGA CAG ATA GA-3′) and gint2 (5′-GTC TAG ACT GGA AGA ATT AGA GAA ACA-3′), digested with *Xba*I and inserted into the *Spe*I site of pKSpL5-Prd and pKSpL5-Pax3 [27] to generate pKSgint-Prd and pKSgint-Pax3, respectively. Subsequently, the 3.4-kb *Xba*I fragments from pKSgint-Prd and pKSgint-Pax3 were cloned into the *Avr*II site of *gsb*-0 to produce *gsb-int*-Prd and *gsb-int*-Pax3. Finally, *gsb*-Prd and *gsb*-Pax3 were constructed by replacing the 5.6-kb *Nhe*I-*Xba*I fragment in *gsb-res* with the corresponding fragments from *gsb-int*-Prd and *gsb-int*-Pax3, respectively.

The *gsb*-Prd and *gsb*-Pax3 constructs were injected together with pUChspΔ2-3 helper plasmid into ry^{506} embryos and ry^{+} transformants were selected.

Immunostaining and *in situ* hybridization of embryos

Embryo collection, fixation, and immunostaining were carried out as described [28]. Polyclonal antibodies against Prd (1:500) [28] Gsb, and Gsbn (1:1000) [9], monoclonal antibody against Wg (1:100) [29], and monoclonal antibody BP102 (1:50), which reveals the patterns of the longitudinal and commissural axons in the CNS [11], have been described. Polyclonal anti-ß-galactosidase antibody (1:1000) was obtained from Cappell.

In situ hybridization with digoxigenin-labeled *Pax3* cDNA was performed essentially as described [25].

Cuticle preparation

Embryos were collected and allowed to develop for 24 h at 25°C before cuticles were prepared as described [1].

Fly strains and rescue experiments

Three *gsb* alleles were used in this work: *Df(2R)IIX62*, a *gsb* null allele that deletes *gsb*, *gsbn*, and five additional genes [17,30]; gsb^{525}, a strong hypomorphic allele in which the first amino acid of the homeodomain is converted to a stop codon [11]; and gsb^{P1155}, a hypomorphic allele with a P-element inserted into the *gsb* promoter region [11]. To rescue the cuticle, CNS, viability, and male fertility functions of *gsb* by the transgenes, we used the following fly stocks: (1) *Df(2R)IIX62/SM1*, (2) gsb^{525}*/SM1*, (3) gsb^{P1155}*/SM1*, (4) *Df(2R)IIX62/SM1*; *gsb-res*, (5) gsb^{525}*/SM1*; *gsb-res*, (6) *Df(2R)IIX62/SM1*; *gsb*-Prd, (7) gsb^{525}*/SM1*; *gsb-Prd*, (8) *Df(2R)IIX62/SM1*; *gsb*-Pax3, and (9) gsb^{525}*/SM1*; *gsb*-Pax3.

Acknowledgments

We are deeply indebted to Markus Noll for his invaluable advice, support, and encouragement throughout this project. We are grateful to Thomas Gutjahr for technical assistance and Fritz Ochsenbein for expert artwork. We thank P. Gruss for a *Pax3* cDNA, S. Cohen for anti-Wg monoclonal antibody, and C. S. Goodman for BP102 monoclonal antibody. We are obliged to Hans Noll for comments on the manuscript.

Author Contributions

Conceived and designed the experiments: LX. Performed the experiments: WL LX. Analyzed the data: WL LX. Wrote the paper: WL LX.

References

1. Nüsslein-Volhard C, Wieschaus E (1980) Mutations affecting segment number and polarity in *Drosophila*. Nature 287: 795–801.
2. Peifer M, Bejsovec A (1992) Knowing your neighbors: Cell interactions determine intrasegmental patterning in *Drosophila*. Trends Genet 8: 243–249.
3. Small S, Levine M (1991) The initiation of pair-rule stripes in the *Drosophila* blastoderm. Curr Opin Genet Dev 1: 255–260.
4. St Johnston D, Nüsslein-Volhard C (1992) The origin of pattern and polarity in the *Drosophila* embryo. Cell 68: 201–219.
5. Bhat KM (1999) Segment polarity genes in neuroblast formation and identity specification during *Drosophila* neurogenesis. Bioessays 21: 472–485.
6. Baylies MK, Bate M, Ruiz Gomez M (1998) Myogenesis: a view from *Drosophila*. Cell 93: 921–927.
7. Whittle JR (1990) Pattern formation in imaginal discs. Semin Cell Biol 1: 241–252.
8. Li X, Noll M (1993) Role of the *gooseberry* gene in *Drosophila* embryos: maintenance of *wingless* expression by a *wingless–gooseberry* autoregulatory loop. EMBO J 12: 4499–4509.
9. Gutjahr T, Patel N, Li X, Goodman CS, Noll M (1993b) Analysis of the *gooseberry* locus in *Drosophila* embryos: *gooseberry* determines the cuticular pattern and activates *gooseberry neuro*. Development 118: 21–31.
10. Bhat KM (1996) The *patched* signaling pathway mediates repression of *gooseberry* allowing neuroblast specification by *wingless* during *Drosophila* neurogenesis. Development 122: 2921–2932.
11. Duman-Scheel M, Li X, Orlov I, Noll M, Patel NH (1997) Genetic separation of the neural and cuticular patterning functions of *gooseberry*. Development 124: 2855–2865.
12. Patel NH, Schafer B, Goodman CS, Holmgren R (1989) The role of segment polarity genes during *Drosophila* neurogenesis. Genes Dev 3: 890–904.

13. Zhang Y, Ungar A, Fresquez C, Holmgren R (1994) Ectopic expression of either the *Drosophila gooseberry-distal* or *proximal* gene causes alterations of cell fate in the epidermis and central nervous system. Development 120: 1151–1161.
14. Deshpande N, Dittrich R, Technau GM, Urban J (2001) Successive specification of Drosophila neuroblasts NB 6-4 and NB 7-3 depends on interaction of the segment polarity genes wingless, gooseberry and naked cuticle. Development 128: 3253–3261.
15. Marie B, Pym E, Bergquist S, Davis GW (2010) Synaptic Homeostasis Is Consolidated by the Cell Fate Gene *gooseberry*, a *Drosophila pax3/7* Homolog. The Journal of Neuroscience 30(24): 8071–8082.
16. Bopp D, Burri M, Baumgartner S, Frigerio G, Noll M (1986) Conservation of a large protein domain in the segmentation gene *paired* and in functionally related genes of *Drosophila*. Cell 47: 1033–1040.
17. Baumgartner S, Bopp D, Burri M, Noll M (1987) Structure of two genes at the *gooseberry* locus related to the *paired* gene and their spatial expression during *Drosophila* embryogenesis. Genes Dev 1: 1247–1267.
18. Treisman J, Harris E, Desplan C (1991) The paired box encodes a second DNA-binding domain in the paired homeodomain protein. Genes Dev 5: 594–604.
19. Goulding MD, Chalepakis G, Deutsch U, Erselius JR, Gruss P (1991) Pax-3, a novel murine DNA binding protein expressed during early neurogenesis. EMBO J 10: 1135–1147.
20. Baumgartner S, Noll M (1990) Network of interactions among pair-rule genes regulating paired expression during primordial segmentation of *Drosophila*. Mech Dev 33: 1–18.
21. Epstein DJ, Vekemans M, Gros P (1991) Splotch (Sp2H), a mutation affecting development of the mouse neural tube, shows a deletion within the paired homeodomain of *Pax-3*. Cell 67: 767–774.
22. Baldwin CT, Hoth CF, Amos JA, Da-Silva EO, Milunsky A (1992) An exonic mutation in the *HuP2* paired domain gene causes Waardenburg's syndrome. Nature 355: 637–638.
23. Tassabehji M, Read AP, Newton VE, Harris R, Balling R, et al. (1992) Waardenburg's syndrome patients have mutations in the human homologue of the *Pax-3* paired box gene. Nature 355: 635–636.
24. Borycki AG, Emerson CP (1997) Muscle determination: another key player in myogenesis? Curr Biol 7: 620–623.
25. Xue L, Noll M (1996) The functional conservation of proteins in evolutionary alleles and the dominant role of enhancers in evolution. EMBO J 15: 3722–3731.
26. Xue L, Noll M (2002) Dual role of the Pax gene paired in accessory gland development of Drosophila. Development 129: 339–46.
27. Xue L, Li X, Noll M (2001) Multiple protein functions of Paired in *Drosophila* development and their conservation in the Gooseberry and Pax3 homologs. Development 128: 395–405.
28. Gutjahr T, Frei E, Noll M (1993) Complex regulation of early *paired* expression: intial activation by gap genes and pattern modulation by pair-rule genes. Development 117: 609–623.
29. Brook WJ, Cohen SM (1996) Antagonistic interactions between *wingless* and *decapentaplegic* responsible for dorsal-ventral pattern in the *Drosophila* Leg. Science 273: 1373–1377.
30. Côté S, Preiss A, Haller J, Schuh R, Kienlin A, et al. (1987) The *gooseberry-zipper* region of *Drosophila*: five genes encode different spatially restricted transcripts in the embryo. EMBO J 6: 2793–2801.
31. Rubin GM, Sprading AC (1982) Genetic transformation of *Drosophila* with transposable element vectors. Science 218: 348–353.
32. Noll M (1993) Evolution and role of *Pax* genes. Curr Opin Genet Dev 3: 595–605.
33. Xue L, Noll M (2000) *Drosophila* female sexual behavior induced by sterile males showing copulation complementation. Proc Natl Acad Sci *USA* 97: 3272–3275.
34. Frigerio G, Burri M, Bopp D, Baumgartner S, Noll M (1986) Structure of the segmentation gene *paired* and the *Drosophila* PRD gene set as part of a gene network. Cell 47: 735–746.
35. Berleth T, Burri M, Thoma G, Bopp D, Richstein S, et al. (1988) The role of localization of *bicoid* RNA in organizing the anterior pattern of the Drosophila embryo. EMBO J 7: 1749–1756.
36. Nutt SL, Busslinger M (1999) Monoallelic expression of *Pax5*: a paradigm for the haploinsufficiency of mammalian *Pax* genes? Biol Chem 380: 601–611.
37. Schedl A, Ross A, Lee M, Engelkamp D, Rashbass P, et al. (1996) Influence of *PAX6* gene dosage on development: overexpression causes severe eye abnormalities. Cell 86: 71–82.
38. Jiao R, Daube M, Duan H, Zou Y, Frei E, et al. (2001) Headless flies generated by developmental pathway interference. Development 128: 3307–3319.
39. Richmond RC, Gilbert DG, Sheehan KB, Gromko MH, Butterworth FM (1980) Esterase 6 and reproduction in *Drosophila melanogaster*. Science 207: 1483–1485.
40. Yamamoto D, Jallon JM, Komatsu A (1997) Genetic dissection of sexual behavior in *Drosophila melanogaster*. Annu Rev Entomol 42: 551–585.
41. Bertuccioli C, Fasano L, Jun S, Wang S, Sheng G, et al. (1996) In vivo requirement for the paired domain and homeodomain of the *paired* segmentation gene product. Development 122: 2673–2685.

Dealing with Food and Eggs in Mouthbrooding Cichlids: Structural and Functional Trade-Offs in Fitness Related Traits

Tim tkint[1]*, Erik Verheyen[2,3], Barbara De Kegel[1], Philippe Helsen[2,3], Dominique Adriaens[1]

1 Department of Biology, Research Group Evolutionary Morphology of Vertebrates, Ghent University, Ghent, Belgium, **2** Vertebrate Department, Royal Belgian Institute of Natural Sciences, Brussels, Belgium, **3** Department of Biology, Research Group Evolutionary Ecology, University of Antwerp, Antwerpen, Belgium

Abstract

Background: As in any vertebrate, heads of fishes are densely packed with functions. These functions often impose conflicting mechanical demands resulting in trade-offs in the species-specific phenotype. When phenotypical traits are linked to gender-specific parental behavior, we expect sexual differences in these trade-offs. This study aims to use mouthbrooding cichlids as an example to test hypotheses on evolutionary trade-offs between intricately linked traits that affect different aspects of fitness. We focused on the oral apparatus, which is not only equipped with features used to feed and breathe, but is also used for the incubation of eggs. We used this approach to study mouthbrooding as part of an integrated functional system with diverging performance requirements and to explore gender-specific selective environments within a species.

Methodology/Principal Findings: Because cichlids are morphologically very diverse, we hypothesize that the implications of the added constraint of mouthbrooding will primarily depend on the dominant mode of feeding of the studied species. To test this, we compared the trade-off for two maternal mouthbrooding cichlid species: a "suction feeder" (*Haplochromis piceatus*) and a "biter" (*H. fischeri*). The comparison of morphology and performance of both species revealed clear interspecific and intersex differences. Our observation that females have larger heads was interpreted as a possible consequence of the fact that in both the studied species mouthbrooding is done by females only. As hypothesized, the observed sexual dimorphism in head shape is inferred as being suboptimal for some aspects of the feeding performance in each of the studied species. Our comparison also demonstrated that the suction feeding species had smaller egg clutches and more elongated eggs.

Conclusions/Significance: Our findings support the hypothesis that there is a trade-off between mouthbrooding and feeding performance in the two studied haplochromine cichlids, stressing the importance of including species-specific information at the gender level when addressing interspecific functional/morphological differences.

Editor: Dirk Steinke, Biodiversity Insitute of Ontario - University of Guelph, Canada

Funding: This study is funded by the Research Foundation Flanders (www.fwo.be)(1.1.A72.10.N.00), FWO project: 3G014911 (http://www.fwo.be), and Belspo project: MO/36/013 (http://www.belspo.be). The funders had no role in study design, data collection and analysis, decision to publish, or preparation of the manuscript.

Competing Interests: The authors have declared that no competing interests exist.

* E-mail: Tim.tkint@Ugent.be

Introduction

Mitochondrial DNA shows that the Lake Victoria super-flock of cichlids has given rise to more than 500 species in less than 200 000 years [1,2]. This unusual high rate of speciation has given rise to numerous lineages that occupy almost every niche available in this young lake. Especially the diversity of trophic adaptations is remarkable, resulting in a wide range of phenotypes which are classified into trophic guilds [3].

As evolutionary processes are known to be constrained at different levels (e.g., [4,5,6,7]), the evolutionary morphospace of the head region of these fishes can be expected to be highly constrained due to the integration of several components that impose different, sometimes conflicting functional demands (e.g. improved performance for speed is inversely related to performance for force, simply because of mechanical constraints). For cichlid species that represent different trophic guilds, these conflicts are especially apparent when the comparison involves a so-called 'suction feeding' species (i.e. a high velocity feeding method) and a so-called 'biting species' (i.e. feeding on hard prey). The differences between these very divergent modes of feeding are even apparent by the straightforward comparison of the components that make up the feeding apparatuses in these fishes, and the inference of their functional properties (for example the mechanical lever ratio's for lower jaw opening and closing in cichlid species characterized as 'biters' and 'suckers') [8]. However, because in most mouthbrooding cichlids, it is either only the female or the male parent that incubates the fertilized eggs in their buccal cavity, phenotypic differences (which may or may not be as large as species-specific differences), and therefore also adaptive peaks of a particular species may be sex-specific. It is obviously important to understand how sex-specific constraints may affect the 'optimal design' of a phenotype that combines various functions such as feeding, respiration and aspects related to reproductive behavior (mouthbrooding, nest building, agonistic display) [9,10,11,12]. Earlier studies have shown that the combination of all these selective pressures yielded a series of morphologically similar cichlid species that repeatedly and independently evolved in the different African lakes [4,13,14, 15,16]. As the shape of the head region of these mouthbrooding cichlid species is mostly defined by the shape and size of the buccal cavity, it can be expected that the evolutionary histories leading to the origin of these ecomorphs are the outcome of the combined, but different trade-offs between the spatial and functional demands imposed on the shape of jaws and related features for feeding and mouthbrooding. In addition, and depending on the gender that incubates the eggs, this difference may also have resulted in sexual dimorphism in the head-region, rather than being the result of sexual selection alone [17]. Besides requiring behavioral and physiological adaptations [3,18] mouthbrooding has a negative impact on the number of offspring per reproductive effort. Despite this consequence, mouthbrooding might be one of the key innovations underlying the success of this group (next to pharyngeal jaw specializations [19,20]). For a mouthbrooder, an increased buccal cavity volume offers several advantages: it increases the reproductive potential for a given egg size [21], it potentially improves the efficiency of oxygen uptake with a mouthful of eggs, and it provides the necessary water volume to churn the eggs for aeration [22]. The prediction that the mouthbrooding sex has a higher buccal volume, has been confirmed for paternal mouthbrooding cardinalfishes [21,23] and a tilapiine cichlid [24], but its effect on feeding efficiency has not yet been established.

The observed diversity in trophic morphologies in cichlids is mainly reflected in the shape of the head-region (and hence of the buccal cavity) which reflects each species' ecomorphology [15,25]. Although the overall morphological features of a functional 'specialist' species are not always unambiguously different from those in a species with a 'generalist' feeding repertoire [26], typical 'suction feeders' tend to have a more elongated and conical head shape, whereas 'biters' are often characterized by a shorter and wider head shape [27]. The corresponding spatial differences in lever systems and muscle organization result in a trophic apparatus that is either kinematically, or force efficient. In both cases, the head shape also determines the shape and size of the buccal cavity (as well as other components of the head that are relevant for mouthbrooding) [28,29]. Obviously, the differences in the head shapes of 'biting' and sucking' species may impose conflicting spatial demands, and may result in a situation where the requirements to optimize the feeding performance may be different or even opposite to the requirements maximizing mouthbrooding performance. It could be logical to assume that the space available for mouthbrooding will be more constrained in a 'biting' species with short jaws and large adductor muscles than in a 'suction feeding' species with less muscled long and slender jaws. Should this working hypothesis be correct, this trade-off between mouthbrooding and feeding performance might be an important factor to consider in the processes of morphological differentiation that occurred during the adaptive radiation of these haplochromine cichlids. Indeed, both functions are predicted to affect two very different aspects of these species' fitness: the number of offspring they can produce per litter and the efficiency by which they can process food to obtain the necessary amount of energy to live and reproduce.

It is the aim of this study to use theoretical capacities as proxies for both performances, by testing whether a difference in this trade-off exists in two mouthbrooding haplochromine species, with different mechanical requirements for feeding: one 'suction feeding' zooplanktivore species (velocity dependent prey capture) and a 'biting' molluscivore species (force dependent prey capture). *Haplochromis piceatus* Greenwood & Gee 1969 is used as a typical suction feeding species (ecomorph) with a long, pointed snout and elongate, gracile jaws. The diet of this zooplanktivorous species includes cladocerans, copepods, insect larvae and pupae [30,31]. In contrast, *Haplochromis fischeri* Seegers 2008 (= *Haplochromis sauvagei* non Pfeffer 1896: Greenwood 1980) is a typical biter with short and stout jaws. Its diet mainly consists of molluscs and only seasonally includes diatoms and copepods [32]. Unlike most other molluscivore haplochromines, this species does not ingest and crush the snail with its pharyngeal jaws, but extracts the snail from its shell by grabbing the exposed soft parts of its prey with its oral jaws, and shaking fiercely [33]. Both species are endemic to Lake Victoria and consequently diverged very recently from a common ancestral phenotype.

This study compares head shape data of both species using geometric morphometrics as well as morphological proxies for their feeding performance (through the inference of kinematic transmission efficiency for jaw protrusion and physiological cross section areas to estimate jaw muscle force). The quantified difference in these performance parameters in both species (ecomorphs) were used to test the following two hypotheses: (1) whether functionally relevant morphological features of 'biting' versus 'suction feeding' ecomorphs cause differences in the mouthbrooding performances of each species (estimated through the number, size and shape of eggs incubated in the buccal cavity), and (2) whether, as the result of gender specific differences related to mouthbrooding, females of both species are morphologically and functionally more constrained than males to perform tasks related to their respective trophic specialization as a 'biter' or a 'suction feeder'. With respect to the first hypothesis, we predict a higher kinematic efficiency for upper jaw protrusion in the 'suction feeding' species (ecomorph), and a higher jaw muscle contraction force in the biter. We also expect to detect differences in the number, size and shape of the eggs that would agree with different strategies for efficient incubation and churning during mouthbrooding, in relation to the buccal cavity shape and size of both species (ecomorphs). With respect to the second hypothesis, we expect to observe smaller jaw muscles and a lower biting force in females of both, with the most important difference between males and females of the 'biting' species (ecomorph).

Results

Morphometric analysis

The geometric morphometric analysis of body shape (figure 1A) shows that overlap between species and sexes is limited. The wild-caught specimens of *H. piceatus* and *H. fischeri* also clustered within the corresponding range of both species.

Differences between the two species are reflected by PC1, whereas sexual differences are represented by PC2. The positive PC1 scores for *H. fischeri* reflect a relatively shortened head, with shorter jaws and a more rostrally positioned opercular region (with respect to the consensus) (figure 1B). The eye is shifted dorsally, resulting in a more rounded head profile, and eye diameter is smaller. The anal fin is longer.

The distribution of males and females on PC 2 indicates that similar sexual dimorphism is present for both species. Females tend to have a longer head, which is mainly due to a more rostral

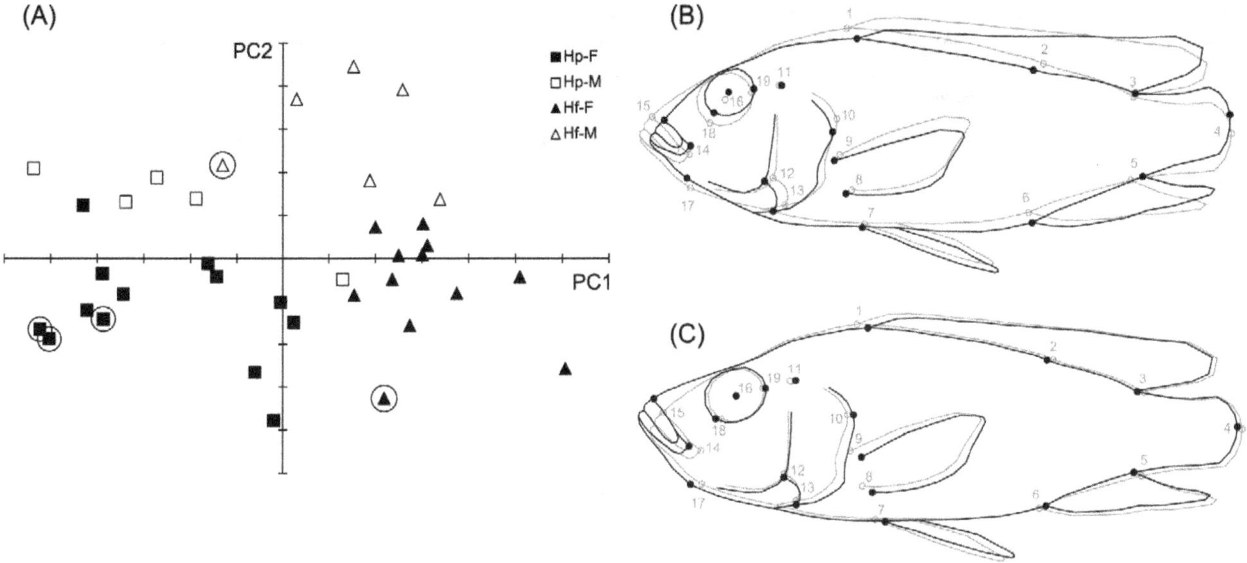

Figure 1. Body shape variation along the first two principal axes. (A) Plot of PC1 versus PC2 (explaining 44% and 17% of the variation, respectively) with indication of species and sex (Legend: Hp-F = *Haplochromis piceatus* females; Hp-M = *Haplochromis piceatus* males; Hf-F = *Haplochromis fischeri* females; Hf-M = *Haplochromis fischeri* males; wild caught specimens are circled). The warped outline drawings represent (B) the positive extreme of PC1 and (C) the negative extreme of PC2 (black outlines) compared to the consensus configuration (gray outline).

and dorsal positioning of the jaws, and an enlargement of the opercular region, without much change in actual length of the oral jaws (figure 1C).

Evaluation of group differences with a permutation test based on Squared Mahalanobis distances showed that both species and the sexes within each species occupied significantly different regions of the morphospace (all p-values for pairwise comparisons <0.002). Measurement of snout width on the dorsal pictures showed that *H. fischeri* had a broader snout than *H. piceatus* ($F_{1,15} = 7.12$, $p = 0.018$). Buccal volume as approximated by elliptical cylinders was equivalent for both species ($F_{1,14} = 0.17$, $p = 0.689$) but differed significantly between sexes ($F_{1,14} = 7.32$, $p = 0.017$) with females having larger buccal cavities than males.

Feeding Performance

Muscle mass of all three parts of the adductor mandibulae differed significantly between species even when standardized for head length, with *H. piceatus* having the lowest relative masses (A_1: $F_{1,15} = 32.51$, $p<0.0001$, A_2: $F_{1,15} = 54.68$, $p<0.0001$, A_3: $F_{1,15} = 25.40$, $p<0.0001$) (Table 1). For the parts A_2 and A_3 of the adductor mandibulae, males also had significantly larger muscles than females (A_2: $F_{1,15} = 7.38$, $p = 0.016$, A_3: $F_{1,15} = 8.41$ $p = 0.011$). Species and sex also had a significant effect on standardized theoretical bite force exerted by A_2 and A_3 (A_2: species: $F_{1,15} = 69.45$, $p<0.0001$, sex: $F_{1,15} = 6.04$, $p = 0.027$, A_3: species $F_{1,15} = 32.01$, $p<0.0001$, sex: $F_{1,15} = 8.94$, $p = 0.009$). *Haplochromis fischeri* bites relatively harder than *H. piceatus*, and males bite harder than females. As a consequence total bite force, which is the sum of bite forces exerted by A_2 and A_3, follows the same pattern. As can be seen from table 1 most factors in the bite model (PCSA, MA and σ) contributed to the significant difference in bite force. Kinematic transmission coefficients of the anterior jaws did not differ between species ($F_{1,16} = 0.15$, $p = 0.71$), although the relative length of the input and output link of the system seems to be shorter for *H. fischeri* (figure 2). Females also had higher KT values than males ($F_{1,16} = 8.50$, $p = 0.01$). *Haplochromis fischeri* did have a significantly lower kinematic efficiency for the opening lever of the lower jaw ($F_{1,16} = 34.56$, $p<0.0001$). Both parameters of the upper jaw (relative length of the ascending process of the premaxilla ($F_{1,15} = 29.75$, $p<0.0001$) and angle between ascending and dentigerous process ($F_{1,15} = 77.82$, $p<0.0001$)) differed between species, where *H. piceatus* had a longer ascending process and a more acutely angled premaxilla.

Egg parameters

Egg size did not differ between species (area: $t_{19} = -0.19$, $p = 0.84$, diameter: $t_{19} = 1.56$, $p = 0.14$, egg volume: $t_{19} = -0.86$, $p = 0.40$), but we did find a highly significant difference in egg shape (Table 2). Eggs of *H. piceatus* had a significantly higher aspect ratio ($t_{19} = 6.83$, $p<0.0001$). A Poisson generalized model showed that *H. piceatus* had smaller clutches, but this difference was only marginally significant after standardization for HL ($\chi^2_{(df=1)} = 3.60$, $p = 0.0579$). The difference in calculated brood volume also became non-significant after standardization ($F_{2,18} = 3.91$, $p = 0.0636$).

Discussion

Structural and functional characterization of the biter versus the sucker

The observed species-specific head shape variation is in accordance with Barel's [27] description of the dichotomy between 'biters' and 'suckers' and with Cooper & Westneat's [34] findings on the morphological differentiation between damselfish herbivores and zooplanktivores. The biter (*H. fischeri*) has a shorter head and a more obtuse head profile, largely due to a shortening of the jaws (figure 1B). Such a shortening has the clear advantage of improving the force transmission (MA) of the jaw, when the input links remain the same. The accommodation of the significantly larger jaw adductors within the head of *H. fischeri*, seems to be associated with the head being broader, whereas the eye is shifted dorsally and reduced in diameter. Albertson & Kocher [8] reported a similar dorsal shift of the eye, associated with a dorsal expansion of the A_1 part of the adductor mandibulae (lying ventral to the eye) for *Labeotropheus fuelleborni* (a Lake Malawi cichlid species

Table 1. Metric data on the jaw muscles and estimates of feeding performance.

	H. piceatus		*H. fischeri*		Main-effect	
	Female (n=5)	**Male (n=5)**	**Female (n=5)**	**Male (n=5)**	**species**	**sex**
Muscle mass (g)						
Pars A1	15.670±4.328	22.320±1.882	39.920±6.847	67.280±19.776	$F_{1,15}=32.5$***	$F_{1,15}=2.4$
Pars A2	8.940±2.809	15.260±1.984	27.450±4.403	46.180±11.525	$F_{1,15}=54.7$***	$F_{1,15}=7.4$*
Pars A3	2.490±0.891	4.420±0.665	6.110±1.078	11.170±2.886	$F_{1,15}=25.4$***	$F_{1,15}=8.4$*
Pars A2						
Fiber length (mm)	4.090±0.629	5.251±0.437	4.579±0.514	5.026±0.275	$F_{1,15}=1.4$	$F_{1,15}=2.4$
PCSA (mm^2)	0.022±0.007	0.029±0.002	0.061±0.013	0.091±0.019	$F_{1,15}=38.5$***	$F_{1,15}=1.8$
MA	0.394±0.025	0.424±0.021	0.484±0.045	0.559±0.043	$F_{1,16}=51.5$***	$F_{1,16}=11.0$**
σ (°)	50.372±2.390	49.962±4.190	54.153±2.639	56.304±2.754	$F_{1,16}=13.6$**	$F_{1,16}=0.4$
Bite force (N)	0.125±0.034	0.178±0.019	0.454±0.108	0.806±0.191	$F_{1,15}=69.5$***	$F_{1,15}=6.0$*
Pars A3						
Fiber length (mm)	3.829±0.651	4.600±0.839	4.555±0.307	5.172±0.981	$F_{1,15}=0.1$	$F_{1,15}=0.0$
PCSA (mm^2)	0.006±0.001	0.010±0.002	0.013±0.002	0.022±0.004	$F_{1,15}=31.5$***	$F_{1,15}=8.9$**
MA	0.249±0.009	0.274±0.010	0.314±0.017	0.297±0.022	$F_{1,16}=39.4$***	$F_{1,16}=0.3$
σ (°)	34.444±6.824	30.377±4.417	29.397±2.671	36.512±5.645	$F_{1,16}=0.06$	$F_{1,16}=0.4$
Bite force (N)	0.017±0.004	0.025±0.004	0.039±0.009	0.072±0.019	$F_{1,15}=32.0$***	$F_{1,15}=8.9$**
Total bite force	0.142±0.033	0.204±0.022	0.493±0.115	0.878±0.208	$F_{1,15}=73.8$***	$F_{1,15}=7.0$*
KT	0.758±0.090	0.607±0.044	0.691±0.076	0.649±0.079	$F_{1,16}=0.2$	$F_{1,16}=8.5$*
KE	5.379±0.717	5.214±0.576	3.653±0.306	3.992±0.563	$F_{1,16}=34.6$***	$F_{1,16}=0.1$
ASC/HL	0.395±0.020	0.360±0.020	0.305±0.036	0.316±0.025	$F_{1,15}=29.8$***	$F_{1,15}=1.0$
β (°)	77.433±4.337	77.914±4.275	93.671±5.577	99.217±4.051	$F_{1,15}=77.8$***	$F_{1,15}=2.0$

PCSA = Physiological cross-sectional area; **MA** = mechanical advantage; **σ** = insertion angle; **KT** = kinematic transmission coefficient; **KE** = Kinematic efficiency of jaw opening; **ASC/HL** = ratio of the length of the ascending arm of the premaxilla and head length; **β** = angle between ascending and dentigerous arm of the premaxilla.
* $p<0.05$, ** $p<0.01$, *** $p<0.001$.
Values are mean ± standard deviation.

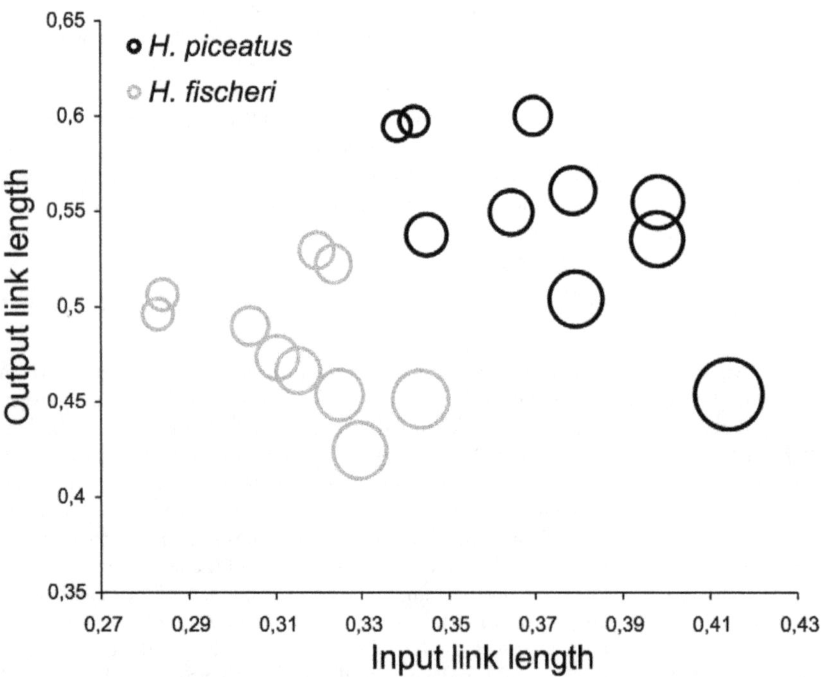

Figure 2. Plot of input link length (lower jaw coronoid processus) versus output link length (maxilla). The size of the circles indicates the value of KT.

Table 2. Egg measurements.

	H. piceatus (n = 10)	*H. fischeri* (n = 11)
Clutch size (# eggs)	33±10	57±20
Aspect ratio	1.37±0.03	1.28±0.03
Area (mm^2)	5.20±0.64	5.21±0.25
Maximum diameter (mm)	3.07±0.21	2.96±0.08
Egg volume (mm^3)	6.94±0.86	6.95±0.33
Brood volume (mm^3)	227.5±82.2	397.3±139.0

Values are mean ± standard deviation.

with a biter morphotype). Through computer modeling, Otten [35] predicted that such a dorsal shift of the eye in cichlids not only increases bite force by providing more space for the muscles, it also allows a more favorable insertion of the A_1 onto the maxilla improving force transmission to the jaws. Furthermore, our observations support the model predictions that a shorter ascending process of the premaxilla placed at an obtuse angle to the dentigerous arm improves force transmission, as this was indeed the case for *H. fischeri*.

The observed morphological differences between the species clearly represent difference in the feeding performance for 'biting' versus 'sucking'. The estimated largest bite forces in *H. fischeri* are not only achieved by an increase in muscle mass, which is a plastic trait that can also be induced by feeding on hard food items [36], but also resulted from an improved force transmission (MA) of the lower jaw and a more favorable insertion (σ) of the jaw adductors.

Although KT is a good predictor of jaw protrusion [37,38,39] and zooplanktivores often have a higher KT than other trophic groups [40], we found no significant difference between *H. piceatus* and *H. fischeri*. This could reflect the fact that neither of the two species is considered as the most specialized within its trophic guild [41]. However, similar KT-values of both species are obtained in different ways (figure 2). The shorter links for *H. fischeri* improve force transmission, but also reduce the extent of jaw protrusion, making suction performance more expensive [27]. As a result of the functional redundancy in the four-bar system involved, different morphologies can result in similar KT values. In fact, the observed KT values fall within the range that has the most theoretically possible morphological solutions [42]. This allowed *H. fischeri* to have an oral jaw system that is more efficient in force transmission, due to its shorter jaws, without compromising its KT. It could be hypothesized that similar KT values in 'biters' and 'suckers' reflect a selective pressure constraining protrusion performance (independent of the morphological configuration to achieve this), similar to that of the ancestral condition of both species (considering their recent common ancestry). However, further comparative studies supported with detailed phylogenetic divergence estimates are required to test this properly.

Sexual dimorphism and possible trade-offs in mouthbrooding females

As traits linked to the buccal cavity in female haplochromine cichlids are related to the efficiency for respiration, feeding and mouthbrooding, it can be expected that sexual dimorphism in the head is due to the added constraint of optimizing buccal incubation (mouthbrooding) performance. Although other aspects undoubtedly play a role as well (e.g. male territorial fighting), the most striking changes we observed result in an increased relative size of the buccal cavity in females. For example, the enlargement of the suspensorial compartment that increases the available volume of the buccal cavity in females is achieved by a longer snout, without much change in length of the jaws. In addition, we observed an enlargement of the opercular compartment of the head by a posterior shift of the base of the pectoral (and pelvic) fins and by a ventral displacement of the interopercle. Interestingly, similar expansions of the lateral aspect of the buccal cavity has also been reported for females of different *Tropheus sp.* populations, a female mouthbrooding cichlid genus that is endemic to Lake Tanganyika [43].

Although the relative size of the buccal cavity did not differ between species, we did observe a trend in clutch size: *Haplochromis piceatus* seemed to have fewer eggs in a buccal cavity of the same size. The eggs of H. piceatus also had a higher aspect ratio, which allows a more efficient packing with less jamming [44], so we might expect that it is easier for this species to churn the eggs in the mouth for aeration. For cardinalfishes it has indeed been found that such a reduction in brood size improves hypoxia tolerance [18]. Physiological performance testing of respiration, in combination with using buccal casts [21,23], would allow a more direct quantification of the trade-off between respiration, feeding and mouthbrooding in females.

Although not all morphological differences between both sexes are statistically significant, most of the determined feeding characteristics suggest that for both species males are better 'biters' than females. Although our model for bite force calculation excluded the A_1 part of the musculus adductor mandibulae (more than 50% of adductor mass) and didn't take the possible sexual difference in muscle physiology into account [45], we found several morphological indicators that support our hypothesis that bite force is higher in males then in females. This does, however, not necessarily imply that reduced theoretical bite performance in females is the exclusive result of an evolutionary trade-off between biting and mouthbrooding capacity. Indeed, as courtship behavior in haplochromine males involves defending a territory [46], it remains possible that higher bite forces in males are the result of other selection factors. Despite the fact that both studied species seem to be typical for their morphotype, a broader species sampling would allow us to determine if the observed patterns in mouthbrooding and feeding performance can be generalized to other species of the same morphotype.

Conclusions

By studying a biting and a sucking ecomorph within haplochromine cichlids, we have shown that observed differences in the head morphology reflect functional demands related to the trophic guild to which they belong. Our data support the hypothesis that the sexual dimorphism in the head region involves an enlargement of the buccal cavity in females to brood eggs, but that this is not without consequences for feeding performance (e.g. bite force). As such, our findings support the hypothesis that a trade-off exists between functional performances that indirectly (feeding) and directly (mouthbrooding) influence fitness in the two species studied. It also suggests that the vast range of selective environments that arose during the explosive radiations of African Lake cichlids may need to be considered at the sex level, rather than the species level in mouthbrooding species. However, a more comprehensive survey in multiple lineages would be required to confirm this.

Materials and Methods

Specimen collection

The specimens used for this study came from laboratory reared stocks at the Royal Belgian Institute of Natural Sciences (Brussels).

These stocks are derived from animals caught in the wild during the 1980's and have been tank bred for approximately 30 generations. The animals were killed with an overdose of MS-222 (*H. piceatus* ♀ n = 10, ♂ n = 5; *H. fischeri* ♀ n = 11, ♂ n = 5). All specimens were sexually mature and females were sacrificed during mouthbrooding. The standard length of the specimens ranged from 57 to 87 mm for *H. piceatus* and 76 to 114 mm for *H. fischeri*. After fixation in 10% formalin for at least two weeks the specimens were transferred to 70% ethanol for preservation. In accordance with the Belgian national law concerning the protection and wellbeing of animals of August 14, 1986, a formal approval from an ethical committee is not required for this kind of project.

As the morphology of tank bred cichlids is known to be variable and might be different from animals in the wild [47], we also included some wild caught specimens, which were provided by the National Natural History Museum (Leiden, The Netherlands) (*H. piceatus*: RMNH 62769 (n = 3) and *H. fischeri*: RMNH 70426 (n = 2)).

Biometry

The left side of all specimens was photographed using a Nikon D40x digital reflex camera, equipped with the standard zoom-nikkor 18–55 mm lens. The fishes were pinned in a dissection board to standardize orientation and spread out the fins. In every picture an individual code and a scale marker was included to allow identification and scaling. Head length (HL) of every specimen was measured on these pictures as defined in Barel et al. [48]. The eggs of the mouthbrooding females were extracted and photographed with a digital camera (Colorview 8, Soft Imaging System) mounted on a dissection microscope (Olympus SZX9). The shape of the eggs was analyzed by taking the following measurements: the length of the long and short axis (assuming it were ellipses), aspect ratio and area. The volume of the eggs was approximated by assuming they were ellipsoids with a long axis and two equal short axes.

For each species five females and five males were dissected to extract the muscles operating the oral jaws. For each individual the A_1, A_2 and A_3 parts of the adductor mandibulae complex were extracted and weighed on an electronic balance (Sartorius BP 121S) to the nearest 0.1 mg (For details on jaw adductor muscle anatomy see [49]). During the dissections photographs were made (with a Nikon D40x and a Sigma 150 mm macro lens) to document the attachment of the muscles to the head skeleton. Snout width was determined on a dorsal picture of these dissected specimens.

The volume of the buccal cavity was approximated using a series of elliptical cylinders as proposed by Drost and Van den Boogaart [50] (see also [51,52]). We measured the width and height of the buccal cavity, which equal the major and minor axis of these ellipses, at 100 equally spaced intervals. These measurements were taken from lateral and ventral pictures of the dissected specimens after clearing and staining [53].

Geometric Morphometrics

The coordinates of 19 landmarks were determined on the habitus pictures of the left side (figure 3) using TPSDIG 2 [54]. Measuring error resulting from variation in positioning of the specimens for photographing and in digitizing of landmarks was quantified, based on two specimens per sex per species (following protocol of Adriaens - http://www.fun-morph.ugent.be/Miscel/Methodology/Morphometrics.pdf). We found that 3.6% of the variation is due to digitization error and 25.4% is due to the combination of orientation and digitization error.

Non-shape variation was removed by performing a Generalized Procrustes Analysis, removing effects of size, position and orientation [55]. To allow the use of traditional multivariate techniques it is also necessary to project the shapes from the non-Euclidean shape space onto a Euclidean tangent space [56]. The correlation between the shape distances in both spaces was checked with TPSSMALL [57] and showed a perfect correlation ($r = 1.0$, slope = 0.9997). Shape variation was analyzed with a PCA using the coordinate data in MorphoJ 1.01a [58]. Due to the limited and unequal sample size we used a permutation test of Squared Mahalanobis distances (10 000 replicates) to test the significance of group differences.

Bite model

The theoretical bite force exerted by the different parts of the adductor mandibulae was estimated using a static bite force model

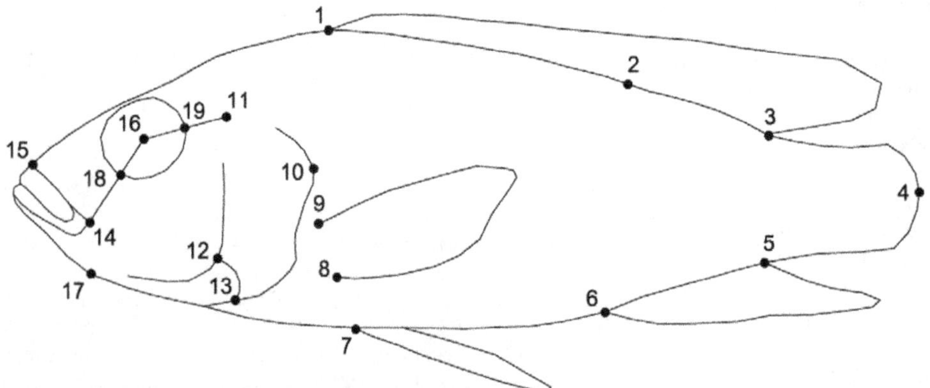

Figure 3. Outline drawing with indication of landmark positions. (1) Anterior insertion of the dorsal fin (2) Insertions of the most caudal spiny fin ray of the dorsal fin (3) Posterior insertion of the dorsal fin (4) Posterior end of the lateral line (5) Posterior insertion of the anal fin (6) Anterior insertion of the anal fin (7) Insertion of the leading edge of the pelvic fin (8) Insertion of the trailing edge of the pectoral fin (9) Insertion of the leading edge of the pectoral fin (10) Posterior extremity of the operculum (11) Center of neurocranial lateral line foramen 5 (12) Dorsal intersection of subopercle and interopercle (13) Ventral intersection of subopercle and interopercle (14) Posterior extremity of the gape (15) Intersection between upper lip and body outline (16) Center of the eye (17) Retroarticular process (18) Intersection of the line connecting landmarks 14 and 16 and the eye outline (19) Intersection of the line connecting landmarks 11 and 16 and the eye outline.

[59]. The output force at the jaw tip was calculated taking into account the maximal force produced by the muscle (based on the physiological cross-sectional area (PCSA) and an estimated unit contraction force – see below) and the geometry of the jaw.

After weighing the muscles (for volume calculation), the average fiber length was determined by immersing the bundles in 30% nitric acid (HNO_3) to dissolve connective tissue holding the muscle fibers together [59]. After about 20 h (depending on muscle size) individual fibers were teased apart and the nitric acid reaction was stopped with an excess of saturated Borax solution (disodium tetraborate). The length of 30 individual fibers per muscle was measured on digital images using analySIS® software (Soft Imaging System) and average fiber length calculated. Muscle density was assumed to be 1 $g.cm^{-3}$ [60], and unit contraction force 19 $N.cm^{-2}$ [61] The contraction force produced by the muscle along its line of action (F_{in}) can then be calculated as: $F_{in} = PCSA*19\ N.cm^{-2}$, where PCSA equals muscle volume divided by average fiber length.

Taking into account the orientation of the line of action of the muscle and the efficiency of force transmission by the lower jaw system, the output force at the jaw tip (F_{out}) was calculated as follows: $F_{out} = F_{in} * \sin \sigma * L_{in}/L_{out}$. The inlever ($L_{in}$) is the distance between the articulation of the lower jaw with the quadrate and the insertion of the muscle onto the lower jaw. Likewise, the outlever (L_{out}) was taken as the distance from the articulation to the jaw tip. The ratio of L_{in} to L_{out} then reflects the mechanical advantage for jaw closing. The σ reflects the angle between the line of action of the muscle and the inlever. All distances and angles were calculated based on the coordinates of four points (jaw tip, jaw articulation with the quadrate, muscle insertion on the jaw and muscle origin) determined on the photos taken during the dissection of the muscles. All calculations from coordinates and muscle fiber lengths to output forces were done in Excel® (Microsoft corporation). We determined the bite force exerted by the A_2 and A_3 part of the adductor mandibulae at an arbitrary gape angle of 20°, the A_1 part was excluded due to its complex pennation and attachment to both the premaxilla and the lower jaw. Consequently, the obtained values are an underestimation of bite force for both species, but still allow meaningful comparison.

Kinematic efficiency

As a measure for efficiency of suction feeding the kinematic transmission coefficient (KT) of the anterior-jaw four-bar linkage was calculated. In cichlids this system consists of the suspensorium as the fixed link, the nasal as coupler, maxilla as output link and coronoid portion of the lower jaw as input link [39]. This linkage describes the amount of maxillary rotation as a result of lower jaw depression, where the KT of this system is defined as the output rotation of the maxilla divided by the input rotation of the lower jaw. Calculations of this coefficient were based on the coordinates of the joints of the linkage determined on the dissection photos and were implemented in Excel® (Microsoft Corporation).

Lever systems and force transmission

The mandible of fishes can be considered as a lever system rotating around the quadrate-articular joint. The outlever, which is the same for jaw opening and closing, is determined as the distance between the joint and the tip of the mandible. The inlever for jaw opening is the bar running from the joint to the tip of the retroarticular process (onto which the interopercular-mandibular ligament attaches). As suction feeding fish rely on fast jaw opening, the kinematic efficiency of this system was calculated as the ratio of outlever to inlever (for jaw opening). Higher values of this ratio represent a kinematically efficient system that more effectively amplifies the input velocity at the retro-articular process. Furthermore two characteristics were quantified that influence force transmission by the upper jaw: the relative length of the ascending process of the premaxilla (for a given head length) and the angle between the ascending and dentigerous arm of the premaxilla [35].

Statistical analyses

Differences in egg metrics between species were analyzed with a t-test. Egg counts were compared using a Poisson generalized model with a log link function [62]. Differences in estimates of feeding performance between species and sexes were statistically evaluated with a glm implementation of a two-way ANOVA with inclusion of HL as covariate for variables that are size-related. All statistical analyses were performed using SAS 9.2.

Acknowledgments

We wish to thank Frans Witte from the National Natural History Museum (Leiden, The Netherlands) for providing the wild-caught specimens and two anonymous reviewers for their suggestions to improve this manuscript.

Author Contributions

Conceived and designed the experiments: TT EV DA. Performed the experiments: TT. Analyzed the data: TT. Contributed reagents/materials/analysis tools: BDK PH. Wrote the paper: TT DA EV.

References

1. Genner MJ, Seehausen O, Lunt DH, Joyce DA, Shaw PW, et al. (2007) Age of cichlids: new dates for ancient lake fish radiations. Molecular Biology and Evolution 24: 1269–1282.
2. Verheyen E, Salzburger W, Snoeks J, Meyer A (2003) Origin of the superflock of cichlid fishes from Lake Victoria, East Africa. Science 300: 325–329.
3. Fryer G, Iles TD (1972) The cichlid fishes of the Great Lakes of Africa: Their biology and evolution. Edinburgh: Oliver & Boyd. 656 p.
4. Barel CDN, Anker GC, Witte F, Hoogerhoud RJC, Goldschmidt T (1989) Constructional constraint and its ecomorphological implications. Acta Morphologica Neerlandica-Scandinavia 27: 83–109.
5. Brakefield PM (2006) Evo-Devo and Constraints on Selection. Trends in Ecology & Evolution 21: 362–368.
6. Gillooly JF, Londono GA, Allen AP (2008) Energetic constraints on an early developmental stage: a comparative view. Biology Letters 4: 123–126.
7. Liem KF (1991) A functional approach to the development of the head of teleosts: implications on constructional morphology and constraints. In: Schmidt-Kittler N, Vogel K, eds. Constructional Morphology and Evolution. Berlin: Springer-Verlag. pp 231–249.
8. Albertson RC, Kocher TD (2001) Assessing morphological differences in an adaptive trait: A landmark-based morphometric approach. Journal of Experimental Zoology 289: 385–403.
9. Balshine-Earn S, Earn DJD (1998) On the evolutionary pathway of parental care in mouth-brooding cichlid fish. Proceedings of the Royal Society of London Series B-Biological Sciences 265: 2217–2222.
10. Duponchelle F, Paradis E, Ribbink AJ, Turner GF (2008) Parallel life history evolution in mouthbrooding cichlids from the African Great Lakes. Proceedings of the National Academy of Sciences of the United States of America 105: 15475–15480.
11. Goodwin NB, Balshine-Earn S, Reynolds JD (1998) Evolutionary transitions in parental care in cichlid fish. Proceedings of the Royal Society of London Series B-Biological Sciences 265: 2265–2272.
12. Schurch R, Taborsky B (2005) The functional significance of buccal feeding in the mouthbrooding cichlid *Tropheus moorii*. Behaviour 142: 265–281.
13. Witte F, Barel CDN, Hoogerhoud RJC (1990) Phenotypic plasticity of anatomical structures and its ecomorphological significance. Netherlands journal of zoology 40(1–2): 278–298.

14. Hulsey CD, Mims MC, Streelman JT (2007) Do constructional constraints influence cichlid craniofacial diversification? Proceedings of the Royal Society B-Biological Sciences 274: 1867–1875.
15. Albertson RC, Kocher TD (2006) Genetic and developmental basis of cichlid trophic diversity. Heredity 97: 211–221.
16. Hoogerhoud RJC, Barel CDN Integrated morphological adaptations in piscivorous and mollusc-crushing *Haplochromis* species; 1978; Wageningen, 52–56.
17. Erlandsson A, Ribbink AJ (1997) Patterns of sexual size dimorphism in African cichlid fishes. South African Journal of Science 93: 498–508.
18. Ostlund-Nilsson S, Nilsson GE (2004) Breathing with a mouth full of eggs: respiratory consequences of mouthbrooding in cardinalfish. Proceedings of the Royal Society B: Biological Sciences 271: 1015–1022.
19. Hulsey CD, de Leon FJG, Rodiles-Hernandez R (2006) Micro- and macroevolutionary decoupling of cichlid jaws: A test of Liem's key innovation hypothesis. Evolution 60: 2096–2109.
20. Liem KF (1974) Evolutionary strategies and morphological innovations: cichlid pharyngeal jaws. Systematic Zoology 22: 425–441.
21. Okuda N, Miyazaki M, Yanagisawa Y (2002) Sexual difference in buccal morphology of the paternal mouthbrooding cardinalfish *Apogon doederleini*. Zoological science 19: 801–807.
22. Hess HC (1993) Male mouthbrooding in jawfishes (Opistognathidae): constraints on polygyny. Bulletin of marine science 52: 806–818.
23. Barnett A, Bellwood DR (2005) Sexual dimorphism in the buccal cavity of paternal mouthbrooding cardinalfishes (Pisces: Apogonidae). Marine Biology 148: 205–212.
24. Oliveira RF, Almada VC (1995) Sexual dimorphism and allometry of external morphology in *Oreochromis mossambicus*. Journal of Fish Biology 46: 1055–1064.
25. Bouton N, De Visser J, Barel CDN (2002) Correlating head shape with ecological variables in rock-dwelling haplochromines (Teleostei: Cichlidae) from Lake Victoria. Biological Journal of the Linnean Society 76: 39–48.
26. Liem KF (1980) Adaptive significance of intraspecific and interspecific differences in the feeding repertoires of cichlid fishes. American Zoologist 20: 295–314.
27. Barel CDN (1983) Towards a constructional morphology of cichlid fishes (Teleostei, Perciformes). Netherlands journal of zoology 33: 357–424.
28. Barel CDN (1984) Form-relations in the context of constructional morphology: the eye and suspensorium of lacustrine cichlidae (Pisces, Teleostei). Netherlands Journal of Zoology 34: 439–502.
29. Devaere S, Adriaens D, Teugels GG, Verraes W (2005) Morphology and Spatial Constraints in a Dorso-Ventrally Flattened Skull, With a Revised Species Description of *Platyallabes Tihoni* (Poll, 1944). Journal of Natural History 39: 1653–1673.
30. Galis F (1993) Interactions between the pharyngeal jaw apparatus, feeding-behavior, and ontogeny in the cichlid fish, *Haplochromis piceatus* - a study of morphological constraints in evolutionary ecology. Journal of Experimental Zoology 267: 137–154.
31. de Jong IML, Witte F, Richardson MK (2009) Developmental stages until hatching of the Lake Victoria cichlid *Haplochromis piceatus* (Teleostei: Cichlidae). Journal of Morphology 270: 519–535.
32. Katunzi EFB (1983) Seasonal variation in the food of a molluscivorous cichlid *Haplochromis sauvagei* Pfeffer 1896. Netherlands journal of zoology 33: 337–341.
33. Slootweg R (1987) Prey selection by molluscivorous cichlids foraging on a schistosomiasis vector snail, *Biomphalaria glabrata*. Oecologia 74: 193–202.
34. Cooper WJ, Westneat MW (2009) Form and function of damselfish skulls: rapid and repeated evolution into a limited number of trophic niches. BMC Evolutionary Biology 9.
35. Otten E (1983) The jaw mechanism during growth of a generalized *Haplochromis* species - *H. elegans* Trewavas 1933 (Pisces, Cichlidae). Netherlands journal of zoology 33: 55–98.
36. Bouton N, Witte F, Van Alphen JJM (2002) Experimental evidence for adaptive phenotypic plasticity in a rock-dwelling cichlid fish from Lake Victoria. Biological Journal of the Linnean Society 77: 185–192.
37. Westneat MW (1994) Transmission of force and velocity in the feeding mechanisms of labrid fishes (Teleostei, Perciformes). Zoomorphology 114: 103–118.
38. Westneat MW (1990) Feeding mechanics of teleost fishes (Labridae, Perciformes) - a test of 4-bar linkage models. Journal of Morphology 205: 269–295.
39. Hulsey CD, De Leon FJG (2005) Cichlid jaw mechanics: linking morphology to feeding specialization. Functional Ecology 19: 487–494.
40. Parnell NF, Hulsey CD, Streelman JT (2008) Hybridization produces novelty when the mapping of form to function is many to one. BMC Evolutionary Biology 8: 122.
41. deVisser J, Barel CDN (1996) Architectonic constraints on the hyoid's optimal starting position for suction feeding of fish. Journal of Morphology 228: 1–18.
42. Alfaro ME, Bolnick DI, Wainwright PC (2005) Evolutionary consequences of many-to-one mapping of jaw morphology to mechanics in labrid fishes. American Naturalist 165: E140–E154.
43. Herler J, Kerschbaumer M, Mitteroecker P, Postl L, Sturmbauer C (2010) Sexual dimorphism and population divergence in the Lake Tanganyika cichlid fish genus *Tropheus*. Frontiers in Zoology 7.
44. Donev A, Cisse I, Sachs D, Variano E, Stillinger FH, et al. (2004) Improving the density of jammed disordered packings using ellipsoids. Science 303: 990–993.
45. Rice AN, Lobel PS (2002) Enzyme activities of pharyngeal jaw musculature in the cichlid *Tramitichromis intermedius*: implications for sound production in cichlid fishes. Journal of Experimental Biology 205: 3519–3523.
46. McKaye KR, Louda SM, Stauffer JR (1990) Bower size and male reproductive success in a cichlid fish lek. American Naturalist 135: 597–613.
47. Witte F (1984) Consistency and functional significance of morphological differences between wild-caught and domestic *Haplochromis squamipinnis* (Pisces, Cichlidae). Netherlands journal of zoology 34: 596–612.
48. Barel CDN, Van Oijen MJP, Witte F, Witte-Maas ELM (1977) An introduction to the taxonomy and morphology of the haplochromine Cichlidae from Lake Victoria, a manual to Greenwood's revision papers. Netherlands journal of zoology 27: 333–389.
49. Anker GC (1978) Morphology of head-muscles of a generalized *Haplochromis* species: *H. elegans* Trewavas 1933 (Pisces, Cichlidae). Netherlands journal of zoology 28: 234–271.
50. Drost MR, Van Den Boogaart JGM (1986) A simple method for measuring the changing volume of small biological objects, illustrated by studies of suction feeding by fish larvae and of shrinkage due to histological fixation. Journal of Zoology (London) 209: 239–249.
51. Van Wassenbergh S, Lieben T, Herrel A, Huysentruyt F, Geerinckx T, et al. (2009) Kinematics of benthic suction feeding in Callichthyidae and Mochokidae, with functional implications for the evolution of food scraping in catfishes. Journal of Experimental Biology 212: 116–125.
52. Van Wassenbergh S, Herrel A, Adriaens D, Aerts P (2007) No trade-off between biting and suction feeding performance in clariid catfishes. Journal of Experimental Biology 210: 27–36.
53. Taylor WR, Van Dyke GC (1985) Revised procedures for staining and clearing small fishes and other vertebrates for bone and cartilage study. Cybium 9: 107–119.
54. Rohlf FJ (2008) tpsDig 2.12, available at http://life.bio.sunysb.edu/morph/.
55. Rohlf FJ, Slice D (1990) Extensions of the Procrustes method for the optimal superimposition of landmarks. Systematic Zoology 39: 40–59.
56. Rohlf FJ (1999) Shape statistics: Procrustes superimpositions and tangent spaces. Journal of Classification 16: 197–223.
57. Rohlf FJ (2003) tpsSmall 1.20, available at http://life.bio.sunysb.edu/morph/.
58. Klingenberg CP (2011) MorphoJ: an integrated software package for geometric morphometrics. Molecular Ecology Resources 11: 353–357.
59. Herrel A, Aerts P, De Vree F (1998) Ecomorphology of the lizard feeding apparatus: a modelling approach. Netherlands journal of zoology 48: 1–25.
60. Westneat MW (2003) A biomechanical model for analysis of muscle force, power output and lower jaw motion in fishes. Journal of Theoretical Biology 223: 269–281.
61. Akster HA, Granzier HLM, Keurs H (1985) A comparison of quantitative ultrastructural and contractile characteristics of muscle fibre types of the perch, *Perca fluviatilis* L. Journal of Comparative Physiology B: Biochemical, Systemic, and Environmental Physiology 155: 685–691.
62. van der Sluijs I, Van Dooren TJM, Seehausen O, Van Alphen JJM (2008) A test of fitness consequences of hybridization in sibling species of Lake Victoria cichlid fish. Journal of Evolutionary Biology 21: 480–491.

PERMISSIONS

All chapters in this book were first published in PLOS ONE, by The Public Library of Science; hereby published with permission under the Creative Commons Attribution License or equivalent. Every chapter published in this book has been scrutinized by our experts. Their significance has been extensively debated. The topics covered herein carry significant findings which will fuel the growth of the discipline. They may even be implemented as practical applications or may be referred to as a beginning point for another development.

The contributors of this book come from diverse backgrounds, making this book a truly international effort. This book will bring forth new frontiers with its revolutionizing research information and detailed analysis of the nascent developments around the world.

We would like to thank all the contributing authors for lending their expertise to make the book truly unique. They have played a crucial role in the development of this book. Without their invaluable contributions this book wouldn't have been possible. They have made vital efforts to compile up to date information on the varied aspects of this subject to make this book a valuable addition to the collection of many professionals and students.

This book was conceptualized with the vision of imparting up-to-date information and advanced data in this field. To ensure the same, a matchless editorial board was set up. Every individual on the board went through rigorous rounds of assessment to prove their worth. After which they invested a large part of their time researching and compiling the most relevant data for our readers.

The editorial board has been involved in producing this book since its inception. They have spent rigorous hours researching and exploring the diverse topics which have resulted in the successful publishing of this book. They have passed on their knowledge of decades through this book. To expedite this challenging task, the publisher supported the team at every step. A small team of assistant editors was also appointed to further simplify the editing procedure and attain best results for the readers.

Apart from the editorial board, the designing team has also invested a significant amount of their time in understanding the subject and creating the most relevant covers. They scrutinized every image to scout for the most suitable representation of the subject and create an appropriate cover for the book.

The publishing team has been an ardent support to the editorial, designing and production team. Their endless efforts to recruit the best for this project, has resulted in the accomplishment of this book. They are a veteran in the field of academics and their pool of knowledge is as vast as their experience in printing. Their expertise and guidance has proved useful at every step. Their uncompromising quality standards have made this book an exceptional effort. Their encouragement from time to time has been an inspiration for everyone.

The publisher and the editorial board hope that this book will prove to be a valuable piece of knowledge for researchers, students, practitioners and scholars across the globe.

LIST OF CONTRIBUTORS

Mario L. Muscedere and James F. A. Traniello
Department of Biology, Boston University, Boston, Massachusetts, United States of America

Marina L. Sardi
División Antropología, Facultad de Ciencias Naturales y Museo, Universidad Nacional de La Plata - CONICET, La Plata, Buenos Aires, Argentina

Fernando V. Ramírez Rozzi
UPR 2147 CNRS, Paris, France

Naoki Morimoto, Christoph P. E. Zollikofer and Marcia S. Ponce de León
Anthropological Institute, University of Zurich, Zurich, Switzerland

Elodie Vercken
Institut Sophia Agrobiotech, UMR 1355 ISA, Institut National de la Recherche Agronomique, Sophia-Antipolis, France

Maren Wellenreuther and Erik I. Svensson
Department of Biology, Lund University, Lund, Sweden

Benjamin Mauroy
Laboratoire J.A. Dieudonné, UMR CNRS 7351, Université de Nice-Sophia Antipolis, Nice, France

James Cotterell
EMBL-CRG Systems Biology Research Unit, Centre for Genomic Regulation (CRG), Barcelona, Spain
Developmental Biology Section, MRC Human Genetics Unit, Edinburgh, United Kingdom

James Sharpe
EMBL-CRG Systems Biology Research Unit, Centre for Genomic Regulation (CRG), Barcelona, Spain
Institució Catalana de Recerca i Estudis Avançats (ICREA) Professor, Centre for Genomic Regulation (CRG), Barcelona, Spain

Caleb D. Phillips, Boyd Butler, Hugo Mantilla-Meluk and Robert J. Baker
Department of Biological Sciences, Texas Tech University, Lubbock, Texas, United States of America

John W. Fondon III
Department of Biology, University of Texas at Arlington, Arlington, Texas, United States of America

Ary A. Hoffmann
Department of Genetics and Bio21 Institute, Melbourne University, Melbourne, Australia

Carla Sgró
School of Biological Sciences, Monash University, Melbourne, Australia

Volker Loeschcke, Trine Bilde and Jesper S. Bechsgaard
Department of Bioscience, Aarhus University, Aarhus, Denmark

Torsten N. Kristensen
Department of Bioscience, Aarhus University, Aarhus, Denmark
Department of Molecular Biology and Genetics, Aarhus University, Tjele, Denmark
NordGen – Nordic Genetic Resource Center, Ås, Norway

Sanna Huttunen
Department of Biology, University of Turku, Turku, Finland

Sanna Olsson
Department of Agricultural Sciences, University of Helsinki, Helsinki, Finland

Volker Buchbender and Dietmar Quandt
Nees-Institute for Biodiversity of Plants, University of Bonn, Bonn, Germany

Johannes Enroth
Department of Biosciences and Botanical Museum, University of Helsinki, Helsinki, Finland

Lars Hedenäs
Department of Cryptogamic Botany, Swedish Museum of Natural History, Stockholm, Sweden

Bradley R. Sturz, Joshua E. Edwards and Ty W. Boyer
Department of Psychology, Georgia Southern University, Statesboro, Georgia, United States of America

Ricard V. Solé
ICREA-Complex Systems Lab, Universitat Pompeu Fabra, Barcelona, Spain
Santa Fe Institute, Santa Fe, New Mexico, United States of America
Institut de Biologia Evolutiva, UPF-CSIC, Barcelona, Spain
European Centre for Living Technology, C Foscari University of Venice, Venice, Italy

Sergi Valverde
ICREA-Complex Systems Lab, Universitat Pompeu Fabra, Barcelona, Spain
Institut de Biologia Evolutiva, UPF-CSIC, Barcelona, Spain
European Centre for Living Technology, C Foscari University of Venice, Venice, Italy

Eli M. Swanson and Kay E. Holekamp
Department of Zoology, Michigan State University, East Lansing, Michigan, United States of America
BEACON Center for the Study of Evolution in Action, Michigan State University, East Lansing, Michigan, United States of America

Barbara L. Lundrigan
Department of Zoology, Michigan State University, East Lansing, Michigan, United States of America
BEACON Center for the Study of Evolution in Action, Michigan State University, East Lansing, Michigan, United States of America
Michigan State University Museum, Michigan State University, East Lansing, Michigan, United States of America

Bradley M. Arsznov
Department of Psychology, Michigan State University, East Lansing, Michigan, United States of America

Sharleen T. Sakai
Neuroscience Program, Michigan State University, East Lansing, Michigan, United States of America
Department of Psychology, Michigan State University, East Lansing, Michigan, United States of America

Miguel Barbosa
Center for Environmental and Marine Studies (CESAM), Departamento de Biologia, Universidade de Aveiro, Campus de Santiago, Aveiro, Portugal
Centre for Biological Diversity and Scottish Oceans Institute, University of St Andrews, St Andrews, Fife, United Kingdom

Joao Pestana
Center for Environmental and Marine Studies (CESAM), Departamento de Biologia, Universidade de Aveiro, Campus de Santiago, Aveiro, Portugal

Amadeu M. V. M. Soares
Center for Environmental and Marine Studies (CESAM), Departamento de Biologia, Universidade de Aveiro, Campus de Santiago, Aveiro, Portugal
Programa de Pós-Graduação em Produção Vegetal, Universidade Federal de Tocantins, Campus de Gurupi, Gurupi, Brasil

Paul R. Martin
Department of Biology, Queen's University, Kingston, Ontario, Canada

Cameron K. Ghalambor
Department of Biology and Graduate Degree Program in Ecology, Colorado State University, Fort Collins, Colorado, United States of America

Marwan El Khoury, José Braga, Jean Dumoncel, Remi Esclassan and Frederic Vaysse
University of Toulouse, Molecular Anthropology and Image Synthesis Laboratory (Centre National de la Recherche Scientifique), Toulouse, France

Javotte Nancy
University of Bordeaux, Faculty of dentistry, Bordeaux, France

April M. Wright and David M. Hillis
Department of Integrative Biology, University of Texas at Austin, Austin, Texas, United States of America

Christiaan V. Henkel, Erik Burgerhout, Ron P. Dirks and Hans J. Jansen
ZF-screens B.V., Leiden, The Netherlands

Daniëlle L. de Wijze, Yuki Minegishi, Herman P. Spaink and Guido E. E. J. M. van den Thillart
Institute of Biology, Leiden University, Leiden, The Netherlands

Sylvie Dufour
UMR BOREA, CNRS 7208, Muse´um National d'Histoire Naturelle, Paris, France

Finn-Arne Weltzien
Norwegian School of Veterinary Science, Oslo, Norway

Katsumi Tsukamoto
Atmosphere and Ocean Research Institute, The University of Tokyo, Kashiwa, Chiba, Tokyo, Japan

Dean C. Adams
Department of Ecology, Evolution, and Organismal Biology, Iowa State University, Ames, Iowa, United States of America

Ryan N. Felice
Department of Biological Sciences, Ohio University, Athens, Ohio, United States of America

Cen Zeng
Department of Biology II, University of Munich, Munich, Germany

Ivan Gomez-Mestre
Ecology, Evolution, and Development Group, Doñana Biological Station, Consejo Superior de Investigaciones Científicas, Seville, Spain

John J. Wiens
Department of Ecology and Evolutionary Biology, University of Arizona, Tucson, Arizona, United States of America

Gareth R. Hopkins, Edmund D. Brodie Jr. and Susannah S. French
Department of Biology and the Ecology Center, Utah State University, Logan, Utah, United States of America

Kalina H. Kapralova, Sigrídur Rut Franzdóttir, Hákon Jónsson, Sigurður S. Snorrason and Zophonías O. Jónsson
Institute of Life- and Environmental Sciences, University of Iceland, Reykjavík, Iceland

Jörg U. Hammel and Michael Nickel
Institut für Spezielle Zoologie und Evolutionsbiologie mit Phyletischem Museum, Friedrich-Schiller-Universität Jena, Erbertstr. 1, 07743, Jena, Germany

Wei Liu
College of Veterinary Medicine, Northwest Agriculture & Forest University, Yangling, Shaanxi, China
Institute for Molecular Biology, University of Zürich, Zurich, Switzerland

Lei Xue
School of Life Science and Technology, Tongji University, Shanghai, China
Institute for Molecular Biology, University of Zürich, Zurich, Switzerland

Tim tkint, Barbara De Kegel and Dominique Adriaens
Department of Biology, Research Group Evolutionary Morphology of Vertebrates, Ghent University, Ghent, Belgium

Erik Verheyen and Philippe Helsen
Vertebrate Department, Royal Belgian Institute of Natural Sciences, Brussels, Belgium
Department of Biology, Research Group Evolutionary Ecology, University of Antwerp, Antwerpen, Belgium

Index

P

R

S

T

V

W